STRUCTURAL PROTEOMICS

and its impact on the

LIFE SCIENCES

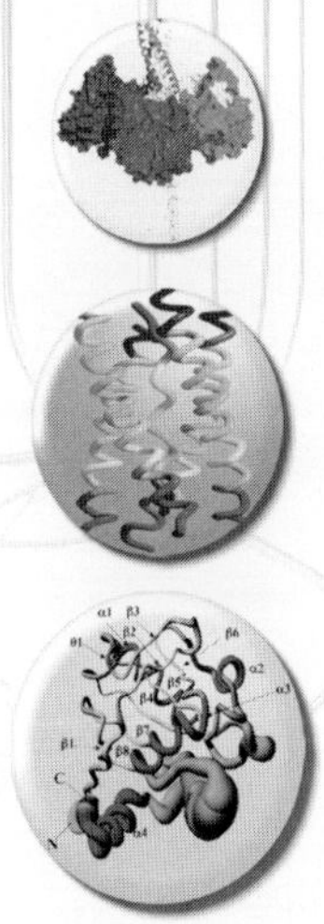

STRUCTURAL PROTEOMICS

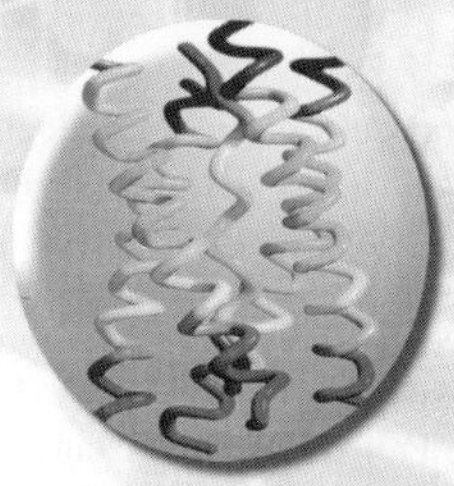

and its impact on the

LIFE SCIENCES

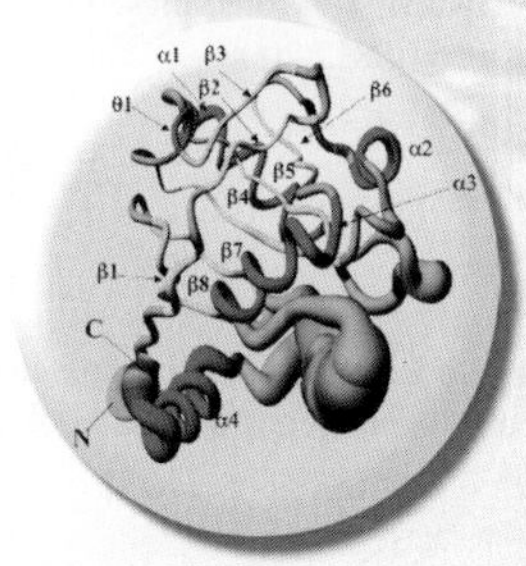

Joel L. Sussman, Israel Silman
Weizmann Institute of Science, Israel
editors

World Scientific
NEW JERSEY · LONDON · SINGAPORE · BEIJING · SHANGHAI · HONG KONG · TAIPEI · CHENNAI

Published by

World Scientific Publishing Co. Pte. Ltd.
5 Toh Tuck Link, Singapore 596224
USA office: 27 Warren Street, Suite 401-402, Hackensack, NJ 07601
UK office: 57 Shelton Street, Covent Garden, London WC2H 9HE

Library of Congress Cataloging-in-Publication Data
Structural proteomics and its impact on the life sciences / editors, Joel L. Sussman, Israel Silman.
p. ; cm.
Includes bibliographical references and index.
ISBN-13: 978-981-277-204-6 (hardcover : alk. paper)
ISBN-10: 981-277-204-9 (hardcover : alk. paper)
1. Proteomics. 2. Proteins--Structure. I. Sussman, Joel. II. Silman, Israel.
[DNLM: 1. Proteomics. QU 58.5 S932 2008]
QP551.S817 2008
572'.6--dc22

2008009897

British Library Cataloguing-in-Publication Data
A catalogue record for this book is available from the British Library.

Typeset by Stallion Press
Email: enquiries@stallionpress.com

Printed by FuIsland Offset Printing (S) Pte Ltd, Singapore

Preface

Joel L. Sussman* and Israel Silman†

The concept of structural genomics (SG) arose in the mid-to-late 1990s in both the USA and Japan, triggered by the success achieved in applying high-throughput (HTP) sequencing methods to whole genomes. It was envisaged that application of a similar HTP approach to obtaining the three-dimensional structures of a substantial fraction of the entire set of proteins of a given organism, the "proteome," would be an efficient way of filling in the gaps in observed "fold-space." The decision to adopt such an approach resulted in the investment of large sums of money, i.e. hundreds of millions of dollars, in large-scale structural genomics projects in both countries. Thus, in Japan, the Protein Research Group was established at the RIKEN Genomic Sciences Center in 1998, and in the USA, the Protein Structure Initiative (PSI), funded by the NIH/NIGMS, commenced at nine major centers in 2000. These projects were characterized by concentration of resources in a small number of large centers, by development of novel, automated technologies which would allow a HTP pipeline approach to structure determination, and a focus on novel folds as a major target criteria.

Europe was slower in implementing HTP approaches to structural biology. Various national efforts such as the Protein Structure Factory

Departments of *Structural Biology and †Neurobiology, Weizmann Institute of Science, Rehovot 76100, Israel

(PSF) in Berlin, the Oxford Protein Production Facility (OPPF), and the Genopoles in France, led the way. But it was only towards the end of 2002 that the first Europe-wide project began, an EC-funded Integrated Project entitled Structural Proteomics in Europe (SPINE). While benefiting from the technological achievements of the US and Japanese programs, and itself also concerned with developing cutting edge technologies as a means of achieving its objectives, SPINE from the outset focused these technologies on biomedically relevant targets. Indeed, as implied by its name, it aimed to establish a pan-European biomedically oriented structural **proteomics** program, placing significant emphasis on **functional aspects** of the target proteins studied. SPINE was followed by a series of EC-funded programs of various scopes and funding, some placing an emphasis on technological development, and others on attacking various classes of targets.

A similar emphasis on the use of the emerging HTP technologies to solve structures of biomedical relevance was adopted by the Structural Genomics Consortium (SGC), established in 2003 with the support of Canadian and British sponsors from both the public and private sectors, with laboratories in Oxford, Toronto and, subsequently, in Stockholm.

From the outset, the scale of funding of the PSI met with considerable criticism, especially in the USA, which has been going through a period during which funding for research by individual PIs has been hard to come by. Many critics have argued that the $270M spent on funding the Pilot Phase of the PSI, over a period of five years, could have been more effectively utilized to fund hypothesis-driven research directed towards targets of fundamental or applied interest. Nevertheless, the achievements of the various SG/SP consortia, viewed in aggregate, and achieved in less than a decade, are impressive. In aggregate, the US PSI centers (September 2000 to June 2005) have determined over 1100 structures (http://www.nigms.nih.gov/Initiatives/PSI/Background/PilotFacts.htm). During PSI-2, which is still ongoing, ~1200 structures have already been solved, of which the vast majority share less than 30% sequence identity with any structure already deposited in the PDB, and many represent novel folds. As a result of the efforts of all the consortia (US, Japanese and

European), 5968 protein structures had been deposited in the Protein Data Bank as of 11-Dec-2007 (http://www.rcsb.org/pdb/static.do?p=general_information/pdb_statistics). Although some of these structures may be redundant, or even appear uninteresting at first sight, many are of the highest technical quality, of fundamental and/or medical importance and, taken overall, provide a valuable database. Moreover, it has been reported that, in 2005, structures arising out of structural genomics and structural proteomics efforts accounted for 44% of the total number of novel structures reported. Although many of the novel structures solved by the SG/SP centers were, on the surface, low-hanging fruit, which filled gaps either in a given proteome or in fold space without yielding novel functional information, other targets have been, as already mentioned, of great fundamental and/or medical importance. Furthermore, filling in fold space provides a robust body of templates for homology modeling, which can rapidly take advantage of these templates as computational techniques become more sophisticated, and computing power increases. Thus, in our view, whatever policy decisions are taken with respect to funding of large-scale SG/SP projects, what has been achieved so far will have a lasting impact on biological and biomedical research. Indeed, the EC, through a Specific Support Action (SSA), established the Forum for European Structural Proteomics (**FESP**, see http://www.ec-fesp.org) to assess the current status and make recommendations for future European infrastructure requirements in the SG/SP area. We feel, therefore, that it is timely to publish a book in which these achievements are presented with a look to their potential impact on biological and biomedical research in general.

In this volume, we have tried to bring together experts capable of addressing all aspects of the SG/SP effort, from target selection, through the various techniques for expressing and purifying proteins and protein complexes and the methodologies for solving their structures, to their impact on drug design and on coping with emerging diseases. In view of the ongoing debate on SG/SP funding, we have also included a special chapter dealing with policy, which includes sections written by several scientists and officials who have been closely associated with the decision-making processes.

Contents

List of Contributors

Enrique E. Abola
Department of Molecular Biology
The Scripps Research Institute
10550 North Torrey Pines Road, La Jolla
California 92037
USA

Mariana Babor
Department of Plant Sciences
Weizmann Institute of Science
Rehovot 76100
Israel

Amos Bairoch
Swiss Institute of Bioinformatics
Centre Medical Universitaire
1, rue Michel-Servet
1211 Geneve 4
Switzerland

Edward N. Baker
School of Biological Sciences
University of Auckland
Private Bag 92019
Auckland
New Zealand

Lucia Banci
Centro Risonanze Magnetiche
University of Florence
Via Luigi Sacconi 6
Sesto Fiorentino
Florence 50019
Italy

Wolfgang Baumeister
Max Planck Institute of Biochemistry
Am Klopferspitz 18a
Martinsried D-82152
Germany

Florence Bedez
Institut de Génétique et de Biologie Moléculaire et Cellulaire
UMR 7104, 1 rue Laurent Fries
BP 10142, 67404 Illkirch Cedex
France

David M. Belnap
Department of Chemistry and Biochemistry
Brigham Young University
Provo, UT 84602
USA

Jeremy Berg
National Institute of General Medical Sciences
National Institutes Health
Bethesda, MD 20B92-6200
USA

Stephen Burley
SGX Pharmaceuticals Inc
10505 Roselle Street
San Diego, CA 92121
USA

Shuai Chen
Institute of Biochemistry
Center for Structural and Cell Biology in Medicine
University of Lübeck
Ratzeburger Allee 160
23538 Lübeck
Germany

A. Chesneau
EMBL Grenoble Outstation
6 rue Jules Horowitz
BP181, 38042 Grenoble Cedex 9
France

Miroslaw Cygler
Department of Biochemistry
McGill University
845 Sherbrooke Street West
Montreal, Quebec H3A 2T5
Canada

Susan Daenke
Division of Structural Biology
Wellcome Trust Centre for Human Genetics
University of Oxford
Roosevelt Drive
Oxford, OX3 7BN
UK

Marvin Edelman
Department of Plant Sciences
Weizmann Institute of Science
Rehovot 76100
Israel

Josefina Enfedaque
European Commission
Research Directorate General
BE-1049 Brussels
Belgium

D. J. Hart
EMBL Grenoble Outstation
6 rue Jules Horowitz
BP181, 38042 Grenoble Cedex 9
France

Udo Heinemann
Max-Delbruck-Center for Molecular Medicine
Robert-Roessle-Str 10
Berlin D-13125
Germany

Rolf Hilgenfeld
Institute of Biochemistry
Center for Structural and Cell Biology in Medicine
University of Lübeck
Ratzeburger Allee 160
23538 Lübeck
Germany

Ursula Hinz
Swiss Institute of Bioinformatics
Centre Medical Universitaire
1, rue Michel-Servet
1211 Geneve 4
Switzerland

Zongchao Jia
Department of Biochemistry
Queen's University
Kingston, ON
K7L 3N6
Canada

Hualiang Jiang
Drug Discovery and Design Center
State Key Laboratory of Drug Research
Shanghai Institute of Materia Medica
Chinese Academy of Sciences
Zuchongzhi
Rd. 555, Shanghai 201203
China

Andrzej Joachimiak
Biosciences Division
Midwest Center for Structural Genomics and
Structural Biology Center
Argonne National Laboratory
9700 S Cass Ave. Argonne
IL 60439

E. Yvonne Jones
Division of Structural Biology
Wellcome Trust Centre for Human Genetics
University of Oxford
Roosevelt Drive
Oxford, OX3 7BN
UK

Christina Kiel
EMBL-CRG Systems Biology partnership Unit
Centre de Regulacio Genomica (CRG)
Dr Aiguader 88
08003 Barcelona
Spain

Saša Jenko Kokalj
Department of Biochemistry and Molecular Biology
Jozef Stefan Institute
Jamova 39
1000 Lijubljana
Solvenija

Seiki Kuramitsu
Department of Biological Sciences
Graduate School of Science
Osaka University
1–1 Machikaneyama-cho
Toyonaka, Osaka 560-0043
Japan

Arthur M. Lesk
Department of Biochemistry and Molecular Biology, and
the Huck Institute for Genomics, Proteomics and Bioinformatics
The Pennsylvania State University
University Park, PA 16802
USA

Ronen Levy
Department of Plant Sciences
Weizmann Institute of Science
Rehovot 76100
Israel

Allan Matte
Biotechnology Research Institute
6100 Royalmount Ave.
Montreal QC H4P 2R2
Canada

Dino Moras
Institut de Génétique et de Biologie Moléculaire et Cellulaire
UMR 7104, 1 rue Laurent Fries
BP 10142, 67404 Illkirch Cedex
France

John Moult
Center for Advanced Research in Biotechnology
University of Maryland
Biotechnology Institute
9600 Gudelsky Drive
Rockville, MD 20850
USA

John Norvell
National Institute of General Medical Sciences
National Institutes Health
Bethesda, MD
USA

Yuan-Ping Pang
Computer-Aided Molecular Design Laboratory
Mayo Clinic
Rochester, Minnesota
USA

Helen Parkinson
European Bioinformatics Institute
Wellcome Trust Genome Campus Hinxton
Cambridgeshire CB10 1SD
United Kingdom

Arnaud Poterszman
Institut de Génétique et de Biologie Moléculaire et Cellulaire
UMR 7104, 1 rue Laurent Fries
BP 10142, 67404 Illkirch Cedex
France

Jacques Remacle
European Commission
BE-1049 Brussels
Belgium

Christophe Romier
IGBMC
1 rue Laurent Fries
B.P. 10142, 67404 Illkirch Cedex

Vineet Sangar
Department of Biochemistry and Molecular Biology, and
the Huck Institute for Genomics, Proteomics and Bioinformatics
The Pennsylvania State University
University Park, PA 16802
USA

Gunter Schneider
Karolinska Institutet
Scheelevägen 2
SE-171 77 Stockholm
Sweden

Luis Serrano
EMBL-CRG Systems Biology partnership Unit
Centre de Regulacio Genomica (CRG)
Dr Aiguader 88
08003 Barcelona
Spain

Xu Shen
Drug Discovery and Design Center
State Key Laboratory of Drug Research
Shanghai Institute of Materia Medica
Chinese Academy of Sciences
Zuchongzhi
Rd. 555, Shanghai 201203
China

Israel Silman
Department of Neurobiology
Weizmann Institute of Sceince
Rehovot 76100
Isreal

Barbara Skene
Formerly Head of Department, Molecules, Genes and Cells
The Welcome Trust
London NW1 2BE
UK

Vladimir Sobolev
Department of Plant Sciences
Weizmann Institute of Science
Rehovot 76100
Israel

Alasdair C. Steven
Laboratory of Structural Biology
National Institute of Arthritis, Musculoskeletal, and Skin Diseases
National Institutes of Health
Bethesda, MD 20892
USA

Raymond C. Stevens
Department of Molecular Biology
The Scripps Research Institute
10550 North Torrey Pines Road, La Jolla
California 92037
USA

David I. Stuart
Division of Structural Biology
Wellcome Trust Centre for Human Genetics
University of Oxford
Roosevelt Drive
Oxford, OX3 7BN
UK

Michael D. Suits
Department of Biochemistry
Queen's University
Kingston, ON
K7L 3N6
Canada

Joel L. Sussman
Department of Structural Biology
Weizmann Institute of Science
Rehovot 76100
Israel

Thomas Szyperski
816 Natural Sciences Complex
Chemistry Department
State University of New York at Buffalo
Buffalo, NY 14260
USA

Jinzhi Tan
Institute of Biochemistry
Center for Structural and Cell Biology in Medicine
University of Lübeck
Ratzeburger Allee 160
23538 Lübeck
Germany

Thomas C Terwilliger
Los Alamos National Laboratory
Los Alamos, NM 87545
USA

Janet M Thornton
EMBL — European Bioinformatics Institute
Wellcome Trust Genome Campus
Hinxton
Cambridge, CB10 1SD
UK

Peter Tompa
Institute of Enzymology
Biological Research Center
Hungarian Academy of Sciences
H-1518 Budapest
P.O. Box 7
Hungary

James D Watson
EMBL — European Bioinformatics Institute
Wellcome Trust Genome Campus
Hinxton
Cambridge, CB10 1SD
UK

James C. Whisstock
Department of Biochemistry and Molecular Biology
Victorian Bioinformatics Consortium
Monash University
Clayton Campus
Melbourne, Victoria 3168
Australia

Ian Wilson
The Scripps Research Institute
10550 N. Torrey Pines Rd
La Jolla, CA 92037
USA

Shigeyuki Yokoyama
Protein Research Group
Genomic Sciences Center
RIKEN Yokohama Institute
1-7-22 Suehiro-cho
Tsurumi, Yokohama 230-0045
Japan

H. Yumerefendi
EMBL Grenoble Outstation
6 rue Jules Horowitz
BP181, 38042 Grenoble Cedex 9
France

Chapter 1

The Importance of Target Selection Strategies in Structural Biology

Enrique E. Abola and Raymond C. Stevens*

Introduction

The industrialization of biology — the large-scale acquisition of biological data — has been pioneered by the sequencing of entire genomes and is being applied to the characterization of other important biological molecules such as the proteome, the interactome (arguably a subset of the proteome), the glycome, and the metabolome. As of October 2007, the complete genomic sequences of 676 organisms have been published and more projects are underway. The DNA sequence data alone is insufficient to generate the level of understanding of biological systems that most biologists seek. Understanding how biological systems operate from the level of single proteins and enzymes, to the level of protein-protein interactions, and finally at the level of intact cellular physiological pathways, a goal of systems biology, will require detailed, quantitative characterization of cellular proteins and their interactions, which is facilitated by access to protein structural information. Thus, the number and types of questions that can now be addressed by structural biologists has increased dramatically. The scope of protein structure space is still too immense for a

*Corresponding author: stevens@scripps.edu. Department of Molecular Biology, The Scripps Research Institute, 10550 North Torrey Pines Road, La Jolla, California 92037, USA.

completely unfocused approach to data acquisition. Therefore, target selection is still a critical step in establishing an industrial-scale protein structure project.

About 50 years after the publication of the first 3-dimensional structure of a protein, that of sperm whale myoglobin completed in John Kendrew's laboratory[1] (a structural proteomics (SP) project in itself as myoglobin from multiple species were pursued), structural biologists have started to explore the possibility of conducting high-throughput (HT) structural studies to permit the structural characterization of proteomes. HT approaches, such as parallel studies of multiple protein targets, are expected to revolutionize the way structures are determined by moving away from one-by-one structural studies. These new approaches are expected to produce important scientific results at reduced costs by using economy of scales and by generating standardized and more generalizable protocols and evaluation metrics for protein expression, purification and structural characterization. Over the past 10 years, the results of successful pilot studies have been reported by the various SP programs. This leads to the important question of what target selection strategies should then be used in the future by both SP and non-SP laboratories in the light of what has been learned from these pilot studies.

This chapter summarizes the strategies in target selection and prioritization used by various SP groups and provides a brief summary of their recent results. We explore the potential of small and medium sized laboratories as well as larger collaborative efforts to make use of the new technologies, protocols and approaches developed by these initial SP initiatives, with a special emphasis on studying biological systems through class-directed target selection approaches.

Global Structural Efforts and their Target Selection Strategies

By 2000, several groups, both academic and for-profit companies, were being setup to establish HT structure determination production lines. Initial efforts were focused on developing new technologies and protocols for each step in the process, from initial cloning to final deposition of coordinate data sets to public databases. The immediate aim was to convert the one-by-one structure determination process, using both single-crystal X-ray diffraction and solution NMR techniques, to work in a HT mode. Although the initial mandate was technology development, there was also a requirement to solve a relatively large number of protein structures to serve both as proof of concept and as a justification of the approach adopted by each center. Major government-sponsored consortium efforts were formed: the Protein Structure Initiative (PSI-1) in the USA, the Structural Proteomics (SPINE) integrated project in Europe, and Project 3000 in Japan, and smaller efforts in other countries. Another major effort was a joint venture between government and industry. The Structural Genomics Consortium, an international project funded by Canada, Sweden, the Wellcome Trust in the UK and industry, with laboratories in Oxford, Stockholm and Toronto. For-profit companies, such as SYRRX, SGX, and ASTEX, were setup with the goal of improving the drug discovery and development process by reducing the risks and cost of getting at the structures of drug targets and their complexes.

Each major consortium had an overarching target selection and prioritization strategy. By and large, all the SP groups mentioned above pursued targets based on general principles which followed the class-directed target strategies outlined in papers published as part of the dialogue on definition and implementation of structural proteomics.[2,3] This is exemplified by the paper by Terwilliger *et al.* (see Table 1; Ref. 3), which put forward a list of protein classes and a scientific rationale for selecting them, and also suggested a protocol for implementation of the target selection strategy. Four classes were suggested: 1) the construction of a database of structural motifs; 2) the study of proteins from microorganisms, including pathogens and thermophiles; 3) a large-scale target class including human targets of biomedical interest, protein assemblies, proteins from plants or animals; and 4) a small-scale target class that is the study of important protein families (e.g. protein kinases, transcription factors). Class 1 attempts to generate structural annotations, while the rest are motivated by the goal of generating functional annotation. These two goals are somewhat related (i.e. fold may

Table 1 Target Classes for a Protein Structure Initiative[†]

Class of Proteins	Importance
(1) Database of protein structure motifs	Prediction of protein structure
(2) Proteins from a microorganism	
Proteins from a pathogen	Potential drug targets
Proteins from a thermophile	Robust enzyme
(3) Large-scale targets	
Human proteins	Medical applications
Plant or animal proteins	Biotechnology
Protein assemblies	Protein interactions
(4) Small-scale targets	
Groups of structurally-similiar proteins	Predicting protein evolution
Proteins from a metabolic pathway	Biocatalysis

[†] From Terwilleger *et al.*[3]

provide clues to function), and one can argue that the target lists generated by some of the PSI-1 centers are more focused on an attempt to functionally annotate an organism's proteome (e.g. the Joint Center for Structural Gemomics (JCSG)'s studies of *Thermotoga maritima*). Below, we summarize the activities of the various SP centers, their initial target selection strategies, and the outcomes. Although the Japanese effort, Protein 3000, has produced more than 2500 structures, accounting for half of the 5000 structures solved in all SP centers worldwide, we have not included an extensive discussion of their efforts. At this time, their target selection strategies remain unpublished, although early descriptions[4] indicate that their main effort was geared towards the generation of structural annotations, viz. looking for new folds. Final statistics indicate that only 34% of their structures have novel sequences and the list of their solved structures appears to indicate that secondary target selection criteria were used.

Selecting and Prioritizing Targets

Initially, most target selection processes centered on choosing proteins based on primary and secondary objectives, which were later supplemented by additional target prioritization schemes. For example, projects that aimed at exploring fold space focused on methodology and approaches to find clusters of related sequences for which 3-dimensional structures of any of its members were not available in the PDB. There was also an interest in developing and using tools for the identification of domains to facilitate the production of expression constructs. This was particularly important for proteins anticipated to be difficult to crystallize and thus more amenable to NMR structure determination. Targets were further prioritized based on the predicted novelty of sequences or on the prediction that structures of new folds would be obtained.

Halfway through the first phase of PSI, it became possible to select and prioritize targets utilizing databases created from the results of the large number of SP experiments carried out in the various PSI laboratories that used different target selection strategies on a number of classes of proteins.[5–7] Data on diverse data sets were, therefore, available to attempt to understand successes (e.g. produces soluble constructs or crystallizes) and failures. Two papers from the JCSG exemplify what can be done. Canaves *et al.*,[6] analyzing the JCSG data for *T. maritima proteins*, and Slabinski, *et al.*[7] analyzing all the available SP data, developed a number of sequence-based metrics, which provides a measure of the difficulty/ease of crystallizing protein (http://ffas.burnham.org/XtalPred-cgi/xtal.pl). Both studies use 12 parameters to arrive at an index. Once crystals are obtained, statistics indicate a 32–38% chance of completing the structure (see Table 2). Interestingly, the Barton laboratory[8] used PDB entries to develop a normalized score, OB_SCORE, based on just two parameters, pI and the Gravy index. This z-score estimates the chances of producing diffraction-quality crystals.[8] These metrics, along with bioinformatics resources established by the various SP centers, were then used to construct prioritized target lists; in the case of the PSI, prioritization was assigned based primarily on the novelty of the sequence. A more integrated approach to target selection is now provided by a web-based system, sgTarget (http://www.ysbl.york.ac.uk/sgTarget/), that produces homology information that measures the uniqueness of the sequence, as well as the calculated physiochemical properties that

Table 2 Success Rates for All Structural Proteomics Centers (October 2007)†

Status	Total Number of Targets	(%) Relative to "Cloned" Targets	(%) Relative to "Expressed" Targets	(%) Relative to "Purified" Targets	(%) Relative to "Crystallized" Targets
Cloned	102306	100	—	—	—
Expressed	64773	63.3	100	—	—
Soluble	27886	27.3	43.1	—	—
Purified	25659	25.1	39.6	100	—
Crystallized	9357	9.1	14.4	36.5	100
Diffraction-quality crystals	4863	4.8	7.5	19	52
Diffraction	4055	4	6.3	15.8	43.3
NMR assigned	1727	1.7	2.7	6.7	—
HSQC	2950	2.9	4.6	11.5	—
Crystal structure	3746	3.7	5.8	14.6	40
NMR structure	1642	1.6	2.5	6.4	—
In PDB	5195	5.1	8	20.2	38
Work stopped	25278	—	—	—	—
Test target	3	—	—	—	—
Other	1	—	—	—	—

† Table downloaded from http://sg.pdb.org/target_centers.html on October 2007.

may affect expression, solubility, and the likelihood that crystals will be obtained.[9] In addition to the target selection activities, statistics derived from data-mining activities on production databases (e.g. TargetDB, PepCDB), as well as process and technology evaluation that measured the performance of the various pipelines, are now able to provide a robust estimate and a better understanding of the risks involved in structural studies which had previously been estimated using anecdotal and *ad hoc* approaches.[10]

The Protein Structure Initiative (PSI)

The PSI project was established by the National Institutes of General Medical Sciences (NIGMS) at the U.S. National Institutes of Health. PSI phase I (PSI-1) studies, initiated in year 2000 and 2001, were conducted in nine centers (Table 3), and were completed in 2005, with over 1100 structures having been deposited in the PDB. Phase II (PSI-2) studies were immediately started, and involved four large-scale production centers, six specialized centers for development, two homology modeling centers, and a research grants program focusing on improving the accuracy of the comparative protein structure modeling. Production centers in PSI-2 were required to produce 4000 new structures within five years, while the specialized centers were given the mission of developing new tools and approaches to handle challenging targets, including eukaryotic proteins, integral membrane proteins, and large macromolecular complexes. Within two years of operation, the four PSI-2 production centers had deposited about 1200 structures in the PDB, thus exceeding the five-year combined production output of the PSI-1 centers.

The focus of the PSI-1 pilot centers was primarily the development of new tools, technologies, and methodology to increase the success rates and lower the costs of structure determination. Each center was responsible for automating protein sample production and the structure determination pipelines, and for meeting production goals. The final production numbers for the PSI-1 centers are presented in Table 4. As the initial goal of the consortium was to set up the pipelines and test their scalability, about 40% of the total number of

Table 3 List of Major Structural Genomics Organizatons

	Center/Consortium	Target Selection Criteria and Target Organism(s)
1.	Berkeley Structural Genomics Center (BSGC), USA http://www.strgen.org	Novel sequences Minimal organisms — *M. genitalium, M. pneumoniae*
2.	Center for Eukaryotic Structural Genomics (CESG), USA http://www.uwstructuralgenomics.org	Novel sequences *Arabidopsis thaliana*
3.	Joint Center for Structural Genomics (JCSG), USA http://www.jcsg.org	Novel sequences *Thermatoga Maritima, mouse*
4.	Midwest Center for Structural Genomics (MCSG), USA http://www.mcsg.anl.gov	Novel sequences Proteins from all three kingdoms of life
5.	Mycobacterium Tuberculosis Structural Genomics Consortium (TBSGC), USA http://www.doe-mbi.ucla.edu/TB/	Novel sequences, *Mycobacterium tuberculosis*
6.	New York Structural Genomics Consortium (NYSGC), USA http://www.nysgrc.org	Novel sequences Disease-related proteins from eukaryotes and bacteria
7.	Northeast Structural Genomics Consortium (NESG), USA http://www.nesg.org	Novel sequences Eukaryotic domain families from *D. melanogaster, S. cerevisiae, C. elegans,* mouse, human
8.	Southeast Collaboratory for Structural Genomics (SECSG), USA http://www.scsg.org	Novel sequences *P. furiosus, C. elegans,* human
9.	Structural Genomics of Pathogenic Protozoa (SGPP), USA http://www.sgpp.org Medical Structural Genomics of Pathogenic Protozoa (MSGPP)	Novel sequences, Pathogenic protozoans — *Leishmania major, Trypanosoma brucei, Trypnasoma cruzi, Plasmodium falciparum, Entemoeba hystolitica, Gardia, Lamblia, Toxomplasma gondii, Cryptosporidium parvum*

(*Continued*)

Table 3 (*Continued*)

	Center/Consortium	Target Selection Criteria and Target Organism(s)
10	Structural Proteomics in Europe (SPINE), UK http://www.spineeurope.org	Bacterial and viral pathogens — *B. antrhacis, M. tuberculosis, SARS-CoV, Herpes virus* Cancer-related proteins Immune defense, neuronal development and neurodegenerative diseases
11.	Structural Genomics Centers, Canada, Sweden, UK http://www.thesgc.com	Human proteins related to diseases and human pathogens
12	Project 3000, Japan http://www.rsgi.riken.jp	Novel sequences and biologically important or human health related

structures were determined in the last year of operation. Overall, these studies now show that, based on the results of PSI-1 activities, there is a 5–10% probability of success for a given target in the class of targets included in the PSI-1 list.

Summary of Target Selection and Results from PSI Centers

The overall scientific goal of the PSI effort was to determine enough structures to completely populate a database which could then be used to construct homology-based models covering most of protein space. Thus, the PSI's primary targets were Class 1 proteins of Table 1. However, target selection efforts were not centralized in PSI-1. All that the centers were required to do was to ensure that a significant number of their targets were unique, i.e. have <30% sequence identity with structures already deposited in the PDB. Target selection was also driven by the informal goal of being able to define the complete fold space of proteins; hence, additional selection criteria were applied by the centers themselves, with higher priority given to proteins for which there was a higher expectation of discovering a new fold.

Table 4 Final Production Numbers for PSI-1 Centers[†]

Center	All targets	Cloned	Crystals	Diffr	Structures (novel) NMR	Structures (novel) X-Ray	In PDB (novel; unique)	Deposits after Oct 1, 2000 (novel; unique)	% Unique	Annual Rate (last 2 months)	Median Length
MCSG	15 565	5730	888	363	0	296 (274)	296 (274; 235)	291 (269; 230)	79	120 (108)	319
NESGC	12 213	5484	163	116	93	116 (97)	198 (169; 138)	186 (157; 128)	68.8	54 (42)	191
NYSGRC	2145	1538	397	196	0	195 (157)	178 (146; 106)	171 (139; 100)	58.5	48 (30)	454
JCSG	6594	3650	1167	268	8	221 (180)	198 (160; 104)	198 (160; 104)	52.5	78 (54)	415
BSGC	911	812	94	65	3	58 (50)	52 (45; 37)	43 (37; 30)	69.8	0 (0)	374
SECSG	14 786	14 378	223	118	2	74 (52)	71 (51; 29)	71 (51; 29)	40.8	0 (0)	214
TB	1758	1547	209	120	2	107 (70)	67 (44; 25)	62 (40; 23)	37.1	12 (12)	611
CESG	6582	4476	104	40	18	34 (22)	47 (33; 27)	47 (33; 27)	57.4	0 (0)	166
SGPP	19 503	10 154	175	45	0	28 (17)	22 (15; 10)	22 (15; 10)	45.5	0 (0)	200
Total PSI	75 104	45 391	3311	1307	125	1114 (919)	1111 (937; 711)	1074 (901; 681)	63.4	312 (246)	358

[†] This table was downloaded from http://www.mcsg.anl.gov/index.html

The centers elected to build their programs around certain organisms and/or classes of organisms, prioritizing their target lists by giving higher preferences to novel sequences within these organisms. Thus, a "spread" of protein classes studied was achieved. Table 5 summarizes the target organisms tackled by the PSI-1 centers. Targets from both Class 2 and Class 3 were chosen by these centers. Each center complemented their lists of proteins from their target organisms with orthologs, thereby increasing the chances of obtaining the desired structures.

In the second phase of the PSI (PSI-2), representatives of the four production centers served as members of a centralized target selection committee responsible for generating and maintaining a target list. Prioritization of targets is being carried out within each center, and is usually based on individual scientific interests and technical capabilities of the center. Overall, the goal of PSI-2 remains the same as that of PSI-1, viz. to attempt to characterize protein space completely. This goal is being approached by coarsely sampling pfam and other large protein families for clusters of sequence-related proteins which lack structural representatives in the PDB. Unlike PSI-1, which provided more latitude to the centers in selecting their targets, a set of clearly defined objectives have been set in order to attain this overall goal (see Table 6). Each center is expected to prioritize the members of a sequence family assigned to it by applying a number of criteria, including: 1) families containing representatives from selected model organisms or groups of organisms; 2) families containing representatives with known or postulated disease associations; 3) families containing representatives with predicted or known biological/biochemical functions; and 4) families containing representatives from all the three kingdoms of life.

PSI Targets from Minimal Organisms

The Berkeley Structural Genomics Center (BSGC) has developed its pipeline to work on minimal organisms (i.e. microbes with the smallest genomes), studying proteins from *M. genitalium,* with 486 ORFs, and *M. pneumoniae,* with 687 ORFs, in their respective genomes. The idea was that by studying the proteomes of these minimal organisms, one might gain insight into the minimal requirements for a viable

Table 5 Statistics on PSI Production Levels Classified by Organism[†]

Organism	Total Number[1]	Work Stopped	Cloned	Expressed	Purified	Crystallized	Crystal Structure	NMR Structure	In PDB[2]
Total Viruses	368	96	204	151	87	15	12	5	17
Archaea	8901	1479	6715	4174	1991	526	211	39	245
Bacteria	66101	9933	50239	34797	13557	4860	1731	130	1768
Total Prokaryotes	**75002**	**11412**	**56954**	**38971**	**15548**	**5386**	**1942**	**169**	**2013**
Yeast	1983	624	1411	733	578	84	33	7	35
Plasmodium	5197	268	2974	1260	196	65	16	0	16
Trypanosoma	6403	62	3953	1909	299	58	10	0	8
Leishmania	9581	288	4557	2202	403	146	21	0	17
Arabidopsis	7525	4118	4122	1663	439	229	38	19	54
Rice	130	94	128	62	12	7	1	0	1
Nematode	15057	3459	12634	5504	417	97	29	3	32
Fly	651	270	142	69	20	1	1	0	1

(*Continued*)

Table 5 (*Continued*)

Organism	Total Number[1]	Work Stopped	Cloned	Expressed	Purified	Crystallized	Crystal Structure	NMR Structure	In PDB[2]
Mouse	1376	660	1179	866	241	152	28	4	32
Human	8714	3466	3751	2171	714	188	47	17	64
Other Eukaryotes	1294	126	1153	962	327	116	36	4	38
Total Eukaryotes	**57911**	**13435**	**36004**	**17401**	**3646**	**1143**	**260**	**54**	**298**
Synthetic	3	0	3	2	3	1	1	2	3
Unknown	1	0	1	1	0	0	0	0	0
Total	133285	24943	93166	56526	19284	6545	2215	230	2331

† This table was downloaded from http://sg.pdb.org/target_centers.html

Table 6 PSI-2 Objective of Improving Coverage of Protein Structural Space

1. To determine at least one structure for each large, hitherto uncharacterized, protein sequence family using coarse sampling (BIG families)
2. To determine representative structures for each branch of very large, diverse protein sequence families (MEGA families) to span the structural and functional diversity within that family (moderate sampling to increase structural coverage and to provide structural coverage of selected families with high biomedical relevance)
3. To determine representative structures for families that are over-represented in the microbiome and metagenome sequence data (moderate sampling of META-family)
4. To determine the structures of biomedical targets and community-proposed targets.

organisms.[11,12] They have been working on 1036 protein targets from these organisms, determining 87 structures of 61 proteins, 52 of which turned out to be novel proteins. Upon completion of their PSI-1 effort, they announced that their efforts had contributed significantly to the almost complete characterization of the *M. genitalium* proteome, which now has fold assignment for 87% of its proteins. The 486 ORFs in *M. genitalium* include 82 integral membrane proteins, 44 soluble globular proteins, and 10 soluble non-globular proteins. Most importantly, they report that their recent efforts in structural biology and SP have succeeded in enabling fold assignments for over ~90% of the soluble globular proteins in five minimal organisms, *Buchnera aphidicola, Blochmannia floridanus, Wigglesworthis glossinidia, Mycoplasma genitalium,* and *Tropheryma whipplei*, thus, providing a rich data set to further drive attempts to understand the minimal requirements for sustaining life.[13]

PSI Targets from Extremophiles

Two centers opted to study extremophile targets. The JCSG worked on *Thermotoga maritima* (*T. maritima*) and the Southeast Collaboratory for Structural Genomics (SECSG) worked on *Pyrococcus furiosus.* Extremophiles were thought to be ideal targets for early SG studies, since their proteins were expected to be more stable and, therefore, amenable to simplified SG sample production protocols.[14]

Full structural and functional characterization of a proteome is now within reach for *Tm*. Although other SP centers have studied proteins from *Tm*, the efforts at the JCSG led to this microbe becoming one of the larger organisms for which a high percentage of its proteome has been structurally characterized. Of its 1877 ORFs, 273 protein structures have been deposited in the PDB, 180 of which were determined by the JCSG. Taking into consideration all of the proteins for which structural information is available, i.e. including those for which structures of homologs are known, it is estimated that 62% of the *T. maritima* proteome now has protein-fold coverage. It should be noted that about 28% of the members of the *T. maritima* proteome have been identified as transmembrane proteins and/or predicted to be of low complexity or to be inherently disordered. Thus, the overall structural coverage from which functional annotation can potentially be generated is quite significant. In addition to the structures in the PDB, over 1000 *Tm* proteins have been purified, about 800 at the JCSG, providing a valuable resource for the community interested in doing further functional and/or biological studies on these proteins. Another important consequence of this work is that questions relating to the putative inherent higher stability of proteins from extremophiles may now be addressed systematically.[15,16]

PSI Targets from Pathogens

The Structural Genomics of Pathogenic Protozoa (SGPP) center selected proteins exclusively from disease-causing protozoans, while the Mycobacterium Tuberculosis Structural Genomics Consortium (TBSGC) studied proteins from *Mycobacterium tuberculosis* (*Mtb*). Their main interests were in developing a better understanding of the biology of these pathogens, as well as in the development of new therapeutics. Upon completion of the PSI-1 program, these two centers continued their studies with funding from the National Institute for Allergy and Infectious Diseases (NIAID), which had decided to fund SP centers focused on protein targets of infectious agents. In 2007, two additional HT centers were funded by the NIAID to determine at least 100 new structures per year, focusing on protein targets from

Table 7 Summary of Progress of the Structure Genomics Consortium (SGC) in Various Target Focus Areas†

Groups	No. of Targets	Targets with Constructs Cloned	Targets Purified	Targets in Crystal Trials	Targets with Crystals	Targets with Crystals Diffracting ≤ 2.8 Å	Approved Structures
TO1	308	254	158	95	74	48	53
TO2	442	401	177	119	56	46	49
TO3	299	219	146	100	55	51	53
Malaria	333	323	246	171	110	52	51
OX1	279	263	137	129	101	66	67
OX2	308	198	108	101	80	55	65
OX3	383	297	144	139	92	63	64
KI	426	376	239	204	108	57	49
Totals	2778	2331	1355	1058	676	438	451

Target focus of research groups:

TO1: Proteases, ubiquitylation pathway and cyclophilins
TO2: Chromatin biology and epigenetics
TO3: ATPases and GTPases
Malaria: Malaria and related diseases caused by apicomplexan parasites

OX1: Oxidoreductaces and metabolic enzymes
OX2: Membrane receptor signaling
OX3: Phosphorylation dependent signaling
KI: Nucleotide and amino acid metabolism, signalling domains in apoptosis and inflammation, phosphoinositol and lipid signaling and RNA helicases

† From http://www.thesgc.com/structures/target_progress.php accessed on October 2007.

organisms implicated in infectious diseases, with the goal of using these new structures for developing new therapeutic protocols.

TBSGC operates as a global structural proteomics effort, with 400 members in 80 institutions.[17,18] The primary goal of the consortium is to study the function of proteins in the pathogen, targeting those which are potential drug targets or are believed to play key role in *Mtb* biology. By now ~200 unique *Mtb* protein structures and ~250 ligand complexes have been deposited in the PDB. Two thirds of these were produced by the TBSGC, the rest by SPINE, the European structural proteomics consortium, and by other laboratories worldwide. Prior to the SP efforts, there were only 8 *Mtb* protein structures in the PDB. Only ~29% of the protein structures solved by the TBSGC have novel sequences, reflecting less emphasis on finding new folds, and more on working on targets of high relevance to the disease. A notable contribution from the TBSGC is the development of a protocol that can be scaled-up on the genome level to identify (using computational approaches), characterize, and determine the crystal structures of protein–protein complexes.

The SGPP center conducted structural studies on proteins from major pathogenic protozoans. These challenging targets included *Plasmodium falciparum*, the causative agent of the most deadly form of malaria, which is responsible for over one million deaths a year, mostly of children. Using the tools of HT SP, they initiated the task of structurally and functionally characterizing the proteome by attempting expression of about 1000 ORFs, leading to the high-level expression of 63 proteins and to solution of 16 structures. Further, protein engineering studies are underway to improve these success rates. The initial pilot study that led to the expression of these targets is now being analyzed so as to lay the ground work for prioritizing and ultimately eliminating barriers to producing protein samples for structural and functional studies.

PSI Targets from Eukaryotic Organisms

Although proteins from eukaryotes remain a difficult class of targets for SP studies, they include important biomedical targets, and a large number of them have been subjected to analysis. As of October 2007, the PSI centers have studied 57 911 eukaryotic proteins, but only about 300 structures of such proteins have been completed. Focused development work is currently underway to produce technologies and processes capable of handling these difficult targets.

In most cases, eukaryotic targets were selected by the PSI centers primarily to solve the structures of proteins with novel sequences; thus, most of the centers had several such targets on their list. However, the plant model organism, *Arabidopsis thaliana,* was the primary target proteome for the Center for Eukaryotic Structural Genomics (CESG). They studied more than 4000 targets from this organism and solved 52. They also worked on 4000 other eukaryotic targets, solving 43. All of the other centers, except for the BCSG, worked on a variety of eukaryotic organisms, including mouse and human. The SECSG extensively studied *C. elegans,* working on almost 12 000 protein targets, but solving only 12 of them. These results again highlight the difficulties associated with working with these eukaryotic protein targets in HT studies. Three centers, the Northeast Structural Genomics Consortium (NESG), the Midwest Center for Structural Genomics (MCSG), and the New York Structural Genomics Consortium (NYSGC), did not focus on particular organisms, but rather worked on novel prokaryotic and eukaryotic targets.

SPINE – Structural Proteomics in Europe Project, Function-based Target Selection

The SPINE project was a three-year project that commenced in 2002, and was funded through the EU FP5 program. In 2006, a special issue of *Acta Crystallographica*, Volume 62, was devoted to a description of the work done by this consortium, and provides a comprehensive discussion of its vision for SP, as well as a description of the development of technologies needed to carry out SP projects. Finally, several of the papers in this volume communicate quite succinctly target selection strategies and their correlation with success rates. SPINE produced 375 structures, of which 305 were unique proteins, and the rest of protein–ligand complexes. It had an overall success rate comparable to that of the PSI-1 centers, determining the structures of 12%

of selected targets. Like the PSI effort, SPINE used HT approaches, and was deliberately named as a Structural Proteomics program, so as to differentiate it from structural genomics efforts. SPINE was driven by the notion of "human health targets," rather than by a bioinformatics-based "fold space" approach.[19] Thus, the criteria of working with novel sequences were not necessarily the primary determinant in selecting targets. Indeed, one early criterion used was to select proteins that could be solved by molecular replacement, which was done primarily to help refine protein production pipelines. The project was subdivided into workpackages, three of which focused on a particular target space of interest. Workpackage 9 focused on proteins from bacterial and viral pathogens, with *B. anthracis* and *Mtb* as the primary bacterial target organisms, and the SARS-CoV and Herpes virus as the viral ones. Workpackage 10 worked on cancer-related proteins (i.e. kinesins, kinases, proteins from the ubiquitin pathway), while Workpackage 11 studied proteins involved in immune defense mechanisms, neuronal development, and on proteins implicated in neurodegenerative diseases. Workpackages 1–8 focused on technology and process development.

New and more challenging targets are being tackled by SPINE2-COMPLEXES, a continuation of the SPINE integrated project funded by the EC within FP6. The project is titled "From Receptor to Gene: Structures of Complexes from Signaling Pathways Linking Immunology, Neurobiology, and Cancer," and reflects the challenging nature of new efforts as well as the establishment of HT centers throughout Europe. The new targets will be protein–protein and protein–nucleic acid complexes that are related to the areas of investigation as shown in Table 8. This new initiative will require the development of new technologies for the HT study of these complexes, ushering in a new era in structural biology.

Target Selection in SPINE

Discussion of target selection strategies used in the SPINE studies are presented in several papers of the *Acta Crystallographica* special issues.[20–23] The overall target selection activity was carried out in two distinct phases, the first involving the identification of targets with

Table 8 SPINE2 Target Selection – Complexes in the Following Areas will be Studied

WP1.1: Complexes in the ubiquitin signaling pathway
- Ubiquitination
- De-ubiquitination
- Proteasome regulation

WP1.2: Cell life, death and integrity
- Cell cycle (including ankyrin repeat protein complexes)
- Apoptotic pathways (p53-dependent and p53-independent)
- Control of cell integrity (e.g. Lon type protease)

WP1.3: Complexes in development and synaptic signaling
- Development and synaptic signaling assemblies
- Protein complexes involving neuronal proteins dependent on copper

WP1.4: Protein kinases
- Signaling and regulatory complexes involving protein kinases

WP1.5: Protein phosphatases
- Protein phosphatases in regulatory cell pathways

WP1.6: Receptors/activators associated with transcription complexes
- Nuclear receptor and transcription factor assemblies
- Chromatin remodelling motors

WP1.7: Innate immune system
- Complexes involved in pathogen recognition and subsequent signaling (including TLRs, NOD/NALP proteins, TIR domains, dectin)

WP1.8: Adaptive immune system
- Cell surface receptors and recognition complexes (including MHC, TCR, NK receptor families)

WP1.9: Viral subversion of cellular signaling and immune modulation
- Proteins modulating host responses (including those from EBV and poxviruses)
- Proteins involved in host interactions such as receptor binding and fusion

possible important biomedical roles, and the second being an assessment of whether the target was amenable to structural studies. The criteria used for the second activity were similar to those described by Canaves *et al.*,[6] using sequence-based predictions of the probability of successfully completing a structural study on a given target. Target selection strategies used by both the SPINE project and the SGC (see below), as well as their success rates, are perhaps more reflective of what could be achieved by those interested in generating functional annotations and/or interested in generating high-resolution structures, in contrast to the bioinformatics-driven, fold-space coverage approach taken by the PSI efforts described above. For example, for drug design applications, targets that can be solved using molecular replacement approaches are not necessarily excluded.

SPINE Studies on *Bacillus anthracis*

A total of 359 proteins from *Bacillus anthracis* were targeted for study resulting in the determination of 46 structures.[23] Two rounds of target selection were carried out. The first was used to identify targets with a high probability of successful completion, primarily to help in developing and establishing the HT pipelines and to help fine tune the target selection process. The criteria used to select these "easy" proteins were (1) sizes of <50 kDa; (2) were possible candidates for molecular replacement; (3) were not part of a complex; (4) did not contain signal peptides or transmembrane regions; and (5) were predicted to be soluble, based on the near absence of disordered regions. The second round of selection relied more on biomedical criteria, selecting the proteins that were predicted to be involved in pathogenesis, and of biomedical interest. Finally, additional challenging targets were included, such as those annotated as hypothetical proteins, or those for which putative molecular replacement models were not available.

The *B. anthracis* studies were carried out in two laboratories at the University of York and at Oxford University. Results from the first round of studies from both laboratories, using 48 targets, were quite encouraging. Oxford, after repeated efforts and refining of protocols, attained a structure solution success rate of 31%. In comparison, York attained a success rate of 21%, but did not carry out as many extensive retrials. These success rates were much reduced when the more challenging targets were studied. It is nevertheless quite gratifying to have learned that interesting questions could be answered using HT approaches through a careful target selection process, and that such a procedure could lead to the solution of a relatively large number of structures at reduced costs.

SPINE Studies on Viral Pathogens

Attempts were also made to study viral pathogens on the SPINE pipelines. The structures of four SARS corona virus (SARS-CoV), four Epstein-Barr virus (EBV), and two vaccinia virus proteins were completed. These studies highlight the difficulties of working with viral proteins in a HT setting. Nevertheless, the results obtained represent significant achievements. The target selection strategy for EBV is discussed in Tarbouriech *et al.*[24] EBV selection was oriented towards enzymes and other proteins that had been predicted to have significant secondary structure, were small in size, and had been calculated to have a high stability index. Proteins were deselected if they were predicted to be parts of a multi-protein assembly and/or had transmembrane domains.

SPINE Studies on Human Proteins of High Biomedical Value

One of the more challenging aspects of the SPINE project was the decision to focus a large portion of the work on human and other eukaryotic proteins that are potentially of high biomedical value. By the start of the SPINE project in 2002, it was clear from early results of the PSI that working with eukaryotic targets required careful attention to target selection in order for some level of success to be achieved. A total of 800 eukaryotic protein targets were selected for study by SPINE, the structures of 170 of which were determined. Biological importance was the primary selection criterion, with calculated physicochemical properties as the secondary criteria. As for the biological targets, preference was given based on the availability of models that could be used in molecular replacement studies.

SGC – Structural Genomics Consortium

The Structural Genomics Consortium (SGC) was organized in 2003 to address industrial and academic pharmaceutical research, and has thus focused on proteins and protein families from human and apicomplexan (e.g. *Plasmodium falciparum*) that are either potential drug targets or have been implicated in human disease processes. The SGC operates with laboratories at the University of Oxford and University of Toronto, and the Karolinska Institute. Unlike other SP efforts, it uses HT pipelines to study protein-ligand (inhibitors, co-factors, substrates and substrate analogs) interactions and protein families. By 2007, the consortium had solved the structures of over 451 out of the 2778 targets selected for study, corresponding to a success rate of 19%, the highest rate for all the SP centers. Table 7 provides a summary of the status of the structural studies, as well as a list of the target focuses of the various member research groups.

Target Selection in the SGC

SGC target selection protocols, including their list of candidates have not been published, as there is a possibility that this information may be used for commercial advantage. What is available are the areas of interests from which these targets are being chosen (Table 7). The main criteria on is relevance to human health and disease. As an example, one of SGC's areas of interest is signaling pathways. The family of protein kinases, the kinome, are an important class of drug targets, and are a prominent category of protein targets in the list. The consortium has solved 21 novel human kinase structures, and along with other SP efforts, was responsible for raising the number of kinase structures from 38 to 93 by the end of 2006.[25] Rather than just solving a unique member of a family, the SGC has elected to attempt total coverage. One important advantage of this approach is that methods and procedures developed for one member of the family could be used for the other members for all the steps in the process, from expression all the way through to crystallization and structure solution. To date, as many as 95% of the targets studied by the SGC have a homologous structure available, simplifying structure solution. Another family that has been extensively studied is that of the human cytosolic sulfur transferases (hSULT).[26]

Proteins in this family are involved in the metabolism of drugs and hormones, the bioactivation of carcinogens, and the detoxification of xenobiotics. Knowledge of the structural and mechanistic basis of substrate specificity and activity is crucial for understanding steroid and hormone metabolism, drug sensitivity, pharmacogenomics, and response to environmental toxins. Thus, they form a class of important targets for the pharmaceutical industry. The SGC has solved the structures of five of the 12 hSULT's; these structures, along with those of six others previously characterized by other groups, have permitted the exploration of local and global structural features of members of this enzyme family. In addition to the structural studies, the enzymes were screened for binding and activity towards a panel of potential substrates and inhibitors, revealing unique "chemical fingerprints" for each protein.[26]

The family-based approach also allows for the exploration of variations among the structures in attempts to develop selective therapeutics. It allows extensive study of small-molecule complexes, thus providing a powerful platform for developing a better understanding of the binding properties of members of the family. For example, in the case of kinases, this has led to the development of inhibitors with picomole potency for PIM kinases and glycogen synthetase kinase 3 (GSK-3) (see Ref. 27).

The SGC is also targeting proteins from *P. falciparum,* the causative organism of malaria, and related apicomplexan organisms. A total of 1008 genes from *P. falciparum* and related organisms have been studied, leading to the determination of 36 structures. This study provides yet another example of a SP survey of a complete organism paralleling those discussed above. The results of the studies are being applied to attempts to develop new vaccines and small molecule therapeutics against the organism.

Expanding the Target List

The ~5000 novel structures that have been produced by the SP efforts within the relatively short time of five years attest to the power of the new HT approaches, since it took almost 40 years to accumulate this number of structures in the PDB using traditional methods. As can be seen from the various target selection approaches discussed above, the range of questions that now can be realistically addressed has increased dramatically. Although SG was initially considered primarily an enterprise whose only goal was to explore protein fold space, activities within the various centers operating within the PSI, as well as in other SP efforts, clearly demonstrate that other important biological questions can be addressed with these new tools of HT structural biology.

The PSI efforts have demonstrated the success of the bioinformatics-based approach by getting close to completing protein-fold coverage for minimal organisms and for a thermophilic organism. These efforts have also shown that this level of coverage of protein-fold space in eukaryotic organisms will require further advances, primarily in protein expression. One of the major lessons of the PSI efforts is that focusing on prokaryotic and thermophilic organisms to cover fold-space and then building homology models for eukaryotic proteins is an efficient, cost-effective route in cases where homologs exist.

The SPINE efforts have demonstrated the success of function-based approaches to target selection, demonstrating that using biological importance for target prioritization can also lead to high success rates. Furthermore, both the PSI and SPINE have demonstrated the value of using sequence-based metrics that provide estimates for success as a further aid to target selection. The SGC effort provides a powerful approach to target selection in which biological questions, which in their case focus on the needs of the biopharmaceutical industry, are addressed through the study of large families of proteins. This effort has generated high resolution structures that can be used for functional and drug design studies. We feel that this approach may be one that makes the best use of HT technologies to address important biomedical questions.

Clearly, there remain important challenges that must be addressed in order to allow for the expansion of target selection strategies to include other classes of important proteins. For example, statistics for the PSI SP efforts reveal very high failure rates for eukaryotic proteins, with only 298 structures being obtained from ~58 000 targets. Membrane proteins and components of large multimeric/multiprotein complexes, remain outside the scope of HT efforts. New initiatives, in the frameworks of PSI-2, SPINE2-COMPLEXES, and the NIH sponsored Roadmap Initiative (http://nihroadmap.nih.gov/structuralbiology/index.asp), are now trying to address these target areas. As mentioned above, the SPINE2 project is focused on working with protein-protein and protein-nucleic acid complexes, and is now developing new tools and methodology to work with these targets using the HT approaches.

References

1. Kendrew JC, Bodo G, Dintizs HM, *et al.* (1958) "A three-dimensional model of the myoglobin molecule obtained by X-ray analysis." *Nature* **181**: 662–66.
2. Brenner S. (2000) "Target selection for structural genomics." *Nature Struct Biol* 7 **Suppl**: 967–69.
3. Terwilliger TC, Waldo G, Peat TS, *et al.* (1998) "Class-directed structure determination: foundation for a protein structure initiative." *Protein Sci* 7: 1851–56.
4. Yokoyama S, Hirota H, Kigawa T, *et al.* (2000) "Structural genomics in Japan." *Nature Struct Biol* 7 **Suppl**: 943–45.
5. Smialowski P, Schmidt T, Cox J, *et al.* (2006) "Will my protein crystallize? A sequence-based predictor." *Proteins* **62**: 343–55.
6. Canaves JM, Page R, Wilson IA, Stevens RC. (2004) "Protein biophysical properties that correlate with crystallization success in the *Thermatoga maritima*: maximum clustering strategy for structural genomics." *J Mol Biol* **344**: 977–91.
7. Slabinski L, Jaroszewski L, Rodrigues AP, *et al.* (2007) "The challenge of protein structure determination lessons from structural genomics." *Protein Sci* **16**: 2472–82.
8. Overton IM, Barton GJ. (2006) "A normalised scale for structural genomics target ranking: the OB-Score." *FEBS Lett* **580**: 4005–09.
9. Rodriques APC, Grant BJ, Hubbard RE. (2006) "SgTarget: a target selection resource for structural genomics." *Nucl Acid Res* **34**: W225–30.
10. Abola E, Carlton DC, Kuhn P, Stevens RC. (2007) "Five years of increasing structural biology throughput — a retrospective analysis." In H Jhoti and A Leach (eds), *Structure-Based Drug Discovery*, Springer, The Netherlands.
11. Kim SH. (2000) "Structural genomics of microbes: an objective." *Curr Opin Struct Biol* **10**: 380–83.
12. Chandonia J-M, Kim S-H, Brenner SE. (2006) "Target selection and deselection at the Berkeley Structural Genomics Center." *Proteins* **62**: 356–70.
13. Chandonia J-M, Kim S-H. (2006) "Structural proteomics of minimal organisms: conservation of protein fold usage and evolutionary implications." *BMC Struct Biol* **6**: 7.

14. Christendat D, Yee A, Dharamsi A, *et al.* (2000) "Structural proteomics of an archaeon." *Nature Struct Biol* **10**: 903–09.
15. Robinson-Rechavi M, Godzik A. (2005) "Structural genomics of *Thermotoga maritima* proteins shows that contact order is a major determinant of protein thermostability." *Structure (Camb)* **13**: 857–60.
16. Robinson-Rechavi M, Alibes A, Godzik A. (2006) "Contribution of electrostatic interactions, compactness and quaternary structure to protein thermostability: lessons from structural genomics of *Thermotoga maritima*." *J Mol Biol* **356**: 547–57.
17. Baker EN. (2007) "Structural Genomics as an approach towards understanding the biology of tuberculosis." *J Struct Funct Genomics* 2007 Aug 1; (Epub ahead of print).
18. Strong M, Sawaya MR, Wang S, *et al.* (2006) "Toward the structural genomics of complexes: crystal structure of a PE/PPE protein complex from Mycobacterium tuberculosis." *Proc Natl Acad Sci USA* **103**: 8060–65.
19. Stuart DI, Jones EY, Wilson KS, Daenke S. (2006) "SPINE: Structural Proteomics IN Europe: the best of both worlds." *Acta Crystallogr D* **62**: preface.
20. Banci L, Bertini I, Cusack S, *et al.* (2006) "First steps towards effective methods in exploiting high-throughput technologies for the determination of human protein structures of high biomedical value." *Acta Crystallogr D* **62**: 1208–17.
21. Fogg MJ, Alzari P, Bahar M, *et al.* (2006) "Application of the use of high-throughput technologies to the determination of protein structures of bacterial and viral pathogens." *Acta Crystallogr D* **62**: 1196–207.
22. Albeck S, Alzari P, Andreini C, *et al.* (2006) "SPINE bioinformatics and data-management aspects of high-throughput structural biology." *Acta Crystallogr D* **62**: 1184–95.
23. Au K, Berrow NS, Blagova E, *et al.* (2006) "Application of high-throughput technologies to a structural proteomics-type analysis of *Bacillus anthracis*." *Acta Crystallogr D* **62**: 1267–75.
24. Tarbouriech N, Buisson M, Géoui T, *et al.* (2006) "Structural genomics of the Epstein-Barr virus." *Acta Crystallogr D.* **62**: 1276–85.
25. Gileadi O, Knapp S, Lee WH, *et al.* (2007) "The scientific impact of the structural genomics consortium: a protein family and ligand-centered approach to medically-relevant human proteins." *J Struct Funct Genomics* 2007 Oct 12; (Epub ahead of print).
26. Allai-Hassani A, Pan W, Dombrovski L, *et al.* (2007) "Structural and Chemical Profiling of the Human Cytosolic Sulfotransferases." *PLoS Biology* **5**: 1063–78.
27. Fedorov O, Sundström M, Marsden B, Knapp S. (2007) "Insights for the development of specific kinase inhibitors by targeted structural genomics." *Drug Discovery Today* **12**: 365–72.

Chapter 2

The Impact of Structural Proteomics on Macromolecular Structure Databases

James D. Watson*,† and Janet M. Thornton†

Introduction

High-throughput protein structure determination has progressed significantly during the last ten years. The development of improved automated technologies has had an immense impact on the procedures involved in structure determination. The scientific goals have varied in the different initiatives, from covering fold space to addressing specific biological questions; however, there is no doubt that the various structural genomics/proteomics initiatives across the globe have produced a vast number of protein structures in the last seven years. With the ramp up to full-scale production as well as investment in new technologies to address the more challenging problems of complexes and membrane proteins, the number of structures deposited

*Corresponding author: watson@ebi.ac.uk

†EMBL — European Bioinformatics Institute, Wellcome Trust Genome Campus, Hinxton, Cambridge, CB10 1SD, UK.

in macromolecular structure databases will rise.[1] In this chapter we discuss the effect of structural proteomics approaches on the structure databases, not only with regard to the number of entries but also the significance of the structures solved. Each new structure can only add to our knowledge and therefore each new deposit has value; however, it is argued that not all proteins are equal in this regard and that the first of a new fold or a biologically significant protein is of greater importance. A number of studies have looked at the impact of structural genomics projects with discussions on fold coverage, Pfam coverage, and functional prediction. We will review these studies and expand on them to discuss structural parameters, procedural influences, and effects on other databases.

Breakdown of Structural Genomics Structures

The RCSB Protein Data Bank (PDB) provides a Structural Genomics Information Portal (http://sg.pdb.org) where it maintains the TargetDB (http://targetdb.pdb.org/statistics/sites/PSI.html),[2] a database tracking the progress of the production and solution of structures from structural genomics initiatives throughout the world. This database was initially set up as part of the first Protein Structure Initiative (PSI-1) to allow the NIH-supported structural genomics initiatives to monitor the progress of each target assigned in that project.[3] However, this was soon expanded to include other worldwide initiatives and to allow non-SG groups to deposit the sequences of their targets to help prevent duplication of effort. In this work, we have used the TargetDB (13th March, 2007) to identify all the SG structures for further examination to investigate their contribution to the macromolecular structure databases in a number of ways.

Number of Structures

The most obvious impact of high-throughput methodology should be seen in the number of structures. Between 1st January, 2000 and the 13 March, 2007, there was a total of 4083 unique structural genomics/proteomics PDB deposits (after removal of multiple PDB deposits for the same target), representing 14.7% of the total deposits during this time. Of the SG deposits, the majority (2697) were solved using X-ray crystallography and a further 1081 were solved using NMR techniques. Information could not be retrieved for the remainder as 288 deposits were still "on hold" and a further 17 entries had been withdrawn.

The number of structures deposited in the PDB has increased steadily, year on year and the proportion of SG vs. non-SG deposits has been rising steadily (see Fig. 1) with the SG initiatives now contributing on average 16% of monthly total deposits.

Despite this rise, and the focus of SG projects on novel folds, it has recently been shown[4] that, overall, the rate of growth of new structural data has actually slowed and the PDB is not growing exponentially as commonly suggested. This conclusion was reached through examination of the growth of novel structures using a number of different measures.

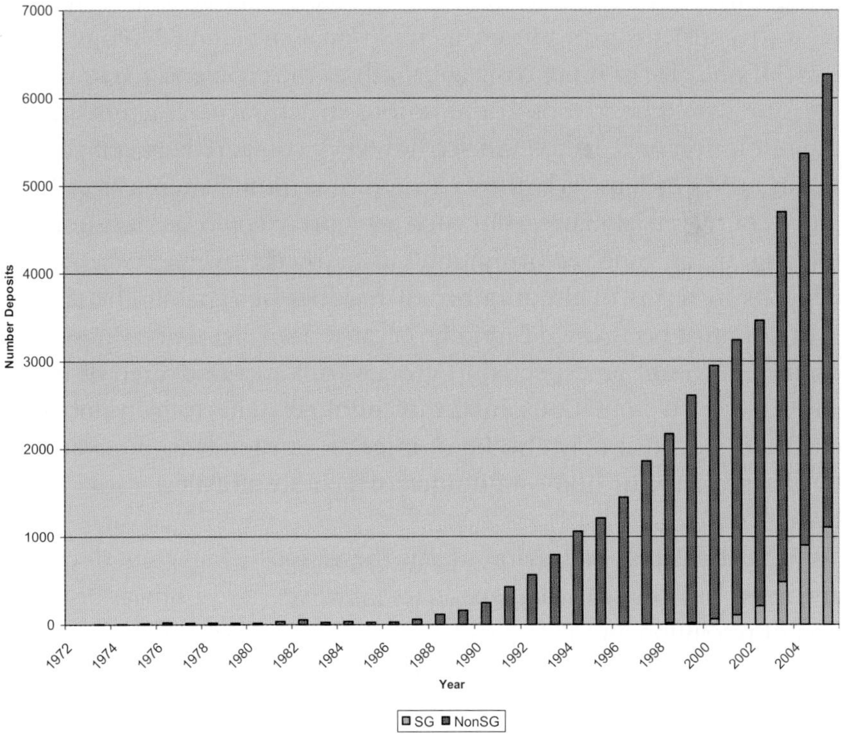

Fig. 1 Chart showing growth of PDB from SG and non-SG groups.

Additional evidence provided by the authors suggests that the SG initiatives (in particular those belonging to the NIH-PSI project) are starting to counter this decline whilst simultaneously producing more structures with low sequence similarity to anything else in the PDB.

Structural Parameters

A common perception is that the structural genomics initiatives have only targeted smaller proteins with structures more amenable to crystallization (the so-called "low hanging fruit") and that high-throughput procedures are likely to lower the quality of depositions. Assessing the quality of a deposition is very difficult with no real consensus on what constitutes "high quality." A recently developed service, the Protein Structure Validation Software suite, aims to do this through the integration of a variety of structure evaluation tools. An analysis using this server suggests that SG proteins exhibit similar quality scores to structures produced in traditional structural biology projects.[5] If one takes a more basic approach to the problem, a very rough measure would be to look at the various structural parameters such as the resolution or R-factor and see how they compare between the SG and non-SG groups. Whether a structure is "low hanging fruit" is a more controversial matter, but such a property could be measured by looking at the number of unique chains or the average size of the deposits in terms of the number of residues per chain. The size of proteins must be assessed carefully, because a smaller than average size of protein would be expected in groups with a large degree of NMR studies due to limitations of the technology. The total number of chains in a deposit may also be of interest, but tends to be swamped by the presence of homo-multimers which are not necessarily more "complex."

The data presented below detail the various parameters discussed and show that the structural genomics initiatives have a similar distribution of resolution and R-factors to the PDB as a whole (Figs. 2 and 3). The data reveal that the SG depositions are slightly smaller in size (based on the number of residues as shown in Fig. 4). Table 1(a) shows a remarkably similar distribution between SG and non-SG

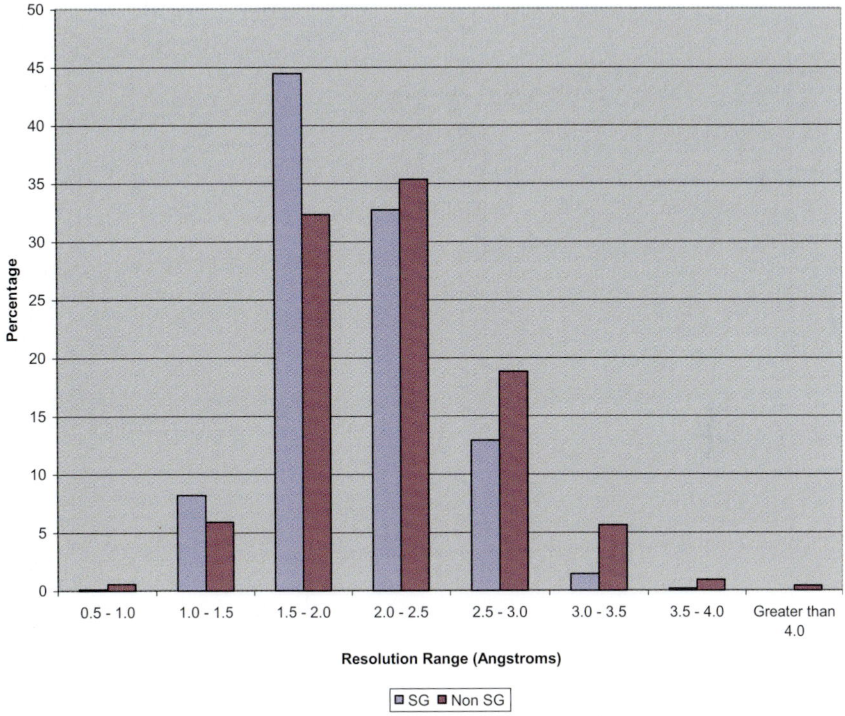

Fig. 2 Resolution distribution for SG and non-SG structures.

deposits as measured by the number of chains. There are slightly more monomers and fewer structures with greater than 10 chains in the SG structures, but the differences are minimal. The differences are more striking if we consider the number of unique chains (Table 1b). It should be noted, however, that the most complex structures (five or more unique chains) only account for about 1% of the entire PDB, and although interesting, are by no means common. It is perhaps unsurprising that the SG structures should have fewer unique chains as the very nature of their target selection means they tend to focus on individual proteins rather than protein–protein complexes.

Therefore, from the perspective of the database providers (wwPDB), one could argue that other than increasing the number of deposits, the structural genomics projects have had no significant

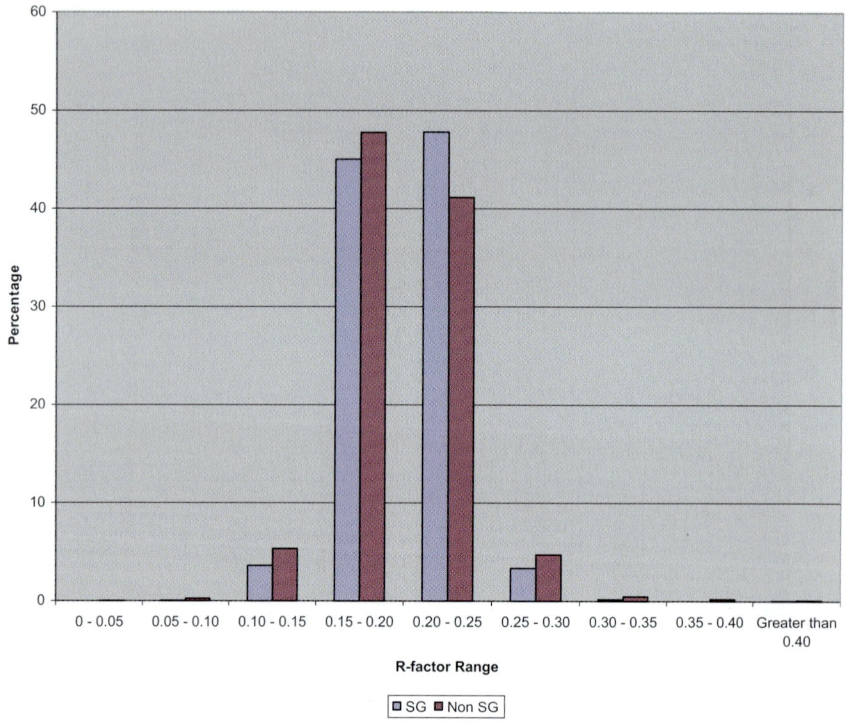

Fig. 3 R-factor distribution for SG and non-SG structures.

impact on the databases, unlike the very large protein complexes solved in traditional laboratories. These very complicated structures have caused a number of problems for the databases; most notably they have revealed the limitations of the original PDB file format and have caused additional problems with the visualization software attempting to display them.[6]

Fold Coverage and Homology Modeling

Although a large number of structures of varying size have been produced by the various initiatives, a recurring question has been how to assess the impact of these proteins. Sequence similarity is one measure of the uniqueness of any given structure. Many of the targets in structural

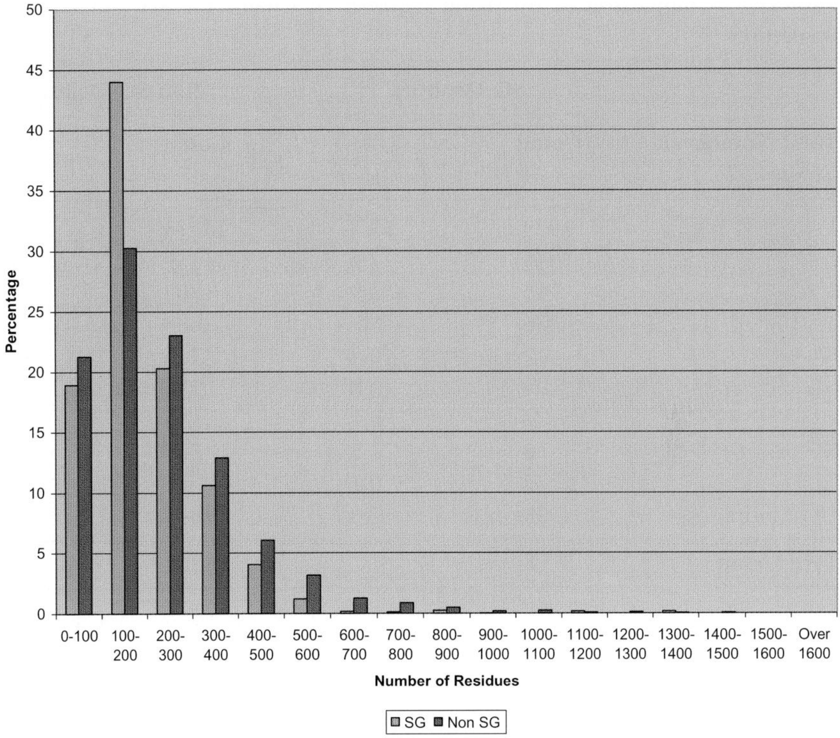

Fig. 4 Number of residues for SG and non-SG structures.

proteomics were initially selected to have less than 30% sequence identity to any other protein in the PDB (although a number of initiatives investigating medically important proteins do not tend to select targets on this criterion). However, in the field of structural proteomics, a protein fold is more commonly hailed as a useful measure of novelty, but how this is measured is not straightforward, and the CATH[7] and SCOP[8] domain classification resources have subtly different assignment protocols, though with a great degree of overlap. The various structural genomics initiatives have often been quick to identify their structures as having a "new fold." In many cases it could be argued that the structure they have deposited resembles a known fold in the core of the structure, but has been embellished outside the core in some way. The increased number of structures with modifications to existing folds

Table 1a Comparison of the Total Number of Chains for SG and non-SG Structures

	SG Deposits		Non-SG Deposits	
Total Number of Chains	**Count**	**%**	**Count**	**%**
1	2029	53.7	19535	50.3
2	1044	27.6	11139	28.7
3	146	3.9	1806	4.6
4	335	8.9	3668	9.4
5	17	0.4	236	0.6
6	104	2.8	950	2.4
7	4	0.1	84	0.2
8	56	1.5	577	1.5
9	5	0.1	41	0.1
10 or more	38	1.0	844	2.2
Total Count	3778		38880	

Table 1b Comparison of the Unique Chains for SG and non-SG Structures

	SG Deposits		Non-SG Deposits	
Number of Unique Chains	**Count**	**%**	**Count**	**%**
1	3522	93.2	31939	82.1
2	218	5.8	4971	12.8
3	23	0.6	1142	3.0
4	5	0.1	440	1.1
5 or more	10	0.3	388	1.0
Total Count	3778		38880	

is leading to a re-evaluation of the actual definition and classification of folds in CATH and SCOP (C. Orengo, University College London, personal communication April 2007).

An early analysis[9] of the status of structural genomics briefly examined the numbers of structures deposited (structural genomics initiatives were responsible for approximately 10% of all deposits at that point); the attrition rate at each stage in the SG pipeline; and the redundancy within each groups target list. The main focus, however, was on the fold and function of the structural genomics structures. Using SCOP definitions, the authors looked at the proportion of new folds resulting from structural genomics and suggested that the proportion of new folds over three six-month periods (Oct 2001–Mar 2002, Apr 2002–Sep 2002, Oct 2002–Mar 2003) was approximately 10%, 18% and 17%, respectively. However, the total number of new folds deposited in the PDB during this timeframe was rather low, so it is possible that the proportions quoted were unusually high. In addition, the study looked at the distribution of all folds from SG as compared to the PDB as a whole and also to some model organisms. Although a degree of bias is hinted at in their analysis, some clear patterns were identified. The most striking observation was that immunoglobulin-like beta sandwiches and TIM barrels were under-represented in the TargetDB and this was seen as a result of successful target selection strategies managing to avoid common folds. On the other hand, RNA/DNA–binding 3 helical bundles, P-loop containing nucleotide triphosphate hydrolases and SAM-dependent methyltransferases were over-represented in TargetDB and this was attributed, in part, to many of these proteins being drug targets. Overall, the structural genomics projects were shown to be producing more structures with a reasonable proportion of new folds.

We have looked at this question again using the CATH database and data collected up to the end of February 2007 (Tables 2a and 2b). Using a simple raw count of the number of CATH codes associated with SG and non-SG structures, we see a similar pattern to the aforementioned analysis, with the most dominant fold in the SG structures being the Rossmann fold. This suggests that the target selection has been successful at avoiding many of the common folds, but that it can still be difficult to identify some folds from sequence similarity.

A more comprehensive analysis[10] of SG targets looked at deposits from 11 consortia (including 8 of the original PSI-1 centers) focusing on their fold space coverage and the contribution of their structures to homology modeling. Unusually for such a study, the authors used both CATH and SCOP in their analysis, acknowledging that fold and superfamily classification is a subjective process.

Table 2a Most Common CATH Terms by Raw Count (non-SG)

CATH code	Count	%	Description
2.60.40.10	3783	7.3	IMMUNOGLOBULINS
2.40.10.10	2238	4.3	SERINE PROTEASES
3.40.50.300	878	1.7	PLOOP NTPases (ROSSMANN)
3.40.50.720	728	1.4	NADP BINDING ROSSMANN-LIKE DOMAINS
3.40.190.10	700	1.4	PERIPLASMIC BINDING PROTEIN LIKE
1.10.490.10	699	1.4	GLOBINS
3.30.70.270	617	1.2	TRANSFERASE/DNA
2.40.70.10	603	1.2	ACID PROTEASES
3.20.20.80	593	1.2	TIM BARREL
3.30.200.20	526	1.0	PHOSPHORYLASE KINASE
1.10.510.10	508	1.0	TRANSFERASE (PHOSPHOTRANSFERASE)
1.10.530.10	501	1.0	LYSOZYME
1.10.238.10	465	0.9	EF HAND
3.20.20.70	457	0.9	TIM BARREL

Table 2b Most Common CATH Terms by Raw Count (SG)

CATH code	Count	%	Description
3.40.50.620	62	3.5	tRNA SYNTHETASE (ROSSMANN FOLD)
3.40.50.720	60	3.4	NADP BINDING ROSSMANN-LIKE DOMAINS
3.20.20.70	34	1.9	TIM BARREL
3.40.50.150	34	1.9	ROSSMANN FOLD
3.40.50.300	33	1.8	ROSSMANN FOLD
1.10.10.10	30	1.7	WINGED HELIX
2.30.42.10	25	1.4	PDZ3 DOMAIN
2.40.50.140	23	1.3	OB FOLD NUCLEIC ACID BINDERS
3.40.50.1000	20	1.1	ROSSMANN FOLD
3.40.30.10	18	1.0	GLUTAREDOXIN
3.10.20.90	16	0.9	UBIQUITIN LIKE
2.60.120.10	15	0.8	JELLY ROLLS
3.40.640.10	15	0.8	TYPE 1 PLP ASP AMIDOTRANSFERASE LIKE
3.90.1150.10	15	0.8	ASP AMIDOTRANSFERASE

At the time of this study, the 11 initiatives provided 15% of the structures deposited in the PDB during 2004. Further analysis suggested that the domain count and mean sequence length of SG structures were comparable to that in the PDB and that structural "quality" had not been compromised with similar average R-factors and resolution for those structures solved by X-ray crystallography.

On in-depth examination of the novelty of the structures, a number of differences were found. In the article, novel structures were clearly defined as not only those forming new folds or new superfamilies in old folds, but also as those where a distant sequence homologue can only be detected using HMMs or the relationship can only be detected from structure. Using these definitions, the results indicated that a total of 67% and 65% of SG domains in CATH and SCOP, respectively were novel (11% and 10% respectively were in the top "novel fold" category), as compared to 21% in non-SG structures deposited during a similar period, 2% of which were "novel folds." Although CATH and SCOP do differ in their construction and definition, the results from this study agree and point towards an increased coverage of fold space.

This large proportion of new folds led to an investigation of the effect that SG structures have had on the reliable construction of homology models in 206 genomes. The results were significant, with over 9000 non-redundant gene sequences identified as being able to be modeled as a direct result of the 316 SG structures deposited by the date of the study. In addition, almost one-fifth of the new models were based on a sequence identity greater than 50% and could therefore be useful for ligand docking and drug design.

Several homology modeling sites are available such as Modbase[11] and Famsbase[12] where users can download homology models for sequences from various genomes. Many of the structural genomics depositions are among the structures used in the modeling process, but there are no readily available statistics on the number of models constructed directly from the SG structures. The Genome Modeling and Model Annotation site (http://www.biochem.ucl.ac.uk/cgi-bin/dlee/GeMMA), developed as a collaborative effort between the University College London (UCL) and the Midwest Centre for Structural Genomics (MCSG), details a number of good quality comparative models generated from each solved MCSG target. The site suggests that the 546 structures listed can be used to generate 99,844 good comparative models. If it could be assumed that this is representative of the modeling potential of the SG structures as a whole, then it would suggest that the 4083 SG structures to date would generate over 746,000 good quality models. Although this is almost certainly an overestimate it highlights the potential of the process as a whole.

A more recent study on the impact of structural genomics[13] looked once again at the structural coverage of protein families by SG structures, but with additional studies into the cost of production of structures. Notably, the investigation compared the results with leading structural biology labs and used several measures of the novelty of structures: Pfam coverage, direct sequence comparison, and SCOP coverage.

By mapping each Pfam family to the SG proteins and the rest of the PDB, the study identified the earliest deposition for each Pfam family. The results showed that the rate of structural characterization of Pfam families has remained constant (20 per month) even though the number of structures has continued to rise. In terms of the first deposition of a new family, the non-SG groups have seen a fall in numbers that appears to have been made up for by the SG groups, which now account for about 50% of all new structurally characterized Pfam families. Using direct sequence comparison (BLAST and PSI-BLAST), the authors find that this figure falls to 44% of the first solved Pfam structures produced by the SG groups.

Through the use of SCOP, the novelty of SG structures and the rest of the PDB was assessed (where one of the protein's domains represented the first structure of a fold, superfamily, family, protein or species). Out of the retrieved 17 654 non-SG domains, 269 (1.5%) were classed as novel at the "fold" level of SCOP. The SG initiatives were described as having solved 64 novel folds from their total of 757 solved domains (8.5%), suggesting a greater level of novelty than non-SG groups. In addition to this, 70% of protein domains solved by non-SG projects are a new experiment on a known structure (e.g. mutations, different bound ligands, etc.) and had these groups used PSI-BLAST[14] in their target selection process, they would have increased their proportion of new folds from under 5% to over 25%.

Functional Annotation

One criticism of the SG centers is that they have published relatively few papers describing their structures and those they have are, on average, cited less than those from non-SG groups. It would appear that publication is rapidly becoming a bottleneck in the high-throughput systems. The number of structures deposited in the PDB with little or no functional information has increased due to the high-throughput structural proteomics programs with many of the functional annotations given as "probable", "possible", "putative", etc. The assignment of function is a time-consuming and difficult process, compounded by the problem of how you actually define the function of a protein.[15] A large number of computational methods to assign function from sequence and structure[16] have been developed to help target experiments and improve function determination. Many of the services provided to the SG groups are open to and used by the entire community and would not have existed without the SG initiatives. Improved protein function prediction servers such as ProFunc,[17] ProKnow[18] and SiteEngine[19] can rapidly return the results of many sequence and structure-based analyses at the same time. As a consequence, the assessment of these predictions by experiment is already the rate-limiting step. To combat this, high-throughput experimental[20] assays are being developed, which will allow mass-screening for specific functions, some of which have already shown success.[21,22]

Ligands

Focusing on the functional importance and elucidating the particulars of enzyme mechanisms has often been a goal of traditional structure determination. To this end a number of identical structures are often deposited with varying ligands bound (often substrate analogues) in an attempt to tease out the fine details of the catalytic mechanism or to identify the possible effects of particular mutations. This is generally speaking not the remit of the various SG groups and it could therefore be expected that structures from these initiatives would not necessarily contain bound ligands. Further examination reveals that this is not the case, as 36% of the SG structures deposited to date have at least one ligand bound (excluding metals). The most commonly bound ligands from the SG structures are shown in Table 3a alongside those for the remaining structures in the PDB. The data show that common ligands tend to be the same in both groups but that the SG structures have a higher proportion of unknown ligands (UNL), presumably due to the high-throughput nature of the experiments and their rapid deposition.

Table 3a Frequency of the Most Common Ligands Bound to SG and non-SG Proteins (excluding Metals)

Non-SG Proteins			SG Proteins		
Ligand	**Count**	**%**	**Ligand**	**Count**	**%**
SO4	4599	10.2	SO4	353	15.8
NAG	2778	6.2	GOL	238	10.7
GOL	1938	4.3	EDO	156	7.0
HEM	1711	3.8	PO4	121	5.4
PO4	1229	2.7	ACT	72	3.2
MAN	956	2.1	UNL	54	2.4
ACT	703	1.6	ACY	45	2.0
FAD	608	1.3	MPD	37	1.7
ADP	559	1.2	FMN	35	1.6
EDO	554	1.2	NAD	35	1.6
GLC	534	1.2	FMT	33	1.5
NAD	498	1.1	SAH	29	1.3
PLP	407	0.9	ADP	28	1.3
FUC	383	0.9	CIT	28	1.3
MPD	335	0.7	HEM	28	1.3
NAP	328	0.7	PLP	28	1.3
ACY	323	0.7	MES	27	1.2
ATP	322	0.7	NAP	26	1.2
FMN	311	0.7	FAD	25	1.1
Total ligands	45057		Total ligands	2230	

Table 3b Frequency of the Most Common Metals Bound to SG and non-SG Proteins

Non-SG Proteins			SG Proteins		
Metal	Count	%	Metal	Count	%
CA	3500	19.0	ZN	309	26.3
ZN	3410	18.5	CL	232	19.8
MG	3023	16.4	MG	207	17.6
CL	2171	11.8	NA	109	9.3
NA	1503	8.1	CA	96	8.2
MN	961	5.2	NI	37	3.2
K	600	3.2	FE	31	2.6
FE	520	2.8	MN	30	2.6
CU	456	2.5	K	17	1.4
CD	360	1.9	HG	15	1.3
HG	274	1.5	IOD	13	1.1
NI	269	1.5	PT	12	1.0
CO	222	1.2	CD	11	0.9
FE2	189	1.0	CO	11	0.9
IOD	134	0.7	BR	9	0.8
BR	91	0.5	FE2	9	0.8
CU1	87	0.5	CU	4	0.3
XE	52	0.3	CU1	3	0.3
SR	47	0.3	ZN2	3	0.3
Total metals	18467		Total metals	1174	

Of the 64% of SG structures without ligands bound, almost one-fifth have some kind of metal bound to them. If the SG and non-SG structures are compared for metal binding (Table 3b), it is evident once again that the most common metals are similar in both except for one striking omission — there are far fewer calcium atoms identified in the SG structures than the PDB as a whole. This could, however, be due to the difficulties in correctly identifying the exact metal from electron density.

Publication

One of the greatest problems with delays to publication is that the public release of the structural data can be held back. This is demonstrated

Table 4 Time from Deposition to Release for SG and non-SG Structures deposited between 01/01/2000 and 31/01/2007

	Non-SG Structures		SG Structures	
Time Difference	**Count**	**%**	**Count**	**%**
30 days or less	4674	16.9	966	25.8
31–90 days	6499	23.5	772	20.6
91–365 days	13 061	47.3	1662	44.3
Greater than 365 days	3401	12.3	349	9.3
Total Structures	27 635		3749	

by the many structures deposited in the PDB that remain "on hold" pending publication.[23] Examination of the difference between the deposition and release dates of structures deposited up to February 1st, 2007 suggests that the structural genomics deposits are actually released sooner than those from traditional laboratories (Table 4). This could, in part, be explained by the fact that the NIH-funded centers are mandated, as part of their grant award conditions, to deposit their data prior to publication. This rapid public data release is in direct contrast to traditional structure determination, where a structure can remain on hold indefinitely until publication, and was designed to prevent data being kept out of the public domain for extended periods of time. It is, therefore, not in the NIH-funded centers' remits to write papers; rather, they are supposed to develop and use automatic annotation pipelines and encourage third parties to perform more detailed analyses. This is reflected in literature searches for structural genomics structures, which indicate 25.3% have a reference as compared with the PDB as a whole, where 80.8% are published.

Impact on Database Protocols

It is clear that the structural proteomics projects have had some effect on the size and content of the macromolecular structure databases. What is not so obvious, yet more significant, are the effects of structural proteomics on the protocols used and the basic make-up of the various databases. The vast number of structures predicted to arrive from structural proteomics led to a re-evaluation of a number of aspects of the macromolecular structure databases, including how the data were stored, the ways users can access the databases and deposition protocols used.

One of the most profound effects has been an increase in the automation of procedures at all steps in the determination and deposition of structural data. At the RCSB, the deposition process was automated with the creation of the AutoDep Input Tool (ADIT) and subsequent development of PDB_Extract to help capture information from commonly used structure determination programs.[24] Harvesters such as this now routinely extract details from depositor's files so as to retrieve information on crystallographic conditions and the various parameters chosen. This type of software has been around for some time, such as the CCP4[25] suite of programs for protein crystallography first set up in 1979, and is commonly used in regular labs. The impact of the structural proteomics programs is that many of the tools are now linked together to a greater degree with improved data management.[26] These types of improvements to the deposition process have resulted in a greater degree of uniformity in the structural databases with fewer pieces of information missing, although it has not totally eradicated errors or omissions.

This theme of integration of data and resources is now more widespread as a direct consequence of the high-throughput processes. In anticipation of a flood of data, many of the major databases have looked to share their annotations through new technologies such as webservices. This secondary annotation has become a much more automated process. A prime example of this is the SIFTS (Structure Integration with Function, Taxonomy and Sequence) initiative started at the European Bioinformatics Institute (EBI) in 2001 (http://www.ebi.ac.uk/msd-srv/docs/sifts). This aims to integrate the EBI's structure database (MSD) with the sequence database UniProt[27] by cleaning up the various taxonomies, allowing accurate cross-referencing and data exchange between the two resources. This is part of a much larger project eFamily (http://www.efamily.org.uk), which aims to extend the integration of MSD and UniProt with Pfam, CATH and SCOP. These resources now directly share access to their services in order to cross-reference each other with the most up-to-date information possible (and to identify possible mistakes in annotation).

Although it is likely that these services would have integrated in due course, the process has been accelerated as a direct response to the problem of updating vast numbers of new entries each week.

Increased automation and integration has had additional knock-on effects to increase efforts in standardization such as the development of Laboratory Information Management Systems (LIMS), Protein Information Management System (PIMS), and ontologies. Gene Ontology (GO) has been around for a number of years now and is the recognized standard of functional ontologies for gene products. Recently, this type of ontology has been extended within the BioSapiens project (http://www.biosapiens.info/page.php?page=home) in order to develop a protein annotation ontology. This has arisen as a direct result of increased integration of resources across Europe which requires each member to use a standard way to describe the various elements of protein sequence, structure and function (G. Reeves, European Bioinformatics Institute, personal communication April 2007).

Conclusions

Structural proteomics has had a number of effects on the macromolecular structure databases in both their content and construction. The number of structures deposited has increased over the years, with the proportion from high-throughput initiatives rising to over 16% of the total monthly deposits worldwide. The structural "quality" of these deposits as measured by various structural parameters has remained high, although it is true to say that the high-throughput structures tend to be smaller on average and less "complex". This does not necessarily mean that the structures deposited by these groups are easier to solve; rather, it reflects the fact that the structural proteomics projects target individual structural domains rather than full-length gene products or protein complexes. One observation that does come to light is that structural proteomics depositions do tend to be released slightly ahead of those from traditional laboratories. The apparent lack of publications by the SG initiatives reflects the "remit" of the projects with their focus on structural determination and rapid release, rather than biological interpretation which is an essential aspect for the publication of structures today.

The promise of vast numbers of structures from high-throughput methods prompted the various databases to reassess how the data are deposited, annotated, and accessed. The first major effect of this has been the development of tools to automatically extract information from the depositors' files, as the experiments are being conducted, for use in automated deposition tools. This has greatly improved the deposition process by speeding up the whole process at the same time as reducing the chances of incorrect or incomplete entries being deposited.

The next most important effect was to improve data exchange between the various online resources. The expected influx of structural data was seen to pose problems for the accurate annotation of sequence and structural properties stored across different sites. Through standardization of terminology and sharing of resources, the sequence and structural databases are now able to more easily cross-reference one another, which has allowed for accurate and automatic updates to their annotations. This process of standardization has been taken up more widely and attempts are now under way to extend the commonly used ontologies to describe structural, sequence and functional features. Overall, the macromolecular structure databases have been transformed as a direct result of the structural proteomics initiatives. In addition, researchers in traditional laboratories have gained from protocols and high-throughput technologies making their way into the mainstream, as well as improved resources with new databases and analytical tools.

References

1. Service R. (2005) "Structural biology. Structural genomics, round 2." *Science* **307**(5715): 1554–58.
2. Chen L, *et al.* (2004) "TargetDB: a target registration database for structural genomics projects." *Bioinformatics* **20**(16): 2860–62.
3. Berman HM, Westbrook JD. (2004) "The impact of structural genomics on the protein data bank." *Am J Pharmacogen* **4**(4): 247–52.

4. Levitt M. (2007) "Growth of novel protein structural data." *Proc Natl Acad Sci USA* **104**(9): 3183–88.
5. Bhattacharya A, Tejero R, Montelione G.(2007) "Evaluating protein structures determined by structural genomics consortia" *Prot Struct Funct Bioinf* **66**(4): 778–95.
6. Dutta S, Berman HM. (2005) "Large macromolecular complexes in the Protein Data Bank: a status report." *Structure* **13**(3): 381–88.
7. Greene LH, *et al.* (2007) "The CATH domain structure database: new protocols and classification levels give a more comprehensive resource for exploring evolution." *Nucleic Acids Res* **35**: D291–97.
8. Hubbard TJ, *et al.* (1999) "SCOP: a Structural Classification of Proteins database." *Nucleic Acids Res* **27**(1): 254–56.
9. Bourne PE, *et al.* (2004) "The status of structural genomics defined through the analysis of current targets and structures." *Pac Symp Biocomput* **9**: 375–86.
10. Todd AE, *et al.* (2005) "Progress of structural genomics initiatives: an analysis of solved target structures." *J Mol Biol* **348**(5): 1235–60.
11. Pieper U, *et al.* (2006) "MODBASE: a database of annotated comparative protein structure models and associated resources." *Nucleic Acids Res* **34**: D291–95.
12. Yura K, Yamaguchi A, Go M. (2006) "Coverage of whole proteome by structural genomics observed through protein homology modeling database." *J Struct Funct Genomics* 7(2): 65–76.
13. Chandonia JM, Brenner SE. (2006) "The impact of structural genomics: expectations and outcomes." *Science* **311**(5759): 347–51.
14. Altschul SF, *et al.* (1997) "Gapped BLAST and PSI-BLAST: a new generation of protein database search programs." *Nucleic Acids Res* **25**(17): 3389–402.
15. Watson JD, *et al.* (2007) "Towards fully automated structure-based function prediction in structural genomics: a case study." *J Mol Biol* **367**(5): 1511–22.
16. Watson JD, Laskowski RA, Thornton JM. (2005) "Predicting protein function from sequence and structural data." *Curr Opin Struct Biol* **15**(3): 275–84.
17. Laskowski RA, Watson JD, Thornton JM. (2005) "ProFunc: a server for predicting protein function from 3D structure." *Nucleic Acids Res* **33**: W89–93.
18. Pal D, Eisenberg D. (2005) "Inference of protein function from protein structure." *Structure* **13**(1): 121–30.
19. Shulman–Peleg, Nussinov AR, Wolfson HJ. (2004) "Recognition of functional sites in protein structures." *J Mol Biol* **339**(3): 607–33.
20. Kuznetsova E, *et al.* (2005) "Enzyme genomics: application of general enzymatic screens to discover new enzymes." *FEMS Microbiol Rev* **29**(2): 263–79.
21. Proudfoot M, *et al.* (2004) "General enzymatic screens identify three new nucleotidases in *Escherichia coli*. Biochemical characterization of SurE, YfbR, and YjjG." *J Biol Chem* **279**(52): 54687–94.

22. Kuznetsova E, *et al.* (2006) "Genome — wide analysis of substrate specificities of the *Escherichia coli* haloacid dehalogenase-like phosphatase family." *J Biol Chem* **281**(47): 36149–61.
23. Rigden DJ. (2006) "Understanding the cell in terms of structure and function: insights from structural genomics." *Curr Opin Biotechnol* **17**(5): 457–64.
24. Yang H, *et al.* (2004) "Automated and accurate deposition of structures solved by X-ray diffraction to the Protein Data Bank." *Acta Crystallogr D Biol Crystallogr* **60(**10): 1833–39.
25. The CCP4 suite: programs for protein crystallography. (1994) *Acta Crystallogr. D Biol Crystallogr* **50**(Pt 5): 760–63.
26. Winn MD, *et al.* (2002) "Ongoing developments in CCP4 for high-throughput structure determination." *Acta Crystallogr D Biol Crystallogr* **58(**11): 1929–36.
27. The Universal Protein Resource (UniProt). (2007) *Nucleic Acids Res* **35**: D193–97.

Chapter 3

The Impact of 3D Structures on a Protein Knowledgebase: From Proteins to Systems

Ursula Hinz*,† and Amos Bairoch†

Introduction

One of the central themes of modern biology is the integration of large amounts of data from different sources, which is no trivial task. More and more protein sequences are available from an ever-increasing range of organisms. Likewise, the number of solved 3D-structures has increased exponentially in the past years, and the number of published papers is simply overwhelming. Information is presented in many different formats, such as nucleic acid and protein sequences or data generated by large-scale transcriptomics and proteomics initiatives. Integrating this information requires a central database that serves both as a repository for protein information and as a hub that provides links to other relevant databases and tools. The Universal Protein Resource KnowledgeBase (UniProtKB) provides the user community with such a resource.[1] It combines protein sequences with functional annotation derived from multiple sources and presents this in a highly structured and user-friendly manner. UniProtKB contains

*Corresponding author: Ursula.Hinz@isb-sib.ch

† Swiss Institute of Bioinformatics, Centre Medical Universitaire, 1, rue Michel-Servet, 1211 Geneve 4, Switzerland.

two sections that together provide access to all published protein sequences. UniProtKB/TrEMBL focuses on the incoming flood of protein data and provides access to highly automated, electronic annotation. UniProtKB/Swiss-Prot gives access to manually annotated protein sequences. The latter strongly emphasizes direct experimental evidence, and displays much more information, with greater detail and depth.[2] UniProtKB/Swiss-Prot is non-redundant, and one entry represents the products of one gene in a given species. Manual annotation is essential to ensure reliability of the data that are extracted from 3D-structures, scientific literature and the results of bioinformatics tools. Rules and controlled vocabularies need to be established to guarantee consistency in the way the data are represented, and to facilitate data retrieval. Constant improvement of annotation tools is crucial to keep pace with the ever-increasing volume of data and the increasingly detailed information that must be made available to the user community. Access to a universal protein resource facilitates the choice of interesting proteins for further structural or biochemical analysis.

Three-dimensional-structures provide a wealth of experimental evidence. Proteins can be assigned to known families, the details of protein-protein interactions can be elucidated and the ligand-binding sites identified. Information derived from 3D-structures is of great interest to the research community, particularly when it is easily accessible and shown in context with a high-quality protein sequence and combined with other types of experimental data. Not everybody feels at ease with 3D-structures, just as other scientists may not feel at ease reading the biomedical literature. Whatever the field of research, it takes time to gather information about a new protein or an unfamiliar metabolic or signaling pathway.

UniProtKB/Swiss-Prot annotators extract data from 3D-structures and from the literature. Information is presented in a clear and highly-structured manner, so it is easily accessible to computer searches, and people working in the lab as well, be they structural biologists, physicians or molecular biologists. The "general annotation" section summarizes the function of a protein, its interactions with other macromolecules, ligands or cofactors, its sub-cellular location and expression pattern, as well as information about possible post-translational modifications. The "sequence annotation" section provides residue-specific annotation, and emphasizes physiological events. Information derived from 3D-structures is shown in the context of a full-length, wild-type sequence. In addition, information is provided on natural sequence variants, such as polymorphisms, disease mutations, or variations arising from alternative splicing, alternative promoter usage or alternative initiation.

With the advent of large-scale sequencing, the number of uncharacterized protein sequences is increasing exponentially. It is not certain whether some proteins that are predicted from genome sequencing really exist in a living cell. Dealing with such "hypothetical proteins" and predicting their function in a meaningful way is a central problem in modern biology. The available knowledge about well-known proteins can be used to try and assign roles to uncharacterized proteins. The validity of these predictions depends on the solidity of the underlying data, and on the care that is exercised in interpreting them. It is the role of a protein knowledgebase to facilitate access to relevant experimental findings that can be used as a basis for meaningful interpretations.

UniProtKB: Facilitating Access to 3D-Structures

In the following section, we will show how a knowledgebase can make use of information gathered from 3D-structures, and integrate this with other information that is available about a protein. We will highlight how the value of 3D-structure information can be increased by complementary data, and how the data, in turn, can augment the value of other types of experimental evidence. We will try in each case to give clear examples. 3D-structures are most interesting when combined with a maximum amount of other types of information, and when they are easily accessible. In the case of UniProtKB/Swiss-Prot, this means facilitating data retrieval by showing the "recommended name" for a protein and using the official nomenclature, whenever this exists. In addition, we strive to provide a list of all the synonyms used in the literature. Likewise, both the recommended gene name and its commonly used synonyms are indicated. Indeed, finding one's way in the jungle of protein and gene names is a complex task. The human mind clearly abhors standard nomenclature, and so, a multitude of common names and variously spelled abbreviations coexist. Thus, the recommended name for the *L. pneumophila* peptidyl-prolyl cis-trans isomerase (Q5ZXE0) is "outer membrane protein MIP," but it is also referred to as "macrophage infectivity potentiator." Worse, non-standard names are redundant, meaning that the same name is used for several different proteins: the name "organic anion transporter 4" (OAT4) has been used in the literature for both SLC22A11 (Q9NSA0) and SLC22A9 (Q8IVM8).

In the case of enzymes, the EC number provides one type of highly informative standard nomenclature. The enzyme commission maintains a database of enzyme-catalyzed reactions and allocates a unique EC number to each type of reaction. In UniProtKB/Swiss-Prot, this number is shown together with other synonyms for a protein name, and a link to the ENZYME database provides access to a wealth of enzyme-specific information.

Once the naming question is solved, UniProtKB provides access to all protein structures, using two complementary approaches. On the one hand, cross-references to PDB give direct access to structural information about the protein itself, while links to HSSP and SMR provide access to 3D-models based on experimental structures for proteins with similar sequences (Refs. 3–5; http://swissmodel.expasy.org/repository/). On the other hand, 3D-structures are used as sources of information for manual annotation in UniProtKB/Swiss-Prot, where the data are displayed in text format. Literature citations indicate the source of the information and increase the visibility of the data. 3D-structures are matched onto UniProtKB by the Macromolecular Structure Database (MSD) group at the European Bioinformatics Institute (EBI) by a semi-automated procedure, using Blast searches, with the sequence corresponding to the structure and information about the taxonomic origins of the proteins as provided by the authors.[6] An experienced annotator manually checks the output. During annotation or when updating entries, the annotators of UniProtKB/Swiss-Prot manually check the links to PDB and give feedback to MSD in case of discrepancies. The manual check is important for maintaining high-quality annotation, and for correcting possible mapping errors. Often, orthologs or paralogs have identical, or almost identical, protein sequences, and it is necessary to rely on the author's indications to know where to map the link to PDB. Regular updates ensure that UniProtKB is well synchronized with PDB and keeps pace with the rapid increase in the number of 3D-structures. For human proteins alone, the number of entries with a link to PDB increased from 1776 to 2487 during the year 2006, and has attained 2950 during 2007.

Database cross-references to PDB display the PDB identifier, the resolution (where applicable), and the chains and extents of the protein that are present in the structure, as this information facilitates comparison between otherwise similar structures. Through the links to PDB, users have several options of accessing tools to analyze and visualize the structures. Additional links to protein structure-related databases and tools can be found at www.ExPASy.org.

UniProtKB provides several different means to identify and retrieve proteins with known 3D-structures to accommodate the specific needs of different types of users. All entries with links to PDB contain the keyword "3D-structure." Keywords are controlled vocabulary, and their definitions are accessible on the ExPASy website (www.expasy. org/txt/keywlist.txt). Similarly, the complete list of all proteins with known 3D-structures in UniProtKB/Swiss-Prot is displayed at the following location: www.expasy.org/cgi-bin/lists?pdbtosp.txt.

Representation of Taxonomic Groups in PDB and in UniProtKB/Swiss-Prot

Traditionally, 3D-structures have been determined for well-characterized proteins from model organisms. Typically, experimental structures are available for 10%–15% of all proteins from a model organism. Thus, out of *ca.* 10 800 entries with experimental 3D-structures in UniProtKB/Swiss-Prot, almost 40% are from mammals, including about 2900 human proteins. Most of the remaining structures concern other model organisms. Amongst bacteria, *E. coli* may be the best characterized: 3D-structures have been solved for almost 20% of the 4931 different proteins encoded on its genome. With the advent of structural genomics projects, numerous structures have been determined for uncharacterized proteins, but their species range is still limited. Recent research efforts have concentrated on proteins with possible biomedical applications, both from humans and from pathogens.[7–12]

UniProtKB gives high priority to the annotation of proteins from model organisms with known 3D-structures. Thus, the vast majority of proteins with experimental 3D-structures are found in the manually annotated UniProtKB/Swiss-Prot section, while a small minority is present in the computer-annotated UniProtKB/TrEMBL section. As new structures are determined, every year a substantial number of such entries are manually annotated or merged with existing entries. UniProtKB/Swiss-Prot Release 54.5 of November 2007 contained 46 410 cross-references to PDB. Often, several structures are available for a single protein, e.g. 375 for phage T4 lysozyme (P00720) alone, while human hemoglobin alpha and beta chains (P69905 and P68871) each have about 170 links to PDB. As a consequence, these 46 410 links correspond to 11 848 individual UniProtKB/Swiss-Prot entries.

Another priority is the annotation of the complete proteome for model organisms. For all species where complete proteome sets are available, the keyword "Complete proteome" appears in each entry that is part of that proteome. Full annotation of all members of a proteome takes time, and while UniProtKB provides access to complete proteome sets for numerous bacteria and archaea, and for some eukaryotes, such as fungi and insects, some of these protein entries are still in the computer-annotated UniProtKB/TrEMBL section. The human proteome is of obvious interest, both for biomedical research groups and for proteomics initiatives. Even though sequencing of the human genome was officially completed in 2001, interpretation is still ongoing; new genes are still being discovered, while gene models undergo important changes. So far, the human gene nomenclature committee (HGNC) has approved 24 405 gene symbols, but this number includes 2594 named pseudogenes. Likewise, gene symbols attributed to RNA genes and to the loci coding for the variable regions of immunoglobulins and T-cell receptors have to be deduced from this number. On the other hand, UniProtKB/Swiss-Prot contains 335 entries for human proteins that are still waiting for an official gene name. All in all, this results in an estimate of *ca.* 21 000 protein-coding human genes. Currently, UniProtKB/Swiss-Prot contains entries for over 17 800 human genes and aims to complete the human proteome by September 2008. Annotation of other mammalian model organisms is another on-going activity that has high priority. UniProtKB/Swiss-Prot currently contain entries for over 14 000 mouse proteins. The numbers for rat and bovine proteins are lower, but are increasing rapidly, as new protein sequences become available. Even when the sequence of the entire genome has been determined for a given species, it still takes time, effort and financing to complete the annotation of the protein sequences and to make them available to the public.

Integrating 3D-Structures with Functional Annotation

The Challenge of Presenting Complex Data I — General Information

In the proteomics field it is important to have rapid access to reliable and detailed information about all known proteins in order to identify truly novel and interesting proteins and select targets for further studies. Similarly, when studying a large complex, one can expect to find already known components of the complex, interesting new components, but also a number of unrelated contaminants.[13] Access to a protein knowledgebase helps to identify interesting candidates, and eliminate others without wasting time and effort.

One of the challenges of data management is finding efficient ways to display ever more complex information, and presenting it in a structured manner that corresponds to the requirements of different types of users. This is achieved by presenting the information, on the one hand, under the form of a "general annotation" section, with specific subsections for each type of data, and on the other hand, by separately presenting residue-specific data in a "sequence annotation" section: the extents of a domain, residues that undergo post-translational modification or are part of the active site of an enzyme, and the like. Links to specialized databases, keywords and ontologies complement the manual annotation, as shown in Fig. 1, where selected parts of the UniProKB/Swiss-Prot entry for human glucosylceramidase (P04062) are displayed. This well-characterized enzyme is associated with lysosomal membranes and is crucial for normal turnover of glucosylceramides. Interaction with saposin C promotes its association with phospholipid membranes and is necessary for optimal enzyme activity. Mutations that alter its activity lead to the accumulation of glucosylceramide and to Gaucher disease, the most prevalent lysosomal storage disease. Depending on the nature and position of the mutation, the disease is more or less severe. Several 3D-structures are available for this protein, and they have been crucial to establish the enzyme mechanism and to understand the effects of the numerous disease mutations and polymorphisms that have been described. Many of these lead to subtle alterations of the protein structure, causing a decrease in activity, and often also enhanced susceptibility to proteolytic degradation.[14,15]

With the advent of whole genome sequences, large scale transcriptomics and proteomics studies and the ever-increasing general volume of data to be integrated, a pressing need for standardized formats and controlled vocabularies has emerged. The development of such vocabularies, and the conversion of the existing annotation to a new format is an arduous and time-consuming task, but it is compulsory to guarantee the consistency of the annotation throughout the database, and to facilitate data retrieval and comparison with other databases. UniProtKB/Swiss-Prot presents more and more of the information using standardized formats and controlled vocabularies, e.g. the section that describes sub-cellular location. Likewise, the section that deals with protein similarity, i.e. identifying the family to which a protein belongs, and the domains it contains, is presented in standardized format. Moreover, this section is complemented by database cross-references to InterPro, PROSITE, Pfam, SMART, PIRSF and related databases that provide additional information about protein domains, their definitions, the organisms in which they are found and the other domains they may be associated with.[16–20]

UniProtKB/Swiss-Prot Entry	**P04062** (GLCM_HUMAN)
Protein names	**Glucosylceramidase** [Precursor] *Also known as:* EC 3.2.1.45

General annotation

Catalytic activity:	D-glucosyl-N-acylsphingosine + H(2)O = D-glucose + N-acylsphingosine.
Subunit:	Interacts with saposin C.
Subcellular location:	Lysosome; lysosomal membrane; peripheral membrane protein; lumenal side. Note=Interaction with saposin C promotes membrane association.
Involvement in disease:	Defects in GBA are the cause of Gaucher disease (GD) MIM:230800]; also known as glucocerebrosidase deficiency. GD is the most prevalent lysosomal storage disease, characterized by accumulation of glucosylceramide in the reticuloendothelial system. Different clinical forms are recognized depending on the presence (neuronopathic forms) or absence of central nervous system involvement, severity and age of onset.
Pharmaceutical use	Available under the names Ceredase and Cerezyme (Genzyme). Used to treat Gaucher's disease.
Sequence similarities	Belongs to the glycosyl hydrolase 30 family.

Sequence annotation

Sites			
Active site	274		Proton donor
Active site	379		Nucleophile
Amino acid modifications			
Glycosylation	58		N-linked (GlcNAc...)
Glycosylation	98		N-linked (GlcNAc...)
Glycosylation	185		N-linked (GlcNAc...)
Disulfide bond	43 with 91		
Disulfide bond	57 with 62		
Natural variations			
Natural variant	409		N → S in GD; common mutation; alters interaction with saposin C and membranes and thereby reduces enzyme activity; mild
Experimental info			
Mutagenesis	43	43	C → S: Loss of activity
Mutagenesis	57	57	C → S: Loss of activity
Mutagenesis	62	62	C → S: Loss of activity

Keywords

Biological process	Lipid metabolism, Sphingolipid metabolism
Cellular component	Lysosome, Membrane
Disease	Disease mutation, Gaucher disease
Molecular function	Glycosidase, Hydrolase
Technical term	3D-structure, Direct protein sequencing, Pharmaceutical

Database crossreferences

PDB	*Structures determined by X-ray crystallography:* 1OGS. 2.00 Angstrom resolution. Chains A/B map to 40-536. 1Y7V. 2.40 Angstrom resolution. Chains A/B map to 40-536. 2F61. 2.50 Angstrom resolution. Chains A/B map to 40-536.
InterPro	IPR001139. Glyco_hydro_30.
Gene3D	G3DSA:3.20.20.80. Glyco_hydro_cat. 1 hit.
Pfam	PF02055. Glyco_hydro_30. 1 hit.
PRINTS	PR00843. GLHYDRLASE30.

Fig. 1 Shows selected parts of the UniProtKB/Swiss-Prot entry for human glucosylceramidase (P04062), illustrating the structured presentation of data in a section for "general annotation," "sequence annotation" and keywords, and complementing links to other databases. For practical reasons, only a part of each section is shown. To show the emphasis on the use of controlled vocabularies and standard formats, information that is presented in a standardized manner is highlighted by using a light blue background.

Functional annotation is further complemented by links to organism-specific databases that provide additional information, such as EcoGene for *E. coli*.[21] Links to Gene Ontologies (GO terms) complement the annotation, and provide a rapid access to highly precise information.[22] Keywords provide another source of "at-a-glance" information and facilitate retrieval of related proteins. Presently, UniProtKB/Swiss-Prot uses close to 900 strictly defined keywords. These include broad functional categories, such as "Hydrolase" or "DNA repair," post-translational modifications, subcellular locations and specific human diseases.

It is, of course, tempting to extrapolate and to propagate information from well-known proteins to other members of a group of proteins that is defined by sequence similarity, or that contain the same type of domains. Such extrapolation has to be done prudently, particularly when it is carried out for organisms that are not closely related at the taxonomic level, or when dealing with multigene families. Even though the outcome is often correct, there are numerous pitfalls, as illustrated by the following examples.

When trying to assign a function to a novel protein, it is preferable, whenever possible, to not only examine an isolated protein from a single organism, but also to compare it with orthologs and paralogs and other family members. Residues that are conserved from bacteria to plants and mammals are likely to be important for the function of the protein, or to be essential for proper folding. 3D-structures permit the identification of the roles of such conserved residues, but this does not remove all of the problems of information propagation, as illustrated by the case of the DNA formamidopyrimidine glycosylases. These enzymes are part of the FPG family that is found from bacteria (P05523) to mammals (Q96FI4) and that plays an important role in base excision repair of DNA that has been damaged by oxidation or by mutagenic agents. The 3D-structure of the *E. coli* DNA formamidopyrimidine-DNA glycosylase (P05523), combined with site-directed mutagenesis and biochemical characterization, has revealed the details of the enzymatic reaction and the basis for substrate specificity. The protein catalyzes two separate reactions. It has DNA glycosylase activity and removes damaged bases from DNA, with a preference for oxidized purines. It then generates a single-strand break at the site of the removed base. Catalytic residues are found both at the N-terminus and in the C-terminal FPG-type zinc finger domain. While the key catalytic residues are conserved from bacteria to humans, the overall sequence similarity is low. In addition, the FPG-type zinc finger domain is not found in members of the mammalian family. Similarly, the substrate specificity of the bacterial enzymes is somewhat different from that of the mammalian enzymes: both act on damaged bases, but contrary to the bacterial family members, the mammalian enzyme has a preference for oxidized pyrimidines. This clearly shows the limits for function prediction and automated annotation, even when proteins are related and all the catalytic residues are conserved.

Most proteins do not function as isolated entities, but are part of biochemical or signaling pathways and need to interact with other proteins to exert their physiological function; others are part of stable complexes, ranging from dimers to complexes such as the ribosome. Most often, these protein-protein interactions are detected by biochemical experiments, such as co-immunoprecipitation, yeast-2-hybrid screens, or isolation of a complex. More recently, mass spectrometry has been used for large scale analysis of proteins that bind to a target peptide.[23,24] Elucidating such signaling cascades and protein-protein interactions is essential to further our understanding of cell biology, but it is inherently difficult to assay by these methods. Fortunately, recent efforts in the structural proteomics field and from individual research groups aim to fill in this information.[25] In UniProtKB/Swiss-Prot annotation of protein-protein interactions has high priority. Information on these interactions is derived from the literature; these interaction are frequently associated with a 3D-structure, and are shown in the "subunit" lines in the "general annotation" section. UniProtKB annotators also contribute to the annotation of binary protein-protein interactions in the IntAct database.[26] IntAct annotation is often derived from large-scale studies, but also from the general literature. The UniProtKB/Swiss-Prot annotation derived from IntAct is shown under the heading "interaction" in the "general annotation" section.

As protein structures are determined for highly concentrated solutions or protein crystals, non-physiological multimers may appear. Likewise, there is always a risk of picking up "contaminants" when constituents of protein complexes are identified by biochemical means, or the risk of non-physiological interactions when proteins are overexpressed in a heterologous system. Thus, protein interactions should be confirmed not only by evidence from 3D-structures, but also by other types of experiments. In a protein knowledgebase, information from different sources is shown side by side, making it easier to evaluate the available data. For example, information about the sub-cellular location of the proposed interaction partners and their expression levels during development or in a particular tissue indicates whether the interaction observed in the 3D-structure is likely to take place in a living cell. On the other hand, it can also indicate that the protein in question may also be found in other tissues, and in other sub-cellular compartments — a hypothesis that can subsequently be tested.

The Challenge of Presenting Complex Data II– 3D-Structures and Sequence Annotation

Topology of membrane proteins

A large number of prediction tools exist for the prediction of transmembrane segments. Most of these tools are based on the assumption that the transmembrane segment is alpha-helical and perpendicular to the plane of the membrane. Hydrophobic segments that are about 20 amino acid residues long are considered potential transmembrane helices. This works very well for proteins that conform to the underlying assumptions, both for type I transmembrane proteins, such as glycophorin A (P02724) and for type II transmembrane proteins, such as the light-harvesting protein B-800/850 alpha chain (P26789). Likewise, signal sequence predictions from different algorithms are usually very reliable.

The situation is radically different for pore-forming proteins, such as a voltage-gated potassium channel (P0A334), where amphipathic transmembrane segments are exposed to a polar environment on one side, and to the membrane lipids on the other side. These pore-forming transmembrane regions contain polar and charged amino acids, and thus escape detection by standard prediction programs. Some membrane proteins, e.g. *E. coli* aquaglyceroporin (P0AER0), present transmembrane helices that cross the membrane at an angle, or contain short helices that are buried within the lipid bilayer, but do not cross the membrane. Similarly, intramembrane proteases, such as the rhomboid-type protease glpG (P09391), cleave their substrates within the membrane. In such cases, prediction programs do not recognize these atypical intramembrane regions, and experimental evidence from 3D-structures is crucial for correct annotation. Moreover, not all integral membrane proteins have alpha-helical transmembrane segments. Proteins, such as maltoporin (P02943) and ompA (P0A910), from the outer membrane of Gram-negative bacteria form typical beta-barrel structures, both with pore-forming and structural roles.[27] There again, 3D-structures have been essential for the determination of the topology and for the subsequent development of prediction tools.[28]

Annotation of ligand binding sites

Proteins can bind to a multitude of ligands, ranging from other polypeptides, DNA or RNA to nucleotides and other organic ligands and ions. These ligands fulfil many roles: they can be essential cofactors, regulate the activity of a protein, or serve as substrates. 3D-structures of proteins with bound ligands show the interactions between a protein and its substrates. They reveal how a transcription factor interacts with DNA, or how a ligand induces a conformational change. In other cases, such structures are not available, but the identity of the ligands has been determined by other means. UniProtKB/Swiss-Prot integrates information from 3D-structures and other sources and gives high priority to the annotation of ligand-binding sites. The "general annotation" section summarizes the catalytic activity and the cofactor requirement of a protein. In the "sequence annotation" section, "binding site" lines indicate residues that form hydrogen bonds or Pi-stacking interactions with organic ligands, while "nucleotide binding" lines are used to indicate short nucleotide-binding regions of the polypeptide. The role of other important residues is shown on "site" lines, e.g. for residues that are critical for interactions with other macromolecules. "Metal binding" lines indicate interactions with simple metal ions, but also with metal ions that are complexed by essential cofactors, such as chlorophyll or heme. This structured presentation of the data facilitates searches and selective retrieval of information. Throughout UniProtKB/Swiss-Prot, bibliographical references indicate the sources of the information.

Numerous organic ligands are seen in 3D-structures, ranging from buffer molecules to physiological substrates and effectors and their synthetic analogs. These analogs can be more or less similar to the physiological ligand they are supposed to represent — for example phosphate or sulfate ions may occupy part of the binding site of ATP or NADP. Other ligands represent substrates and transition state analogs. The question that arises is how should one represent these ligands for maximum accuracy and clarity? PDB files contain a multitude of names for such ligands, reflecting the manifold nature of the analogs and inhibitors that are used. Besides, several different names are commonly found in the literature to design a single chemical entity. In a database, the use of a standardized vocabulary is essential to facilitate searches. In UniProtKB/Swiss-Prot, priority is given to the physiological situation, and the ligand names reflect their presumed physiological role. Thus, whenever possible, the term "substrate" instead of the full chemical name, is used. When an enzyme acts on two substrates, common names are used to indicate the ligands. A typical example is the spermidine/spermine synthase family. It groups proteins with clear sequence similarity but different function, such as spermidine synthase (EC 2.5.1.16) and putrescine N-methyltransferase (EC 2.1.1.53), and these enzymes bind similar, but distinct ligands. Spermidine synthase is a highly conserved enzyme involved in polyamine biosynthesis that is found in all kingdoms of life. It utilizes S-adenosylmethioninamine (decarboxylated SAM) to convert putrescine into spermidine. Several structures are available, but most of these show the apoprotein. Still, the human enzyme (P19623) has been crystallized with bound substrates and *T. maritima* spermidine synthase (Q9WZC2) has been crystallized and complexed with a multisubstrate adduct inhibitor. The functionally important residues are conserved throughout the spermidine/spermine synthase family, making it possible to propagate the ligand binding sites to other family members, including the putrescine N-methyl transferases. These enzymes utilize S-adenosyl-L-methionine to convert putrescine into N-methylputrescine and S-adenosyl-L-homocysteine, and the UniProtKB/Swiss-Prot entries for these proteins all show the physiological ligands.

Metal ions are essential for the function of numerous proteins, and they often fulfill both structural and catalytic roles, as shown for the rattlesnake protease adamalysin-2 (P34179). Metal ions stabilize small structural elements, such as the zinc fingers that are present in the drosophila protein ush (Q9VPQ6) or the EF hands in chicken skeletal muscle troponin C (P02588). They also promote protein-protein interactions, e.g. between the yeast superoxide dismutase 1 copper chaperone and the copper and zinc-dependent superoxide dismutase SOD1 (P00445 and P40202). Other metal ions provide a link between the protein and essential cofactors, such as the heme in cytochrome P450 2C5 (P00179) or the chlorophyll in the light-harvesting protein B-800/850 alpha chain (P26789). Such residue-specific information is shown in the "sequence annotation" section. Interactions via amino acid side chains are considered the norm, and interactions involving backbone amine nitrogen or carbonyl oxygen atoms are indicated explicitly, as can be seen in the entries for adamalysin-2 and quinoprotein glucose dehydrogenase-B (P34179, P13650). When several metal ions are bound, as in *Acinetobacter* quinoprotein glucose dehydrogenase-B, they are numbered, starting with the ion that binds to the most N-terminal residue.

In UniProtKB/Swiss-Prot, the situation in a healthy living cell is emphasized. Care is taken to show the physiological ligand, provided that this information is available. For novel proteins, biochemical characterization is often not yet done, and 3D-structures solved by structural proteomics groups are not always accompanied by a publication. Buffer-derived metal ions may occupy physiological binding sites, but do not give information about the physiological ligand. In such cases, annotation merely indicates the presence of divalent metal ions without further details, e.g. for the structural proteomics target UPF0135 protein ybgI (P0AFP6).

Active site residues

Often 3D-structures permit the formulation of a hypothesis about the catalytic mechanism of an enzyme — a hypothesis that has to be tested by other types of experiments. Mutagenesis of catalytic residues is expected to lead to a complete loss of activity. Conversely, loss of activity can also be caused by mutation of an essential residue that is far from the active site. Therefore, hypotheses about proposed active site residues based on mutagenesis studies should be verified by 3D-structures. Again, combining the results from complementary analyses enhances the value of each.

In the literature the meaning of the term "active site residue" is quite variable. In the broadest sense it is used to indicate all residues that line the active site pocket. This groups, without distinction, catalytic residues with those that have other roles in the enzyme reaction, e.g. by binding the substrate or cofactor, and those that are simply within a certain range. In a knowledgebase, it is preferable to provide detailed information on the roles of individual residues. Therefore, in UniProtKB/Swiss-Prot, the term "active site" is reserved for amino acid residues that are directly involved in catalysis, e.g. as a proton donor or acceptor, as in the case of human glucosylceramidase or *H. pylori* 3-dehydroquinate dehydratase (P04062, Q48255), or as a nucleophile, as in the case of *P. aeruginosa* GDP-mannose 6-dehydrogenase (P11759). When a large protein, such as the genome polyprotein from human Coxsackievirus (Q65900), has several domains with enzymatic activities, the type of activity is indicated in the sequence annotation. Residues that play important, but not essential roles in catalysis are indicated under "site." The term "site," therefore, includes residues that lower the pK of the active site residue, as shown for human NADP-dependent alcohol dehydrogenase (P14550), or that stabilize the transition state, as in the case of the *E. coli* protein ushA or *H. pylori* 3-dehydroquinate dehydratase (P07024, Q48255).

3D-structures have permitted the identification of covalent reaction intermediates with the active site nucleophile, as shown for human PTPN1 or *P. angusta* peroxisomal copper amine oxidase (P18031, P12807), as well as the elucidation of the reaction mechanism and mode of action of proteins, e.g. for *S. aureus* coenzyme A disulfide reductase (EC 1.8.1.14) (O52582), where the reaction is mediated by a redox-active cysteine residue. Numerous protein families contain catalytic metal ions, e.g. alcohol dehydrogenases, metalloproteases and dioxygenases, such as the bacterial protein alkB (P05050) and its mammalian paralogs ALKBH2 (Q6NS38) and ALKBH3 (Q96Q83). While the overall sequence similarity is low, sequence analysis suggests that these proteins all function as alpha-ketoglutarate-dependent dioxygenases. 3D-structure analysis combined with site-directed mutagenesis and biochemical characterization has elucidated the reaction mechanism, for both the *E. coli* protein and the human orthologs. While both alkB and its mammalian paralogs ALKBH2 and ALKBH3 are important for the repair of alkylated DNA, the bacterial and mammalian enzymes have slightly differing substrate specificities. Another mammalian family member, ALKBH1 (Q13686), has apparently no such activity, and the role of the other mammalian alkB homologs still remains to be elucidated. This inability to fully elucidate reaction mechanisms again shows the limitations in the propagation of information from one family member to another, and especially the risks associated with extrapolations across a wide taxonomic range.

Covalent modifications

Post-translational modifications are an important source of protein variety, and play important roles in the proper functioning or targeting of proteins. Protein 3D-structures can provide solid evidence for numerous types of covalent modifications, e.g. the disulfide bonds and glycosylations found in secreted proteins such as the snake venom L-amino-acid oxidase (Q6STF1), and in the extracytoplasmic part of transmembrane proteins. Numerous proteins contain covalently bound cofactors, such as the heme in human cytochrome c (P99999) or pyridoxal phosphate in ornithine aminotransferase (P04181), while for others the cofactor is derived from the post-translational modification of a residue, as it is the case for the topaquinone present in peroxisomal copper amine oxidase (P12807). While modifications with relatively small groups can often be confirmed by protein crystallography, other biologically important modifications, such as sumoylation or mono-ubiquitination, need to be determined by biochemical analyses and mutagenesis studies, as in the case of PARK7 (Q99497), a multifunctional protein where sumoylation is required for normal activity. As discussed, the importance of combining information from different sources is evident.

Combining 3D-Structures with Protein Sequences

It is important to look at a 3D-structure both in the context of the sequence that was used for the experiment and of the full-length, wild-type sequence, and combine these with what is known about natural sequence variation and the results of *in vitro* mutagenesis studies. In UniProtKB/Swiss-Prot, one entry corresponds to the product(s) of a single gene from a given organism. The product may be a protein sequence based on a gene model from a microbial genome, or a mammalian protein sequence based on numerous cDNAs and ESTs. When several sequences have been submitted for a gene product, these entries are merged, and the sequence differences are carefully annotated. Cross-references to EMBL and GenBank give access to the original data, and further information is provided by links to additional genome and organism-specific databases. Entries for human proteins are based on an average of 6.3 nucleotide sequence submissions, thus ensuring high sequence quality. Alternative splicing is an important source of sequence variation for eukaryotic proteins, and this is documented in the "alternative sequence" lines. Each alternatively spliced isoform has its own identifier and links to specific information. Unexplained sequence differences are indicated as "sequence conflict."

Within a population, individuals differ from each other due to natural sequence variation. A wealth of information is available about human sequence variants, particularly at the nucleotide level. Such information can be found in databases, such as dbSNP or HGMD, and of course in the scientific literature.[29,30] It is important for researchers to have access to a central repository that lists mutations leading to altered amino acid sequences, and in addition describes the effects of these mutations, as far as this is known. In UniProtKB/Swiss-Prot, annotation of natural sequence variation has high priority. Each human variant has its own identifier and its own web page that summarizes the available information for that particular variant. When a 3D-structure is available for the protein itself or an ortholog with sufficient sequence similarity, the mutated amino acid is shown in the context of a 3D-structure model.[31]

It is essential (but not always easy) to distinguish between disease mutations, polymorphisms and sequencing errors. Mutations found in cancerous tissues are not necessarily a cause of the disease. For many genetically determined diseases, the severity of the symptoms and the age of onset of the disease can vary considerably, making it difficult to draw a line between polymorphisms and disease mutations. Environmental factors, infections and general ill-health, age and genetic predisposition, all contribute to the clinical outcome. Disease mutations can be validated by their segregation pattern, and by biochemical characterization of the mutant protein, provided that an appropriate test is available. From the structural point of view, polymorphisms, defined as neutral missense mutations that do not lead to a clinical phenotype, are not expected to lead to dramatic changes in protein structure. Polymorphisms are likely to be conservative substitutions, or to affect residues that are not particularly important for proper folding and protein-protein interactions. It is of obvious interest to elucidate the 3D-structures of proteins that are linked to human disease. 3D-structures can explain why some mutations have no apparent effect on the function of a protein, while others lead to a clinical phenotype. Moreover, determining the molecular basis for disease may help to develop adequate treatments.

Missense mutations are the most common defect associated with human disease.[32] Diseases can also be caused by the absence of a protein, due to deletions or mutations leading to premature stop codons, or due to altered expression levels. Some diseases are caused by the inactivation of an essential biological process, such as peroxisome biogenesis. The peroxisomal proteins are synthesized in the cytoplasm and are then transported into the peroxisome, and thus mutations interfering with peroxisomal protein import lead to peroxisome biogenesis disorders of variable severity. Cytoplasmic PEX5 (P50542) recognizes the C-terminal targeting sequence that is present in the majority of peroxisomal proteins, and the PEX5-cargo complex then binds to a docking complex on the peroxisomal membrane. This binding step is essential for protein import and biogenesis of functional peroxisomes. A nonsense mutation that abolishes PEX5 expression leads to Zellweger syndrome, a fatal peroxisome biogenesis disorder. Missense mutations that permit some degree of protein import give rise to less severe disease phenotypes. Mutation of Asn-526 to Lys in the sixth TPR repeat causes neonatal adrenoleukodystrophy (NALD). Asn-526 interacts with the C-terminal end of the targeting peptide, and mutation to Lys may interfere with proper binding due to the loss of hydrogen bonds and steric clash. The mutation of Ser-600 to Trp gives a milder phenotype. The residue is at the end of a loop, and is not in proximity to the bound target peptide. The mutation of Ser-600 to Trp probably affects local protein folding.

Typical examples of complex and multifactorial diseases are high blood pressure and obesity, and this is also a subject where the underlying physiology is highly complex and the literature on this is particularly abundant. It is the role of a protein knowledgebase to assemble key findings and present them in a way that will also enable non-specialists to quickly become familiar with the subject. Literature citations indicate the sources of the information and provide a starting point for further reading. This is exemplified by the mineralocorticoid receptor (MCR; P08235) and its role in the regulation of plasma sodium levels and blood pressure.[33] The MCR is part of the nuclear receptor superfamily. In the presence of bound ligands, it translocates from the cytoplasm to the nucleus and acts as a transcription factor. Its physiological ligand is aldosterone. It stimulates sodium reabsorption and potassium secretion in the kidney, and thus regulates plasma sodium homeostasis and blood pressure. The 3D-structure of its ligand-binding domain has recently been determined.[34,35]

Aldosterone binds to the receptor via interactions with Asn-770, Gln-776, Arg-817 and Thr-945. Numerous polymorphisms and disease mutations have been described, such as mutations that affect aldosterone-binding residues, which lead to apparent mineralocorticoid resistance and type 1 pseudohypoaldosteronism (PHA1). On the other hand, the mutation of Ser-810 to Leu alters the conformation and the hydrogen-bonding network of the ligand-binding site.[36] As a result, compounds that are normally weak activators or antagonists of the receptor, such as the drug spironolactone, are able to activate the receptor, leading to early-onset hypertension with exacerbation during pregnancy.

Annotation of natural sequence variation has high priority in UniProtKB/Swiss-Prot. Out of the *ca.* 17 800 entries for human proteins in UniProtKB/Swiss-Prot, about 7600 mention sequence variants (polymorphisms and disease mutations). This corresponds to over 37 000 individual sequence variants. Typically, about 200 individual variants are listed for well-characterized proteins, where mutations are involved in human disease, such as the cystic fibrosis transmembrane conductance regulator CFTR (P13569), the hemoglobin beta subunit (P68871) and fibrillin-1 (P35555). Over 460 individual variants are shown for coagulation factor VIII (P00451), a 2332-amino acid protein. On the average, there are *ca.* 2 sequence variants per human entry.

Outlook: Function Prediction for Uncharacterized and Hypothetical Proteins

These days, many protein sequences are derived from gene predictions, particularly for microbial proteins. Although matching ESTs or cDNAs are frequently present in eukaryotes, it can still be difficult to determine the correct reading frame. Also, one should bear in mind that some RNA molecules function as transcripts, and that the existence of transcribed RNA is not a proof of the existence of a protein. This means that there are two distinct problems: a predicted protein may simply not exist in the living cell, or the predicted sequence may present a certain number of errors. In UniProtKB/Swiss-Prot, we have recently introduced a Protein Existence (PE) line to indicate the level of confidence that a protein really exists in a living cell. In the ideal case, direct evidence is available at the protein level for the existence of a given protein. Examples of such experimental confirmation are protein sequencing (Edman), a clear identification by mass spectrometry (MS), or an *in vivo* post-translational modification (PTM). 3D-structures provide additional evidence about the likelihood of the existence of an otherwise uncharacterized protein. Of particular interest are cases where the structures of similar proteins from distantly related organisms have been determined, e.g. yeaZ, where the structure has been determined for the protein from *E. coli* (P76256), *S. typhimurium* (Q7CQE0) and for *T. maritima* (Q9WZX7). This proves that the underlying sequences have properties that are necessary for assuming a similar stable fold, even when there are amino acid substitutions, insertions and deletions. Nevertheless, the physiological role of yeaZ is still not at all clear.[37]

Traditionally, 3D-structures have been used to elucidate the exact mode of action of well-known proteins. Now, with structural proteomics, 3D-structures are used as a basis to predict the functions and interaction sites of novel proteins.[38] For this type of prediction, it is essential to take into consideration not only the overall sequence similarity and the fold of the protein, but also the presence of known key residues that are needed for protein function, e.g. active site residues, or residues necessary for ligand binding.[39,40] The value of even the vaguest prediction crucially depends on the validity of the underlying data set, and on the care that is invested in the predictions.

In the case of uncharacterized proteins, it is sometimes possible to predict a function when the protein can be assigned to a well-known family based on sequence similarity, or clear similarity at the level of the 3D-structure, even when sequence similarity is low. The most valuable structures are those that contain fortuitously bound ligands from the cloning organism and thus provide an indication of the likely function of the protein. Examples are the putative oxidoreductase ydhF (P76187) and the alcohol dehydrogenase yqhD (Q46856). The structure of both shows bound NADP, and for yqhD, catalytic zinc ions are also shown, in agreement with the proposed alcohol dehydrogenase activity. However, the nature of the physiological substrate remains unknown.[41] In other cases, proteins can be assigned to a family based on their overall sequence similarity and domain structure, but they lack key functional residues, indicating that their function must be different. This is often observed after gene duplications and for some members of multigene families. For example, the isochorismatase family protein yecD (P0ADI7) is clearly a member of the isochorismatase family, but it has a glycine at the position of the conserved active site cysteine and is thus not expected to have isochorismatase activity.

While for some protein families, all members have the same type of function, this is by no means the rule. Even for groups of clearly related proteins, function may not be conserved, as exemplified by the citrate lyase beta subunit. Citrate lyase activity is found in archaea, bacteria and eukaryotes. While in eukaryotes an ATP-dependent citrate lyase converts citrate to acetyl-coenzyme A and oxaloactetate, in prokaryotes the corresponding reaction takes place during anaerobic fermentation of citrate and is catalyzed by an ATP-independent enzyme that requires magnesium as cofactor. Bacterial citrate lyase is a complex containing six copies each of the catalytic alpha and beta subunits, and of the gamma subunit that acts as acyl-carrier protein. The alpha subunit replaces the acyl group of acetyl-coenzyme A with a citryl group. Subsequently, the beta subunit cleaves citryl-coenzyme A and releases oxaloacetate, with concomitant regeneration of acetyl-coenzyme A. In microbes with citrate lyase activity, such as *E. coli*, these genes are grouped in an operon. In others, e.g. *M. tuberculosis* (O06162), a protein with clear similarity to citrate lyase beta subunit is present, but the genes for the alpha and gamma subunits are absent. Recently a human gene (CLYBL; Q8TDH8) corresponding to the bacterial citrate lyase beta subunit has been identified, but again, the genes for the other subunits are missing. The 3D-structure of the *M. tuberculosis* citrate lyase beta subunit-like protein complexed with oxaloacetate and magnesium has been determined. The residues that mediate the interaction with the ligands are conserved in the whole family, from *E. coli* (P0A9I1) to *M. tuberculosis* (O06162) and humans (Q8TDH8), and one can presume that they fulfil analogous functions. However, one expects that these citE homologs must be involved in some other metabolic pathway, and that their enzymatic activity must be somewhat different in organisms that lack the other subunits of the magnesium-dependent citrate lyase complex. Clearly, annotation that is based exclusively on sequence similarity can easily go wrong, even when all the key residues are conserved. Manual annotation is essential to reach a meaningful and accurate result. For the future success of automated annotation, it will be crucial to find ways to integrate a maximum of information.

Information from 3D-structures and evidence from genetic and physiological analyses, data showing protein-protein interactions, expression patterns, regulation, sub-cellular location and post-translational modifications, are all complementary pieces of a puzzle. Modern biology is highly specialized, and it is a challenge for individual scientists and for databases to combine all the available information from different large-scale proteomics and transcriptomics efforts, and from individual research groups to arrive at a global picture. They need to find ways to deal with ever more, and always more detailed, information, and present it in a way that corresponds to the needs of different types of users. Meeting these challenges requires efforts towards standardization, establishment of collaborations and exchange of data, to facilitate access to all the available information by members of the scientific community and avoid duplication of efforts. Another challenge is to enhance public awareness of the need for adequate funding so that databases can keep pace with the information flow and new developments, and continue to fulfill their role as freely-accessible sources of reliable and up-to date information based on experimental evidence.

Acknowledgments

Sincere thanks to Lydie Bougueleret for her suggestions and for critically reading the manuscript. Many thanks also to Marie-Claude Blatter and Salvo Paesano for help with the illustrations.

References

1. The UniProt Consortium. (2007) "The Universal Protein Resource (UniProt)." *Nucleic Acids Res* **35**: D193–97.

2. Boeckmann B, Blatter MC, Famiglietti L, *et al.* (2005) "Protein variety and functional diversity: Swiss-Prot annotation in its biological context." *C R Bio.* **328**: 882–99.
3. Berman H, Henrick K, Nakamura H, Markley JL. (2007) "The worldwide Protein Data Bank (wwPDB): ensuring a single, uniform archive of PDB data." *Nucleic Acids Res* **35**: D301–03.
4. Schneider R, Sander C. (1996) "The HSSP database of protein structure-sequence alignments." *Nucleic Acids Res* **24**: 201–05.
5. Kopp J, Schwede T. (2006) "The SWISS-MODEL Repository: New features and functionalities." *Nucleic Acids Res* **34**: D315–18.
6. Velankar S, McNeil P, Mittard-Runte V, *et al.* (2005) "E-MSD: an integrated data resource for bioinformatics." *Nucleic Acids Res* **33**: D262–65.
7. Badger J, Sauder JM, Adams JM, *et al.* (2005) "Structural analysis of a set of proteins resulting from a bacterial genomics project." *Proteins* **60**: 787–96.
8. Marsden RL, Lewis TA, Orengo CA. (2007) "Towards a comprehensive structural coverage of completed genomes: a structural genomics viewpoint." *BMC Bioinformatics* **8**: 86–86.
9. Abergel C, Coutard B, Byrne D, *et al.* (2003) "Structural genomics of highly conserved microbial genes of unknown function in search of new antibacterial targets." *J Struct Funct Genomics* **4**: 141–57.
10. Albeck S, Burstein Y, Dym O, *et al.* (2005) "Three-dimensional structure determination of proteins related to human health in their functional context at the Israel Structural Proteomics Center (ISPC)." *Acta Crystallogr D* **61**: 1364–72.
11. Busso D, Poussin-Courmontagne P, Rose D, *et al.* (2005) "Structural genomics of eukaryotic targets at a laboratory scale." *J Struct Funct Genomics* **6**: 81–88.
12. Fogg MJ, Alzari P, Bahar M, *et al.* (2006) "Application of the use of high-throughput technologies to the determination of protein structures of bacterial and viral pathogens." *Acta Crystallogr D Biol Crystallogr* **62**: 1196–207.
13. Rappsilber J, Ryder U, Lamond AI, Mann M. (2002) "Large-scale proteomic analysis of the human spliceosome." *Genome Res* **12**: 1231–45.
14. Premkumar L, Sawkar AR, Boldin-Adamsky S, *et al.* (2005) "X-ray structure of human acid-beta-glucosidase covalently bound to conduritol-B-epoxide. Implications for Gaucher disease." *J Biol Chem* **280**: 23815–19.
15. Liou B, Kazimierczuk A, Zhang M, *et al.* (2006) "Analyses of variant acid beta-glucosidases: effects of Gaucher disease mutations." *J Biol Chem* **281**: 4242–53
16. Mulder NJ, Apweiler R, Attwood TK, *et al.* (2007) "New developments in the InterPro database." *Nucleic Acids Res* **35**: D224–28.
17. Hulo N, Bairoch A, Bulliard V, *et al.* (2006) "The PROSITE database." *Nucleic Acids Res* **34**: D227–30.
18. Finn RD, Mistry J, Schuster-Bockler B, *et al.* (2006) "Pfam: clans, web tools and services." *Nucleic Acids Res* **34**: D247–51.
19. Letunic I, Copley RR, Pils B, *et al.* (2006) "SMART 5: domains in the context of genomes and networks." *Nucleic Acids Res* **34**: D257–60.

20. Wu CH, Nikolskaya A, Huang H, *et al.* (2004) "PIRSF: Family classification system at the Protein Information Resource." *Nucleic Acids Res* **32**: D112–14.
21. Rudd KE. (2000) "EcoGene: a genome sequence database for Escherichia coli K-12." *Nucleic Acids Res* **28**: 60–64.
22. Harris MA, Clark J, Ireland A, *et al.* (2004) "The Gene Ontology (GO) database and informatics resource." *Nucleic Acids Res* **32**: D258–61.
23. Sali A, Glaeser R, Earnest T, Baumeister W. (2003) "From words to literature in structural proteomics." *Nature* **422**: 216–25.
24. Cho S, Park SG, Lee DH, Park BC. (2004) "Protein-protein interaction networks: from interactions to networks." *J Biochem Mol Biol* **37**: 45–52.
25. Romier C, Ben Jelloul M, Albeck S, *et al.* (2006) "Co-expression of protein complexes in prokaryotic and eukaryotic hosts: experimental procedures, database tracking and case studies." *Acta Crystallogr D Biol Crystallogr* **62**: 1232–42.
26. Kerrien S, Alam-Faruque Y, Aranda B, *et al.* (2007) "IntAct — open source resource for molecular interaction data." *Nucleic Acids Res* **35**: D561–65.
27. Schulz GE. (2002) "The structure of bacterial outer membrane proteins." *Biochim Biophys Acta* **1565**: 308–17.
28. Bagos PG, Liakopoulos TD, Hamodrakas SJ. (2005) "Evaluation of methods for predicting the topology of beta-barrel outer membrane proteins and a consensus prediction method." *BMC Bioinformatics* **6**: 7.
29. Sherry ST, Ward MH, Kholodov M, *et al.* (2001) "dbSNP: the NCBI database of genetic variation." *Nucleic Acids Res* **29**: 308–11.
30. Stenson PD, Ball EV, Mort M, *et al.* (2003) "Human Gene Mutation Database (HGMD): 2003 update." *Hum Mutat* **21**: 577–81.
31. Yip YL, Scheib H, Diemand AV, *et al.* (2004) "The Swiss-Prot variant page and the ModSNP database: a resource for sequence and structure information on human protein variants." *Hum Mutat* **23**: 464–70.
32. Antonarakis SE, Cooper DN. (2003) Mutations in human genetic disease. In: DN Cooper (ed), Encyclopedia of the Human Genome, pp. 227–53. Nature Publishing Group, London.
33. Fuller PJ, Young MJ. (2005) "Mechanisms of mineralocorticoid action." *Hypertension* **46**: 1227–35.
34. Bledsoe RK, Madauss KP, Holt JA, *et al.* (2005) "A ligand-mediated hydrogen bond network required for the activation of the mineralocorticoid receptor." *J Biol Chem* **280**: 31283–93.
35. Fagart J, Huyet J, Pinon GM, *et al.* (2005) "Crystal structure of a mutant mineralocorticoid receptor responsible for hypertension." *Nat Struct Mol Biol* **12**: 554–55.
36. Nichols CE, Johnson C, Lockyer M, *et al.* (2006) "Structural characterization of Salmonella typhimurium YeaZ, an M22 O-sialoglycoprotein endopeptidase homolog." *Proteins* **64**: 111–23.

37. Watson JD, Laskowski RA, Thornton JM. (2005) "Predicting protein function from sequence and structural data." *Curr Opin Struct Biol* **15**: 275–84.
38. George RA, Spriggs RV, Bartlett GJ, *et al.* (2005) "Effective function annotation through catalytic residue conservation." *Proc Natl Acad Sci USA* **102**: 12299–304.
40. Rigden DJ. (2006) "Understanding the cell in terms of structure and function: insights from structural genomics." *Curr Opin Biotechnol* **17**: 457–64.
41. Sulzenbacher G, Alvarez K, Van Den Heuvel RH, *et al.* (2004) "Crystal structure of *E. coli* alcohol dehydrogenase YqhD: evidence of a covalently modified NADP coenzyme." *J Mol Biol* **342**: 489–502.

Chapter 4

Bioinformatics of Protein Function

Arthur M. Lesk*,†, Vineet Sangar†, Helen Parkinson‡ and James C. Whisstock§

Introduction

A genome sequence contains the plans of the potential life histories of an organism, but the implementation of genetic information depends on the functions of the encoded proteins and nucleic acids. Bioinformatics uses databases and algorithms to trace the logical progression: protein sequence → protein structure → protein function. However, function is a more elusive property than sequence or structure.

The success of high-throughput methods of sequencing, and structure determination and modeling, has given rise to copious experimental sources of functional information, joined by microarrays and chromatin immunoprecipitation techniques for the study of DNA-binding proteins. However, the assignment of function to gene products in the absence of direct experimental information remains an important challenge for computational molecular biology.[1–4] Pressing

*Corresponding author. Email: aml25@psu.edu

†Department of Biochemistry and Molecular Biology, and the Huck Institute for Genomics, Proteomics and Bioinformatics, The Pennsylvania State University, University Park, PA 16802, USA.

‡European Bioinformatics Institute, Wellcome Trust Genome Campus, Hinxton, Cambridgeshire CB10 1SD, United Kingdom.

§Department of Biochemistry and Molecular Biology, Victorian Bioinformatics Consortium, Monash University, Clayton Campus, Melbourne, Victoria 3168, Australia.

examples arise when genes responsible for diseases are identified but their specific functions are unknown.

Whole-genome sequencing projects, and metagenomic data, constitute a major source of proteins of unknown function. Annotation of a genome involves assignment of functions to gene products, in most cases on the basis of amino acid sequence alone. Three-dimensional structure can aid the assignment of function, and the challenge of structural genomics projects is to make structural information available for novel uncharacterized proteins.[5–11]

A common practice, supported by tools of bioinformatics, is to transfer annotation from a previously annotated homologous protein.[12–14] This approach is based on the assumptions that: a) as homologous proteins have similar sequences and structures, they have similar functions; and b) the annotation of the homologue is correct. Often, but certainly not always, these assumptions are valid.

In this chapter, we survey the field of "functional bioinformatics," focussing on the question of how to assign protein function. Some methods attempt to infer function directly from a single protein sequence or structure. Other methods try to infer function by applying what is known about the function of close relatives. Still other methods apply contextual information, of many types, including but not limited to neighbors in the genome, metabolic pathway reconstruction, and phylogenetic distribution patterns.

Different methods achieve functional inferences at different degrees of detail. Sometimes it is possible to predict only a general class of function; for instance: likely hydrolytic enzyme. Other methods have the aim of predicting, for a protein for which the general function is known, the precise substrate. Combinations of these methods may be required. However, even if it is possible to ascribe a particular function to a gene product, the protein may have multiple functions. A fundamental problem is that function is in many cases an ill-defined concept.

We shall describe some of the underlying difficulties, consider what tools are available, how to calibrate them, and how well they work. This will require us to consider more general questions, of what we mean by function, how functions can be classified, and how differences in function can be measured.

Evolution of Protein Sequence, Structure and Function

At the molecular level, evolution proceeds according to the cascade:

> gene sequence determines amino acid sequence,
> amino acid sequence determines protein structure,
> protein structure determines protein function,
> selection acts on function to modify allele frequencies in populations (to close the loop).

During protein evolution, sequences, structures and functions all diverge. However, the relationship between sequence and structure is substantially simpler than that between either sequence or structure, and function.

Sequences and structures of orthologous proteins show coordinated evolutionary divergence. As sequences progressively diverge, structures progressively deform.[15] Typically, a core of the structure, including the major elements of secondary structure and, usually, the active site, retains its folding pattern. Other, peripheral, regions of the structure can refold. In this process, structure changes more conservatively than sequence. This is why, in many families of proteins, we can recognize structural similarity in relatives so distant that there is no easily visible sign of the similarity in the sequence.

One general reason for the retention of protein conformation in general, and the structure of the active site in particular, is selection for the maintenance of function. A need to retain function imposes constraints on protein structural change during evolution. In contrast, when a protein evolves to change its function, many of these constraints are released — or, more precisely, are replaced by alternative constraints required by the new function — and the sequence and structural changes are correspondingly greater.

It is easier to see the effects of these constraints than to understand their mechanism. In some cases certain specific residues are directly involved in function; an example is the iron-linked proximal histidine of the globins, and these are immutable. In contrast, constraints that maintain the overall folding pattern are dispersed around the sequence, and only the patterns of residue conservation in large-scale alignments of homologous proteins can elucidate them.

The relationship between sequence and function is even more complex. Small changes in sequence, during evolution, usually make only small changes in structure; this is why homology modeling is so successful. Often they make only small changes in function too. But changes in function do not necessarily require large changes in sequence or structure — function can jump. Indeed, proteins can change function without any sequence changes at all. Many examples are known of "recruitment" or "moonlighting" by proteins — adapting to a novel function with relatively little sequence change. For instance, in the duck, an active lactate dehydrogenase and an enolase serve as crystallins in the eye lens, although they do not encounter the substrates *in situ* (see below).

Conversely, proteins with very different sequences and structures can have the *same* function. For instance, many families of proteinases differ in sequence and structure, sharing only a common general catalytic activity. Figure 1 summarizes, in a schematic way, the topology of protein space with respect to the relations between sequence, structure, and function.

All features of proteins — sequence, structure and function — are potentially useful in interpreting new genome sequences.

We expect that many regions of any new genome encode proteins similar to known relatives in other species. We identify them from similar patterns in the sequences. We can expect that the structures will be similar, and indeed the expected difference in structure can be calibrated from the extent of divergence in the sequence. If a protein in the new genome is related sufficiently closely to a homologue of known structure, it is possible to build a model of the new protein.

Assigning function to the proteins encoded in the new genome is a more difficult problem. The study of protein function arises in two general contexts.

1. In the past, the paradigm was for a focussed research group to assemble a thick dossier of detailed experimental information.

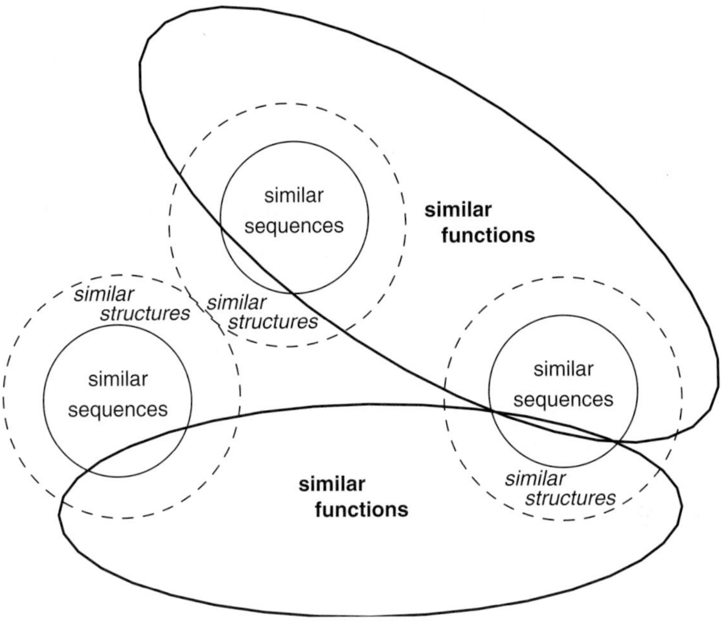

Fig. 1 The topology of protein sequence, structure and function:

- *similar sequences produce similar protein structures,* with divergence in structure increasing progressively with the divergence in sequence.[15]
- conversely, *similar structures are often found with very different sequences.* For instance, many proteins form TIM barrels with no easily detectable relationship between their sequences.[104,105]
- *similar sequences and structures sometimes produce proteins with similar functions,* but exceptions abound.[106,107]
- conversely, *similar functions are often carried out by proteins with dissimilar structures;* examples include the many different families of proteinases, sugar kinases, and lysyl-tRNA synthetases.[108,109]

Such studies might include identification of cofactors and post-translational modifications, and even a structure determination and a check on the phenotypic effect of a knockout. These demonstrated and characterized the function.

2. However, more and more commonly, we must deal with much sparser information. The largest sources of proteins of unknown function are complete genome sequences or metagenomic data,

giving us the challenge of annotating them.[7,16,17] Identification of genes provides the amino acid sequences of an organism's proteins. In many cases, the data about specific proteins are limited to their amino acid sequences.

In analyzing a novel genome, how well do we understand nature's rules in proceeding from DNA sequence to amino acid sequence to protein structure to function?

- Starting from a genome sequence, gene identification is still problematic, especially in eukaryotes where alternative splicing patterns compound the difficulty.[18]
- From sequence to structure is the safest step. Nature has strict rules for determining protein structure uniquely from amino acid sequence, with only a few exceptions — notably the prion proteins,[19] and the serpins.[20] This generalization is among the most robust we have. Although as yet we do not understand the physical basis of nature's folding algorithm in sufficient detail to predict structure from sequence, progress is being made.[21,22] The observation that similar sequences determine similar structures gives us general confidence in homology modeling — the "differential form" of the folding problem.[23]

However, the assumption that homologues share function is less and less safe as the sequences progressively diverge. Moreover, even closely-related proteins can change function, either through divergence to a related function or by recruitment for a very different function.[24] In such cases, assignment of function on the basis of homology, in the absence of direct experimental evidence, will give the wrong answer, leading to misannotations in databanks. Many authors have called attention to annotation "howlers."[25–32] Iyer *et al.*[33] have collected cases in which prediction and experiment agree, but both are likely to be wrong!

Indeed, the situation can be even worse. An often-asked question is "How much must a protein change its sequence before its function changes?" The answer is: Not at all! There are numerous examples of proteins with multiple functions:

1. We have mentioned the eye lens proteins in the duck that are identical in sequence to active lactate dehydrogenase and enolase in other tissues. They have been recruited to provide a completely unrelated function based on the optical properties of their assembly. Several other avian eye lens proteins are identical or similar to enzymes. In some cases, residues essential for catalysis have mutated, proving that the function of these proteins in the eye is not an enzymatic one.[34] Note that the coexistence in some species of mutated inactive enzymes in the eye, and active enzymes in other tissues, implies that the gene must have been duplicated.
2. Certain proteins interact with different partners to produce oligomers with different functions. In *Escherichia coli,* a protein that functions on its own as lipoate dehydrogenase is also an essential subunit of pyruvate dehydrogenase, 2-oxoglutarate dehydrogenase and the glycine cleavage complex.[35]
3. Proteinase Do functions as a chaperone at low temperatures and as a proteinase at high temperatures. The logic, apparently, is that under conditions of moderate stress it attempts to salvage misfolded proteins; under conditions of higher stress it "gives up" and recycles them.[36]
4. Phosphoglucose isomerase (= neuroleukin = autocrine motility factor = differentiation and maturation mediator) functions as a glycolytic enzyme in the cytoplasm, but as a nerve growth factor and cytokine outside the cell.[37,38] The structural origin of the extracellular receptor function is obscure.

Because evolution has so assiduously pushed the limits in its exploration of sequence-structure-function relationships, many procedures described in the literature on function prediction do not specify function exactly, but provide general hints. For instance, a protein known to be TIM barrel is likely to be a hydrolytic enzyme. It is not useless to predict such general aspects of function, even if the details remain obscure. Such hints are very useful in guiding experimental investigations of function, or specialized computational tools such as *in silicio* screening. An example from the *Haemophilus influenzae* structural genomics project illustrates the point.

HI1679[20] has an α/β-hydrolase fold, with a putative remote homology, based on sequence analysis, to members of the L-2-haloacid dehydrogenase family, the P-domain of Ca^{2+} ASPase and phosphoserine phosphatase. It was the first structure of a protein in the L-2-haloacid dehydrogenase family to be determined, and one of the motives for selecting it for investigation was the goal of learning about the structure and the mechanism of function of this family. The structure was consistent with a phosphatase, this being confirmed in the laboratory where a variety of potential substrates were used. The protein cleaved 6-phosphogluconate and phosphotyrosine. To achieve the goal of elucidating the functions of this family of proteins, selected substrates were modeled into the binding pocket so as to determine how sequence variation in the active site might affect specificity.[39]

Natural Mechanisms of Development of Novel Protein Functions[40]

Observed mechanisms of protein evolution that produce altered or novel functions include: a) *divergence;* b) *recruitment;* and c) *"mixing and matching" of domains.*

(a) *Divergence*

Among closely-related proteins, mutations usually conserve function but modulate specificity. For example, the trypsin family of serine proteinases contains a specificity pocket: a surface cleft complementary in shape and charge distribution to the sidechain adjacent to the scissile bond. Mutations tend to leave the backbone conformation of the pocket unchanged but affect the shape and charge of its lining, altering the specificity.

The change in specificity of the proteases illustrates a common theme: although homologous proteins show a general drifting apart of their sequences as they accumulate mutations, often a few specific mutations account for functional divergence,[41] as initially proposed by Perutz[42] for hemoglobin. The malate and lactate dehydrogenase (MDH/LDH) family is a good example. Malate and lactate dehydrogenases are related enzymes catalyzing related reactions. Wilks *et al.*[43] showed by site-directed mutagenesis, that a single residue change could switch the activity. Their paper may have been read by a *Trichomonad* which developed an MDH much more similar to LDH molecules than to other MDHs, and appears to have arisen by convergent evolution.[44]

The enolase superfamily, which exhibits a folding pattern very closely related to TIM-barrels, contains several enzymes that catalyze different reactions with shared features of their mechanisms.[45] These includes enolase itself, mandelate racemase, muconate lactonizing enzyme 1, and D-glucarate dehydratase. From the point of view of sequence similarity, these enzymes are fairly close relatives. Mandelate racemase and muconate lactonizing enzyme 1 have 25% sequence identity. However, looking only at sequence and structure runs the risk of overlooking a more subtle similarity. What these enzymes share is a common feature of their *mechanism:* each acts by abstracting a proton adjacent to a carboxylic acid to form an enolate intermediate. The stabilization of a negatively-charged transition state is conserved. In contrast, the subsequent reaction pathway, and the nature of the product, vary from enzyme to enzyme. These enzymes have not only a similar overall structure, a variant of the TIM-barrel fold, but each requires a divalent metal ion, bound by structurally equivalent ligands. Different residues in the active site produce enzymes that catalyze different reactions.

An aspect of divergence important for its implications on function is the distinction between orthologues and paralogues. Any two proteins that are related by descent from a common ancestor are homologues. Two proteins in different species descended from the same protein in an ancestral species are orthologues. Two proteins related via a gene duplication within one species (and the respective descendants of the duplicates) are paralogues. After gene duplication, one of the resulting pairs of paralogoues can continue to provide its customary function, releasing the other to diverge to develop new functions.

Therefore, inferences of function from homology are more secure for orthologues than for paralogues.

The database, Clusters of Orthologous Groups (COG), is a collection of proteins encoded in fully-sequenced genomes, organized into families.[46] The COG database has been applied to the analysis of function and genome annotation.

Comparative analyses of known structures in such families of enzymes illustrate the kinds of structural features that change and those that stay the same. In some cases, the catalytic atoms occupy the same positions in molecular space, although the residues that present them are located in different context within the fold. In other cases, the positions in space of the catalytic residues are conserved even though the identities and functions of the catalytic residues vary. In these cases, there appears to be a set of conserved "functional positions" relative to the molecular framework.

However, several enzyme families show an even greater degree of divergence, including variation in the residues responsible for mediating catalysis. For example, the Apurinic/Apyrimidinic Endonuclease superfamily is a large diverse family of phosphoesterases. The family includes members that cleave nucleic acids (both DNA and RNA). However, the family has diverged to include lipid phosphatases. The essential catalytic residues vary between different subfamilies; for example, an essential His in the DNA repair enzyme, DnaseI, is not conserved in Exonuclease 111. In these cases, the conservation patterns from which we could hope to identify function have disappeared.

(b) *Recruitment*

The application of enzymes as lens crystallins illustrated another route of evolution: a novel function *preceding* divergence. It is more difficult to distinguish divergence and recruitment than it might first appear. Divergence and recruitment are at the ends of a broad spectrum of changes in sequence and function. Aside from cases of "pure" recruitment such as the duck eye lens proteins or phosphoglucose isomerase, in which a protein adopts a new function with no sequence change at all, there are examples of relatively small sequence changes correlated with very small function changes (which most people would think of as relatively pure divergence); relatively small sequence changes with quite large changes in function (which most people would think of as recruitment), as well as many cases in which there are large changes in both sequence and function.

(c) *"Mixing and matching" of domains, including duplication/oligomerization, and domain swapping or fusion*

Many large proteins contain tandem assemblies of modules which appear in different contexts and orders in different proteins. Censuses of genomes suggest that many proteins are multimodular. Stein[17] reports that of 4401 genes in *E. coli,* 287 correspond to proteins containing 2, 3 or 4 modules. Teichmann *et al.*[47] have analyzed, for enzymes involved in metabolism of small molecules, the distribution and redistribution of domains. The structural patterns of 510 enzymes could be accounted for in total or in part by 213 families of domains. Of the 399 which could be entirely divided into known domains, 68% were single-domain proteins, 24% comprised two domains, and 7%, three domains. Only 4 of the 399 had 4, 5 or 6 domains. Wu *et al.*[48] reported a new scheme for classification of domains of proteins. They arranged proteins in a hierarchy of superfamilies to subfamilies, depending upon the domain structure of the proteins. Families are homologous and homeomorphic arrangements of domains. These families form directed acyclic graphs based on the following relationship: protein families are children of a protein superfamily when they have the same ancestor as well as the same domain content but a different domain arrangement. (With the same domain arrangement, they would be in the same family.) Subfamilies distinguish specialized functions or structural variations. UniProt provides this information, which can be used to infer different functions for different domains when they appear in different combinations.

Multidomain proteins present particular problems for functional annotation, because domains may possess independent functions, modulate one another's function, or act in concert to provide a single function. On the other hand, in some cases the presence of a particular domain or combination of domains is associated with a specific function. For example, NAD-binding domains appear almost exclusively in dehydrogenases.

To follow changes in protein function during evolution is to aim at a moving target. Can we pin down the definition of protein function?

Definition and Classification of Protein Function

Proteins constitute most of the "executive branch" of the cellular nation, and as such are involved in many different types of functions. It is far from trivial to write a set of "job descriptions," as many proteins are involved in several activities, alone and in interaction. Different observers, from different points of view, may describe the function of a protein in different terms. Indeed unlike genes or protein structures, which are physical objects, a protein function is a description of a process or processes: much less tangible. Indeed, our definitions of protein functions have been derived from explicit enumerations, or ontologies, of categories of protein functions. These contain different levels of detail, from very general — "enzyme" — to very specific — "hexokinase."

There are many such classifications (reviewed by Ouzounis *et al.*[49]). Probably the most widely known schemes are those of the Enzyme Commission, limited of course to that class of functions, and of the Gene Ontology consortium. Other protein function classification schemes have been proposed, many in connection with individual organisms or individual families of proteins. However, a scheme appropriate for one organism is not necessarily appropriate for others. Indeed, even for very well understood proteins, there are different legitimate points of view about what aspects of function to focus on. The biochemist looks for the process mediated by the isolated protein in dilute solution. The molecular biologist looks for the significance, in the overall scheme of the life of the cell, of the process or processes in which the protein participates.

Definition of protein functions in most dictionaries include classifications, or clusterings of similar functions. A more difficult problem is to define a *quantitive* measure of function divergence. Given three *sequences,* it is possible to decide which of the three possible pairs is the most closely related, based on alignments of the sequences. Given three *structures,* methods are also available to measure and compare the similarity of the pairs. But in general, given three protein *functions,* it would be harder to choose the pair with the most similar function. That is, although it is possible to define metrics for quantitative comparisons of different protein sequences and structures, this is more difficult for different protein functions. One aim of this chapter is to review methods that have been proposed for inferring protein function from amino acid and three-dimensional structure, and, as far as possible, to evaluate them. Without a metric on function, it is difficult to state the criteria for the success or failure of such methods.

Classification Schemes for Protein Function

General Schemes

Several schemes for the classification of protein functions have been proposed. We begin with some fairly general categories.

Andrade *et al.*[50] distinguished functional classes of proteins involved in Energy, Information, and Communication and Regulation. Within these general classes they offered the subdivisions as shown in Table 1. These categories comprise fairly general activities rather than

Table 1 General Classification of Protein Functions[50]

- **Energy**
  - — Biosynthesis of cofactors, amino acids
  - — Central and intermediary metabolism
  - — Energy metabolism
  - — Fatty acids and phospholipids
  - — Nucleotide biosynthesis
  - — Transport
- **Information**
  - — Replication
  - — Transcription
  - — Translation
- **Communication and Regulation**
  - — Regulatory functions
  - — Cell envelope/cell wall
  - — Cellular processes

Table 2 Functional Groups of Proteins for *E. coli*[51]

Regulatory function
Putative regulatory proteins
Cell structure
Putative membrane proteins
Putative structural proteins
Phage, transposons, plasmids
Transport and binding proteins
Putative transport proteins
Energy metabolism
DNA replication, recombination, modification, and repair
Transcription, RNA synthesis, metabolism, and modification
Translation, posttranslational protein modification
Cell processes (including adaptation, protection)
Biosynthesis of cofactors, prosthetic groups, and carriers
Putative chaperones
Nucleotide biosynthesis and metabolism
Amino acid biosynthesis and metabolism
Fatty acid and phospholipid metabolism
Carbon compound catabolism
Central intermediary metabolism
Putative enzymes
Other known genes (gene product or phenotype known)
Hypothetical, unclassified, unknown

individual protein functions. For example, biosynthesis of an amino acid often involves a sequence of reactions catalyzed by unrelated enzymes. Despite the differences in the precise function of these enzymes and in their structure and mechanism, all would fall into a single class in this scheme.

Other classifications have appeared in connection with genome sequencing projects. It is interesting to compare an analysis of functional categories suggested for a prokaryotic *(E. coli)* (Table 2) with those suggested for a eukaryote *(Saccharomyces cerevisiae)* (Table 3).

There is a good deal more overlap in these two schemes than would appear at a casual glance. The *E. coli* classes contain a much more precise subdivision of metabolic reactions than the Yeast scheme. Perhaps this is an example of the difference in point of view among biochemists, molecular biologists and cell biologists. Nevertheless, for purposes of

Table 3 Functional Categories suggested for Yeast[52]

Metabolism
Energy
Cell cycle and DNA processing
Transcription
Protein synthesis
Protein fate (folding, modification, destination)
Cellular transport and transport mechanisms
Cellular communication/signal transduction mechanism
Cell rescue, defense and virulence
Regulation of/interaction with cellular environment
Cell fate
Transposable elements, viral and plasmid proteins
Control of cellular organization
Subcellular localization
Protein activity regulation
Protein with binding function or cofactor requirement (structural or transport facilitation)
Classification not yet clear-cut
Unclassified proteins
This hierarchy contains a total of 261 classes and subclasses

annotating a genome, most people would hope for more specific assignments of function than any of these categories.

Given the goal of mapping a functional classification onto sequence and structure classifications, several problems associated with the current functional categorizations are generally recognized. One is that the function is defined without reference to homology in general and structure in particular. The EC, for instance, merges nonhomologous enzymes that catalyze similar reactions.

Gerlt and Babbitt,[29] who are among the most thoughtful writers on the subject, pointed out that "no structurally contextual definitions of enzyme function exist." They proposed a general hierarchical classification of function better integrated with sequence and structure. For enzymes they define:

- *Family*: homologous enzymes that catalyze the same reaction (same mechanism, same substrate specificity). These can be hard to detect at the sequence level if the sequence similarity becomes very low.

- *Superfamily*: homologous enzymes catalyzing similar reaction with either a) different specificity, or b) different overall reactions with common mechanistic attribute (partial reaction, transition state, intermediate) that share conserved active-site residues.
- *Suprafamilies*: different reactions with no common feature. Proteins belonging to the same suprafamily would not be expected to be detectable from sequence information alone.

There is also the "culture clash": the traditional biochemist's view of function arises from the study of isolated proteins in dilute solutions, in the presence of carefully controlled concentrations of substrates. The molecular biologist knows that an adequate definition of function must recognize the biological role of a molecule in the context of a living cell (or intracellular compartment) or the complete organism, on the one hand, and its role in a network of metabolic or control processes, on the other.[53,54] (In addition to the fundamental point of providing a more appropriate definition of function, information about context is often useful in assigning of function.) As a result, there is a generic problem with all attempts to force functional classifications into a hierarchical format. (See comments of Riley.[55])

The Enzyme Commission Classification

The origin of the EC classification was the action taken by the General Assembly of the International Union of Biochemistry (IUB), in consultation with the International Union of Pure and Applied Chemistry (IUPAC), in 1955, to establish an International Commission on Enzymes. The Enzyme Commission published its classification scheme, first on paper and now on the web: http://www.chem.qmul.ac.uk/iubmb/enzyme/.

EC numbers (looking suspiciously like IP numbers) contain four fields, corresponding to a four-level hierarchy. For example, EC 1.1.1.1 corresponds to alcohol dehydrogenase, catalyzing the general reaction:

an alcohol + NAD = the corresponding aldehyde or ketone + $NADH_2$

Note that several reactions, involving different alcohols, would share this number; but that the same dehydrogenation of one of these alcohols by an enzyme using the alternative cofactor NADP would be assigned EC 1.1.1.2.

The first number shows the division, out of the six main divisions (classes), to which the enzyme belongs:

Class 1. Oxidoreductases
Class 2. Transferases
Class 3. Hydrolases
Class 4. Lyases
Class 5. Isomerases
Class 6. Ligases

The significance of the second and third numbers depends on the class. For Oxidoreductases, the second number describes the substrate and the third number the acceptor. For Transferases, the second number describes the class of the item transferred, and the third number describes either more specifically what they transfer or in some cases, the acceptor. For Hydrolases, the second number signifies the kind of bond cleaved (e.g. an ester bond) and the third number, the molecular context (e.g. a carboxylic ester or a thiolester). (Proteinases are treated slightly differently, with the third number including the mechanism: serine proteinases, thiol proteinases and acid proteinases are classified separately.) For Lyases, the second number signifies the kind of bond formed (e.g. C–C or C–O), and the third number, the specific molecular context. For Isomerases, the second number indicates the type of reaction and the third number, the specific class of reaction. For Ligases, the second number indicates the type of bond formed and the third number, the type of molecule in which it appears. For example, EC 6.1 for C–O bonds (enzymes acylating tRNA), EC 6.2 for C–S bonds (acyl-CoA derivatives), etc. The fourth number gives the specific enzymatic activity.

Specialized classifications are available for some families of enzymes; for instance, the MEROPS database by N. D. Rawlings and A. J. Barrett provides a structure-based classification of peptidases and proteinases, http://www.merops.sanger.ac.uk/.

The Gene Ontology Consortium

In 1999, M. Ashburner and others faced the problem of annotating the soon-to-be-completed *Drosophila melanogaster* genome sequence. As a classification of function, the EC classification was unsatisfactory, if only because it was limited to enzymes. Ashburner organized the Gene Ontology Consortium to produce a standardized scheme for classifying function, as described in his memoir.[56]

The Gene Ontology Consortium has adopted a more general approach to the *logical* structure of a functional classification.[57] (http://www.geneontology.org.) Its goal is a systematic attempt to classify function, by creating a dictionary of terms and their relationships for describing molecular functions, biological processes and the cellular context of proteins and other gene products. It supports the annotation efforts by providing a set of terms that individual annotators or databases may adopt.

Organizing concepts of the Gene Ontology project include the distinctions between:

- *Molecular function*: a function associated with what an individual protein or a RNA molecule does in itself; it involves either a general description such as "enzyme," or a specific one such as "alcohol dehydrogenase." This is function from the biochemists' point of view.

and

- *Biological process*: a component of the activities of a living system, mediated by a protein or RNA, possibly in concert with other proteins or RNA molecules; either a general term such as signal transduction, or a particular one such as cyclic AMP synthesis is involved. This is function from the cell's point of view.

Because many processes are dependent on location, Gene Ontology also tracks:

- *Cellular component*: the assignment of site of activity or partners; this can be a general term such as nucleus or a specific one such as ribosome.

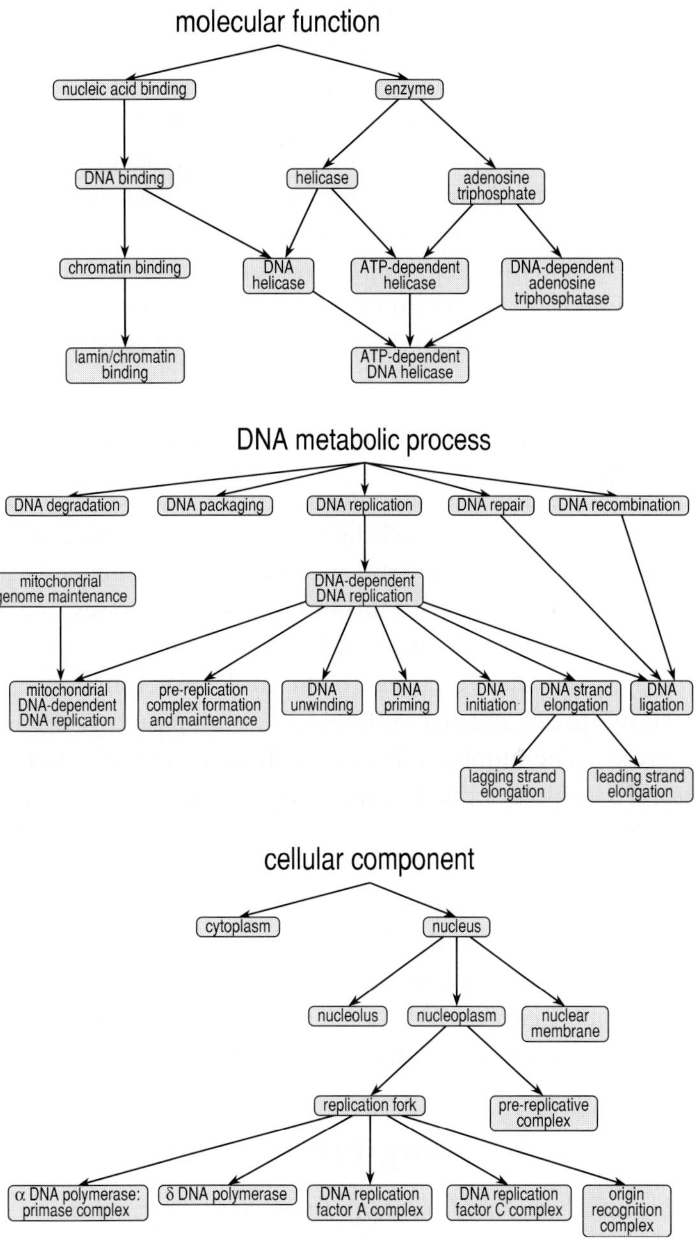

Fig. 2 Examples of fragments of GO classifications in three categories: molecular function, biological process and cellular location.

Figure 2 shows a fragment of the GO classifications in the three different categories. Note that the GO classification is not a strict hierarchy, or tree, but a more general structure called a directed acyclic graph.

It is very important to emphasise that neither the EC nor the GO classification is an assignment of function to individual proteins. The EC emphasized that: "*It is perhaps worth noting, as it has been a matter of long-standing confusion, that enzyme nomenclature is primarily a matter of naming reactions catalyzed, not the structures of the proteins that catalyze them.*" Assigning EC or GO numbers to proteins is a separate task. Such assignments appear in protein databases such as PIR or SWISS-PROT.[58]

Comparison of Enzyme Commission and GO Classifications

Enzyme Commission identifiers form a strict four-level hierarchy, or tree. For example, isopentenyl-diphosphate Δ–isomerase is assigned EC number 5.3.3.2. The initial 5 specifies the most general category; 5 = isomerases; 5.3 comprises intramolecular isomerases; 5.3.3 those enzymes that transpose C=C bonds; and the full identifier 5.3.3.2 specifies the particular reaction. In the molecular function ontology, GO assigns the identifier 0004452 to isopentenyl-diphosphate Δ–isomerase. (The numbers themselves have no specific significance.)

Figure 3 compares the EC and GO classifications of isopentenyl-diphosphate Δ–isomerase. The figure shows a path from GO:0004452 to the root node of the molecular function graph, GO:0003674. In this case, there are four intervening nodes, progressively more general categories as we move up the figure. Note that the GO description of this enzyme as an oxidoreductase is inconsistent with the EC classification, in which a committed choice between oxidoreductase and isomerase must be made at the highest level of the EC hierarchy.

Methods for Assigning Protein Function

Detection of Protein Homology from Sequence, and its Application to Function Assignment

If there is a "default" method for predicting protein function, in the absence of experimental information, it is the detection of similarity of

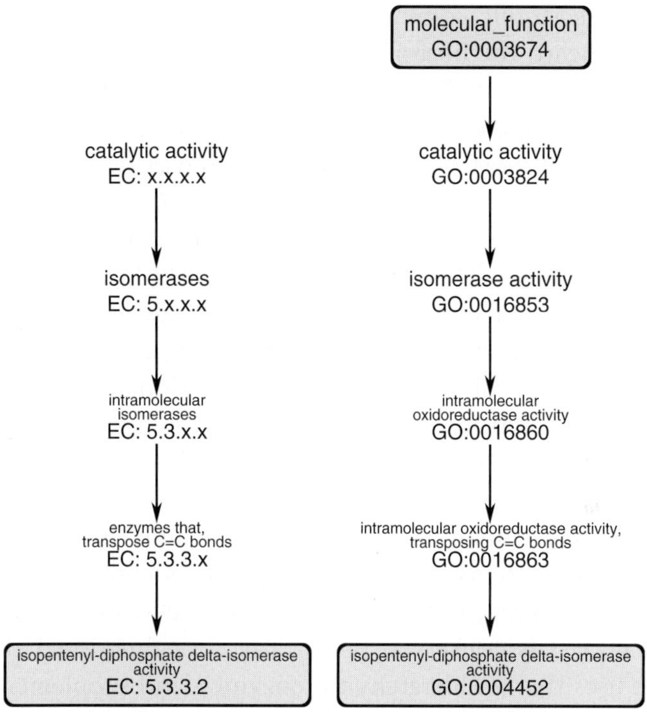

Fig. 3 Comparisons of Enzyme Commission and Gene Ontology classifications of isopentenyl-diphosphate Δ–isomerase.

amino acid sequence by database searching, and assuming that the molecules identified are homologues with similar functions. Search engines such as PSI-BLAST pull out sequences similar to a query sequence, from general protein sequence databases. The most favorable result is to find that the query sequence is identical or very closely-related to that of a well-characterized protein. However, even in these cases the assignment of function may not be correct or complete. The problem of assigning function becomes significantly more complex in cases where similarity between the unknown sequence and its (putative) homologue failed to be found. In some cases, however, specific sequence signature patterns identify active sites, even in proteins with little overall sequence similarity to homologues of known function.

Several groups have tested the correlations between sequence similarity and functional similarity. The feeling, as sensed by those in the relevant scientific community, could be roughly stated as, "Yes, we have all heard the horror stories about proteins with very closely related sequences but different functions, but those are rare exceptions, and the inference of function from similarity in sequence works fairly well most of the time." Does the evidence support this assumption?

Shah and Hunter[59] determined the sequence similarity of proteins within any EC class. They used a sample of 1327 classes and 15208 proteins, and tested various similarity thresholds. Their conclusions were that the errors were mostly false positives, and that it would be better to carry out this kind of analysis at the domain level.

Wilson, Kreychman and Gerstein,[60] Todd, Orengo and Thornton,[61] and Devos and Valencia[30] reached similar (although not identical) optimistic conclusions.

Wilson, Kreychman and Gerstein[60] concluded that for pairs of single-domain proteins, at levels of sequence identity ≥40%, precise function is conserved, and for levels of sequence identity ≥25%, broad functional class is conserved (according to a functional classification that uses the EC hierarchy for enzymes, and supplements it for material from FLYBASE[62] for nonenzymes). Todd, Orengo and Thornton[61] found that for pairs of proteins, both known to be enzymes, slightly less than 90% of pairs with sequence identity ≥40% conserve all four EC numbers. Even at ≥30% sequence identity, they found conservation of three levels of the EC hiererchy for 70% of homologous pairs of enzymes. Devos and Valencia[30] reached very similar conclusions; they also reported the ability to predict correctly the agreement of FSSP categories[63] and SWISS-PROT keywords, as a function of the level of sequence similarity.

Sangar, Blankenberg, Altman and Lesk[64] have extended these studies by determining how sequence divergence is related to functional divergence as measured using the GO functional classification. They find a threshold at about 50% sequence identity beyond which the divergence of function is accelerated.

Function prediction from sequence similarity can take advantage of multiple sources of information to back up the prediction from levels of sequence identity alone, and to improve the results in cases of lower sequence similarity.

Putative homologues having been identified, multiple sequence alignments enable identification of conserved residues; the literature may provide crucial information about the family as a whole and the role of conserved residues; and phylogenetic trees can provide information as to whether an unknown clusters with a particular functional grouping.[65,66] In general, if an unknown protein shares significant sequence similarity with a family of known function, and possesses the "right essential conserved residues" (e.g. active site residues), then a prediction as to function (proteinase, exonuclease, etc.) can reasonably be proposed. In addition, if the unknown also forms part of a well-supported functional cluster or clade within a phylogenetic tree, then a more detailed level of functional prediction may be possible.

Hannenhalli and Russell[67] examined nucleotidyl cyclases. Changing the specificity between an ATP cyclase and a GTP cyclase requires mutations of only two residues, E937K and C1018D. From a common alignment of ATP and GTP cyclases, they were able to identify residues correlated with the change in specificity, including the two crucial positions. Given the sequence of a new enzyme in this family, it could be identified as a family member by overall sequence similarity, and its specificity could be inferred from the residues occupying the selected positions. Hannenhalli and Russell[66] also showed that a similar analysis permitted prediction of specificity of protein kinases. (Motifs were already known that were able to distinguish Ser/Thr from Tyr kinases.)[67,68]

As a control, an illustration of a negative inference: an evolutionary tree of myotubularin-related proteins permitted Nandurkar *et al.*[69] to infer that their protein, although related to active phosphatases, lacks the essential catalytic residues and acts as an adapter rather than an enzyme.

The situation for multidomain proteins is even more complex. Although it may be relatively straightforward to predict the role of some of the domains, others may prove more challenging. Thus, a complete functional description of a multidomain protein of unknown function may be limited if it contains one or more domains that cannot be accurately annotated. Furthermore, the possibility of domains acting in concert with one another to modulate the behavior of the complete molecule is difficult to predict.

Detection of Structural Similarity, and Structure/Function Correlations

Several groups have attempted to correlate protein structure and function.[70,71] Hegyi and Gerstein[71] correlated the enzymes in the yeast genome between their fold classification in SCOP[72] and their EC functional categories, via the annotations in SWISS-PROT. They identified 8937 single-domain proteins that could be assigned both a fold and a function.

The broadest categories of structure were from the top of the SCOP hierarchy, including the all–α, all–β, α/β, $\alpha + \beta$, multidomains, and small classes. The broadest categories of function were from the top of the EC hierarchy: oxidoreductases, transferases, hydrolases, lyases, isomerases, and ligases; plus an additional category, non-enzymes. There are therefore 6 (structural classes) × 7 (functional classes) = 42 possible combinations of the highest-level correlates. By using finer classifications of structure and function (down to the third level of EC numbers) a total of 21 068 potential fold-function combinations are obtained. Only 331 of these are observed, among the 8937 proteins analyzed.

The observed distribution is highly nonrandom. Nonenzymatic functions account for 59% of the sequences of which well over half are in the all–α or all–β fold category. Of the enzymes, the most popular combinations were α/β folds among the oxidoreductases and transferases, and all–β and $\alpha + \beta$ among the hydrolases.

When the structure of a domain is known, what can be inferred about its function? Many folds are compatible with very different activities. The five most "versatile" folds are the TIM barrel, α–β hydrolase, NAD-binding fold, P-loop-containing NTP hydrolase fold, and ferredoxin fold. Conversely, the functions carried out by the most varied types of structure are glycosidases and carboxylases. These two functions are carried out by seven different fold types, from three different fold classes.

What we are looking for, however, are cases where structure provides reliable clues to function. In their cross table, Hegyi and Gerstein[71] show several folds that appear in combination with only one function. These appear to have predictive significance for function. Of course one cannot tell whether this is just because they are rare folds, and whether the correlation will hold up as the databases grow.

Shapiro and Harris[73] and Teichmann, Murzin and Chothia[74] illustrate the power of structure, including but not limited to identification of distant relationships not derivable from sequence comparisons.

- Identification of structural relationships unanticipated from sequence can suggest similarity of function. The crystal structure of AdipoQ, a protein secreted from adipocytes, showed a similarity of folding pattern to that of tumor necrosis factor. The inference that AdipoQ is a cell signaling protein was subsequently verified.
- The histidine triad proteins are a broad family with no known function. Analysis of their structures indicated a catalytic centre and nucleotide binding site, identifying them as a nucleotide hydrolase. Note that this did *not* depend on detection of a distant homology.
- Structural similarity of a gene product of unknown function from *Methanococcus jannischii* and other proteins containing nucleotide-binding domains led to experiments showing it to be a xanthine or inosine triphosphatase.[75]

Several groups have attempted to determine the common functionally active site of a family of proteins.

Lichtarge, Bourne and Cohen[76] have developed an evolutionary trace method to define binding surfaces common to protein families. They extract functionally important residues from sequence conservation patterns and map them onto the protein surface to identify functional clusters.

Given a set of homologous sequences, and at least one structure, the goal of the evolutionary trace method is to identify surface sites implicated in function. The assumptions of the method are:

1. The set of proteins has a common surface-exposed active site.
2. The homologous sequences produce similar structures that retain the location in that molecular space of the active site.
3. The functional site is less subject to mutation that the average surface sites.
4. Those mutations in the functional site that do occur are not random but create discrete sets of structures with shifts in function (also see Refs. 41 and 47).

The method begins by forming a multiple sequence alignment from which the molecules are hierarchically clustered into a tree. By choosing different levels in the hierarchy, clusters of different sizes may be extracted. If different functions are known in the family, the clusters are chosen to reflect the subgroups with different functions. By choosing larger or smaller clusters, grosser or finer resolution in function distinction may be made. For each cluster in the partition, form a consensus sequence alignment. Then coalign all the consensus sequences. The residues can be divided into: a) those that are absolutely conserved; b) those that are conserved within clusters but differ between clusters; and c) unconserved positions. By mapping the conserved residues onto the structure, a pattern is observed that defines a surface patch predicted to correspond to the active site.

Lichtarge, Bourne and Cohen[76] applied their method to SH2 and SH3 signaling domains, and the DNA-binding domain of nuclear hormone receptors. Their results correctly identified the known functional sites in these molecules. If the Evolutionary Trace method depended on a classification induced by known functional divergence, as is in these test cases, it would be arguable that it was really a method for assigning structure to function rather than function to structure. However, it can be applied using trees from other sources, and the classifications they induce; for instance, those based solely on multiple sequence alignments.

Successful predictions by the Evolutionary Trace method include identification of the functional surface in families of G protein α subunits[79] and regulators of G protein signaling.[79,80]

Both cases were *blind* predictions as subsequently verified by experiment. The success of the Evolutionary Trace method has led to its being taken up and developed by a number of groups.[81–84]

Irving, Whisstock and Lesk[85] applied the principle that active sites tend to be among the structurally best conserved parts of a protein-using superposition methods to extract regions of the lowest root-mean-square deviation of Cα atoms in a pair of proteins of known structure.

Prediction of Ligands

Once a binding site is targeted, the identification of a ligand involves, computationally, the same principle as in drug design, for which a great deal of mature algorithms and software exist.[86] Knowing, or hypothesizing, an active site in a known protein structure, modeling of ligands can predict what might bind to the site, which might permit inference of function.

Song *et al.*[87] have studied proteins of the enolase family, which share a modified TIM-barrel structure, and a common mechanism involving formation of an enolate intermediate, but having divergent specificity with respect to substrate and product. They attempted to determine the substrate specificity of a *Bacillus cereus* protein by *in silicio* screening.

Building a homology model of the *B. cereus* protein from a parent of known structure (with 35% sequence identity), in the liganded conformation, docking of molecules from a library of 420 dipeptides and N-succinyl-amino acids correctly showed the specificity for peptides with C-terminal arginine or lysine.

Motifs as Functional Signatures

Despite the progress in structural genomics projects, most proteins encoded in newly-sequenced genomes are known from their amino acid sequences alone. A major problem in genome annotation is that of assigning their functions. Note that not only are the three-dimensional structures in most cases unknown, there is also generally no information even about cofactors or post-translational modifications, which are often essential for function.

We have already discussed the standard method based on the assumption that in at least many cases, evolutionary divergence is slow enough to permit recognition of homologues that may have the same or at least similar structures and functions. Often the general similarity of sequences reflects a similarity in overall folding pattern, and particular residues within the fold may form a localized active site. Clearly the conservation of active site residues is important in reasoning from sequence similarity to functional similarity. Indeed, in some cases it is possible to shortcut the reasoning and to recognize the residues comprising the active site from a specific signature pattern or motif within the sequence. However, although many motifs do reflect functional active sites, others reflect positions for post-translational modification (e.g. glycosylation sites), or structural signals (e.g. N and C caps of α-helices), or signal sequences, with no direct functional implications.

Attwood[88] has described general methods for deducing sequence patterns. All these methods start with (or produce) a multiple sequence alignment, and seek to identify common distinctive features of particular positions of the sequence. These features may involve:

1. A motif describing a single consecutive set of residues.
2. Multiple motifs — a combination of several motifs involving separate consecutive sets of residues.
3. Profile methods, based on entire sequences and weighting different residue positions according to the variability of their contents. Extensions and generalizations of profile methods, including Hidden Markov models, are among the most sensitive detectors of distant homology based entirely on sequence data that we have.

Databases of single motifs

Motifs may be expressed in terms of uniquely defined sequences, such as:

GWTLNSAGYLLGP,

which characterizes the neuropeptide galanin. Or, motifs may contain alternative residues; for instance [LIVMF]-T-T-P-P-[FY], the signature of N-4 cytosine-specific DNA methylases. Here [LIVMF] means that that first position may contain *any* of the amino acids L, I, V, M, or F, followed by the unique sequence TTPP, followed by a position that may contain either F or Y.

PROSITE contains a collection of motifs covering a wide range of groups of proteins, together with retrieval software to check a submitted sequence for the presence of one or more motifs.[89] The motifs are calibrated to indicate the number of false negatives and positives to be expected. The [LIVMF]-T-T-P-P-[FY] motif detects all N-4 cytosine-specific DNA methylases, but also picks up false positives. The L-x(6)-L-x(6)-L-x(6)-L motif is least specific, missing one known leucine zipper (L-myc, which contains a methionine instead of one of the leucines) and promiscuously picking up hundreds of other sequences from many different types of proteins.

Several authors have sought to extend motif searching to three dimensions. Given that motifs tend to correspond to regions of conserved structure linked to function, Wallace, Laskowski and Thornton[90] searched known protein structures for the Ser-His-Asp catalytic triad of trypsin-like serine proteinases. They identified all known serine proteinases in their dataset, an in addition, triglycerol lipases which share the catalytic triad.

de Rinaldis[91] derived three-dimensional profiles from a single protein structure or a set of aligned structures. They applied their results to identifying proteins with matching surface patches. Analysis of the three-dimensional profiles of ATP and GTP binding P-loop proteins led to the identification of a positively charged phosphate-binding residue (Arg or Lys) in a position conserved in space but not in sequence.

Using a similar approach, Jackson and Russell[92] have identified regions with conformations similar to those of PROSITE motifs, but not necessarily sharing sequence similarity with them. They were able to identify serine proteinase inhibitors that contain regions similar in conformation to the loops in known inhibitors that have a common structure that docks to the proteinase.

Function Identification from Sequence by Feature Extraction

Brunak and colleagues have examined an alternative intermediate between sequence and function.[93] They reasoned that information about function should be contained in a spectrum of features of proteins, including secondary structure, post-translational modifications, protein sorting, and general properties of the amino acid composition such as the isoelectric point. Using neural networks, they predicted the following features from protein sequences, and correlated the results with functional classes.

- Extinction coefficient
- Grand average hydrophobicity
- Number of negative residues
- Number of positive residues
- O-glycosylation
- Serine/Threonine phosphorylation
- Tyrosine phosphorylation
- N-glycosylation
- PEST rich regions
- Secondary structure
- Subcellular location
- Low complexity regions
- Signal peptides
- Transmembrane helices

They recognized that the predictions of the features would be imperfect, but this need not fatally degrade their prediction of function.

The combined networks were trained to recognize a general set of functional classes based on categories originally defined by Riley,[94] and, within the proteins predicted to be enzymes, the Enzyme Commission classification. As a measure of the quality of the results for the general categories, at a level of thresholding giving 70% correct predictions, the range of false positives varied between below 10% up to below 40%, with most of the categories giving about 20% false positives.

By analyzing the networks, Jensen *et al.*[93] were also able to analyze which particular combinations of features were most effective signals for specific functional types.

Applications of Full-Organism Information: Inferences from Genomic Context and Protein Interaction Patterns

For proteins encoded in complete genomes, approaches to function prediction making use of contextual information and intergenomic comparisons are useful.[95–98]

1. *Gene fusion.* A composite gene in one genome may correspond to separate genes in other genomes. The implication is that there is a relationship between the functions of these genes.
2. *Local gene context.* It makes sense to coregulate and cotranscribe components of a pathway. In bacteria, genes in a single operon are usually functionally linked.
3. *Interaction patterns.* As part of the development of full-organism methods of investigation, data are becoming available on the patterns of protein interactions.[99] The network of interactions reveals the function of a protein.
4. *Phylogenetic profiles.* Pellegrini *et al.*[100] have exploited the idea that proteins in a common structural complex or pathway are functionally linked and expected to coevolve. For each protein encoded in a known genome, they construct a phylogenetic profile that indicates which organisms contain a homologue of the protein in question. Clustering the profiles identifies sets of

proteins that co-occur in the same group of organisms. Some relationship between their functions is expected.

For instance, the *E. coli* ribosomal protein, RL7 has homologues in 10 out of 11 eubacterial genomes, but no homologue appears in an archaeal genome.[100] Most of the *E. coli* proteins that share the phylogenetic profile of RL7 have ribosome-associated functions. If the function of RL7 were unknown, one could infer that it is associated in some way with the ribosome. Comparison of keywords in SWISS-PROT annotations affords a general test of this approach. Sets of nonhomologous proteins with similar phylogenetic profiles had, on average, 18% of SWISS-PROT keywords in common.

There need be no sequence or structural similarity between the proteins that share a phylogenetic distribution pattern. One unusual and very welcome feature of this method is that it is one of the few that derives information about the function of a protein from its relationship to *nonhomologous* proteins.[95,100]

Inference of Function from Metabolic Pathways

Assignment of proteins to metabolic pathways can be useful for functional assignment.

Many proteins are known only as amino acid sequences translated from genes. Although it is often possible to assign their function on the basis of similarity to proteins of known function in other organisms, sometimes a protein of unknown function may show weak similarities to several other proteins and it is unclear which is the true orthologue. Conversely, sometimes an organism has a metabolic pathway but no annotated enzyme for an essential step. Confronting the unannotated proteins with the unassigned functions can sometimes identify a protein that fills the gap in a pathway.

If an enzyme needed for a pathway cannot be identified even by weak sequence similarity, it may be that the organism has evolved a non-homologous enzyme for the task. For example, the archaeon *Methanococcus jannaschii* has a pathway for biosynthesis of chorismate from 3-dehydroquinate. Enzymes for most of the steps have known homologues in bacteria and/or eukarya. However, shikimate kinase was not identifiable from sequence similarity. *M. jannaschii* must have *some* protein with this function. How can it be found?

In bacteria, genes consecutive in pathways are often consecutive in operons in the genome. This is not true of the 3-dehydroquinate → shikimate pathway in *M. jannaschii.* However, the genes for successive steps of the chorismate biosynthesis pathway *are* clustered and consecutive in another archaeon, *Aeropyrum pernix.* From the gene order, it was possible to propose a gene for a shikimate kinase in *A. pernix*, and to identify a homologue of that gene in *M. jannaschii.*[101]

Experiment confirmed the prediction that the *M. jannaschii* gene so identified, *(MJ1440)*, encodes a shikimate kinase. It has no sequence similarity to bacterial or eukaryotic shikimate kinases. Archaea have recruited a protein from a different family to catalyze this step.

Microarrays

The most familiar type of microarray contains immobilized oligonucleotides, and is used to identify oligonucleotides in a mixcture. It is also possible to create protein arrays in which individual proteins are distributed on a chip. The proteins can be screened for protein-protein interaction, enzymatic activity, or ligand binding. Many other methods of high-throughput identification of protein-protein interactions have been developed (see Lesk,[102] pp. 382–385).

Chromatin Immunoprecipitation

An important class of protein functions involves DNA binding, in many cases for the regulation of gene expression. A high-throughput method for identifying the sequence specificities of DNA-binding proteins is chromatin immunoprecipitation. By cross-linking proteins to their DNA targets, followed by fragmenting the chromatin and pulling down the proteins with specific antibodies, the DNA ligand of the protein can be determined. Microarrays permit finding all the binding sites across the genome.

These represent direct measurements of protein function, not inferences in the absence of experimental data. Bioinformatics can, in these cases, sit back and analyze the data, rather than work out ways of living without them. Less glibly, bioinformatics can develop methods whereby array-based data can be integrated with other measurements and theoretical analysis, to yield the most complete and reliable description of protein function.

Conclusions

Harrington *et al.*[103] have assessed the state of the art of protein function prediction using metagenomics shotgun sequences, concluding that it is possible to infer specific functions for 76% of the 1.4 million predicted ORFs. This suggests that we are three-quarters of the way to a satisfactory solution of the problem of prediction of function from amino acid sequence and protein structure.

Some problems are difficult, while others are both difficult and messy. The prediction of protein structure from amino acid sequence is difficult, but we know that nature has an algorithm and all we have to do is find it, and given any procedure we can easily decide whether the answer is correct or not. The prediction of protein function is messy, partly because function is a fuzzy and multifaceted concept, and partly because very small (or even no) changes in amino acid sequence are compatible with large changes in function.

Many of the methods that have been applied to function prediction work part of the time but none is perfect. Morever, the more expert analysis of the results is applied, the better the predictions are. This makes it difficult to envisage a purely "black-box" automatic annotation machine for new whole-genome sequences. In most cases, predictions suggest, but do not determine, the general class of function. Their most useful effect is to guide investigations in the laboratory to confirm, or refute, the prediction, and, even if correct, to define the function in greater detail.

We conclude that predictions are useful but are no substitute for work in the laboratory. Indications from theory may indict, but only experimental evidence can convict.

References

1. Laskowski RA, Watson JD, Thornton JM. (2003) "From protein structure to biochemical function?" *J Struct Funct Genom* **4**: 167–77.
2. Whisstock JC, Lesk AM. (2003) "Prediction of protein function from protein sequence and structure." *Quart Revs Biophys* **36**: 307–40.
3. Roberts RJ. (2004) "Identifying protein function-A call for community action." *PLoS Biol* **2**: e42.
4. Sivashankari S, Shanmughavel P. (2006) "Functional annotation of hypothetical proteins — a review." *Bioinformation* **1**: 335–38.
5. Brenner SE. (1999) "Errors in genome annotation." *Trends Gen* **15**: 132–33.
6. Skolnick J, Fetrow JS, Kolinski A. (2000) "Structural genomics and its importance for gene function analysis." *Nat Biotechnol* **18**: 283–87.
7. Eisenstein E, Gilliland GL, Herzberg O, *et al.* (2000) "Biological function made crystal clear — annotation of hypothetical proteins via structural genomics." *Curr Opin Biotechnol* **11**: 25–30.
8. Brenner SE. (2001) "A tour of structural genomics." *Nat Rev Genet* **2**: 801–09.
9. Gilliland GL, Teplyakov A, Obmolova G, *et al.* (2002) "Assisting functional assignment for hypothetical *Haemophilus influenzae* gene products through structural genomics." *Curr Drug Targets Infect Disord* **2**: 339–53.
10. Chance MR, Bresnick AR, Burley SK, *et al.* (2002) "Structural genomics: a pipeline for providing structures for the biologist." *Protein Sci* **11**: 723–38.
11. Zhang C, Kim SH. (2003) "Overview of structural genomics: from structure to function." *Curr Opin Chem Biol* **7**: 28–32.
12. Andrade MA, Sander C. (1997) "Bioinformatics: from genome data to biological knowledge." *Curr Opin Biotechnol* **8**: 675–83.
13. Camon E, Magrane M, Barrell D, *et al.* (2003) "The Gene Ontology Annotation (GOA) project: implementation of GO in SWISS-PROT, TrEMBL, and InterPro." *Genome Res* **13**: 662–72.
14. Conesa A, Gotz S, Garcia-Gomez JM, *et al.* (2005). "Blast2GO: a universal tool for annotation, visualization and analysis in functional genomics research." *Bioinformatics* **18**: 3674–76.
15. Chothia C, Lesk AM. (1986) "The relation between the divergence of sequence and structure in proteins." *EMBO J* **5**: 823–26.
16. Smith TF. (1998) "Functional genomics — bioinformatics is ready for the challenge." *Trends Gen* **14**: 291–93.
17. Stein L. (2001) "Genome annotation: from sequence to biology." *Nat Rev Genet* **2**: 493–503.
18. Jones J, Field JK, Risk JM. (2002) "A comparative guide to gene prediction tools for the bioinformatics amateur." *Int J Oncol* **20**: 697–705.
19. Cohen FE, Prusiner SB. (1998) "Pathologic conformations of prion proteins." *Annu Rev Biochem* **67**: 793–819.

20. Peretz D, Williamson RA, Legname G, *et al.* (2002) "A change in the conformation of prions accompanies the emergence of a new prion strain." *Neuron* **34**: 921–32.
21. Schonbrun J, Wedemeyer WJ, Baker D. (2002) "Protein structure prediction in 2002." *Curr Opin Struct Biol* **12**: 348–54.
22. Tramontano A. (2003) "Of men and machines." *Nat Str Biol* **10**: 87–90.
23. Tramontano A. (2006) *Protein Structure Prediction: Concepts and Applications.* Wiley-VCH, Weinheim, Baden-Württemberg, Germany.
24. Ganfornina MD, Sanchez D. (1999) "Generation of evolutionary novelty by functional shift." *Bioessays* **21**: 432–39.
25. Smith TF, Zhang X. (1997) "The challenges of genome sequence annotation or "the devil is in the details." *Nat Biotechnol* **15**: 1222–23.
26. Bork P, Koonin EV. (1998) "Predicting functions from protein sequences — where are the bottlenecks?" *Nat Genet* **18**: 313–18.
27. Karp R. (1998) "What we do not know about sequence analysis and sequence databases." *Bioinformatics* **14**: 753–54.
28. Doerks T, Bairoch A, Bork P. (1998) "Protein annotation: detective work for function prediction." *Trends Genet* **14**: 248–50.
29. Gerlt JA, Babbitt PC. (2000) "Can sequence determine function?" *Genome Biol* **1**: REVIEWS0005.
30. Devos D, Valencia A. (2000) "Practical limits of function prediction." *Proteins: Struct Funct Genet* **41**: 98–107.
31. Devos D, Valencia A. (2001) "Intrinsic errors in genome annotation." *Trends Genet* **17**: 429–31.
32. Jeong SS, Chen R. (2001) "Functional misassignment of genes." *Nat Biotechnol* **19**: 95.
33. Iyer LM, Aravind L, Bork P, *et al.* (2001) "Quod erat demonstrandum? The mystery of experimental validation of apparently erroneous computational analyses of protein sequences." *Genome Biol* **2**: research0051.1-research0051.11.
34. Wistow G, Piatigorsky J. (1987) "Recruitment of enzymes as lens structural proteins." *Science* **236**: 1554–56.
35. Riley M. (1997) "Genes and proteins of *Escherichia coli* K-12 (GenProtEC)." *Nucl Acids Res* **25**: 51–52.
36. Spiess C, Beil A, Ehrmann M. (1999) "A temperature-dependent switch from chaperone to protease in a widely conserved heat shock protein." *Cell* **97**: 339–47.
37. Jeffery CJ. (1999) "Moonlighting proteins." *Trends Biochem Sci* **24**: 8–11.
38. Jeffery CJ, Bahnson BJ, Chien W, *et al.* (2000) "Crystal structure of rabbit phosphoglucose isomerase, a glycolytic enzyme that moonlights as neuroleukin, autocrine motility factor, and differentiation mediator." *Biochemistry* **8**: 955–64.
39. Parsons JF, Lim K, Tempczyk A, *et al.* (2002) "From structure to function: YrbI from *Haemophilus influenzae* (HI1679) is a phosphatase." *Proteins: Str Funct Genet* **46**: 393–404.

40. Piatigorsky J. (2007) *Gene Sharing and Evolution.* Harvard University Press, Cambridge, MA, USA.
41. Golding GB, Dean AM. (1998) "The structural basis of molecular adaptation." *Mol Biol Evol* **15**: 355–69.
42. Perutz MF. "Species adaptation in a protein molecule." *Mol Biol Evol* **1**: 1–28.
43. Wilks HM, Hart KW, Feeney R, *et al.* (1988) "A specific, highly active malate dehydrogenase by redesign of a lactate dehydrogenase framework." *Science* **242**: 1541–44.
44. Wu G, Fiser A, ter Kuile B, *et al.* (1999) "Convergent evolution of *Trichomonas vaginalis* lactate dehydrogenase from malate dehydrogenase." *Proc Nat Acad Sci USA* **96**: 6285–90.
45. Hasson MS, Schlichting I, Moulai J, *et al.* (1998) "Evolution of an enzyme active site: the structure of a new crystal form of muconate lactonizing enzyme compared with mandelate racemase and enolase." *Proc Nat Acad Sci USA* **95**: 10396–401.
46. Natale DA, Shankavaram UT, Galperin MY, *et al.* (2003) "Towards understanding the first genome sequence of a crenarchaeon by genome annotation using clusters of orthologous groups of proteins (COGs)." *Genome Biol Res* 0009.1–0009.19.
47. Teichmann SA, Rison SC, Thornton JM, *et al.* (2001) "The evolution and structural anatomy of the small molecule metabolic pathways in *Escherichia coli.*" *J Mol Biol* **311**: 693–708.
48. Wu CH, Nikolskaya A, Huang H, *et al.* (2004) "PIRSF: family classification system at the Protein Information Resource." *Nucl Acids Res* **32**: D112–14.
49. Ouzounis CA, Coulson RMR, Enright AJ, *et al.* (2003) "Classification schemes for protein structure and function." *Nature Rev Genet* **4**: 508–19.
50. Andrade MA, Ouzounis C, Sander C, *et al.* (1999) "Functional classes in the three domains of life." *J Mol Evol* **49**: 551–57.
51. Blattner FR, Plunkett GR, Bloch CA, *et al.* (1997) "The complete genome sequence of *Escherichia coli* K-12." *Science* **277**: 1453–74.
52. http://mips.gsf.de/proj/yeast/catalogues/funcat/.
53. Lan N, Jansen R, Gerstein M. (2002) "Towards a systematic definition of protein function that scales to the genome level: defining function in terms of interactions." *Proc IEEE* **90**: 1848–58.
54. Lan N, Montelione GT, Gerstein M. (2003) "Ontologies for proteomics: towards a systematic definition of structure and function that scales to the genome level." *Curr Opin Struct Biol* 7: 44–54.
55. Riley M. (1998) "Systems for categorizing functions of gene products." *Curr Opin Struct Biol* **8**: 388–92.
56. Ashburner M. (2006) *Won for All: How the* Drosophila *Genome Was Sequenced.* Cold Spring Harbor Laboratory Press, Cold Spring Harbor-New York.
57. The Gene Ontology Consortium. (2000) Gene Ontology: tool for the unification of biology. *Nature Genet* **25**: 25–28.

58. Couto FM, Silva MJ, Lee V, *et al.* (2006) "GOAnnotator: linking protein GO annotations to evidence text." *J Biomed Discov Collab* **1**: 19.
59. Shah I, Hunter L. (1997) "Predicting enzyme function from sequence: a systematic appraisal." *Proc Int Conf Intell Syst Mol Biol* **5**: 276–83.
60. Wilson CA, Kreychman J, Gerstein M. (2000) "Assessing annotation transfer for genomics: quantifying the relations between protein sequence, structure and function through traditional and probabilistic scores." *J Mol Biol* **297**: 233–49.
61. Todd AE, Orengo CA, Thornton JM. (2001) "Evolution of protein function, from a structural perspective." *J Mol Biol* **307**: 1113–43.
62. Ashburner M, Drysdale R. (1994) "Flybase: the *Drosophila* genetic database." *Development* **120**: 2077–79.
63. Holm L, Sander C. (1999) "Protein folds and families: sequence and structure alignments." *Nucl Acids Res* **27**: 244–47.
64. Sangar V, Blankenberg DJ, Altman N, Lesk, AM. (2007) "Quantitative sequence-function relationships in proteins based on gene ontology." *BMC Bioinforma* **8**: 294.
65. Hannenhalli SS, Russell RB. (2000) "Analysis and prediction of functional sub-types from protein sequence alignments." *J Mol Biol* **303**: 61–76.
66. Gu X, Vander Velden K. (2002) "DIVERGE: phylogeny-based analysis for functional-structural divergence of a protein family." *Bioinformatics* **18**: 500–01.
67. Hanks SK, Hunter T. (1995) "Protein kinases 6. The eukaryotic protein kinase superfamily: kinase (catalytic) domain structure and classification." *FASEB J* **9**: 576–96.
68. Hanks SK, Quinn AM, Hunter T. (1998) "The protein kinase family: conserved features and deduced phylogeny of the catalytic domains." *Science* 241: 42–52.
69. Nandurkar HH, Caldwell KK, Whisstock JC, *et al.* (2001) "Characterization of an adapter subunit to a phosphatidylinositol (3)P 3-phosphatase: identification of a myotubularin-related protein lacking catalytic activity." *Proc Nat Acad Sci USA* **98**: 9499–504.
70. Thornton JM, Orengo CA, Pearl FM. (1999) "Protein folds, functions and evolution." *J Mol Biol* **293**: 333–42.
71. Hegyi H, Gerstein M. (1999) "The relationship between protein structure and function: a comprehensive survey with application to the yeast genome." *J Mol Biol* **288**: 147–64.
72. Lo Conte L, Brenner SE, Hubbard TJP, *et al.* (2002) "SCOP database in 2002: refinements accommodate structural genomics." *Nucl Acid Res* **30**: 264–67.
73. Shapiro L, Harris T. (2000) "Finding function through structural genomics." *Curr Opin Biotech* **11**: 31–35.

74. Teichmann SA, Murzin AG, Chothia C. (2001) "Determination of protein function, evolution and interactions by structural genomics." *Curr Opin Struct Biol* **11**: 354–63.
75. Hwang KY, Chung JH, Kim S-H, *et al.* (1999) "Structure-based identification of a novel NTPase from *Methanococcus jannaschii.*" *Nature Struct Biol* **6**: 691–96.
76. Lichtarge O, Bourne HR, Cohen FE. (1996) "An evolutionary trace method defines binding surfaces common to protein families." *J Mol Biol* **257**: 342–58.
77. Gu X. (1999) "Statistical methods for testing functional divergence after gene duplication." *Mol Biol Evol* **16**: 1664–74.
78. Lichtarge O, Bourne HR, Cohen FE. (1996) "Evolutionarily conserved $G_{\alpha\beta\gamma}$ binding surfaces support a model of the G protein-receptor complex." *Proc Natl Acad Sci USA* **93**: 7507–11.
79. Sowa ME, He W, Wensel TG, Lichtarge O. (2000) "A regulator of G protein signaling interaction surface linked to effector specificity." *Proc Natl Acad Sci USA* **97**: 1483–88.
80. Sowa ME, He W, Slep KC, *et al.* (2001) "Prediction and confirmation of a site critical for effector regulation of RGS domain activity." *Nat Struct Biol* **8**: 234–37.
81. Aloy P, Querol E, Aviles FX, Sternberg MJ. (2001) "Automated structure-based prediction of functional sites in proteins: applications to assessing the validity of inheriting protein function from homology in genome annotation and to protein docking." *J Mol Biol* **311**: 395–408.
82. Madabushi S, Yao H, Marsh M, *et al.* (2002) "Structural clusters of evolutionary trace residues are statistically significant and common in proteins." *J Mol Biol* **316**: 139–54.
83. Lichtarge O, Sowa ME. (2002) "Evolutionary predictions of binding surfaces and interactions." *Curr Opin Struct Biol* **12**: 21–27.
84. Yao H, Kristensen DM, Mihalek I, *et al.* (2003) "An accurate, sensitive, and scalable method to identify functional sites in protein structures." *J Mol Biol* **326**: 255–61.
85. Irving JA, Whisstock JC, Lesk AM. (2001) "Protein structural alignments and functional genomics." *Proteins: Struct Funct Genet* **42**: 378–82.
86. Finn PW, Kavracki LE. (1999) "Computational approaches to drug design." *Algorithmica* **25**: 347–71.
87. Song L, Kalyanaraman C, Fedorov AA, *et al.* (2007) "Prediction and assignment of function for a divergent N-succinyl amino acid racemase." *Nature Chem Biol* **3**: 486–91.
88. Attwood TK. (2000) "The quest to deduce protein function from sequence: the role of pattern databases." *Int J Biochem Cell Biol.* **32**: 139–55.
89. Falquet L, Pagni M, Bucher P, *et al.* (2002) "The PROSITE database, its status in 2002." *Nucl Acids Res* **30**: 235–38.

90. Wallace AC, Laskowski RA, Thornton JM. (1996) "Derivation of 3D coordinate templates for searching structural databases: application to Ser-His-Asp catalytic triads in the serine proteinases and lipases." *Protein Sci* **5**: 1001–13.
91. de Rinaldis M, Ausiello G, Cesareni G, *et al.* (1998) "Three-dimensional profiles: a new tool to identify protein surface similarities." *J Mol Biol* **284**: 1211–21.
92. Jackson RM, Russell RB. (2000) "The serine protease inhibitor canonical loop conformation: examples found in extracellular hydrolases, toxins, cytokines and viral proteins." *J Mol Biol* **296**: 325–34.
93. Jensen LJ, Gupta R, Blom N, *et al.* (2002) "Prediction of human protein function from post-translational modifications and localization features." *J Mol Biol* **319**: 1257–65.
94. Riley M. (1993) "Functions of gene products of *Escherichia coli*." *Microb Rev* **57**: 862–952.
95. Marcotte EM, Pellegrini M, Thompson MJ, *et al.* (1999) "A combined algorithm for genome-wide prediction of protein function." *Nature* **402**: 83–86.
96. Huynen MA, Snel B. (2000) "Gene and context: integrative approaches to genome analysis." *Adv Prot Chem* **54**: 345–79.
97. Kolesov G, Mewes HW, Frishman D. (2001) "SNAPping up functionally related genes based on context information: a colinearity-free approach." *J Mol Biol* **311**: 639–56.
98. Kolesov G, Mewes HW, Frishman D. (2002) "SNAPper: gene order predicts gene function." *Bioinformatics* **18**: 1017–19.
99. Xenarios I, Salwinski L, Duan XJ, *et al.* (2002) "DIP, the Database of Interacting Proteins: a research tool for studying cellular networks of protein interactions." *Nucl Acids Res* **30**: 303–05.
100. Pellegrini M, Marcotte EM, Thompson MJ, *et al.* (1999) "Assigning protein functions by comparative genome analysis: protein phylogenetic profiles." *Proc Nat Acad Sci USA* **96**: 4285–88.
101. Daugherty M, Vonstein V, Overbeek R, Osterman A. (2001) "Archaeal shikimate kinase, a new member of the GHMP-kinase family." *J Bacteriol* **183**: 292–300.
102. Lesk AM. (2007) Introduction to Genomics, Oxford University Press, Oxford, UK.
103. Harrington ED, Singh AH, Doerks T, *et al.* (2007) "Quantitative assessment of protein function prediction from metagenomics shotgun sequences." *Proc Nat Acad Sci USA* **104**: 13913–18.
104. Copley RR, Bork P. (2000) "Homology among $(\beta\alpha)_8$ barrels: implications for the evolution of metabolic pathways." *J Mol Biol* **303**: 627–41.
105. Nagano N, Orengo CA, Thornton JM. (2002) "One fold with many functions: the evolutionary relationships between TIM barrel families based on their sequences, structures and functions." *J Mol Biol* **321**: 741–65.

106. Ponting CP. (2001) "Issues in predicting protein function from sequence." *Brief Bioinform* **2**: 19–29.
107. Rost B. (2002) "Enzyme function less conserved than anticipated." *J Mol Biol* **318**: 595–608.
108. Doolittle RF. (1994) "Convergent evolution: the need to be explicit." *Trends Biochem Sci* **19**: 15–18.
109. Galperin MY, Walker DR, Koonin EV. (1998) "Analogous enzymes: independent inventions in enzyme evolution." *Genome Res* **8**: 779–90.

Chapter 5

Comparative Modeling in Structural Genomics

John Moult

Structural genomics around the world now has many themes, but a common thread is the idea of greatly increased coverage of protein space by three-dimensional structure. Since the number of known sequences will always be much much larger than the number of experimentally determined structures (currently the ratio is of the order of 1000 to 1), large-scale coverage strategies implicitly or explicitly invoke the use of modeling, using templates derived from experimental sampling of structure space. The US NIH Protein Structure Initiative (http://www.nigms.nih.gov/Initiatives/PSI/) is most directly focused on that goal, with the stated objective of the PSI2 project being "to make the three-dimensional atomic-level structures of most proteins easily obtainable from knowledge of their corresponding DNA sequences." This and related strategies build on the fact that when two proteins have a detectable sequence relationship, the structures are similar. Thus, when the structure of one protein of an evolutionary family is determined experimentally, it becomes possible to build some kind of at least partial model of all other detectable members.

Center for Advanced Research in Biotechnology, University of Maryland Biotechnology Institute, 9600 Gudelsky Drive, Rockville, MD 20850, USA. Email: jmoult@tunc.org

The success of this large-scale coverage strategy depends on three factors: the nature and size of evolutionary families, the quality of models required for intended applications, and the effectiveness of modeling methods in delivering appropriately accurate structures. We examine each of these below.

The Nature and Size of Evolutionary Families

Understanding of evolutionary relationships among protein domains has changed rapidly in the last few years, with new insights provided by the increased number of known structures and particularly by the data from fully sequenced genomes. It is now clear that there is an approximately power-law distribution of the apparent size of both sequence[1] and fold[2] families. That is, from both a sequence and structure perspective, there are a small number of large families and many small ones. The evolutionary implications of these observations are not yet clear; it may be that not only sequence families but also fold family definitions are artificial, reflecting the limitations of the methods for identifying evolutionary relationships. In particular, contrary to the prevailing wisdom, folds are not generally fully conserved over long evolutionary distances. The implications for structural genomics are clear, though — a relatively small sample of structures will provide coverage of a very large fraction of sequence space, while beyond that point there will be diminishing returns as sequence family size decreases. For example, in an analysis of all protein sequence families in a set of 67 prokaryotic genomes, Yan and Moult[1] found that about 6000 structures would be required to obtain one representative for each domain family with three or more members, thus providing modeling templates for 88% of the 250 000 sequences, but a further 26 000 structures would be required for the remaining 12% of sequences. The number of families is still increasing as more genomes are sequenced. Nevertheless, these and other similar estimates do show that a strategy of obtaining one experimental structure for each of the larger families leads to the possibility of producing model structures for a very large fraction of proteins. As discussed later, the more remote the sequence relationship between a structural representative of a family and a protein of interest, the less accurate the model; so not all models will be equal.

Model Accuracy Requirements for Typical Applications

Although a detectable sequence relationship between two proteins implies that knowledge of the structure of one will allow the production of at least some kind of partial model of the other, the quality of these models varies widely. As has been pointed out before,[3] different qualities of model are useful for different purposes. We define four levels of model quality, and consider which applications are appropriate at each.

Level 1, Very-high-accuracy models: Those that are competitive in accuracy with a moderate-resolution X-ray structure. The main chain is accurate to within about 0.5 Å; most non-surface side chains are correctly oriented; and there are no or at worst a few short, poorly defined regions of backbone. Typically, these models require a very high level of sequence identity to a known structure, in excess of 50%.

Level 2, High-accuracy models: Those with a backbone accuracy of better than 1 Å RMSD on Cα atoms, some internal side chains incorrectly oriented, small or no alignment errors, and some short regions of chain poorly defined. These models are of comparable accuracy to a typical NMR structure, and usually require a sequence identity to a known structure of 30% or higher, or successful model refinement.

Level 3, Medium-accuracy models: Those with an accuracy of up to about 2 Å RMSD on Cα atoms, perhaps with more extensive alignment errors, with unreliable side chain orientations, and with some larger regions of backbone poorly defined. Typically, this level of accuracy applies to an unrefined model produced from a sequence relationship at the limit of detectability by PSI_BLAST.

Level 4, Low-accuracy models: Those where the overall topology is correct, but with substantial parts of the structure not present, substantial alignment errors, and significant backbone errors in the main chain. Such models are worst-case outcomes when a template can be detected only by the most sophisticated means (well beyond the PSI-BLAST limit), and there has been no successful refinement or further modeling.

Given this accuracy framework, we can now examine the relationship between model accuracy and usefulness for particular applications.

Low-Accuracy Models Are Useful for:

Assignment of approximate molecular function: Identification of a fold often reveals that a protein is a member of a known super-family, and so allows an approximate assignment of function. These partial function assignments often rely on the fact that chemistry is usually conserved within enzyme superfamilies.[4] There are many cases of superfamily-based function assignment in structural genomics, for example.[5]

Determining the position of structural and functional domain boundaries: For expression of domains, and for dividing up complete proteins into functional units, the most effective methods make use of structural information to identify domain boundaries.[6] Low-accuracy models will not give exactly the same boundaries as might be obtained from an experimental structure, but the boundaries will be close enough for parsing functional units and for allowing limited variability in the constructs needed to test expression.

Assessing the impact of alternative splicing on protein structure and function: Alternatively spliced messages often insert, remove, modify or exchange exons. A low-resolution model is usually adequate to obtain an approximate idea of the likely role of these changes in function.[7] There is a database of models of alternatively spliced human protein structures (AS3D.umbi.umd.edu), in many cases based on remote evolutionary relationships.

Selection of epitopes: Knowledge of protein structure is frequently used to choose short stretches of polypeptide to be used to produce a vaccine.[8] The best choice is generally regions of the structure that form flexible surface loops. Provided the fold of the relevant domain structure has been identified almost any model will be accurate enough to produce a short list of candidates.

Medium-Resolution Models Add Use for:

Identification of specific aspects of function, such as protein–protein interaction sites: This aids in Assisting the design of experiments, particularly the selection of mutants to probe function.[9] Protein–protein interaction sites may be identified by mapping sequence conservation onto the surface of a model,[10] or by inferring interaction sites from the template protein or proteins. Likely small ligand binding sites can also often be identified from low-resolution surface topology.[11]

General location of non-synonymous SNPs and disease-causing mutations: Monogenic disease[12] and driver mutations for cancer are often single amino acid substitutions.[13] It is likely that the SNPs contributing most to susceptibility to common human diseases such as Alzheimer's, disease stroke heart disease, asthma and diabetes act in the same way.[14] A medium-resolution model is often sufficient to identify clustering substitutions in particular regions, thus allowing likely mechanisms of action to be identified.[15]

High Accuracy Models Add Use for:

Assignment of orthologs within protein families: As discussed further below, assigning exact molecular function (beyond a super-family designation) is generally difficult, even with a high-accuracy experimental structure. But the problem of determining whether two proteins from different species have the same detailed function is in principle more tractable, relying on the conservation of key structural features. Although examples are so far scarce, tests with experimental structures[16] suggest that a medium-resolution model will suffice in favorable cases.

Assessing the detailed impact of non-synonymous SNPs: As noted above, most disease-related genetic variants are single base changes that result in an amino acid substitution affecting protein function *in vivo*.[17] While it is possible to identify likely high-impact non-synonymous SNPs from sequence profiles,[14,18,19] interpretation of the mode of impact on protein function requires examination of the amino acid substitution in a structural context. Benchmarking of one method has shown that models based on sequence identity down to 40% are as reliable as experimental structures for this type of interpretation.[14]

Very-High-Accuracy Models May Be Useful for:

Drug design: Proposing potential lead compounds and analyzing the reasons why a particular candidate molecule does or does not bind tightly is one of the most demanding uses of structure. Generally, high-resolution structures of not only the apo-molecule, but also of the various complexes with candidates, are required. Nevertheless, there have been a number of cases reported where models have been sufficient, for example.[20]

Assignment of detailed molecular function: One of the conclusions from structural genomics projects has been that it is often very difficult to provide a complete assignment of molecular function, even given a high-resolution experimental structure.[5] For non-enzymes, the identification of protein binding partners or the specificity of RNA or DNA binding is not generally possible from structure alone, although in some cases it is possible to identify the nature of binding partners.[21] For enzymes, aspects of substrate specificity may be deduced, particularly by comparison of structural features with those of other members of the family that have been previously functionally characterized.[22] However, perhaps because many enzymes have a range of substrates, of which not all of these them have biological significance, definitive assignment of biological specificity is often difficult. The same restrictions apply to even the most accurate models. Nevertheless, a number of computational tools which aid in functional assignment are now available, for example,[23] and can be used in favorable cases.

Models of All Accuracy Are Useful for:

Modeling in conjunction with experimental structure determination techniques: Structure modeling techniques are finding increasing application in augmenting both established experimental methods of structure determination and newly emerging ones. In X-ray crystallography, a number of studies have demonstrated that comparative modeling of the target structure can provide greatly improved search structures in molecular replacement,[24,25] sometimes spectacularly so.[26] It is likely that the generation of many trial models will become a routine procedure, gradually diminishing the need for other methods of solving the phase problem. It has recently been conclusively demonstrated that methods developed for the refinement of comparative models are able to improve the accuracy of NMR structures, as measured by the standard of similarity to the corresponding X-ray structure.[26] A particularly powerful application is in increasing the detail in structures devised from cryo-electron microscopy, by the use of envelope constraints together with modeling.[27] A growing area of application is in fitting folds to solution X-ray scattering data,[28] and the combination of modeling methods with constraints from chemical crosslinking[29] and surface labeling.[30]

Aiding intuition and the development of hypotheses: This is a rather vague, but unquestionably real, value of every structure model. Experience working with experimentalists shows that they are often very excited about even the most approximate structure, and these often do provide a new dimension for thinking about a problem.

In summary, although a high-quality experimental structure is always the ideal, most applications do not need a structure that has an accuracy competitive with a high-quality X-ray structure, and even the poorest-quality models have many uses.

Factors Determining Model Accuracy

As discussed above, substantially different accuracy is required of models, depending on the application. But how accurate are models, and how fast is accuracy increasing as better methods and more data become available? These factors have been tracked by the CASP experiments for the past 14 years, providing a pool of information with which to assess the relationship between the experimental structural coverage of protein space and the usefulness of models.

The basic principle behind CASP is simple — the capabilities of any computational method can be best measured by *bona fide* blind prediction. To implement this principle, every two years CASP solicits a set of modeling targets from the experimental structural biology community. These are structures which are about to be solved, or which have been solved but not yet made public. In recent CASPs, the NIH PSI project (http://www.nigms.nih.gov/Initiatives/PSI/) and the Structural Genomics Consortium (http://www.sgc.utoronto.ca/) have provided almost all of the targets. Details of the most recent CASP experiment and the general procedure may be found in Moult *et al.*[31] Here, we focus on the current quality of comparative models,[32] and recent progress in improved accuracy and usefulness of these models.[33]

Four major factors determine the accuracy of any comparative model: 1) the success in identifying possible templates on which to base the model, 2) the accuracy of mapping (aligning) the target protein sequence onto the template or templates, 3) the ability to model parts of the structure not included in the best template, and 4) the effectiveness of techniques for refining an initial model towards the correct structure.

1) *Fold recognition:* Recognition of a fold relationship between two proteins utilizes several principles: sequence similarity, comparison of predicted target and template secondary structure, and compatibility of the target sequence with aspects of the template fold. Knowledge of many sequences, both related to the protein of interest and to each potential template, greatly improves the sensitivity of detection of remote evolutionary relationships. Methods that compare the sequence profiles of target and template families,[34] as well as hidden Markov models of structures and partial structures,[35] are currently the most effective approaches. Both methods can also incorporate secondary structure information. Aspects of three-dimensional fold and sequence compatibility have not proven as powerful a signal as once thought, but still play a role in the most comprehensive methods, often expressed in the form of long-range contact restraints.[36] There has been steady progress in this area over the CASP experiments such that, in the most recent one, templates were correctly identified by at

least one team for all but two of the relevant target domains (see Fig. 4 in Kryshtafovych *et al.*[33]).

2) *Alignment:* Early CASP experiments showed that even when it was possible to recognize a correct fold for use as a template, the resulting alignment was surprisingly poor. This factor too has improved greatly over the CASPs; in the most recent one, 60% of the best models of each target had at least as many residues correctly positioned as were present in a single best template.[33] That is, the mapping between every residue in the target protein with a structurally equivalent position in the principal template had been correctly made. Alignment values greater than 100% reflect the use of techniques to add information not present in the principal template, discussed further below. The same methods outlined above for fold recognition have also contributed most of the alignment improvement, though an additional contribution is now made from methods that explore the quality of models incorporating alternative alignment and other features.[37]

3) *Modeling of regions not present in the principal template:* As sequence similarity decreases, the fraction of residues that are part of the common fold between two proteins decreases, though to varying degrees: sometimes folds are well conserved down to low or even undetectable levels of sequence similarity; while in others, the structure diverges more rapidly. The fraction of the fold considered common between two proteins also depends on the exact metric used. With the fairly generous measure used in CASP,[38] it is quite usual for a small fraction of residues not to be equivalent between two proteins with 30% sequence identity, and this fraction may rise to 40% or 50% of non-equivalent residues for a pair of proteins with remotely related sequences. To generate a complete model, these residues must be added from some other source.

These non-principal template regions may be modeled in two ways. First, other members of the family may provide templates for some regions. Where multiple templates are available and provide differing versions of parts of the structure, the task of choosing which template to use for which part of the target is non-trivial. For a long time, there was little evidence that this could be done; but in the two most recent CASPs, there are striking examples of successful multi-template combinations (see Fig. 5 in Kryshtafovych *et al.*[33] for examples). If no alternative template can be identified for a missing region, other methods must be used. There is an extensive history of development of such "loop building" methods. Although these have long been capable of rebuilding regions of chain removed from an experimental structure, they have generally performed less well against the environment of a model: Even moderate errors in the surrounding structure tend to result in an incorrect choice of conformation. Once again, though, examples can be found in the two most recent CASPs, where non-template approaches have been successful (see also Fig. 5 in Kryshtafovych *et al.*[33]). Although there is encouraging progress, at present, it is the exception rather than the rule that the methods work completely.

4) *Refinement:* Even in regions where a good template is available, there will be small conformational differences from the target structure. The size of these differences generally increases with decreasing sequence similarity. To correct these, and to clean up initial models of non-template regions, some form of refinement method is needed. Attempts to apply conventional molecular dynamics procedures for this purpose have not been effective, and for a long time this seemed a very difficult problem. Results from the most recent CASP have shown some thing of a breakthrough, with several spectacular cases of refinement of quite poor initial models towards the experimental structure (see, for example, Qian *et al.*[26]). At present, these methods seem to work best for fairly inaccurate starting models (say, 4–5 Å RMSD on Cα atoms) and can refine down to around 2 Å RMSD, rather than achieving very high accuracy, and also work only on small proteins in some instances. Nevertheless, these results are extremely encouraging. As noted earlier, recent work has shown that the new methods are effective at improving initial NMR structures, for example. Current refinement methods use a combination of alternative local conformations, discrete local moves, and gradient/dynamics methods.[36,39]

Overall Model Accuracy: The progress in all four primary areas of modeling has resulted in a dramatic cumulative increase in overall model quality: Almost 80% of template-based best models received in CASP7 are more accurate at the main chain level than could be obtained by optimally copying the single best template[33] (itself a very non-trivial level of accuracy, requiring essentially perfect sequence alignment). The additional structural information is contributed by the processes outlined above — through combining information from multiple templates, using non-template methods for parts of the structure, successfully refining the model, or some combination of these techniques.

The accuracy of side chain orientations has also improved substantially over the course of the CASP experiments. Two factors have contributed here. First, side chain accuracy is dominated by main chain accuracy — side chain building methods perform well when rebuilding side chains on experimentally determined backbone, but performance deteriorates markedly as backbone accuracy decreases[40]; therefore, better main chain selection and alignment have automatically improved side chain results. Second, new sampling procedures that vary the backbone and side chain in an integrated manner[39] deliver better results.

Estimating Model Accuracy: As outlined above, models are now of much better quality for a given level of difficulty. However, quality is still very variable, and it is obviously desirable to provide estimates of overall accuracy as well as detailed accuracy at the main chain level. In the most recent CASP, the ability to assign accuracy was carefully and thoroughly assessed. The best current method for doing this utilizes algorithms originally devised for building consensus models, taking input from several model servers.[41] In favorable cases, the results are already quite impressive,[42] although there is still a long way to go. Fortunately, there is now great interest in this topic, and there will likely be considerable progress in the short term.

In summary then, although models are still far from perfect, there has been enormous progress in the quality of comparative models over the course of the CASP experiments. Furthermore, recent progress in identifying remote fold relationships, using multiple templates, refining initial models, and estimating model accuracy shows every sign of continuing.

References

1. Yan Y, Moult J. (2005) "Protein family clustering for structural genomics." *J Mol Biol* **353**: 744–59.
2. Coulson AF, Moult J. (2002) "A unifold, mesofold, and superfold model of protein fold use." *Proteins* **46**: 61–71.
3. Baker D, Sali A. (2001) "Protein structure prediction and structural genomics." *Science* **294**: 93–96.
4. Todd AE, Orengo CA, Thornton JM. (2001) "Evolution of function in protein superfamilies, from a structural perspective." *J Mol Biol* **307**: 1113–43.
5. Gilliland GL, Teplyakov A, Obmolova G, *et al.* (2002) "Assisting functional assignment for hypothetical *Heamophilus influenzae* gene products through structural genomics." *Curr Drug Targets Infect Disord* **2**: 339–53.
6. Tress M, Cheng J, Baldi P, *et al.* (2007) "Assessment of predictions submitted for the CASP7 domain prediction category." *Proteins* **69(Suppl 8)**: 137–51.
7. Wang P, Yan B, Guo JT, *et al.* (2005) "Structural genomics analysis of alternative splicing and application to isoform structure modeling." *Proc Natl Acad Sci USA* **102**: 18920–25.
8. Nassal M, Leifer I, Wingert I, *et al.* (2007) "A structural model for duck hepatitis B virus core protein derived by extensive mutagenesis." *J Virol* **81**: 13218–29.
9. Krasley E, Cooper KF, Mallory MJ, *et al.* (2006) "Regulation of the oxidative stress response through Slt2p-dependent destruction of cyclin C in *Saccharomyces cerevisiae*." *Genetics* **172**: 1477–86.
10. Glaser F, Pupko T, Paz I, *et al.* (2003) "ConSurf: identification of functional regions in proteins by surface-mapping of phylogenetic information." *Bioinformatics* **19**: 163–64.
11. Laskowski RA, Luscombe NM, Swindells MB, Thornton JM. (1996) "Protein clefts in molecular recognition and function." *Protein Sci* **5**: 2438–52.
12. Stenson PD, Ball EV, Mort M, *et al.* (2003) "Human Gene Mutation Database (HGMD): 2003 update." *Hum Mutat* **21**: 577–81.
13. Sjoblom T, Jones S, Wood LD, *et al.* (2006) "The consensus coding sequences of human breast and colorectal cancers." *Science* **314**: 268–74.
14. Yue P, Moult J. (2006) "Identification and analysis of deleterious human SNPs." *J Mol Biol* **356**: 1263–74.
15. Wood LD, Parsons DW, Jones S, *et al.* (2007) "The genomic landscapes of human breast and colorectal cancers." *Science* **318**: 1108–13.
16. Najmanovich RJ, Allali-Hassani A, Morris RJ, *et al.* (2007) "Analysis of binding site similarity, small-molecule similarity and experimental binding profiles in the human cytosolic sulfotransferase family." *Bioinformatics* **23**: e104–09.
17. Yue P, Li Z, Moult J. (2005) "Loss of protein structure stability as a major causative factor in monogenic disease." *J Mol Biol* **353**: 459–73.

18. Ramensky V, Bork P, Sunyaev S. (2002) "Human non-synonymous SNPs: server and survey." *Nucleic Acids Res* **30**: 3894–900.
19. Ng PC, Henikoff S. (2003) "SIFT: predicting amino acid changes that affect protein function." *Nucleic Acids Res* **31**: 3812–14.
20. Becker OM, Dhanoa DS, Marantz Y, *et al.* (2006) "An integrated *in silico* 3D model-driven discovery of a novel, potent, and selective amidosulfonamide 5-HT1A agonist (PRX-00023) for the treatment of anxiety and depression." *J Med Chem* **49**: 3116–35.
21. Murray D, Honig B. (2002) "Electrostatic control of the membrane targeting of C2 domains." *Mol Cell* **9**: 145–54.
22. Teplyakov A, Liu S, Lu Z, *et al.* (2005) "Crystal structure of the petal death protein from carnation flower." *Biochemistry* **44**: 16377–84.
23. Watson JD, Sanderson S, Ezersky A, *et al.* (2007) "Towards fully automated structure-based function prediction in structural genomics: a case study." *J Mol Biol* **367:** 1511–22.
24. Schwarzenbacher R, Godzik A, Grzechnik SK, Jaroszewski L. (2004) "The importance of alignment accuracy for molecular replacement." *Acta Crystallogr D Biol Crystallogr* **60**: 1229–36.
25. Raimondo D, Giorgetti A, Giorgetti A, *et al.* (2007) "Automatic procedure for using models of proteins in molecular replacement." *Proteins* **66**: 689–96.
26. Qian B, Raman S, Das R, *et al.* (2007) "High-resolution structure prediction and the crystallographic phase problem." *Nature* **450**: 259–64.
27. Chiu W, Baker ML, Jiang W, Zhou ZH. (2002) "Deriving folds of macromolecular complexes through electron cryomicroscopy and bioinformatics approaches." *Curr Opin Struct Biol* **12**: 263–69.
28. Putnam CD, Hammel M, Hura GL, Tainer JA (2007) "X-ray solution scattering (SAXS) combined with crystallography and computation: defining accurate macromolecular structures, conformations and assemblies in solution." *Q Rev Biophys* **40**: 191–285.
29. Sinz A. (2006) "Chemical cross-linking and mass spectrometry to map three-dimensional protein structures and protein–protein interactions." *Mass Spectrom Rev* **25**: 663–82.
30. Xu G, Chance MR. (2007) "Hydroxyl radical-mediated modification of proteins as probes for structural proteomics." *Chem Rev* **107**: 3514–43.
31. Moult J, Fidelis K, Kryshtafovych A, *et al.* (2007) "Critical assessment of methods of protein structure prediction — Round VII." *Proteins* **69**: 3–9.
32. Kopp J, Bordoli L, Battey JN, *et al.* (2007) "Assessment of CASP7 predictions for template-based modeling targets." *Proteins* **69(Suppl 8)**: 38–56.
33. Kryshtafovych A, Fidelis K, Moult J. (2007) "Progress from CASP6 to CASP7." *Proteins* **69**: 194–207.
34. Ohlson T, Wallner B, Elofsson A. (2004) "Profile–profile methods provide improved fold-recognition: a study of different profile–profile alignment methods." *Proteins* **57**: 188–97.

35. Karplus K, Karchin R, Shackelford G, Hughey R. (2005) "Calibrating E-values for hidden Markov models using reverse-sequence null models." *Bioinformatics* **21**: 4107–15.
36. Zhang Y. (2007) "Template-based modeling and free modeling by I-TASSER in CASP7." *Proteins* **69 (Suppl 8)**: 108–17.
37. Venclovas C, Margelevicius M. (2005) "Comparative modeling in CASP6 using consensus approach to template selection, sequence-structure alignment, and structure assessment." *Proteins* **61 (Suppl 7)**: 99–105.
38. Zemla A, Venclovas C, Moult J, Fidelis K. (2001) "Processing and evaluation of predictions in CASP4." *Proteins* (Submitted).
39. Das R, Qian B, Raman S, *et al.* (2007) "Structure prediction for CASP7 targets using extensive all-atom refinement with Rosetta@home." *Proteins* **69 (Suppl 8)**: 118–28.
40. Chung SY, Subbiah S. (1995) "The use of side-chain packing methods in modeling bacteriophage repressor and cro proteins." *Protein Sci* **4**: 2300–09.
41. Wallner B, Elofsson A. (2007) "Prediction of global and local model quality in CASP7 using Pcons and ProQ." *Proteins* **69 (Suppl 8)**: 184–93.
42. Cozzetto D, Kryshtafovych A, Ceriani M, Tramontano A. (2007) "Assessment of predictions in the model quality assessment category." *Proteins* **69**: 175–83.

Chapter 6

The Contribution of Structural Proteomics to Understanding the Function of Hypothetical Proteins

Michael D. Suits[†], Allan Matte[‡], Zongchao Jia[†] and Miroslaw Cygler[*,‡,§]

Developments in structural genomics have led to the automation of protein production and given rise to the structure determination pipeline, resulting in a rapid increase in the number of new protein structures. The next challenge will be to infer the functions for many of these proteins. Here, we describe the efforts of the Montreal-Kingston Bacterial Structural Genomics Initiative in the utilization of structural information for protein functional assignment as illustrated by several examples.

The Challenge of Assigning Protein Functions to Hypothetical Proteins Using High-Throughput Methods

A key role of proteomics research, including structural proteomics, is to gain an insight into the function of previously uncharacterized proteins

---

*Corresponding author: Mirek Cygler, Tel: 1 514 496 6321, Fax: 1 514 496 5143 Email: mirek@bri.nrc.ca

[†]Department of Biochemistry, Queen's University, Kingston, ON Canada.

[‡]Biotechnology Research Institute, 6100 Royalmount Ave., Montreal QC H4P 2R2 Canada.

[§]Department of Biochemistry, McGill University, Montreal, QC, Canada.

at the level of whole genomes. The high-throughput cloning and expression of open-reading frames provides simultaneous confirmation that the predicted genes, have in fact, been translated into proteins. This combined genomic and proteomic information will be critical, for example, in the construction of complete metabolic networks for a bacterial cell. A considerable number of proteins of viral, microbial and eukaryotic origin remain un-annotated at the functional level. For many others, the annotation is based on weak sequence similarity to proteins with known functions. While genetic screens and various high-throughput functional assays play a key role in assigning protein function, these methods cannot be adapted to all proteins and are best used to characterize enzymes.[1] Protein structure often provides functional clues that go beyond what sequence analysis alone can provide.[2] Function is often not preserved in proteins which have diverged to a level of sequence identity below ~40%,[3,4] a much higher level of sequence identity than in proteins with significant structural similarity. Many proteins whose functional assignment is based on sequence comparisons require experimental validation, part of which can be achieved through structural analysis. The structure of a protein often leads directly to the design of tailor-made experiments to test the predicted function using activity or binding assays *in vitro* or *in vivo*. The availability of a structure also permits the full complement of structural bioinformatics tools to be employed, sometimes automatically, in analyses ranging from sequence-structure relationships to the identification of possible ligand-binding sites.[5–8] In addition, functional clues are often provided by the presence of small molecules identified in the crystal structure, e.g. cofactors, ions, etc.

Although *Escherichia coli* is the bacterial model organism, there still remain a large number of open reading frames (ORFs), either of unknown function or of questionable functional assignment based only on weak sequence similarity. This is applicable to even the best investigated and annotated non-pathogenic K-12 strain.[9] Pathogenic strains of *E. coli* contain a large number of ORFs that are not present in the K-12 strain and therefore may be related to their pathogenicity. The genome of the enterohemorrhagic strain O157:H7 of *E. coli* has 1.4 Mb of DNA that is not present in the K-12 strain.[10,11] Much of this O157-specific DNA is thought to have been acquired through horizontal transfer from foreign species and contains many bacteriophage elements that mediate genetic reorganization and exchange, contributing to its pathogenesis through frequent genetic change.[10] Sequence analysis of this O157-specific DNA indicates that it encodes 1632 unique ORFs and 20 unique tRNAs. At least 131 of the O157-encoded proteins are associated with virulence.[10] These unique proteins are very important, as understanding of their function and interplay with other protein partners will contribute to our overall understanding of bacterial pathogenesis, in addition to the principles and mechanisms of infection.

Here, we provide three examples from our structural genomics program,[12] where either functional assignments or mechanistic insight were revealed from clues provided initially by the crystal structure. The examples highlight a quinate/shikimate dehydrogenase, a family I CoA transferase and proteins present in pathogenic *E. coli* O157:H7 that are involved in bacterial heme metabolism.

Defining Protein Function by Experiment and Structural Bioinformatics

A Quinate/Shikimate Dehydrogenase from Escherichia coli — YdiB

The *ydiB* gene was selected as one of the targets for the structural proteomics study as a result of sequence analysis, which indicated that no significant sequence homologues of known structure had been determined. The sequence was annotated as a putative shikimate dehydrogenase, as it possessed 28% sequence identity to *E. coli* AroE, a characterized shikimate dehydrogenase.[13,14] For a number of years, AroE has been characterized as a NADPH-dependent dehydrogenase that converts 3-dehydroskikimate to shikimate, the fourth step in chorismate biosynthesis; with chorismate being a precursor to vitamins and aromatic amino acids.[15] The central role of these enzymes in intermediary metabolism and their absence in metazoans led to the almost simultaneous determination of structures of this enzyme from

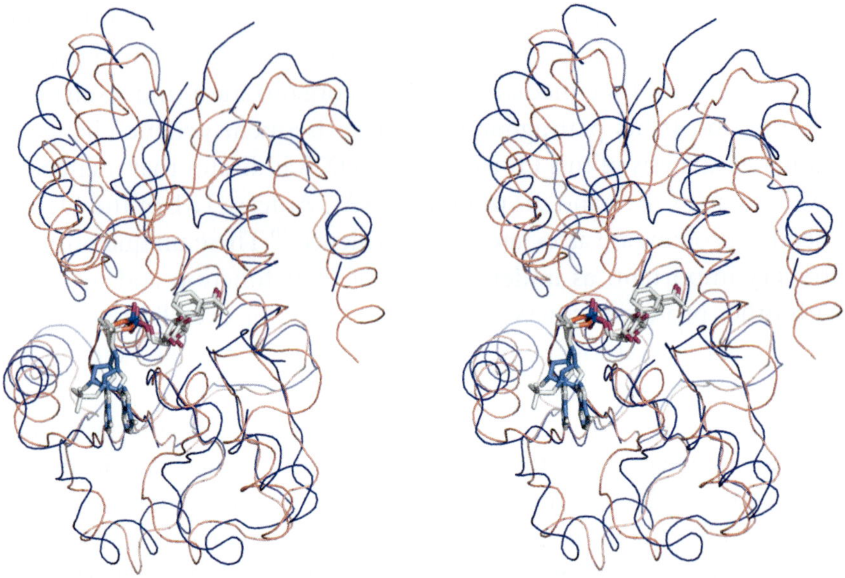

Fig. 1 Superposition of the structures of *E. coli* YdiB (PDB 1O9B, blue) and *E. coli* AroE (PDB 1NYT, orange) based on the NAD/NADP binding domains. The molecules are shown as ribbon tracings and the cofactors in stick representation. The N- and C-terminal domains show some orientational flexibility both when compared different molecules in the asymmetric unit (2 molecules in YdiB, 4 molecules in AroE) and between the two proteins.

several bacteria[16–20] and has sparked interest in these enzymes as potential antimicrobial targets.[17,21]

The structure of *E. coli* YdiB revealed interesting similarities as well as differences to *E. coli* AroE (Fig. 1).[16] Both enzymes share similar folds, contain signature motifs indicative of dinucleotide (NAD^+, $NADP^+$) binding, and while AroE is a monomer, the crystal structure of *E. coli* YdiB revealed a dimer.[16] The mode of dimerization of YdiB was unusual, as it involved a relatively small surface of each monomer, but was consistent with the molecular weight as determined by size exclusion chromatography. Both enzymes consist of two α/β domains, with the C-terminal domain adopting a canonical Rossmann fold topology, consistent with dinucleotide binding.

Crystallographic analysis of AroE revealed a bound molecule of NADP, the expected cofactor, while co-crystallization of YdiB with NADH revealed clear density for this cofactor at the active site. Based on this result, the cofactor requirements of YdiB were investigated in more detail. It was found that using shikimic acid as a substrate, both the K_M and k_{cat} for either $NADP^+$ or NAD^+ were similar. AroE did not exhibit competitive inhibition with respect to $NADP^+$, indicating that NAD^+ did not bind to the enzyme. Unlike AroE, YdiB also displayed activity with quinic acid as a substrate, in addition to shikimic acid, with a clear preference for NAD^+ as the electron acceptor. On this basis, YdiB was characterized as a quinate/shikimate dehydrogenase.[16]

A detailed analysis of the cofactor binding sites in the two enzymes revealed clues as to their respective cofactor preference. In YdiB, Val^{206} replaced Ser^{190} of AroE, forming a hydrophobic interaction with the adenine ring. In particular, two residues of YdiB, Asp^{158} and Phe^{160}, were found to substitute for the corresponding residues Thr^{151} and Arg^{154} in AroE. The Phe-for-Arg substitution is important, as it creates a more neutral environment suitable for binding both cofactors.

To better define residues involved in substrate binding and catalysis for YdiB, a series of site-specific mutations were prepared in the active site region, and were kinetically characterized and compared with the wild-type enzyme.[22] These residues were selected primarily on the basis of the available crystal structures, although their level of sequence identity in related enzymes was also evaluated. The kinetic results indicated critical roles for Lys^{71} and Asp^{107}, with both residues contributing towards substrate binding in the Michaelis complex. The residue Thr^{106} also contributed towards substrate affinity. To our surprise, no single residue was identified as catalytically essential in these studies, although the observed decrease in k_{cat} for the K71G mutant was consistent with a role for this residue in transition-state stabilization. Similar mutations of the equivalent lysine and aspartate in the shikimate dehydrogenase from *Hemophilus influenza*, which according to phylogenetic analysis belongs to a different functional class, did not lead to any measurable activity.[20] These results leave unresolved the question of which residue in the active site functions as the general base in catalysis. Other possible mechanisms, including quantum-mechanical tunneling.[23] Substrate-assisted catalysis should be considered in explaining this issue. Nevertheless, these studies have made it possible to distinguish between different proposed models for the mode of substrate binding in the ternary complex.[16,18,19] The observed role for Lys^{71} and Asp^{107} in substrate binding is most consistent with the ternary complex model of Michel *et al.*[16] The proposed orientation of the substrate has been confirmed recently in the structure of *Arabidopsis* dehydroquinate dehydratase-shikimate dehydrogenase, which has been determined with bound shikimate.[24]

Crystallographic Trapping of the γ-Glutamyl-CoA Intermediate of E. coli YdiF, a Family-I CoA Transferase

Coenzyme A is a ubiquitous cofactor, used by an enormous number of different enzymes for a variety of metabolic functions. CoA-transferases catalyze the reversible transfer of a CoA moiety from a CoA-thioester to a carboxylic acid acceptor. The *E. coli* enzyme YdiF was annotated as a putative CoA-transferase of unknown function, and was selected for structural and enzymatic characterization. Enzymatic assays and ESI-MS were employed to evaluate the activity profile of YdiF using several combinations of CoA-thioester and carboxylic acid acceptor, revealing that the enzyme had a preference for acceptors containing short acyl chains.[25] The highest activity for YdiF was observed in the presence of acetoacetyl-CoA with acetate as an acceptor.

The crystal structure of YdiF revealed a tetrameric enzyme, composed of a dimer-of-dimers. Each monomer consisted of two α/β domains, separated by a linker region. The structural similarity of these two domains indicated a gene duplication event. While the two domains can be superimposed with an rmsd of 1.6 Å for 85 Cα pairs, the amino acid sequence identity between them was very low (~6%), suggesting that the gene duplication occurred in the distant past. A similar ancient gene duplication event has been proposed for the α- and β-subunits of *Acidaminococcus fermentens* glutaconate CoA transferase, which together form the active heterodimer.[26]

In order to define the substrate binding site and residues involved in catalysis, we co-crystallized YdiF with a number of CoA-thioesters in the absence of the carboxylic acid acceptor and in each case trapped the same γ-glutamyl CoA-thioester intermediate, covalently linked to Glu^{333}. The extent of electron density observed for CoA varied in the different sub-units obtained from various data sets. In the crystal structure of YdiF co-crystallized with butyryl-CoA, electron density corresponding to that of a covalent thioester between Glu^{333} and CoA was observed in sub-units A, B and C but not D. This intermediate could also be detected by ESI-MS when YdiF was incubated in the presence of butryl-CoA in solution. While this intermediate had been previously identified in solution, in some cases almost 40 years earlier with other family I CoA-transferases,[27,28] it had never been structurally characterized. CoA was found to bind in a cleft between the N- and C-terminal domains, with the cofactor-binding interactions contributed by the C-terminal domain. No large conformational changes were observed accompanying CoA binding. Comparison of residues in the neighborhood of Glu^{333} in individual subunits across all the crystal structures of complexes and the apo form allowed us to postulate local structural changes in the YdiF active site along the reaction pathway.[25]

Escherichia coli Proteins Associated with Heme Metabolism

Effective iron acquisition and utilization are essential for colonization of invading Gram-negative bacteria, and contribute to their pathogenesis. One mechanism of iron acquisition used by enterohemorrhagic pathogens involves the induction of hemolytic lesions to gain access to heme, the most abundant source of invertebrate iron.[29] It had been previously established through gene knockout experiments in *Yersinia enterocolitica* that five gene products, homologous to those in a similar *E. coli* operon, were required for heme uptake and utilization. The five gene products are the following: the outer membrane receptor, ChuA; the periplasmic heme binding protein, ChuT; the heme permease protein, ChuU; an ATP-binding hydrophilic protein; and ChuV. Four other proteins, ChuS, ChuW, ChuX and ChuY, which are conserved across Gram-negative heme utilization loci, were poorly characterized in terms of function and structure. In *Shigella dysenteriae*, the gene homologous to *chuS* can be found 48-nucleotides downstream of the outer membrane receptor and is separated by an intergenic region homologous to the Fur box, a promoter element responsive to low iron levels.[30] Sequence homology analysis using Blastp[31] provided no functional insight. At the beginning of our investigation, ChuS was known only as a protein upregulated under low iron conditions, responsible for the prevention of heme toxicity.

In *Shigella dysenteriae, shuW, shuX*, and *shuY*, the three remaining genes clustered within the heme utilization locus, are poorly characterized but appear to be co-transcribed with the periplasmic heme transport protein *shuT*. While expression of these three genes is not required for heme utilization as an iron source,[32,33] the fate of the cytoplasmic heme and the proteins involved in modulating heme utilization are not known. In the heme uptake locus from *Y. pestis*, genes homologues to *chuX* and *chuY* (*orfX* and *orfY*, respectively) were found to be located immediately downstream of the heme receptor *hmuR* and are therefore likely co-expressed.[34] Similarly, in *S. dysenteriae*, the initiation codon of *shuY* overlaps the stop codon of *shuX*, which may indicate that these two smaller genes are co-transcribed.[30] In *Vibrio anguillarum*, which can use heme and hemoglobin as sources of iron, it was shown using a *lacZ* reporter system that *huvX*, a homologue of *chuX*, was co-transcribed with the uncharacterized gene *huvZ* under iron limiting conditions, but that *huvX* was not essential for heme utilization.[35,36] An examination of the −10 and −35 elements downstream of *huvX* revealed sequence similarity to σ^{70}-promoters and the presence of a Fur element, suggesting that HuvX expression is upregulated under iron-limiting conditions where heme utilization is stimulated.[36] Although a review of the current literature did not suggest an integral role for ChuX, it was initially selected as a target for functional and structural investigation to bridge a gap in our understanding of the process of iron uptake.

In cases where sequence and structure homology fail to reveal functional insight, certain functional information can still be obtained from a variety of other sources. Experimental clues contributing to the functional annotation of ChuS were revealed even prior to its structure determination. During overexpression of ChuS, a blue-green pigment developed and persisted in fractions containing ChuS following affinity purification.[37] As ChuS is found in an operon responsible for heme uptake and utilization, the formation of this pigment may have been the result of the production of biliverdin or bilirubin, the end product of heme degradation that has reduced cytotoxicity compared with biliverdin. However, the pigment did not co-purify with ChuS during subsequent purification steps, indicating that the pigmented compound was neither tightly nor covalently bound by ChuS. This result suggested that while bilirubin or biliverdin may be responsible for the observed color, neither was the preferred substrate or ligand for ChuS.

Concurrent with efforts to determine the structure of ChuS, and in view of its presence in a heme-utilization operon, we tested heme binding to ChuS by spectral analysis. When hemoproteins coordinate heme, a characteristic absorbance spectrum is produced that is dependent on the nature of the protein-heme interaction. Based on the differences in the spectra between free heme and heme bound to ChuS, together with those reported for other hemoproteins, we determined that the spectra of ChuS-heme resembled those of heme oxygenases. A Soret maximum was evident at 408 nm, with a smaller set of characteristic peaks at 545 nm and 580 nm, suggesting that the ChuS-heme complex formed was ferric hexacoordinate and that a histidine residue was involved in this coordination.[38] Heme oxygenases use oxygen together with an electron donor such as cytochrome P450 reductase-NADPH or ascorbic acid to degrade heme, producing biliverdin, carbon monoxide, and free iron. When either cytochrome P450 reductase-NADPH or ascorbic acid was added, the spectra of ChuS-heme changed over time. The change was similar to those characterized for other heme oxygenases, i.e. the disappearance of the Soret peak. Furthermore, we used sensitive gas chromatography to first detect, and then measure, carbon monoxide production during heme degradation, further confirming ChuS as a novel heme oxygenase. To eliminate the possibility of coupled oxidation or non-specific heme degradation, catalase and superoxide dismutase were added prior to the addition of either electron donor where similar spectral changes and CO production were observed.

When the structure of ChuS was determined, it was analyzed for structural similarity to other proteins of known structure. A number of bioinformatics web-based servers have been developed to assist in extracting functional information from protein structures, including CATH, CE, DALI, DEJAVU, MATRAS, PALI, SCOP, SUPFAM, and VAST. A recent evaluation of these computational tools suggests that CE, DALI, MATRAS, and VAST showed the best performance, but understandably, none of the programs achieved a 100% success rate.[39] An advantage of *in silico* analysis is that a large amount of information is gathered in a short period of time. Subsequently, however, the information must be analyzed to identify true positives from false positives. In the case of ChuS, the 3-D structure-based homology comparison indicated that the structure represents a novel fold and, therefore, no new functional information was obtained from this analysis.

The structure of ChuS has a central core of two, nine-stranded anti-parallel β-sheets, each flanked at their N-termini by a pair of parallel α-helices and at its C-terminus by a set of three α-helices in a helix-loop-helix-loop-helix configuration (Fig. 2). The structure clearly indicated that ChuS arose by gene duplication, followed by divergence in sequence, as the N- and C-terminal halves of the protein share only 18% sequence identity. Notably, the structure of ChuS shows no structural homology to any heme degrading enzymes that have been characterized so far. Normally, these enzymes are mainly α-helical, with the heme sandwiched between a histidine residue and a glycine residue, with human heme oxygenases the best-characterized representatives of this group.[40] Due to the appearance of the Soret peak following ChuS reconstitution with heme, we postulated that an axial histidine side chain may be responsible for heme coordination. Sequence alignment of ChuS with its homologues highlighted four conserved histidine residues at positions 73, 87, 193 and 277. The structure of ChuS shows two large clefts delineated on opposite sides of the central β-pleated sheets core, with the other side of the clefts flanked by a set of three α-helices. Interestingly, three of the four conserved histidine residues are adjacent to, or point into, either of these

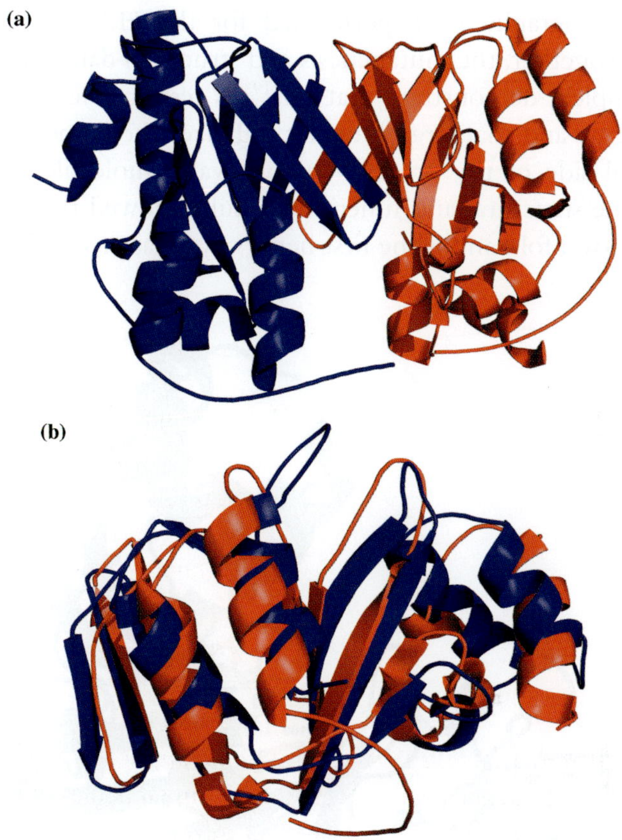

Fig. 2 (**a**) Structure of apo-ChuS. The N- and C-terminal domains are colored red and blue, respectively; (**b**) superimposition of the N- and C-terminal halves of ChuS. The two domains can be superimposed with a root-means-squares deviation of 2.1 Å.

two clefts. The exception is His^{73}, which points towards the interior of the structure.

In an attempt to disrupt the interaction of ChuS with heme, we undertook mutagenesis of these conserved histidine residues. In parallel, we obtained two new crystal forms of the ChuS-heme complex, which, upon structure determination, revealed the structural basis for heme coordination. Intriguingly, we observed only one heme molecule bound to ChuS. Both spectral analysis and a CO release assay using gas chromatography were performed for the H73A and H193N mutants. Together, the mutagenesis and structural data on the ChuS-heme complex, demonstrate that His[193] is responsible for axial ligation of heme and that Arg[100] further stabilizes the heme group from the medial side of the protein via two water molecules (Fig. 3).[41] Overall, the structural and functional results showed that ChuS represents a novel fold and is the first heme oxygenase to be identified in

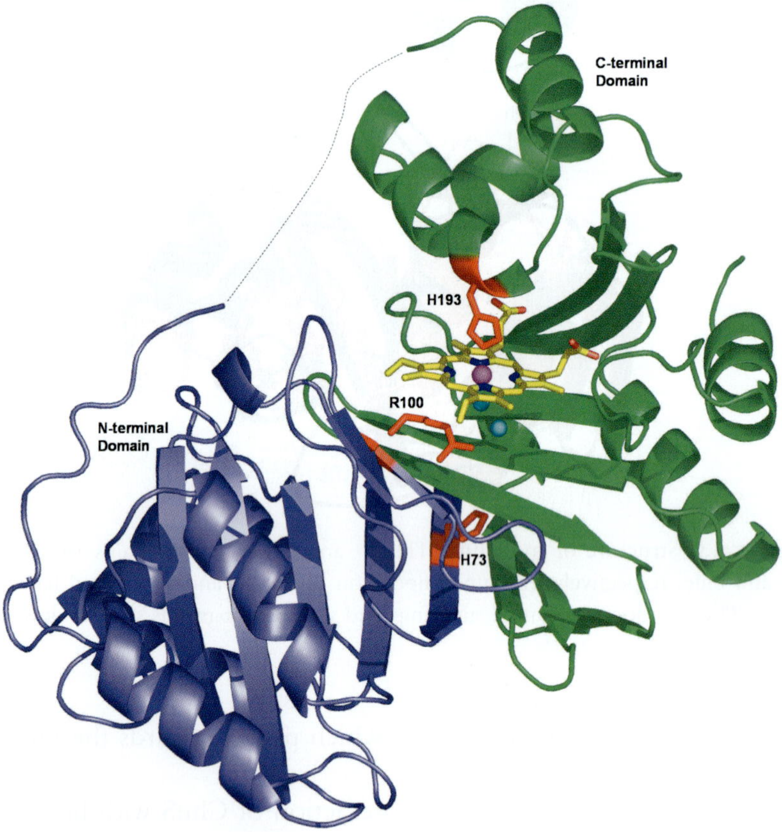

Fig. 3 Ribbon diagram of ChuS with bound heme. The heme-degrading enzyme ChuS displays a unique mode of heme coordination compared to other heme oxygenases identified to date. ChuS employs domain swapping between the two halves of the monomer structure to coordinate heme between a central set of β-sheets via Arg[100] and His[198] that originate at the base of an α-helix.

any strain of *E. coli*. Further, its mode of heme coordination is unique compared to that of other heme-degrading enzymes. Recently, the structure of another protein with the same fold has been determined, i.e. *Yersinia enterocolitica* HemS with bound heme.[42]

The structural characterization and functional annotation of ChuX was different from that of ChuS, although the entire process was accelerated as a direct result of our previous experience with ChuS. The structure of ChuX forms a dimer, which displays a surprising degree of similarity to the monomer structures of two other heme utilization operon proteins, *E. coli* ChuS and *Y. enterocolitica* HemS, both of which have an internal domain repeat (Fig. 4). A recent search for related structures in the PDB identified a protein from *Erwinia carotovora* Q6D2T7_ERWCT that has a single repeat structure similar to that of ChuX and also forms dimers (unpublished, PDB code 2PH0).

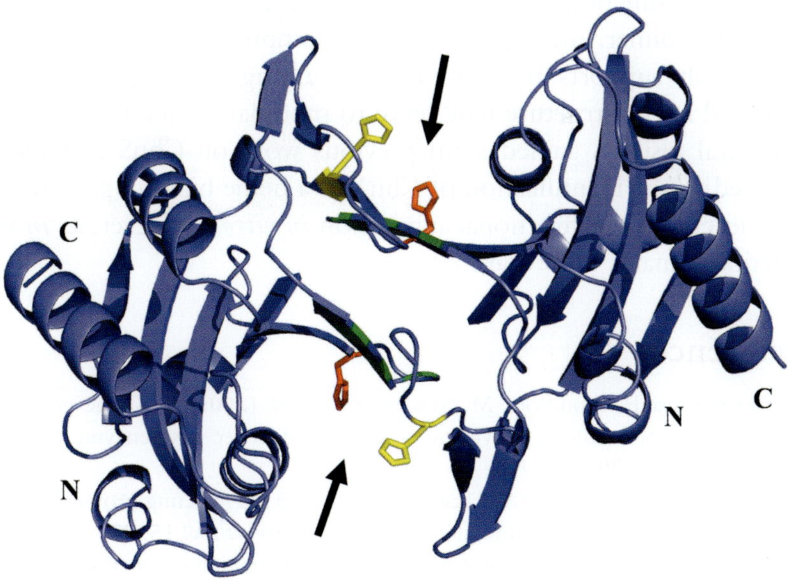

Fig. 4 Ribbon structure of ChuX highlighting the conserved histidines and the putative heme binding clefts. Despite the low sequence identity between ChuX and ChuS the topology of ChuX is similar to that of the domains of ChuS. The conserved His^{65} and His^{98} (yellow and red respectively), have been shown to contribute to heme binding via site-directed mutagenesis.

Despite having low sequence similarity, all these structures have very similar overall topologies. The ChuS monomer structure resembles that of the homodimer of both ChuX and AGR_C_4470p. This structural similarity suggests an evolutionary relationship where ChuS and its homologues have evolved from an ancestral *chuX* gene through duplication. Absorption spectra analysis of ChuX reconstituted with heme, demonstrates that ChuX binds heme in a 1:1 molar ratio, implying that each homodimer coordinates two heme molecules. This is in contrast to ChuS and HemS where only one heme molecule is bound in the crystal structures. Based on its structural similarity to ChuS, we designed a number of site-directed mutations in ChuX to probe the putative heme binding sites and the dimer interface. Spectral analysis of ChuX and its mutants suggested that mutations of both His^{65} and His^{98} are required to remove the heme binding capacity of ChuX. We also demonstrated that heme was transferred from ChuX to ChuS, which then functioned to degrade heme. Furthermore, mutations at the ChuX dimer interface resulted in compromised heme binding. This result supports the notion that varying ChuX juxtapositions observed in the structure may serve to modulate heme binding. The structural insights gained from previous work on ChuS and HemS have led to the identification of ChuX as a heme binding protein and have accelerated its functional annotation *in vitro*. However, its *in vivo* function remains to be clarified.

References

1. Kuznetsova E, Proudfoot M, Sanders SA, *et al.* (2005) "Enzyme genomics: application of general enzymatic screens to discover new enzymes." *FEMS Microbiol Rev* **29**: 263–79.
2. Watson JD, Laskowski RA, Thornton JM. (2005) "Predicting protein function from sequence and structural data." *Curr Opin Struct Biol* **15**: 275–84.
3. Todd AE, Orengo CA, Thornton JM. (2001) "Evolution of function in protein superfamilies, from a structural perspective." *J Mol Biol* **307**: 1113–43.
4. Tian W, Skolnick J. (2003) "How well is enzyme function conserved as a function of pairwise sequence identity?" *J Mol Biol* **333**: 863–82.
5. Laskowski RA, Watson JD, Thornton JM. (2005a) "ProFunc: a server for predicting protein function from 3D structure." *Nucleic Acids Res* **33**: W89–93.

6. Laskowski RA, Watson JD, Thornton JM. (2005b) "Protein function prediction using local 3D templates." *J Mol Biol* **351**: 614–26.
7. Glaser F, Morris RJ, Najmanovich RJ, *et al.* (2006) "A method for localizing ligand binding pockets in protein structures." *Proteins* **62**: 479–88.
8. Watson JD, Sanderson S, Ezersky A, *et al.* (2007) "Towards fully automated structure-based function prediction in structural genomics: a case study." *J Mol Biol* **367**: 1511–22.
9. Riley M, Abe T, Arnaud MB, *et al.* (2006) "*Escherichia coli* K-12: a cooperatively developed annotation snapshot–2005." *Nucleic Acids Res* **34**: 1–9.
10. Hayashi T, Makino K, Ohnishi M, *et al.* (2001) "Complete genome sequence of enterohemorrhagic *Escherichia coli* O157:H7 and genomic comparison with a laboratory strain K-12." *DNA Res* **8**: 11–22.
11. Ogura Y, Kurokawa K, Ooka T, *et al.* (2006) "Complexity of the genomic diversity in enterohemorrhagic *Escherichia coli* O157 revealed by the combinational use of the O157 Sakai OligoDNA microarray and the Whole Genome PCR scanning." *DNA Res* **13**: 3–14.
12. Matte A, Sivaraman J, Ekiel I, *et al.* (2003) "Contribution of structural genomics to understanding the biology of *Escherichia coli*." *J Bacteriol* **185**: 3994–4002.
13. Chaudhuri S, Coggins JR. (1985) "The purification of shikimate dehydrogenase from *Escherichia coli*." *Biochem J* **226**: 217–23.
14. Anton IA, Coggins JR. (1988) "Sequencing and overexpression of the *Escherichia coli* aroE gene encoding shikimate dehydrogenase." *Biochem J* **249**: 319–26.
15. Herrmann KM, Weaver LM. (1999) "The shikimate pathway." *Annu Rev Plant Physiol Plant Mol Biol* **50**: 473–503.
16. Michel G, Roszak AW, Sauve V, *et al.* (2003) "Structures of shikimate dehydrogenase AroE and its Paralog YdiB. A common structural framework for different activities." *J Biol Chem* **278**: 19463–72.
17. Padyana AK, Burley SK. (2003) "Crystal structure of shikimate 5-dehydrogenase (SDH) bound to NADP: insights into function and evolution." *Structure* **11**: 1005–13.
18. Ye S, von DF, Brooun A, *et al.* (2003) "The crystal structure of shikimate dehydrogenase (AroE) reveals a unique NADPH binding mode." *J Bacteriol* **185**: 4144–51.
19. Benach J, Lee I, Edstrom W, *et al.* (2003) "The 2.3-A crystal structure of the shikimate 5-dehydrogenase orthologue YdiB from *Escherichia coli* suggests a novel catalytic environment for an NAD-dependent dehydrogenase." *J Biol Chem* **278**: 19176–82.
20. Singh S, Korolev S, Koroleva O, *et al.* (2005) "Crystal structure of a novel shikimate dehydrogenase from *Haemophilus influenzae*." *J Biol Chem* **280**: 17101–108.
21. Han C, Wang L, Yu K, *et al.* (2006) Biochemical characterization and inhibitor discovery of shikimate dehydrogenase from *Helicobacter pylori*. *FEBS J* **273**: 4682–92.

22. Lindner HA, Nadeau G, Matte A, *et al.* (2005) "Site-directed mutagenesis of the active site region in the quinate/shikimate 5-dehydrogenase YdiB of *Escherichia coli.*" *J Biol Chem* **280**: 7162–69.
23. Kohen A, Cannio R, Bartolucci S, Klinman JP. (1999) "Enzyme dynamics and hydrogen tunnelling in a thermophilic alcohol dehydrogenase." *Nature* **399**: 496–99.
24. Singh SA, Christendat D. (2006) "Structure of Arabidopsis dehydroquinate dehydratase-shikimate dehydrogenase and implications for metabolic channeling in the shikimate pathway." *Biochemistry* **45**: 7787–96.
25. Rangarajan ES, Li Y, Ajamian E, *et al.* (2005) "Crystallographic trapping of the glutamyl-CoA thioester intermediate of family I CoA transferases." *J Biol Chem* **280**: 42919–28.
26. Jacob U, Mack M, Clausen T, *et al.* (1997) "Glutaconate CoA-transferase from *Acidaminococcus fermentans*: the crystal structure reveals homology with other CoA-transferases." *Structure* **5**: 415–26.
27. Hersh LB, Jencks WP. (1967) "Isolation of an enzyme-coenzyme A intermediate from succinyl coenzyme A-acetoacetate coenzyme A transferase." *J Biol Chem* **242**: 339–40.
28. Selmer T, Buckel W. (1999) "Oxygen exchange between acetate and the catalytic glutamate residue in glutaconate CoA-transferase from *Acidaminococcus fermentans*. Implications for the mechanism of CoA-ester hydrolysis." *J Biol Chem* **274**: 20772–78.
29. Law D, Kelly J. (1995) "Use of heme and hemoglobin by *Escherichia coli* O157 and other Shiga-like-toxin-producing *E. coli* serogroups." *Infect Immun* **63**: 700–02.
30. Wyckoff EE, Duncan D, Torres AG, *et al.* (1998) "Structure of the *Shigella dysenteriae* haem transport locus and its phylogenetic distribution in enteric bacteria." *Mol Microbiol* **28**: 1139–52.
31. Altschul SF, Madden TL, Schaffer AA, *et al.* (1997) "Gapped BLAST and PSI-BLAST: a new generation of protein database search programs." *Nucl Acids Res* **25**: 3389–402.
32. Stojiljkovic I, Hantke K. (1992) "Hemin uptake system of *Yersinia enterocolitica*: similarities with other TonB-dependent systems in gram-negative bacteria." *EMBO J* **11**: 4359–67.
33. Stojiljkovic I, Hantke K. (1994) "Transport of haemin across the cytoplasmic membrane through a haemin-specific periplasmic binding-protein-dependent transport system in *Yersinia enterocolitica*." *Mol Microbiol* **13**: 719–32.
34. Thompson JM, Jones HA, Perry RD. (1999) "Molecular characterization of the hemin uptake locus (hmu) from *Yersinia pestis* and analysis of hmu mutants for hemin and hemoprotein utilization." *Infect Immun* **67**: 3879–92.
35. Mourino S, Osorio CR, Lemos ML. (2004) "Characterization of heme uptake cluster genes in the fish pathogen *Vibrio anguillarum*." *J Bacteriol* **186**: 6159–67.

36. Mourino S, Osorio CR, Lemos ML, Crosa JH. (2006) "Transcriptional organization and regulation of the *Vibrio anguillarum* heme uptake gene cluster." *Gene* **374**: 68–76.
37. Suits MD, Pal GP, Nakatsu K, *et al.* (2005) "Identification of an *Escherichia coli* O157:H7 heme oxygenase with tandem functional repeats." *Proc Natl Acad Sci USA* **102**: 16955–60.
38. Chu GC, Katakura K, Zhang X, *et al.* (1999) "Heme degradation as catalyzed by a recombinant bacterial heme oxygenase (Hmu O) from *Corynebacterium diphtheriae*." *J Biol Chem* **274**: 21319–25.
39. Novotny M, Madsen D, Kleywegt GJ. (2004) "Evaluation of protein fold comparison servers." *Proteins* **54**: 260–70.
40. Poulos TL. (2005) "Structural and functional diversity in heme monooxygenases." *Drug Metab Dispos* **33**: 10–18.
41. Suits MD, Jaffer N, Jia Z. (2006) "Structure of the *Escherichia coli* O157: H7 heme oxygenase ChuS in complex with heme and enzymatic inactivation by mutation of the heme coordinating residue His-193." *J Biol Chem* **281**: 36776–82.
42. Sharp KH, Schneider S, Cockayne A, Paoli M. (2007) "Crystal structure of the heme-IsdC complex, the central conduit of the Isd iron/heme uptake system in *Staphylococcus aureus*." *J Biol Chem* **282**: 10625–31.

Chapter 7

Intrinsically Disordered Proteins

Peter Tompa

The traditional structure-function paradigm equates protein function with a well-defined three-dimensional structure. This view is based on tens of thousands of high-resolution structures in the Protein Data Bank, which provide the foundation for our understanding of protein function and malfunction. Rapid progress in the classical field of structural biology over the past decades, however, has led to many observations that apparently did not to fit into this unifying scheme. There has been a steady increase in the number of proteins/protein domains that resembled highly denatured proteins, although they were studied under native conditions. By the end of the 1990s, the exceptions that violated the rule had become so extensive that their existence could no longer be ignored. This increase called for a re-assessment of the structure-function paradigm. The feverish activity that followed has brought disordered proteins into the spotlight, and transformed our basic views of protein structure and function. It has been realized that structurally disordered proteins provide many functional advantages and enable functions completely out of the reach of globular proteins. Protein disorder was found to prevail in regulatory and signaling functions, and its frequency increased sharply in evolution from prokaryotes to eukaryotes. The structure-function characterization of these proteins now poses a major challenge to proteomic and small-scale studies alike, and has already reached epic proportions. In this review the history, recent focus and possible future directions of IDP research are surveyed.

Institute of Enzymology, Biological Research Center, Hungarian Academy of Sciences, 1113 Budapest, Karolina ut 29, tel: +361-279-3143, fax: +361-466-5465, e-mail: tompa@enzim.hu

Introduction

Arguably, the current most exciting development in protein structure-function studies is the general recognition that many full-length proteins and protein domains lack a well-defined three-dimensional structure under native, physiological conditions.[1–4] These intrinsically disordered, or unstructured, proteins (IDPs/IUPs) can be best approximated as fluctuating ensembles of alternative conformations, where they resemble the unfolded states of globular proteins attained under denaturing conditions, such as low pH or high concentrations of denaturants. Unlike denatured globular proteins, however, IDPs carry out basic functions most often in signal transduction and transcription regulation.[5–7] Structural disorder provides functional advantages, such as separation of specificity from binding strength, adaptability to various partners and frequent involvement in post-translational modifications. These advantages explain the occurrence of structural disorder, sometimes in high proportions in functionally important proteins, such as p53,[8] prion protein,[9] or BRCA1.[10]

The major repository of our knowledge on protein disorder is the DisProt database, which currently contains about 460 proteins, with roughly 1100 disordered regions, collected from the literature reporting serendipitous observations.[11] However, DisProt only contains a negligible fraction of all IDPs, as most bioinformatic predictions suggest that about 5–15% of proteins are fully disordered, and 30–50% of proteins contain at least one long disordered region in the higher organisms.[5–7]

The apparent wide gap between predicted and actually characterized disorder raises basic conceptual issues and suggests a variety of studies for delineating the structure-function relationship of this potentially important class. While detailed studies on individual proteins provide important information toward re-formulating the structure-function paradigm, large-scale high-throughput studies are also required for faster development in the field. Large-scale studies also provide data generated under controlled and comparable conditions, which might enable the formation of well-founded concepts on the structure and function of these proteins. In this chapter the most important recent developments in the field are covered, which will hopefully provide the reader with a comprehensive view of IDPs. Key references are also given to direct the interested reader to the primary sources of information.

What Do We Mean By Disorder: A Short Historical Overview

Although the traditional structure-function paradigm dominated all literature and thought on how proteins exist and function, observations that could not be interpreted within its confines have always been made. These have generally been treated as slightly disturbing outliers that did not comply with the rule and were not numerous enough to break into mainstream thought on proteins. At the very end of the last millennium, however, different interpretations of protein structures became more common. Here we present a couple of interesting examples, and further cases can be found in some excellent reviews.[12,13]

The formulation of the classical structure-function paradigm, that a well-defined 3D conformation is required for protein function revealed that certain enzymes had the capacity to act on structurally diverse substrates, which required *configurational adaptability*, or *induced fit*, of the active site. In retrospect, this observation could be considered as the prelude to protein disorder, and similar examples followed after the disorder of functionally important regions of proteins were described some 30 years ago.[12,13] Since then, the number of examples increased steadily until they reached a critical mass that called for the re-assessment of the structure-function paradigm.[14,15] For example, the inability to crystallize myelin basic protein, now known as an IDP, sent a warning sign that problems of crystallization may have an underlying cause more serious than the absence of the right conditions.[16] On similar grounds, it was considered abnormal when the trans-activator domain of transcription factors did not appear among structurally-characterized proteins. Sigler termed these regions "acid blobs" and "negative noodles", long before protein disorder as a general phenomenon was suggested.[17] Further, individual proteins have been reported for their unusual, random coil-like behavior, as observed by heat-resistance, FTIR and CD (tau protein[18]), SAXS and CD (prothymosin alpha[19]), gel-filtration and electron microscopy (caldesmon[20]), and NMR (FnBP[21] and p21^{Cip1} [22]). These observations led to the systematic analysis of the phenomenon when the first predictor, PONDR was developed. Subsequently, it was recognized that protein disorder was of general occurrence and importance.[14,15,23]

The revolution in the protein structure-function paradigm is a result of the widespread acceptance of this notion. For about eight years now, we have witnessed an exponential growth in the number of references to IDPs, and a similar increase in the description of novel cases. In the light of these discoveries, it is perhaps very surprising to both experts and students entering the field that a consensus on the very concept underlying this field, i.e. protein disorder, has not yet been reached. The expression "protein disorder" was originally used in contrast to order, and was thought to cover everything that differed from globular proteins (excluding fibrous and transmembrane proteins, of course). Today, we think of disorder as the native structural state of proteins, which corresponds to an ensemble of rapidly interconverting conformations, spanning structures from the fully extended, random-coil-like conformations to compact but disordered, molten-globule-like states.[12,24] Arguably, there is a continuum of structural states from full disorder to folded structures with various amounts of secondary/tertiary residual structure, fluctuating on various time-scales. Further variation results from the fact that disorder in some cases extends to only a few residues, but can in other cases cover the whole protein.[11,25] Many large proteins have a complex modular organization, composed of several globular domains connected by flexible linkers, but globular domains can also be interspersed with domain-sized disordered regions.[4] A further level of ambiguity is caused by the fact that many experimental techniques are probably responsive to proteins that are "mostly" disordered, i.e. have more disordered than ordered residues. However, no consensus has been reached on how to define this feature either. In essence, the diversity in the type and proportion of disorder underscores uncertainties in the field, and may be responsible for the several conceptually different approaches of identifying and studying IDPs.

Proteomic Identification of IDPs

The number of known IDPs in the DisProt database is approximately 460 proteins with about 1100 regions,[11] but predictions suggest that several thousand IDPs/IUPs exist in the human proteome alone.[6,26] The wide gap between known and suggested IDPs promoted proteomic approaches for their large-scale identification from cellular extracts.

Proteomic approaches generally rely on two-dimensional electrophoresis (2DE) and subsequent mass-spectrometric (MS) analysis, in combination with some pre-treatment that enriches solutions for IDPs. A caveat to these approaches, however, is that they treat proteins as single structural units, and thus separate and identify proteins that are dominated by disorder (mostly disordered proteins). Proteomic identification of regions of disorder requires prior bioinformatic filtering followed by large-scale cloning and structural characterization. However, no such effort has so far been published to our knowledge. As a matter of fact, only a few large-scale studies have been directed toward identifying even fully (or mostly) disordered proteins.

The primary goal in all these studies is to initially enrich solutions for proteins, which are mostly disordered, i.e. proteins that probably have more residues in regions of disorder than order. The protocol for enrichment usually relies on the noted resistance of IDPs to denaturing conditions, i.e. low pH and high temperature.[1] The methods take advantage of the fact that IDPs do not lose solubility under denaturing conditions, because they contain a low level of hydrophobic residues, which cause precipitation of ordered proteins.[12] In one study, acid-induced precipitation was used to enrich *E. coli* and *S. cerevisiae* extracts,[27] enriched extracts were subsequently analyzed by 2DE/MS, and the proteins thus identified were assessed for disorder by sequence-based prediction. Mostly unstructured proteins made up about 60% of acid-enriched extracts. In another study, heat treatment was applied to the much larger proteome of the mouse,[28] with an enrichment in IDPs from 12% to 42%. The treatment resulted in the enrichment of regulatory, signaling and structural proteins, and the depletion of proteins of metabolic functions. The extract was also enriched in proteins of increased proteolytic susceptibility, in perfect agreement with the enhanced proteolytic sensitivity of IDPs.[1] A related study addressed the stress-related phosphoproteome of *A. thaliana*,[29] in which heat treatment was combined with a subsequent step of phospho-affinity chromatography, to take advantage of the expected prevalence of phosphorylated forms in stress-related plant proteins (dehydrins, LEA proteins). Accordingly, the phospho-affinity chromatography preserved about half of the heat-resistant fraction, which fully belonged to the LEA/storage categories. The combination of enrichment, separation and subsequent identification illustrates the functional insight that can be gained from proteome-level IDP studies.

The enrichment and separation steps were combined in a completely different way in a recent study.[30] In this approach, an initial step of heat treatment was applied to precipitate most globular proteins, and to provide the first line of evidence for the structural status of proteins. The subsequent 2D electrophoretic step combined a native electrophoresis in the first dimension and an 8M-urea electrophoresis in the second dimension. In the native gel, proteins were separated according to their charge/mass ratios, whereas in the denaturing second dimension IDPs behaved similar to native conditions, which caused them to line up along the diagonal of the gel. Heat-resistant globular proteins, which unfold in urea, occupied positions above the diagonal. The conceptual novelty of this approach stems from the fact that it enables the identification of novel IDPs, and it also provides direct evidence for their structural status, i.e. disorder.

Prediction of Disorder from Sequence

Proteomic approaches cannot provide sequence-specific information on disorder. Consequently, bioinformatics predictors need to be invoked in the case of novel sequences that arise from genome sequencing or proteomic studies. Although IDPs are rather heterogeneous in structural details, they appear to share some common attributes, which enable their primary characterization based on sequence alone. IDPs are depleted in hydrophobic amino acids (WCFIYVL, generally denoted as order-promoting) and are enriched in hydrophilic/charged amino acids and proline (KEPSQRA, generally denoted as disorder promoting).[12] Although disordered regions show some correlation with low complexity segments,[31,32] no unequivocal correspondence exists between the two. Thus, predictors are based on the above differences in amino acid composition and sequence, but they rely on rather diverse implementations of this principle. At present, there are about 20 different predictors, most of them available via the Internet (Table 1).

The simplest approaches rely on various measures of amino acid propensity, such as the classical approach suggested by Uversky.[33] The approach plots the net charge of the protein as a function of its net hydrophobicity. IDPs occupy the high net charge — low mean hydrophobicity region, and are separated from globular proteins in this 2D plane. Although this approach provides a clear physical picture of the reasons for disorder (or order) of a protein, its major limitation is

Table 1 Predictors for Intrinsic Disorder of Proteins

Predictor	URL	Principle
PONDR VSL2	http://www.ist.temple.edu/disprot/predictorVSL2.php	SVM
DISOPRED2	http://bioinf.cs.ucl.ac.uk/disopred/	SVM + NN for smoothing
IUPred	http://iupred.enzim.hu/	estimated interaction energy
FoldIndex	http://bip.weizmann.ac.il/fldbin/findex	AA propensities
DisEMBL	http://dis.embl.de	neural network
GlobPlot	http://globplot.embl.de/	single AA propensity
DISpro	http://www.ics.uci.edu/~baldig/dispro.html	1-D recursive NN with profiles
Spritz	http://protein.cribi.unipd.it/spritz/	SVM with non-linear kernel
FoldUnfold	http://skuld.protres.ru/~mlobanov/ogu/ogu.cgi	single AA propensity
PreLink	http://genomics.eu.org/spip/PreLink	AA propensity + hydrophobic cluster analysis

that it can only be applied to full-length proteins. The limitation has been corrected by calculating the distribution of these features within a pre-defined sequence window (FoldIndex[34]), while other predictors apply different and more sophisticated amino acid propensity scales (e.g. GlobPlot[35] and FoldUnfold[36]), or combine the compositional bias with the distance to the closest hydrophobic cluster (PreLink[37]).

The largest class of predictors, which rely on machine learning approaches, i.e. neural networks and support vector machines (SVM), can capture more complex associations of structure with sequence in implicit ways. These predictors are always trained on datasets of disorder and order, which enable the prediction of disorder to be reduced to a simple binary classification problem. PONDR® was the first such predictor, which has now been developed into a family of predictors (VLXT, VL3[38]). The latest implementation, VSL2, integrates two specific predictors for short- and long disorder.[25] Some methods do not define protein disorder directly, and use sequence attributes instead. Among them fractional composition, hydropathy, and sequence complexity may form the basis of prediction. The input data may be a profile generated by sequence alignment, as in the case of DISOPRED2,[6] which relies on an SVM algorithm. Spritz uses a very similar approach, but its SVM is mounted with a non-linear kernel.[39] The DISpro method uses a 1D recursive neural network for the same problem.[40] These methods can readily accommodate other factors, such as predicted secondary structure or solvent accessibility. The numerous algorithms differ not only in the representation of the input data and in the learning algorithm, but also in the type of disorder used for training. A limitation of these predictors occurs because of the uncertainties of the underlying databases.

An algorithm relying on different principles has been developed recently. IUPred[41,42] has not been trained on any database of disorder, but uses low-resolution force-fields deduced from the structures of globular proteins, and estimates the total pair-wise interaction energy of a sequence (or its predefined segment). IDPs are distinguished from globular proteins by their significantly lower overall potential for inter-residue interactions (Fig. 1). As IUPred does not rely on a potentially

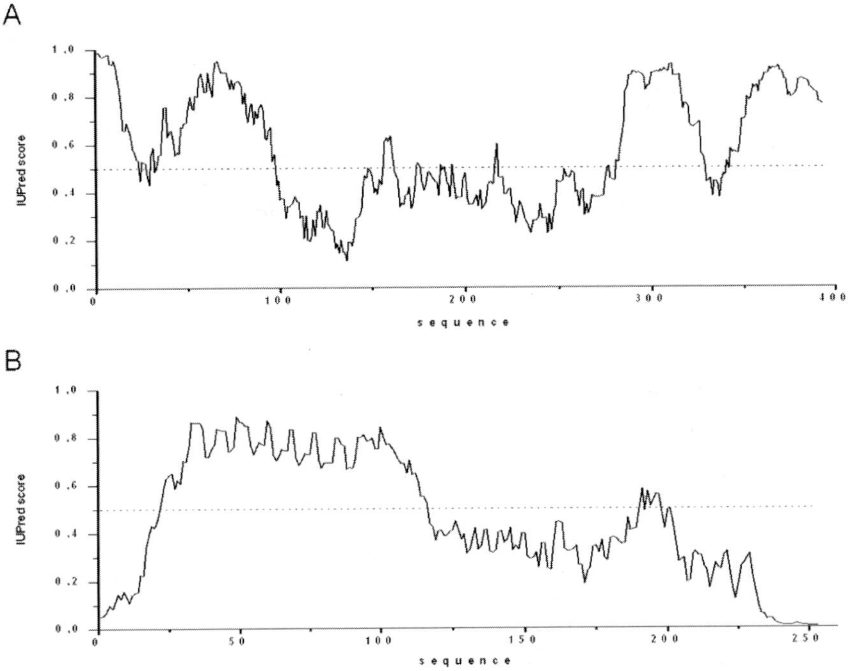

Fig. 1 IUPred prediction of disorder. Disorder of human p53 (A) and human prion protein (B) has been predicted by the IUPred predictor (http://iupred.enzim.hu/). The score above 0.5 indicates disorder, while a score below 0.5 indicates order. **A**) The N-terminal 100 amino acids (transactivator domain) and C-terminal 50 amino acids (regulatory domain) of p53 tend to be disordered, while the middle region (res. 100–300), i.e. the DNA binding domain, tends to be ordered. The region around res. 300, i.e. the tetramerization domain is predicted to be disordered, and is known to attain its structure only in a tetrameric form. **B**) The prion protein is composed of an entirely disordered N-terminal half and an ordered C-terminal half.

erroneous database of protein disorder, its assessment of the structural status of a (region of a) protein may be considered as an unbiased and independent assessment of the structural status of an IDP. Essentially, this assessment provides independent evidence for the existence of intrinsic protein disorder.

A natural question with respect to disorder prediction is, which algorithm is the best? Unfortunately, comparison of the performance of disorder predictors depends critically on the type of data used for testing, and also on the evaluation criteria, which practically preclude an unbiased assessment. This task was nevertheless undertaken in the CASP6 experiment,[43] on a dataset of short disorder, i.e. residues missing from globular proteins. It is of note that various criteria resulted in different scores, but most of the time the predictors on top of the list in Table 1 performed the best.

An important application of disorder predictor(s) is related to structural genomics/proteomics programs, which aim to solve all possible structures, or at least all representative folds. A serious bottleneck to these efforts is practically countless potential structures to be solved, which makes selection of targets a critical issue.[44] Traditionally, the preference is to select targets that show little sequence similarity to proteins of known structures and/or which are of particular biological/biomedical importance. Irrespective of these preferences, however, the initial removal of targets with a significant level of disorder might be expected to increase the success rate, because disorder is highly detrimental to solving structures by X-ray crystallography or NMR. Such an improvement has been demonstrated by the retrospective application of bioinformatic filtering to the targets of the Center for Eukaryotic Structural Genomics,[45] and the TargetDB database.[46]

Experimental Characterization of IDPs

Identification of proteins as disordered by high- or low-throughput experiments and/or by bioinformatics analysis only represents the first step towards re-assessing the structure-function paradigm. Re-assessing the paradigm requires not only recognizing the high frequency and probable functional importance of protein disorder, but also the detailed structure-function characterization of a large number of IDPs. This ultimately leads to generalizations that match those already established in the field of ordered proteins. Fortunately, the physical character of IDPs enables characterization with a wide range of techniques, which provide a detailed description of the structural ensemble that can be related to function (Table 2). As this issue far

Table 2 Major Experimental Techniques for Studying IDPs

Type	Method	Information Gained
Indirect	Heat/acid-resistance	Low level of hydrophobic residues
	SDS-PAGE mobility	Unusual amino-acid composition, low SDS binding
	Proteolytic sensitivity	Exposure of chain, possible identification of most exposed sites
	Differential scanning calorimetry (DSC)	Lack of cooperative structure
Hydrodynamic	Gel filtration	Hydrodynamic radius
	Dynamic light scattering (DLS)	Hydrodynamic radius
	Analytical ultracentrifugation (AU, UC)	Hydrodynamic properties, solution molecular mass, stoichiometry
	Small-angle X-ray scattering (SAXS)	Hydrodynamic properties, low-resolution model of shape
Spectroscopic	Förster resonance energy transfer (FRET)	Distance constraints, dynamics of structure
	UV fluorescence	Exposure of aromatic side-chains, quenching of residues
	Circular dichroism (CD)	Ratio of secondary structural elements
	Raman optical activity (ROA)	Secondary structural elements, including PPII, dynamics of chain
	Nuclear magnetic resonance (NMR)	Sequence-specific information on local structure and dynamics

exceeds the limits of not only this review article, but easily the entire volume, the article only gives a flavor of the multiplicity of methods already at our disposal, and directs the reader to the several excellent reviews for further information.[1,24,47,48]

A range of unrelated and indirect techniques may provide the first line of evidence for the IDP status of a protein. IDPs have been noted for their heat-resistance, which enables the rapid identification of a potentially disordered protein. Acid resistance is also indicative of the lack of a well-defined stable fold. Their aberrant electrophoretic mobility on SDS-PAGE also provides a positive sign for the unusual structural status. The extreme proteolytic sensitivity of IDPs, which is a result of their largely exposed polypeptide chain, also enables their rapid identification. Degradation by the 20S proteasome, apparently without prior ubiquitination is a special attribute that provides information on the disordered state of IDPs both *in vitro* and *in vivo*.[49] Furthermore, when proteolysis is applied under limiting conditions, it can provide details on the structure of the IDP, as exposed and sensitive sites may correspond to the actual partner-binding regions of IDPs. Differential scanning calorimetry (DSC) can also be used for detecting the lack of a compact, cooperative structure of IDPs.

Techniques sensitive to the hydrodynamic behavior of proteins may enable us to gain a more detailed insight into the structural character of IDPs, because their hydrodynamic behavior is a sensitive measure of their (type of) disorder. Due to their extended and heterogeneous nature, IDPs display anomalously large apparent molecular weight in gel-filtration, which results from their exclusion from the molecular sieving medium. Calibration of the column by globular standards enables us to determine the apparent M_w, which, upon comparison with its absolute mass, provides a measure of the compactness of the protein. The hydrodynamic dimension can also be addressed by two less-frequently used techniques, dynamic light scattering (DLS), and analytical ultracentrifugation (AU); however, it can be most thoroughly characterized by small-angle X-ray scattering (SAXS). SAXS not only enables us to determine the overall dimensions, such as the radius of gyration (Rg) or Stokes radius (Rs), of the protein, but also allows a more sophisticated analysis of scattering intensities to build a low-resolution structural model of the molecule, which can be filled in with high-resolution structural details. In this sense, hydrodynamic techniques cannot only prove disorder, but also provide a gross estimate of its structural sub-type and overall structural topology.

Spectroscopic techniques, of course, provide even more details on the structural ensemble of IDPs. Their rapid identification may simply rely on recording the UV fluorescence spectrum, which shows the character of an exposed Trp residue, with a maximum of around 340 nm. A variant of this technique relies on applying a quencher of fluorescence, such as acrylamide, because IDPs are far more susceptible to quenching than globular proteins. Circular dichroism (CD) spectroscopy is another often-used technique, which can report the absence of repetitive secondary structural elements in IDPs, with a characteristic large negative peak of around 200 nm. The proportion of residual structure, i.e. transient secondary structural elements, such as α-helix and β-strand, can also be determined by the deconvolution of CD spectra. Generation of deletion mutants or fragments of the IDP may further permit us to characterize the actual location of these potentially important secondary structural elements. A rather recent development in the field of IDPs is the use of Raman optical activity (ROA) measurement, which can provide information on the structure and dynamics of IDPs, particularly the presence of the often mentioned polyproline II (PPII) helix conformation. The single most powerful technique for studying IDPs, however, is NMR, which can provide atomic-level resolution on the structural ensemble of IDPs. When full resonance assignment is achieved, various NMR measures such as secondary chemical shifts, relaxation rates and residual dipolar coupling values can be determined and associated with particular residues. These measures enable both structural and dynamic information to be extracted, which reveals the enormous potential of atomic-scale description of the structure-function relationship of IDPs. Furthermore, as mentioned in the next section, NMR also has the capacity for in-cell structural studies, which enable characterization of IDPs under truly physiological conditions.

Disorder *In Vivo*

A somewhat neglected issue is the *in vivo* structural state of proteins that appear disordered under artificial, *in vitro* conditions. The primary reason that their disorder does not necessarily apply to conditions in the living cell is that crowding elicited by extreme macromolecular concentrations may actually make them fold.[50] This issue should be examined seriously, not only because it challenges the very concept of disorder, but also because it has been shown that crowding elicited *in vitro* basically affects the folding state of denatured/mutated globular proteins, such as staphylococcal nuclease[51] or apomyoglobin.[52] Only a few experiments have so far addressed this question in the case of IDPs, with rather mixed results. Crowding mimicked by dextran or Ficoll had no effect on the structural state of the KID domain of p27^{Kip1}, and the TAD domain of c-Fos,[53] but high concentrations of glucose brought α-synuclein to a compact, albeit not ordered, structural state.[54] TMAO, a bacterial osmolyte sometimes erroneously used as a crowding agent, induced some folding of α-synuclein,[55] but the limitation of *in vitro* approaches is best demonstrated by the fact that α-synuclein apparently remains mostly unfolded *in vivo*, as shown by in-cell NMR studies.[56] In all, an array of experiments suggest that IDPs probably undergo limited ordering *in vivo*, but do not become folded and fully ordered under conditions encountered in the cell. Some indirect considerations further corroborate this picture, i.e. that intrinsic disorder is the physiological state of these proteins.

The extended binding mode of IDPs, which is compatible with their actual functions, strongly argues against a folded state *in vivo*. For several IDPs, their structures in the partner-bound state have been solved and they have been deposited in the Protein Data Bank (Fig. 2). These structures invariably suggest an open, extended mode of binding.[57] A folded, compact, structural state prior to binding is hardly compatible with this picture, because it would require unfolding of IDP prior to binding.[1,2] An additional argument against a stable, folded state *in vivo* arises from the observed adaptability in binding (termed binding promiscuity,[22] or moonlighting[58]) of IDPs, which assumes a structural malleability incompatible with a stable structure prior to binding. The high observed evolutionary rate of IDPs is probably also indicative of this structural state.[59] As amino acid replacements in evolution are limited by both functional and structural constraints, such an accelerated evolutionary rate argues against a compact, folded structure with extended structural constraints. These different observations discredit any claims that disorder observed *in vitro* would be an artifact, which corroborates our contention that

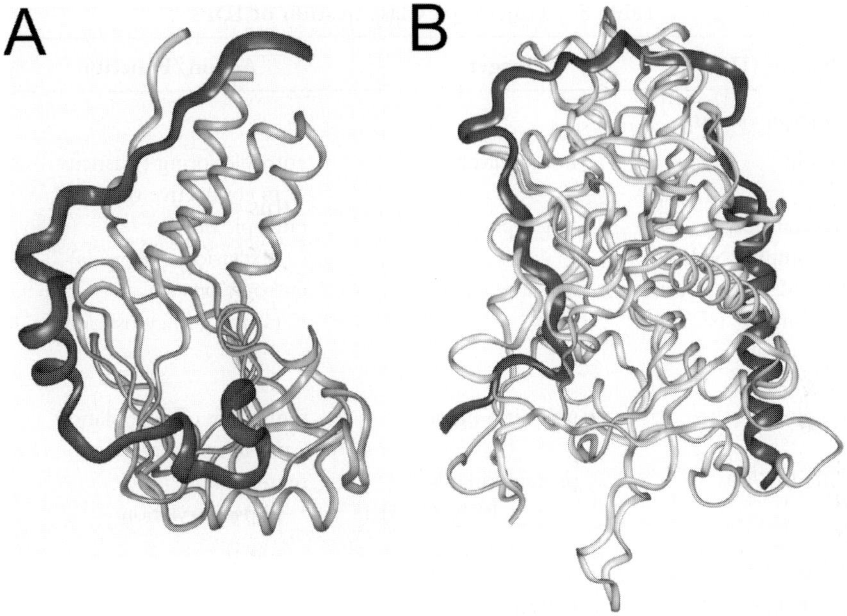

Fig. 2 IDPs bind their partners in an extended conformation. Two IDPs in the bound form are shown, SARA SBD bound to SMAD MH2 domain (pdb 1DEV) and SNAP25 bound to BoNT/A (pdb 1XTG). The partner is shown in light gray, while the IDPs (SARA SBD and SNAP25) are rendered in a darker, wider ribbon. The figure demonstrates that IDPs attain open, extended conformations when they bind to their partners.

disorder is the *in vivo* state of many IDPs. As already discussed, an unrelated argument for the *in vivo* structural state of IDPs stems from their predictability from sequence by IUPred, the algorithm not trained on IDP sequences.[41] An additional point to note is that a significant number of extracellular IDPs[1,2] do not exist in a crowded physiological niche, i.e. their disorder observed *in vitro* must reflect their behavior *in vivo*. A final solid argument for disorder under physiological conditions comes from the entropic chain functions of many IDPs (Table 3), which are not compatible with a folded state.[1,13] For example, disorder has been observed for certain components of the nuclear pore complex,[60] the function of which relies on entropic exclusion, which is incompatible with a folded, stable structural state.[61]

Table 3 Functional Classification of IDPs

Protein (IUP)	Target	Action/Function
Entropic chains		
elastin	not applicable	entropic spring (elasticity in connective tissue)
nuclear pore complex NUPs	not applicable	entropic bristle (exclusion of objects)
tau/MAP2 projection domain	not applicable	entropic bristle (spacing in cytoskeleton)
Display sites		
cyclin B	ubiquitin ligase	regulation of degradation by ubiquitination
CREB TAD	protein kinases (e.g. PKA, CaMKIV)	regulation by phosphorylation
Chaperones		
crystallins	protein	protein chaperones
hRNP A1	RNA	RNA chaperone
Effectors		
Inhibitor 2 (I2)	protein phosphatase 2	inhibitor of phosphatase
securin	separase	inhibitor of separase in anaphase initiation
$p21^{Cip1}/p27^{Kip1}$	cyclin-dependent kinases	inhibitors of Cdks in cell cycle regulation
Scavengers		
caseins	calcium-phosphate	preventing calcium-phosphate precipitation in milk
dehydrins	water	retaining water to prevent desiccation of plants
Assemblers		
BRCA1	various	assembly of complexes in DNA repair
$p21^{Cip1}/p27^{Kip1}$	cycline D – Cdk 4	assembly of functional CycD-Cdk4 complex

(*Continued*)

Table 3 (*Continued*)

Protein (IUP)	Target	Action/Function
prions		
Sup35	subunits of eukaryotic release factor	suppression of nonsense mutations
CPEB	mRNA	polyadenylation of dormant mRNA, memory in *Aplysia*

Protein Disorder and Function

All the previous points culminate in probably the single most exciting issue of IDPs, i.e. their functions. In fact, the sweeping change in the structure-function paradigm has come about only after the general recognition and acknowledgment that protein disorder has very important functional roles. Protein disorder enables functions or functional modes not really accessible to folded, globular proteins. These may be viewed from the aspect of either the functional advantages associated with disorder, or the actual functional modes they represent. Here we will examine both aspects.

In terms of the actual advantages, we should be aware that these are applicable only in comparison to globular proteins. Protein disorder is often mentioned in conjunction with adaptability in binding, which, in principle, enables binding to different partners, with potentially different functional outcomes. Such a binding promiscuity was first suggested in the case of $p21^{Cip1}$.[22] Later, it was established that such promiscuity actually enables completely different, even opposing activities by the protein, which has been termed moonlighting.[58] If proven to be a general characteristic, moonlighting may fundamentally increase the complexity of the protein networks of the cell. Directly related to this phenomenon is that IDPs often use only a small segment of their sequence for binding, which often is the site of post-translational modification.[3] Phosphorylation sites,[62] binding sites of SH3 domains or 14-3-3 domains,[63] just to mention a few, have been directly associated with disorder, and a recent study has shown that short recognition motifs (Linear Motifs, LMs) in proteins correlate strongly with disorder.[64] It also follows from this binding mode that IDPs use much less of their surface for protein-protein interactions than globular proteins, which enables them to interact with more partners, and basically serve to organize the interactome. In fact, in a range of recent studies it was found that proteins with multiple interacting partners in the interactome (hub proteins) tend to have more disorder than other proteins.[65–67] The extended and dynamic conformational state and binding by short recognition elements probably also result in much faster interactions than in the case of globular proteins. This was first observed in the case of DNA renaturation,[68] where the time required for strand re-annealing was significantly shortened by non-specific polyamines, such as spermidine. In the case of IDPs, this effect is qualitatively and quantitatively rendered in the protein fishing[69] and fly-casting[70] mechanisms. Perhaps the most often cited advantage of disorder is that regions of IDPs involved in binding undergo significant disorder-to-order transition,[71] which contributes unfavorably to binding strength due to a decrease in configurational entropy. This is thought to separate binding strength from specificity, which is of prime importance in regulation enabling highly specific interactions to be weak and readily reversible. Although the magnitude of this effect has hardly been analyzed in the case of IDPs,[72] an instructive and influential case study has been published in the case of DNA-binding proteins.[73] A final point to be noted is that IDPs often carry out functions directly, by the multiplicity of the conformational states of the protein, that are not realized by binding . The existence of these entropic chain functions may be considered a significant functional advantage of protein disorder.[1,2]

The special functional capacities that result from protein disorder and the potential advantages discussed manifest themselves in special functional modes that can be classified into seven general functional categories (Table 3). These modes provide the basic toolkit by which IDPs integrate into the functioning proteome, and an overview of these will also shed light on the evolutionary rise of IDPs. In previous reviews we considered only five[1] or six[2] categories, but recent observations on the involvement of IDPs in the formation of amyloids[74] add the ability to form prions/amyloids to the functional arsenal of IDPs. Although we have not systematically screened all entries in DisProt,[11] this classification scheme appears to cover the functions of most IDPs known and characterized so far.

As mentioned, IDPs have unique functions that directly result from disorder, termed *entropic chains* in the literature. These functions are usually characterized by passive force generation (elasticity) or determination of the orientation/localization of bound proteins or attached domains (linker/spacer functions).[13] The elastic (entropic spring) function of elastin,[75] gating in nuclear pore complex,[61] or the entropic spacer function provided in the cytoskeleton by microtubule-associated proteins[76] are adequate examples of this function.

Further functions in the classification scheme invoke mechanisms that rely on molecular recognition, i.e. binding of other macromolecule(s) or small ligand(s). In the case of permanent binding, we may solve the structure of the IDP complexed with its partner(s), but in the case of transient binding the structure is probably too loose to be pinned down by any technique. In transient binding, *display sites* are sites of post-translational modification, carried out by specific enzymes, which require flexible and adaptable regions for transient but productive interaction within the substrate. The improvement in the prediction accuracy of phosphorylation sites by considering disorder,[62] or the occurrence of ubiquitination sites in locally disordered segments of proteins[77] demonstrate the possible general correlation of the two phenomena. This generality has been underlined by our prediction study, which has shown that short recognition elements of proteins found in the ELM (Eukaryotic Linear Motif) database[78] almost always reside in a locally disordered sequential environment.[64]

Chaperones also function by transient partner binding. In a comprehensive study,[79] we found a high frequency of disorder in RNA chaperones, with 40% of their residues falling into long disordered regions. The number is 15% in the case of protein chaperones, still among the most disordered functional classes of proteins. Disordered regions in both chaperone classes are often found to be directly involved in chaperone function, which enabled the formulation of an "entropy transfer" model of structural disorder in chaperone function.[79]

Other disordered proteins function by permanent partner binding, where structures can be solved and functions can be more readily interpreted in terms of the classical structure-function paradigm. *Effectors* bind and modify the activity of their partner, primarily an enzyme.[1] We found that the action of effectors is inhibitory in most cases, but activation has also been described in the literature. To cite two examples, we refer to securin, the inhibitor of separase[80] and I2, the inhibitor of protein phosphatase 2.[81] Demonstrating the potential of these proteins to adapt to the structures of different partners, or even the same partner in different modes, securin has been shown recently to be able to both inhibit and activate separase.[82] Several such examples point to the fact that structural disorder may permit multiple, often opposing activities of certain proteins, as formulated in the model of moonlighting.[58]

The open and extended structure of IDPs has also been suggested to permit binding of more partners for a given length than globular proteins,[83] which enables these proteins to organize interaction networks, i.e. to function as *assemblers* of complexes.[1] In fact, a high level of disorder has been reported in some scaffolding proteins, such as Ste5 and BRCA1,[10,84] and also in hub proteins that interact with many partner proteins.[65–67] These results are in line with a recent analysis that showed an increase of structural disorder with increasing number of components of protein complexes (Hegyi and Tompa, unpublished).

Protein disorder may be instrumental not only in binding other proteins, but also small ligand molecules. This class of permanent binders is termed *scavengers*, which can store and/or neutralize ligand molecules. Casein(s), which bind small calcium phosphate seeds and prevent calcium phosphate precipitation in milk,[85] or plant dehydrins, which avert dehydration stress conditions by their large-capacity water binding[86,87] are two prime examples of this kind of IDP function.

An unusual, and probably not yet generally appreciated function of IDPs, is their capacity to aggregate *in vivo*, which results in the loss of function of attached domains, and also occasionally in the gain of some new function not manifest in the soluble state of the protein. Such *prions* or *amyloids* have been traditionally considered as pathogens, mostly because of the conditions caused by the misfolded state of the mammalian prion protein[88] or other amyloidogenic proteins, such as α-synuclein or Aβ peptide.[89] The prion/amyloid phenomenon, however, was shown to serve physiological functions as well, in either yeast or higher organisms.[74] Invariably, the prion domain of these physiological prions is intrinsically disordered,[90] which suggests that they represent a separate functional class within IDPs.

Protein Disorder in Disease

As shown above, IDPs are prevalent in the proteomes of higher organisms, and they correlate with functions related to signal transduction and transcription regulation.[6,7] Disorder has been observed in important regulatory proteins, such as p53, p21$^{\text{Cip1}}$, securin and BRCA1, to mention just a few. It directly follows from these observations that mutations and subsequent malfunctioning of these proteins may cause severe disorders. In the bioinformatic analysis of whole proteomes it was found that disorder correlates with cancer-associated proteins.[5] Therefore, it is no surprise that many proteins, such as those mentioned above, are known proto-oncogene products, i.e. their mutations have been directly linked with oncogenic transformations. Disorder is also often found in proteins implicated in neurodegenerative disorders, such as Alzheimer's (tau[18]) and Parkinson's (α-synuclein[56]) disease, and in proteins with repetitive regions that may undergo repeat expansion, causing the formation of amyloid-type of aggregates (Q-repeat or trinucleotide-repeat diseases.[91]) Furthermore, significant enrichment of disorder has also been reported in proteins involved in cardiovascular disease.[92] These correlations, together with the observation that IDPs usually bind their partner via short recognition motifs, often at a hydrophobic crevice of the partner, have recently led to the suggestion that the binding mode of IDPs and their involvement in disease may open new avenues for rational drug design.[93] In short, interfaces of globular proteins complexes are relatively large and flat, and are not usually amenable to interference by small-molecule inhibitors. On the other hand, binding of IDPs in a peptide-like fashion in the binding crevice of the partner enables small molecules to bind and interfere, as convincingly demonstrated by inhibiting the p53-MDM2 interaction by Nutlins.[94]

Conclusion

The importance of intrinsic structural disorder is underlined by the functional advantages it imparts to proteins in signaling and regulatory functions, which explains its prevalence in various proteomes. Further, IDP malfunction often results in debilitating diseases, such as neurodegeneration and cancer. A detailed understanding of IDPs will thus not only lead to the extension of the structure-function paradigm, but also to the discovery of new remedies for a range of serious diseases. Thus, identifying and characterizing the disordered complement of the proteome, i.e. the "unfoldome" or "disorderome" holds the promise of fundamentally advancing both basic science and biomedical research. Although many of the concepts and technological advances are hard to anticipate, it is of little doubt that we will see many interesting and exciting results in the coming years in the field of IDPs.

Acknowledgments

Work in the laboratory of the author is supported by grants OTKA K60694 from the Hungarian Scientific Research Fund, ETT 245/2006 from the Hungarian Ministry of Health, and International Senior Research Fellowship ISRF 067595 from the Wellcome Trust.

References

1. Tompa P. (2002) "Intrinsically unstructured proteins." *Trends Biochem Sci* **27**: 527–33.
2. Tompa P. (2005) "The interplay between structure and function in intrinsically unstructured proteins." *FEBS Lett* **579**: 3346–54.
3. Uversky VN, Oldfield CJ, Dunker AK. (2005) "Showing your ID: intrinsic disorder as an ID for recognition, regulation and cell signaling." *J Mol Recognit* **18**: 343–84.
4. Dyson HJ, Wright PE. (2005) "Intrinsically unstructured proteins and their functions." *Nat Rev Mol Cell Biol* **6**: 197–208.
5. Iakoucheva L, Brown C, Lawson J, *et al.* (2002) "Intrinsic Disorder in Cell-signaling and Cancer-associated Proteins." *J Mol Biol* **323**: 573–84.

6. Ward JJ, Sodhi JS, McGuffin LJ, *et al.* (2004) "Prediction and functional analysis of native disorder in proteins from the three kingdoms of life." *J Mol Biol* **337**: 635–45.
7. Tompa P, Dosztányi Z, Simon I. (2006) "Prevalent structural disorder in E. coli and S. cerevisiae proteomes." *J. Proteome Res* **5**: 1996–2000.
8. Bell S, Klein C, Muller L, *et al.* (2002) "p53 contains large unstructured regions in its native state." *J Mol Biol* **322**: 917–27.
9. Lopez GF, Zahn R, Riek R, *et al.* (2000) "NMR structure of the bovine prion protein." *Proc Natl Acad Sci USA* **97**: 8334–39.
10. Mark WY, Liao JC, Lu Y, *et al.* (2005) "Characterization of segments from the central region of BRCA1: an intrinsically disordered scaffold for multiple protein-protein and protein-DNA interactions?" *J Mol Biol* 345: 275–87.
11. Sickmeier M, Hamilton JA, LeGall T, *et al.* (2007) "DisProt: the database of disordered proteins." *Nucl Acids Res* **35**: D786–93.
12. Dunker AK, Lawson JD, Brown CJ, *et al.* (2001) "Intrinsically disordered protein." *J Mol Graph Model* **19**: 26–59.
13. Dunker AK, Brown CJ, Lawson JD, *et al.* (2002) "Intrinsic disorder and protein function." *Biochemistry* **41**: 6573–82.
14. Romero P, Obradovic Z, Kissinger CR, *et al.* (1998) "Thousands of proteins likely to have long disordered regions." *Pac Symp Biocomput* **3**: 437–48.
15. Wright PE, Dyson HJ. (1999) "Intrinsically unstructured proteins: re-assessing the protein structure-function paradigm." *J Mol Biol* **293**: 321–31.
16. Sedzik J, Kirschner DA. (1992) "Is myelin basic protein crystallizable?" *Neurochem Res* **17**: 157–66.
17. Sigler PB. (1988) "Transcriptional activation. Acid blobs and negative noodles." *Nature* **333**: 210–12.
18. Schweers O, Schonbrunn-Hanebeck E, Marx A, Mandelkow E. (1994) "Structural studies of tau protein and Alzheimer paired helical filaments show no evidence for beta-structure." *J Biol Chem* **269**: 24290–97.
19. Gast K, Damaschun H, Eckert K, *et al.* (1995) "Prothymosin alpha: a biologically active protein with random coil conformation." *Biochemistry* **34**: 13211–18.
20. Lynch WP, Riseman VM, Bretscher A. (1987) "Smooth muscle caldesmon is an extended flexible monomeric protein in solution that can readily undergo reversible intra- and intermolecular sulfhydryl cross-linking. A mechanism for caldesmon's F-actin bundling activity." *J Biol Chem* **262**: 7429–37.
21. Penkett CJ, Redfield C, Dodd I, *et al.* (1997) "NMR analysis of main-chain conformational preferences in an unfolded fibronectin-binding protein." *J Mol Biol* **274**: 152–59.
22. Kriwacki RW, Hengst L, Tennant L, *et al.* (1996) "Structural studies of p21Waf1/Cip1/Sdi1 in the free and Cdk2-bound state: conformational disorder mediates binding diversity." *Proc Natl Acad Sci USA* **93**: 11504–09.

23. Garner E, Cannon P, Romero P, *et al.* (1998) "Predicting disordered regions from amino acid sequence: common themes despite differing structural characterization." *Genome Inform Ser Workshop Genome Inform* **9**: 201–213.
24. Uversky VN. (2002) "Natively unfolded proteins: a point where biology waits for physics." *Protein Sci* **11**: 739–56.
25. Peng K, Radivojac P, Vucetic S, *et al.* (2006) "Length-dependent prediction of protein intrinsic disorder." *BMC Bioinformatics* 7: 208.
26. Dunker AK, Obradovic Z, Romero P, *et al.* (2000) "Intrinsic protein disorder in complete genomes." *Genome Inform Ser Workshop Genome Inform* **11**: 161–71.
27. Cortese MS, Baird JP, Uversky VN, Dunker AK. (2005) "Uncovering the unfoldome: enriching cell extracts for unstructured proteins by acid treatment." *J Proteome Res* **4**: 1610–18.
28. Galea CA, Pagala V, Obenauer JC, *et al.* (2006) "Proteomic studies of the intrinsically unstructured mammalian proteome." *J Proteome Res* **5**: 2839–48.
29. Irar S, Oliveira E, Pages M, Goday A. (2006) "Towards the identification of late-embryogenic-abundant phosphoproteome in Arabidopsis by 2-DE and MS." *Proteomics* **6**: S175–85.
30. Csizmok V, Szollosi E, Friedrich P, Tompa P. (2006) "A novel two-dimensional electrophoresis technique for the identification of intrinsically unstructured proteins." *Mol Cell Proteomics* **5**: 265–73.
31. Wootton JC. (1994) "Sequences with 'unusual' amino acid compositions." *Curr Opin Struct Biol* **4**: 413–21.
32. Romero P, Obradovic Z, Li X, Garner EC, *et al.* (2001) "Sequence complexity of disordered protein." *Proteins* **42**: 38–48.
33. Uversky VN, Gillespie JR, Fink AL. (2000) "Why are "natively unfolded" proteins unstructured under physiologic conditions?" *Proteins* **41**: 415–27.
34. Prilusky J, Felder CE, Zeev-Ben-Mordehai T, *et al.* (2005) "FoldIndex: a simple tool to predict whether a given protein sequence is intrinsically unfolded." *Bioinformatics* **21**: 3435–38.
35. Linding R, Russell RB, Neduva V, Gibson TJ. (2003) "GlobPlot: exploring protein sequences for globularity and disorder." *Nucl Acids Res* **31**: 3701–08.
36. Garbuzynskiy SO, Lobanov MY, Galzitskaya OV. (2004) "To be folded or to be unfolded?" *Protein Sci* **13**: 2871–77.
37. Coeytaux K, Poupon A. (2005) "Prediction of unfolded segments in a protein sequence based on amino acid composition." *Bioinformatics* **21**: 1891–900.
38. Obradovic Z, Peng K, Vucetic S, *et al.* (2003) "Predicting intrinsic disorder from amino acid sequence." *Proteins* **53** Suppl 6: 566–72.
39. Vullo A, Bortolami O, Pollastri G, Tosatto SC. (2006) "Spritz: a server for the prediction of intrinsically disordered regions in protein sequences using kernel machines." *Nucl Acids Res* **34**: W164–68.

40. Cheng J, Sweredoski M, Baldi P. (2005) "Accurate prediction of protein disordered regions by mining protein structure data." *Data Min Knowl Disc* **11**: 213–22.
41. Dosztanyi Z, Csizmok V, Tompa P, Simon I. (2005) "The pairwise energy content estimated from amino acid composition discriminates between folded and instrinsically unstructured proteins." *J Mol Biol* **347**: 827–39.
42. Dosztanyi Z, Csizmok V, Tompa P, Simon I. (2005) "IUPred: web server for the prediction of intrinsically unstructured regions of proteins based on estimated energy content." *Bioinformatics* **21**: 3433–34.
43. Jin Y, Dunbrack RL, Jr. (2005) "Assessment of disorder predictions in CASP6." *Proteins* **61** Suppl 7: 167–75.
44. Brenner SE. (2000) "Target selection for structural genomics." *Nat Struct Biol* 7: 967–69.
45. Oldfield CJ, Ulrich EL, Cheng Y, *et al.* (2005) "Addressing the intrinsic disorder bottleneck in structural proteomics." *Proteins* **59**: 444–53.
46. Dosztanyi Z, Sandor M, Tompa P, Simon I. (2007) "Prediction of protein disorder at the domain level." *Curr Protein Pept Sci* **8**: 161–71.
47. Dyson HJ, Wright PE. (2004) "Unfolded proteins and protein folding studied by NMR." *Chem Rev* **104**: 3607–22.
48. Ferron F, Longhi S, Canard B, Karlin D. (2006) "A practical overview of protein disorder prediction methods." *Proteins: Structure, Function, Bioinformatics* **65**: 1–14.
49. Tsvetkov P, Asher G, Paz A, *et al.* (2007) "Operational definition of intrinsically unstructured protein sequences based on susceptibility to the 20S proteasome." *Proteins* in press.
50. Ellis RJ. (2001) "Macromolecular crowding: obvious but underappreciated." *Trends Biochem Sci* **26**: 597–604.
51. Baskakov I, Bolen DW. (1998) "Forcing thermodynamically unfolded proteins to fold." *J Biol Chem* **273**: 4831–34.
52. McPhie P, Ni YS, Minton AP. (2006) "Macromolecular crowding stabilizes the molten globule form of apomyoglobin with respect to both cold and heat unfolding." *J Mol Biol* **361**: 7–10.
53. Flaugh SL, Lumb KJ. (2001) "Effects of macromolecular crowding on the intrinsically disordered proteins c-Fos and p27(Kip1)." *Biomacromolecules* **2**: 538–40.
54. Morar AS, Olteanu A, Young GB, Pielak GJ. (2001) "Solvent-induced collapse of alpha-synuclein and acid-denatured cytochrome c." *Protein Sci* **10**: 2195–9.
55. Uversky VN, Li J, Fink AL. (2001) "Trimethylamine-N-oxide-induced folding of alpha-synuclein." *FEBS Lett* **509**: 31–35.
56. McNulty BC, Young GB, Pielak GJ. (2006) "Macromolecular Crowding in the Escherichia coli Periplasm Maintains alpha-Synuclein Disorder." *J Mol Biol* **355**: 893–97.

57. Fuxreiter M, Simon I, Friedrich P, Tompa P. (2004) "Preformed structural elements feature in partner recognition by intrinsically unstructured proteins." *J Mol Biol* **338**: 1015–26.
58. Tompa P, Szasz C, Buday L. (2005) "Structural disorder throws new light on moonlighting." *Trends Biochem Sci* **30**: 484–89.
59. Brown CJ, Takayama S, Campen AM, *et al.* (2002) "Evolutionary rate heterogeneity in proteins with long disordered regions." *J Mol Evol* **55**: 104–10.
60. Denning DP, Uversky V, Patel SS, *et al.* (2002) "The Saccharomyces cerevisiae nucleoporin Nup2p is a natively unfolded protein." *J Biol Chem* **277**: 33447–55.
61. Patel SS, Belmont BJ, Sante JM, Rexach MF. (2007) "Natively unfolded nucleoporins gate protein diffusion across the nuclear pore complex." *Cell* **129**: 83–96.
62. Iakoucheva LM, Radivojac P, Brown CJ, *et al.* (2004) "The importance of intrinsic disorder for protein phosphorylation." *Nucl Acids Res* **32**: 1037–49.
63. Bustos DM, Iglesias AA. (2006) "Intrinsic disorder is a key characteristic in partners that bind 14-3-3 proteins." *Protein* **63**: 35–42.
64. Fuxreiter M, Tompa P, Simon I. (2007) "Local structural disorder imparts plasticity on linear motifs." *Bioinformatics* **23**: 950–56.
65. Dosztanyi Z, Chen J, Dunker AK, *et al.* (2006) "Disorder and sequence repeats in hub proteins and their implications for network evolution." *J Proteome Res* **5**: 2985–95.
66. Ekman D, Light S, Bjorklund AK, Elofsson A. (2006) "What properties characterize the hub proteins of the protein-protein interaction network of Saccharomyces cerevisiae?" *Genome Biol* 7: R45.
67. Haynes C, Oldfield CJ, Ji F, *et al.* (2006) "Intrinsic disorder is a common feature of hub proteins from four eukaryotic interactomes." *PLoS Comput Biol* **2**: e100.
68. Pontius BW. (1993) "Close encounters: why unstructured, polymeric domains can increase rates of specific macromolecular association." *Trends Biochem Sci* **18**: 181–86.
69. Dafforn TR, Smith CJ. (2004) "Natively unfolded domains in endocytosis: hooks, lines and linkers." *EMBO Rep* **5**: 1046–52.
70. Shoemaker BA, Portman JJ, Wolynes PG. (2000) "Speeding molecular recognition by using the folding funnel: the fly-casting mechanism." *Proc Natl Acad Sci USA* **97**: 8868–73.
71. Dyson HJ, Wright PE. (2002) "Coupling of folding and binding for unstructured proteins." *Curr Opin Struct Biol* **12**: 54–60.
72. Ferreon JC, Hilser VJ. (2004) "Thermodynamics of Binding to SH3 Domains: The Energetic Impact of Polyproline II (P(II)) Helix Formation." *Biochemistry* **43**: 7787–97.
73. Spolar RS, Record MT, Jr. (1994) "Coupling of local folding to site-specific binding of proteins to DNA." *Science* **263**: 777–84.

74. Fowler DM, Koulov AV, Balch WE, Kelly JW. (2007) "Functional amyloid — from bacteria to humans." *Trends Biochem Sci* **32**: 217–24.
75. Rauscher S, Baud S, Miao M, *et al.*(2006) "Proline and glycine control protein self-organization into elastomeric or amyloid fibrils." *Structure* **14**: 1667–76.
76. Mukhopadhyay R, Hoh JH. (2001) "AFM force measurements on microtubule-associated proteins: the projection domain exerts a long-range repulsive force." *FEBS Lett* **505**: 374–78.
77. Cox CJ, Dutta K, Petri ET, *et al.* (2002) "The regions of securin and cyclin B proteins recognized by the ubiquitination machinery are natively unfolded." *FEBS Lett* **527**: 303–08.
78. Puntervoll P, Linding R, Gemund C, *et al.* (2003) "ELM server: a new resource for investigating short functional sites in modular eukaryotic proteins." *Nucl Acids Res* **31**: 3625–30.
79. Tompa P, Csermely P. (2004) "The role of structural disorder in the function of RNA and protein chaperones." *FASEB J* **18**: 1169–75.
80. Waizenegger I, Gimenez-Abian JF, Wernic D, Peters JM. (2002) "Regulation of human separase by securin binding and autocleavage." *Curr Biol* **12**: 1368–78.
81. Yang J, Hurley TD, DePaoli-Roach AA. (2000) "Interaction of inhibitor-2 with the catalytic subunit of type 1 protein phosphatase. Identification of a sequence analogous to the consensus type 1 protein phosphatase-binding motif." *J Biol Chem* **275**: 22635–44.
82. Hornig NC, Knowles PP, McDonald NQ, Uhlmann F. (2002) "The dual mechanism of separase regulation by securin." *Curr Biol* **12**: 973–82.
83. Gunasekaran K, Tsai CJ, Kumar S, *et al.* (2003) "Extended disordered proteins: targeting function with less scaffold." *Trends Biochem Sci* **28**: 81–85.
84. Bhattacharyya RP, Remenyi A, Good MC, *et al.* (2006) "The Ste5 scaffold allosterically modulates signaling output of the yeast mating pathway." *Science* **311**: 822–26.
85. Holt C, Sawyer L. (1993) "Caseins as rheomorphic proteins: interpretation of primary and secondary structures of the alpha(s1)-, beta- and kappa-caseins." *J Chem Soc Faraday Trans* **89**: 2683–92.
86. Bokor M, Csizmok V, Kovacs D, *et al.* (2005) "NMR relaxation studies on the hydrate layer of intrinsically unstructured proteins." *Biophys J* **88**: 2030–37.
87. Tompa P, Banki P, Bokor M, *et al.* (2006) "Protein-water and protein-buffer interactions in the aqueous solution of an intrinsically unstructured plant Dehydrin: NMR Intensity and DSC aspects." *Biophys J* **91**: 2243–49.
88. Prusiner SB. (1998) "Prions." *Proc Natl Acad Sci USA* **95**: 13363–83.
89. Uversky VN, Fink AL. (2004) "Conformational constraints for amyloid fibrillation: the importance of being unfolded." *Biochim Biophys Acta* **1698**: 131–53.

90. Pierce MM, Baxa U, Steven AC, Bax A, Wickner RB. (2005) "Is the prion domain of soluble Ure2p unstructured?" *Biochemistry* **44**: 321–28.
91. Masino L, Kelly G, Leonard K, *et al.* (2002) "Solution structure of polyglutamine tracts in GST-polyglutamine fusion proteins." *FEBS Lett* **513**: 267–72.
92. Cheng Y, LeGall T, Oldfield CJ, *et al.* (2006) "Abundance of intrinsic disorder in protein associated with cardiovascular disease." *Biochemistry* **45**: 10448–60.
93. Cheng Y, Legall T, Oldfield, CJ, *et al.* (2006) "Rational drug design via intrinsically disordered protein." *Trends Biotechnol* **24**: 435–42.
94. Klein C, Vassilev LT. (2004) "Targeting the p53-MDM2 interaction to treat cancer." *Br J Cancer* **91**: 1415–19.

Chapter 8

Metalloproteins: Structure, Conservation and Prediction of Metal Binding Sites

Marvin Edelman*, Mariana Babor, Ronen Levy and Vladimir Sobolev*

The composition and structure of transition or "soft" metal binding sites and their conservation and flexibility are described. Algorithms for soft metal-binding-site prediction from sequences, holo structures and apo structures are presented, with emphasis on the latter. The chapter ends with a discussion on the importance of modeling for future predictive algorithms.

Introduction

Currently, about 20 novel protein structures are resolved each week by the Structural Genomics Initiative, a worldwide effort having as one of its goals the creation of a catalog of all protein folds. Functional information for Structural Genomics Initiative targets is often limited or nonexistent; thus, there is a growing need for procedures to deduce the information directly from the resolved structure.[1]

*Corresponding authors.
Department of Plant Sciences, Weizmann Institute of Science, Rehovot 76100, Israel.
Emails: marvin.edelman@weizmann.ac.il and vladimir.sobolev@weizmann.ac.il

In such instances, initial clues to biochemical function can be sought from ligands and cofactors that often accompany the protein during crystallization.

No cofactor group is more prevalent than metal ions. More than one third of the known proteins require metal ions to perform their functions.[2,3] Metal binding sites in proteins serve a great variety of purposes — among them, increasing the structural stability of the protein in a conformation required for biological function, participation in catalytic processes and playing regulatory roles. Certain metal ions can also undergo redox reactions.[4] A large fraction of metal binding proteins are resolved in the Protein Data Base (PDB)[5] in a pre-bound, or "apo", state with respect to their metal ion cofactors.

Metal-binding sites in proteins are very diverse, varying in coordination numbers, geometries, amino acid composition and metal ion preferences. They include backbone carbonyl oxygens, side chain groups — mainly of Asp, Glu, His and Cys — and water molecules. It has been observed that many metal binding sites in proteins are centered in a shell of hydrophilic ligands surrounded by a hydrophobic area.[6]

A number of algorithms are available for predicting metal binding sites[6–12] and metal binding proteins.[13] Mostly, they are based on information derived from holo forms. However, metal binding is the result of dynamic processes and examples of conformational changes upon metal binding are rife. Babor *et al.*[14] analyzed such changes at a database level in order to develop an effective metal binding site prediction algorithm for apo proteins, especially for novel protein structures of unknown function, such as those emanating from the Structural Genomics Initiative.

This chapter describes the composition and structure of transition or "soft" metal binding sites, their conservation and flexibility. Algorithms for soft metal-binding-site prediction from sequences, holo structures and apo structures are presented, with emphasis on the latter. The chapter ends with a discussion on the importance of modeling for future predictive algorithms.

Amino Acid Composition and Structure of Metal Binding Sites

The atoms (and their residues) surrounding the metal ion and interacting directly with it by donating electron pairs are called ligands. Pre-transition metal ions (e.g. Ca and Mg) are "hard" and interact through electrostatic forces mainly with oxygen atoms (often, backbone carboxyls). On the other hand, most of the transition metal ions relevant to biological systems (e.g. Fe, Zn, Cu, Ni, Co) are "soft" and have intermediate properties. Soft metal ions form stronger complexes and ligate mainly with nitrogen and sulfur atoms.[3] In numerous studies it was shown that soft-metal first shell residues fall almost exclusively into four amino acid types: Cys, His, Asp and Glu.[15–18] Statistics from current PDB structures fully agree with this (Fig. 1).

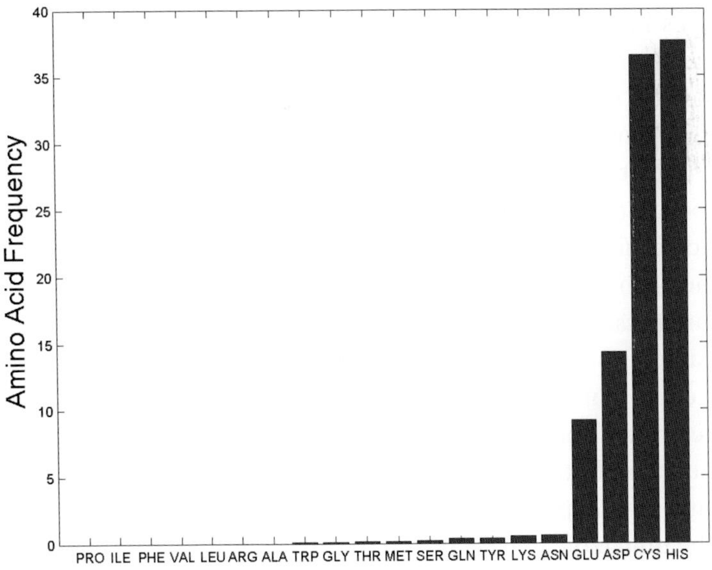

Fig. 1 Amino acid composition of Zn ion binding sites in PDB structures. The vast majority of first shell residues are His, Cys, Asp, and Glu. These residues bind Zn via nitrogen, oxygen, or sulfur donors. Only in rare cases are other amino acids involved. 348 non-redundant (<30% identity) sequences of resolution equal to or better than 2.5 Å were analyzed.

Metal-binding site geometry depends upon the number of ligands, termed the coordination number, and their stereochemical arrangement. The coordination number strives to be as large as possible, with the ligands atoms arranged so as to minimize repulsive interactions among them. Several coordination geometries are possible; for example, the arrangement of atoms can be tetrahedral if the coordination number is four, square pyramidal or trigonal bipyramidal if it is five, octahedral if it is six. The principal coordination numbers for metal ions found in proteins range from four to eight, with six being the most common. Mg and Mn each has a main coordination number of six,[19] whereas Zn and Ca show greater coordination flexibility, with values from four to six and six to eight, respectively. While Ca almost always, binds oxygens, the preference of Zn, a softer metal ion, depends upon the ligation geometry. When the coordination number is low, Zn binds nitrogen and sulfur; when it is six, Zn prefers oxygen.[15,20,21] Changes in coordination number occur because the energy barriers to pass from one to the other are low. Coordination number flexibility is also related to the role these metals play in biological systems: Mg ions generally have a structural role while Zn ions are more likely to participate directly in catalytic processes, with possible changes in the coordination number during the reaction.[21]

Side chain carboxylates, sulfur and the imidazole groups dominate soft metal coordination in proteins, although main-chain carbonyl oxygens are present in some binding sites.[3] Generally, metal cations lie in the plane of the carboxylic group. Direct bonding, in which the metal ion is equidistant from both oxygens of the carboxylate ion, seems to be preferred. These results agree with those obtained by analyzing protein crystal structures, where liganded carboxylate groups often belong to the Asp and Glu side chains.[20,22,23]

The imidazole groups in His residues bind a variety of metal ions, especially Zn, Cu and Fe cations. Metal ions bound to the imidazole group nearly always lie in the plane of the ring along the lone-pair direction of the nitrogen. This observation is independent of the identity of the metal ion and is consistent in both small molecules and protein structures.[20,24] In aqueous solution, the NE2-protonated tautomeric form is preferred. When bound to a cation, the

ND2-protonated tautomer is the most stable, with the NE2 atom bound to the metal ion and the ring oriented through a hydrogen bond between the ND1 atom and a polar side-chain of a neighbor.[20,24]

Conservation and Flexibility of Soft Metal Binding Sites

Sequence Conservation

The majority of Zn binding site residues is positionally conserved. In Fig. 2, it can be seen that Zn binding site positions are clearly distinguished from the random positions of these same amino acid types, and the overlap between the two is very low. These results illustrate that metal binding residues are far more conserved than randomly picked residues of the same amino acid type along the protein sequence. Indeed, it is commonly accepted that there is a tendency for functional residues to be conserved.[8,25]

In attempts to recognize residues forming metal binding sites within a protein sequence, it is important to determine whether a metal binding signature, or a pattern by which the residues are spaced along the sequence, exists. Figure 3 presents a statistical analysis of the sequence spacing between Zn binding residues for sites with three and four residue ligands. Similar to Auld,[16] we find that at least one spacer is short (5 residues or less). However, at least one spacer in both cases varies significantly in length. This could be an important limitation when trying to extract conserved spacing patterns for predictive purposes.

Structural Conservation

Protein function is the result of dynamic processes, which may involve conformational changes affecting protein structure at a variety of levels: motion of a few atoms or amino acid side chains, movements affecting a few secondary structural elements, movements of whole subunits within a domain, and those affecting the orientation of a domain relative to another.[26] In line with this, metal binding to an apo protein may trigger changes at different scales. For example, upon

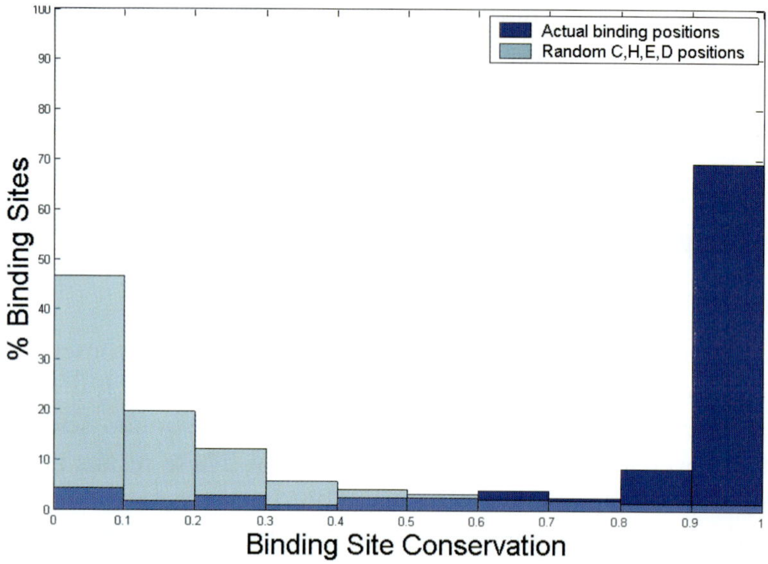

Fig. 2 Binding site conservation analysis. 270 non-redundant (<30% identity), high resolution (equal to or better than 2.5 Å) PDB Zn binding proteins served as templates for structural alignment against homologous sequences (having 95%–30% identity to template) using the HSSP database.[40] The fraction of multiply aligned sequences with *all* residues of the binding site positionally conserved was determined (allowing substitution among the 4 CHED residue types themselves). The dark blue color represents the percentage distribution of conservation for these binding sites. In almost 70% of cases ≥ 90% of the HSSP sequences in the homologous set had all residues of the Zn binding site positionally conserved. This distribution was compared to those of other C, H, E, D residues in the template that are *not* part of the Zn binding site (cyan color). A set of these, equal in number to the number of Zn binding residues, was randomly chosen, 100 times in each case, and analyzed for positional conservation of all the residues in the set. As can be seen, in almost 70% of these cases, < 20% of the HSSP sequences had all the residues of the randomly chosen, non-binding sets positionally conserved. The intermediate blue color represents the overlap between the two distributions. The comparison of distributions demonstrates that for Zn ions, CHED binding site residues are significantly more conserved than the other C, H, E, D residues along the sequence.

metal binding, carboxypeptidase A undergoes only small structural rearrangements. This enzyme is a Zn-dependent hydrolase, in which the metal is pentacoordinated by two histidine residues (His 69, His 196), a glutamic acid residue (Glu 72, in a bidentate form) and a water

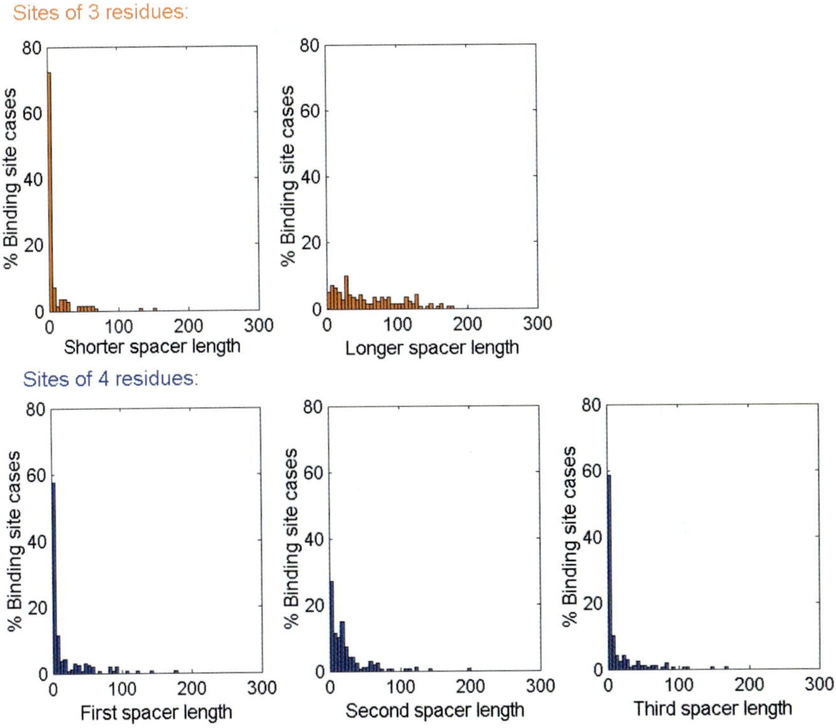

Fig. 3 Statistical analysis of sequence spacing between Zn binding residues. Biologically relevant binding sites with three or four ligand residues are most common (cf., Table 1). The sequence length of the spacers between residue binding positions belonging to the same site was measured. Non-redundant (< 30% sequence identity) sets of high-resolution (equal to or better than 2.5 Å) Zn binding sites were analyzed. **Sites of 3 residues** (141 entries): a short (≤ 5 residue) spacer separating two of the three residues is accompanied by a spacer of highly variable length separating the third residue from the other two. There is no directionality to the data. **Sites of 4 residues** (165 entries): going from N-terminal to C-terminal of the protein, the first spacer (between binding residues 1 and 2) and the third spacer (between binding residues 3 and 4) are short (≤ 5 residues), while the intermediate spacer (between binding residues 2 and 3) is variable in length.

molecule. Structural comparison of apo and holo carboxypeptidase A showed that they share a very similar structure. However, in the apo form, His 196 is rotated 110° about the torsion angle χ^2 (around the $C\beta$-$C\gamma$ bond) with respect to the holo form to make a salt bridge with

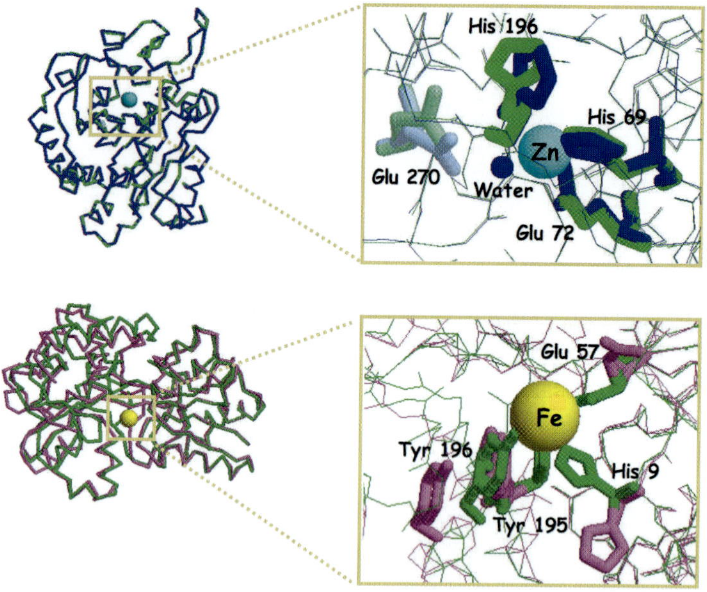

Fig. 4 Structural rearrangements upon metal binding. *Upper panel*: Structural alignment of Carboxipeptidase A holo (1yme; blue) and apo (1arl; green) forms. Enlarged on the right, metal binding site ligands (blue) and apo-form amino acid residues analogous to the binding site ligands (green). *Lower panel*: structural alignment of Ferric binding protein holo (1mrp; green) and apo (1d9v; magenta) forms. Enlarged on the right, metal binding site ligands (green) and analogous apo amino acids (magenta).

the side chain of Glu 270[27] (Fig. 4, upper panel). On the other hand, the ferric binding protein, a transport protein with considerable intrinsic flexibility, undergoes major conformational changes involving hinge motions during the uptake and release of iron (Fig. 4, lower panel).[28]

Babor *et al.*[14] analyzed at a database scale the putative metal binding sites in the apo protein state and their rearrangements upon actual metal binding. A high resolution, non-redundant dataset containing all available apo-holo pairs for the most populated metals in the PDB was created. Structural rearrangements upon metal binding were found to be mainly restricted to the first and second shell residues. While backbone motions occurred in less than 15% of cases, side chain reorientations were more frequent. This was first pointed out by Chakrabarti,[29] who showed that in apo proteins, cysteine thiol side chains in non-α/non-β structures prefer the g^+ conformation, while upon metal binding the t conformer is the most populated, with g^- increasing substantially and g^+ rarely found.

Babor *et al.*[14] showed that the frequency of side chain reorientation is directly correlated with metal ligand flexibility, measured as B factor, and solvent accessibility, calculated using CSU software,[30] in the unbound state (Fig. 5). Side chain reorientation occurred in more than 40% of binding sites having three amino acid ligands. However,

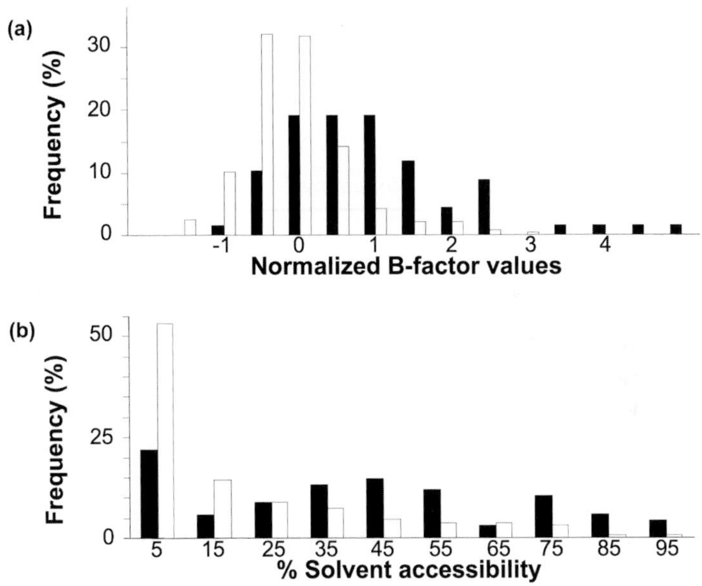

Fig. 5 Metal binding-site residue flexibility and solvent accessibility in apo forms. Adapted from Babor *et al.*[14] A high resolution, non-redundant dataset of 71 apo-holo pairs containing three or more amino acid ligands for the most populated metal ions (excluding Na and K) in the PDB was analyzed. Black bars: ligand side chains undergoing conformational changes; White bars: ligand side chains not reorienting. (**a**) *Normalized B-factor distribution*. B-factor is a measure of atomic disorder in crystal structures. The mean values for ligand side chains that do, or do not, reorient are +1.0 and −0.1, respectively. Thus, on average, ligand residues that reorient upon metal binding are more disordered in the pre-bound state than those that do not move. (**b**) *Side chain reorientations as a function of solvent accessibility*. Ligands whose side chains reorient upon metal binding, also tend to be more accessible to solvent in the pre-bound state.

Table 1 Side Chain Rearrangements in Soft Metal Binding Sites upon Metal Binding. A high-resolution, non-redundant PDB set of 93 apo-holo protein pairs having soft metal binding sites populated with 3 or more amino acid ligands was analyzed

Characteristic	No. of Amino Acid Ligands per Binding Site	
	Three	Four or More
Total binding sites	47	46
Total reoriented side chains	20	14
One side chain reoriented	15	10
Two side chains reoriented	2	3
Three side chains reoriented	3	1
Probability of reorientation	0.43	0.30
Probability of one or no side chains reorienting	0.89	0.91

in 90% of rigid-backbone binding sites, on average not more than one side chain moved (Table 1). These findings revealed that, in general, a significant part of the metal-ion binding site is already structured in the unbound state and transition from the apo to the holo state can be approximated by rearrangement of a single amino acid ligand.

Metal Binding Sites Prediction From Sequence

Metal binding predictions based on 3D structural properties of proteins are often not applicable since the great majority of the proteins studied to date are derived from translated DNA sequences rather than resolved structures. In an attempt to bypass this, new methods have been devised.

Metal binding is encoded in the primary sequence of proteins. As primary sequence is related to 3-D structure, it also affects the arrangement of the amino acid ligands forming the metal binding site. One approach taken to harvest this information systematically determines all the possible metal-binding signatures present in the PDB. These signatures include the binding residues and their spacing along the sequence. The method[8] was applied to copper proteins only, and a library of *exact* metal binding patterns was built. Each metal-binding pattern is used together with the primary sequence of the corresponding metalloprotein to browse any ensemble of sequences of interest. The level of confidence of this method is variable, ranging between 50% to over 90%, depending on the lengths of the local alignments identified around each binding residue. It is not clear as to what extent the method is applicable to metals other than Cu. Moreover, a limitation of this work is that the PDB contains only a fraction of all the possible metal-binding patterns.

Lin *et al.*[10] took fragments of proteins as input and assumed that the surrounding environment influences the metal binding residues. The amino acid at the center of the fragment is the target amino acid, whereas the others are the "neighbors." The fragment sequence is encoded to a feature vector, which contains information on the occurrence probability of the amino acid, the propensities of the secondary structure, and the metal-binding propensity of the amino acid. This process is repeated by shifting one position at a time along the protein sequence, resulting in a series of new fragments. The feature vectors are fed into a neural-network learning machine, which decides whether the target amino acid binds metal or not. The binding residues are identified with higher than 90% sensitivity. A limitation of this approach is that it predicts the probability of each putative binding residue individually, instead of taking into consideration the combined context of all the residues belonging to one unified site. In addition, a protein chain can include more than one binding site, with residues of multiple sites intertwined.[31] We note that 14% of Zn binding sites have an additional same-chain metal binding partner. The method is blind to such sites if they are intertwined.

Predicting Metal Binding Sites Based on HOLO Structures

The first metal-binding-site prediction algorithm described[6] made use of holo structures. It is based on the fact that metal binding sites are often centered in an area of hydrophilic ligands, surrounded by a hydrophobic shell. However, this method also selects regions of high contrast that are not associated with metal binding, such as charged surface residues and buried, positively charged residues.[7]

More recently, Sodhi *et al.*[9] calculated the likelihood of a given residue to be a metal ligand by considering multiple sequence alignment of homologous proteins as well as approximate structural information. This algorithm performed satisfactorily for SCOP[32] superfamilies where large sets of evolutionary related proteins are available. The algorithm was developed considering 190, 18, 11 and 49 superfamilies for Zn, Fe, Cu and Mn, respectively, while valuation of performance was applied to five, four, one and one cases, respectively. For the Zn, Cu and Fe cases, the algorithm succeeded in capturing 77%, 62.5% and 92% of the ligand residues, with a selectivity of 43%, 44% and 27%, respectively. This algorithm suffers from the drawbacks discussed above for the individual residue approach: it is often difficult to identify the location of a metal binding site by inspecting the distribution of predicted individual residues within the protein structure.

A third algorithm, developed by Schymkowitz *et al.*,[11] was aimed at pinpointing the location of the metals within the proteins. The algorithm uses a library containing the most common metal-ion spatial positions relative to the corresponding ligating atoms of the protein. In the first step, the library is used for searching for possible metal positions within the protein structure. Then, an optimization step is performed to find the best position for the predicted metal using the Fold-X force field.[33] The resulting position is used to estimate the energy of binding. A hydration step is also included to discriminate water ligands. This algorithm is geared to, and is reported to perform well in identifying the metal binding sites in holo forms. It captured about 90% of the metal binding sites for Zn, Cu and Mn.[11] The highest selectivity value was achieved for Zn (88%), while slightly lower values were obtained for Cu (79%) and Mn (78%).

Predicting Metal Binding Sites Based on APO Structures

Babor *et al.* (2007) recently published a prediction algorithm based on the apo forms that considers commonalities in soft metal binding modes. The algorithm was derived using the Zn ion, the archetype for soft metals,[34] and involves two steps: a search for binding site candidates founded on a geometric definition of the pre-bound state, followed by a machine-learning filtering step to reduce the number of false positives.

Geometric Considerations

The algorithm searches for a 3D constellation of three accepted amino acid residues (a "triad"), whose metal-ligating atoms satisfy a distance criterion. All the possible triads from the four most common ligands for zinc residues (Cys, His, Glu and Asp; referred to by us as "CHED"), were retrieved whose collective Cβ atom distances are less than 13.0 Å. If distances d_1, d_2 and d_3 among the ligand atoms from separate CHED residues satisfied the cutoff criterion $d_1 \leq 4.7$ Å, $d_2 \leq 5.1$ Å and $d_3 \leq 5.4$ Å, the set was retained. The cutoff values were chosen by analyzing a redundant set of more than 1000 holo forms, and refined using a set of 28 non-redundant apo structures that have holo PDB partners. In addition, if one or two of the three inter-ligand distances were not initially satisfied, alternative side chain conformations of the relevant residues were built, one at the time, using a backbone-independent rotamer library.[35] If no clashes were eventually observed, and d_1, d_2 and d_3 now satisfied the cutoff distances, then the built-up triad was retained. Clashing was considered to occur if the distance between any two atoms from two different residues ($\text{dist}_{a,b}$) is $< 0.7 * (Ra + Rb)$, where Ra and Rb are Van der Waals radii of atoms a and b, as defined by Bondi.[36] The value 0.7 was determined by statistically analyzing inter atomic distances in PDB structures. The distances between atoms separated by four or more bonds were analyzed to check for clashing.

We note that the optimal distance between a Zn ion and a ligand atom is 2.0–2.3 Å. Therefore, the minimal distance between two ligated atoms would be less than 3 Å when all the four ligand atoms and the metal ion are situated in one plane, or about 3.3 Å when in a tetrahedral ligand arrangement. The maximal distance would then be 4.6 Å when two ligand atoms are located opposite to each other. For the geometric search, increments beyond this maximum distance can result from different ligand atom positions between apo and holo structures and from an inaccurate atom position when an alternative side chain position is built based on the rotamer library.

At the binding site level, a triad of residues is considered "true," "partially-true" or "false" if all three, some or none of its residues are correct. A site is "true" if it includes at least a partially-true triad; it is "false" if it does not. More than one true triad is possible if more than three amino-acid-ligating residues are involved. Using the above geometric procedure, at least one true triad was identified in 24 out of 28 apo binding sites (distributed among 27 apo chains), and at least one partially-true-triad in 3 of the 4 remaining ones (Table 2). In the one case that failed (1bec), one of the three ligands was a rare non-CHED ligand type (Gln). However, in the cases when there are four ligated residues with one rare residue (Tyr for 1ial, Gly for 1e65), soft metal binding sites were predicted correctly based on the three remaining CHED residues. There are also cases of three ligated residues that include a rare non-CHED type with the binding site correctly predicted. In these instances, the geometric search picked up a CHED type residue from the second sphere. Such cases involves the Mn binding sites in 1dck, 1lv5 and 3tmy.

Filtering Out the False Positives

Two filters were generated: a mild one, yielding high sensitivity (i.e. maximum true positives), and a stringent one, yielding high selectivity (i.e. minimum false positives). The mild filter is based on an empirical observation that sites composed of a relatively large number of triads tend to be true. Therefore, in cases where a site is found to contain at least five triads, all the other putative sites with three or fewer triads are discarded. Predictions for four different non-redundant sets (mononuclear training set; mononuclear, multinuclear and apo testing sets) of Zn binding sites using the mild filter captured over 95% of the experimental sites in each case with a selectivity of 63% or higher (Fig. 7a).

The stringent filter makes use of two machine-learning techniques: a decision tree classifier[37] and a support vector machine classifier.[38,39]

Table 2 Results of the Search Procedure for Apo Forms

PDB ID	Number of Triads			Number of Residues			Number of Sites		
	True	Partial True	False	Exp.[a]	Predict True	Predict False	Exp.[a]	Predict	Correct
1arl_	1	1	0	3	3	1	1	1	1
1dq0A	4	0	1	4	4	3	1	2	1
1et9A	1	0	0	3	3	0	1	1	1
1qovH	1	0	0	3	3	0	1	1	1
1qlpA[b]	1	1	1	6	4	3	2	2	2
1oztM	4	4	0	4	4	3	1	1	1
1fnuA	4	0	1	4	4	3	1	2	1
1xlaA	8	7	2	7	7	8	1	3	1
1om6A	0	1	2	3	2	7	1	3	1
1i60A	4	9	2	4	4	8	1	2	1
1iad_	1	2	3	4	3	7	1	2	1
1lt7A	1	0	3	3	3	4	1	2	1
1k0fA	4	3	1	4	4	4	1	2	1
1e65A	1	1	0	4	3	2	1	1	1
3enl_	1	5	4	3	3	14	1	4	1

(Continued)

Table 2 (*Continued*)

PDB ID	Number of Triads			Number of Residues			Number of Sites		
	True	Partial True	False	Exp.[a]	Predict True	Predict False	Exp.[a]	Predict	Correct
1pta_	0	5	1	4	2	5	1	1	1
1emvB	1	1	0	3	3	1	1	1	1
1rdzA	1	0	1	3	3	3	1	2	1
2cbe_	1	6	2	3	3	6	1	2	1
1et6A	1	0	0	3	3	0	1	1	1
1ilwA	1	4	0	3	3	3	1	1	1
1kspA	1	0	5	3	3	12	1	4	1
1bi1_	1	1	1	3	3	4	1	2	1
1c3pA	1	3	5	3	3	11	1	2	1
1k6kA	1	1	0	3	3	1	1	1	1
1h4uA	1	0	0	3	3	0	1	1	1
1bec_	0	0	0	3	0	0	1	0	0
Totals	46	55	35	96	86	113	28	47	27

[a] Experimentally determined.
[b] Contains two experimental binding sites.

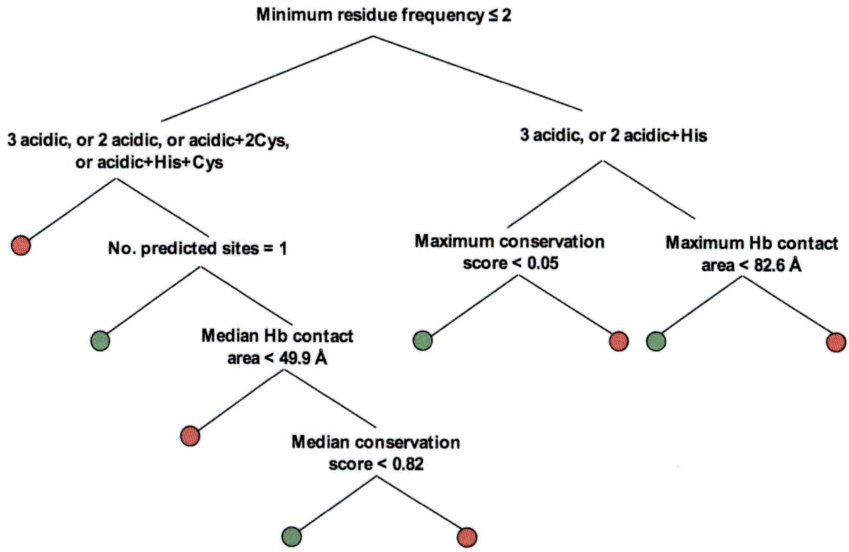

Fig. 6 Decision tree filtration. The filter was created based on a CART tree algorithm. The tree, pruned using a 10-fold cross-validation test, contains nine terminal nodes. See text for explanation of attributes. A triad of metal binding residues is sorted along the tree according to its attribute values until it reaches a terminal node (shown as circles). Terminal nodes that retain the triads are shown as green and discarded ones as red.

The decision tree (Fig. 6) considers: *Number of predicted sites* — number of sites predicted by the geometric search; *Minimum residue-position frequency* — the frequencies of all the positions within a site are first tallied, then the position with the lowest frequency is scored for each triad; *Amino acid composition* — the number of acidic, His and Cys residues present in a given triad; *Conservation score* — the HSSP database[40] was used to obtain a multiple sequence alignment. The conservation score was then calculated by a modification of Mirny and Shakhnovich,[41] in which CHED residues replaced the acidic group. Median and maximum conservation scores were defined as the intermediate and highest conservation score values among the three residues of the triad; *Hydrogen bond contact surface area* — contact surface area between each residue in a triad and its neighbors was calculated using CSU software[42] and a distance cutoff of 3.5 Å, and summed. The median and maximum areas were defined as the intermediate and highest hydrogen bonding contacting surface values among the three residues of the triad. The support vector machine includes, in addition, the *number of triads per predicted site* and *relative solvent accessible surface*. A triad was retained if found by at least one of the classifiers and removed if excluded by both of them. For each of the four data sets, the stringent filter captured at least 79% of the experimental sites while increasing site selectivity to at least 94% (Fig. 7a).

The strength of the algorithm resides in its search for a triad binding site motif, a set of three residues with a particular spatial relationship, and not for the likelihood of a given residue to be a metal ligand. Nevertheless, predictions at the individual residue level (Fig. 7b) show that for all four Zn sets analyzed, at least 90% sensitivity and 49% selectivity were obtained with the mild filter, while for the stringent filter the values are at least 77% sensitivity and 71% selectivity.

The performance of the CHED algorithm (http://ligin.weizmann.ac.il/ched) was evaluated in holo and apo sets of soft metals other than Zn (viz., Fe, Mn, Cu, Ni, Co) (Fig. 8). The mild filter identified 70–100% of the sites with a selectivity range of 50–70%, depending on the metal type. For the stringent filter, high selectivity values of 84–100% were achieved in all the sets analyzed. Concerning sensitivity, values were around 80% except for Mn sites, where sensitivity was reduced to about 50%. Mn is the hardest of the soft metals analyzed and the stringent filter captures hard metals poorly (e.g. Mg and Ca).

General Application to Structures from Proteomics Initiatives and the PDB

The high selectivity (i.e. very low level of false positives) obtained with the CHED algorithm suggests that it can be useful in assigning soft metal binding capabilities to the growing list of unidentified protein structures emanating from the Structural Genomics Initiative. Frequently, little is known about the function of the protein targets chosen for high-throughput crystallization; furthermore, procedures

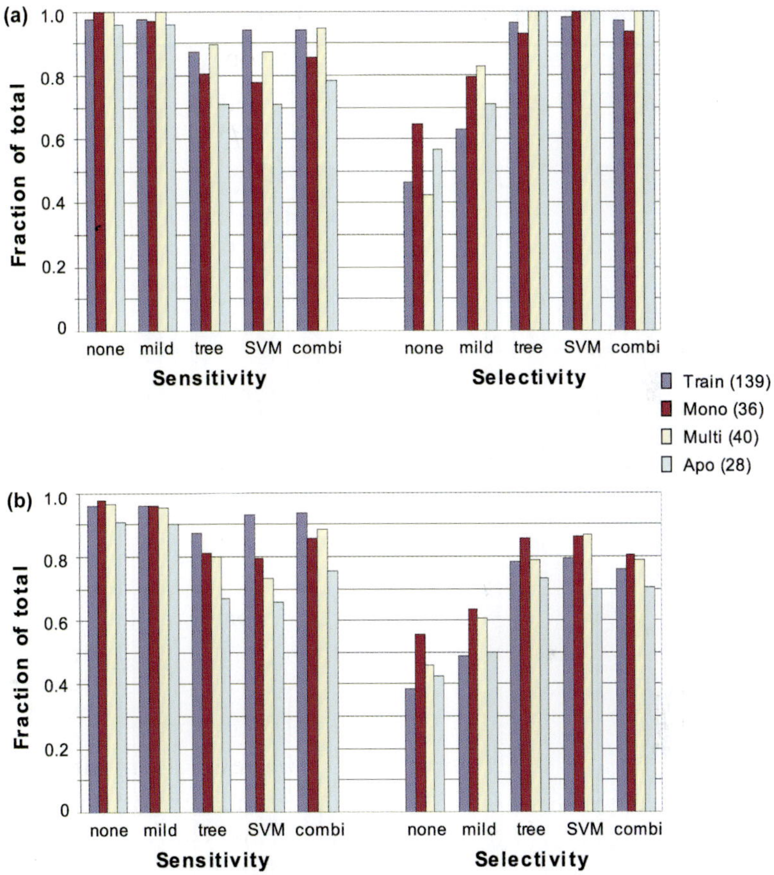

Fig. 7 Predicting experimental Zn binding sites and binding residues using the CHED algorithm. Results for four independent high-resolution, non-redundant Zn binding site data sets are shown (color coded), with the number of sites analyzed given in parenthesis. The datasets consist of a large training set, and three testing sets containing mononuclear Zn sites, multinuclear sites and apo structures from a set of apo-holo pairs. An experimental site is scored if at least one correct binding site residue is identified. On average, 4.0–4.5 residues were predicted per mononuclear site and 6–8 per multinuclear site. (**a**) *Binding sites*; (**b**) *Binding residues.*

currently applied (e.g. heterologous overexpression, chelating agents) may result in loss of cofactors such as metals. Babor *et al.*[12] analyzed a non-redundant set of 230 crystallized apo protein chains extracted from such a list. Of the 33 putative, soft metal binding sites predicted

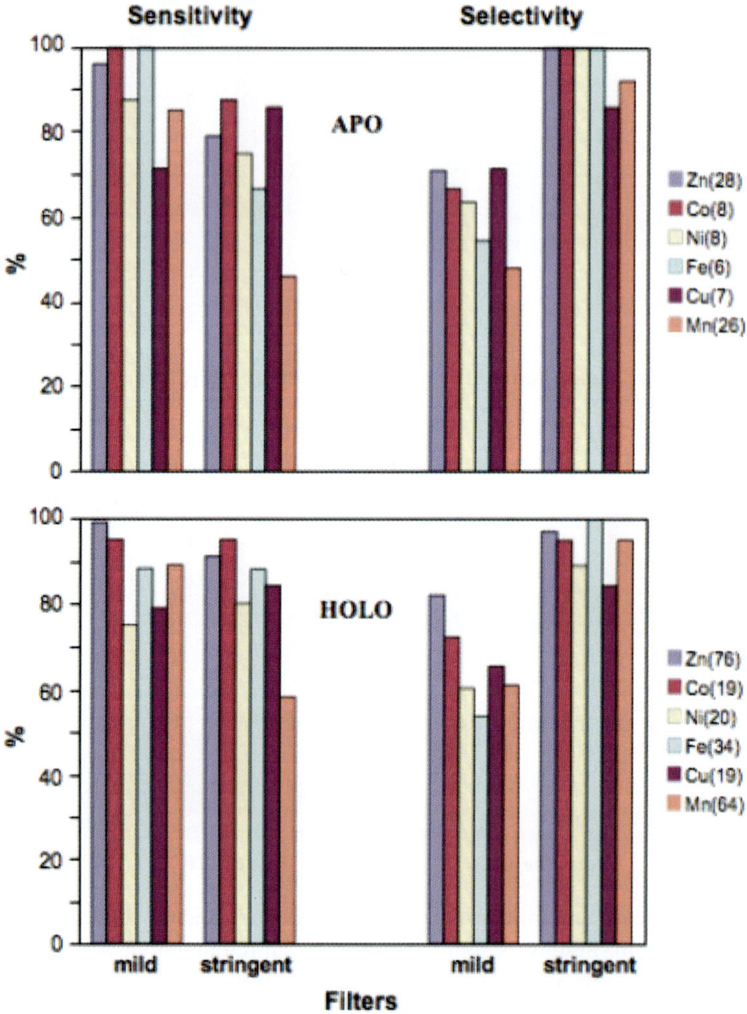

Fig. 8 Sensitivity and selectivity predictions for soft metal ions using the CHED algorithm (from Babor *et al.*[12]). Non-redundant datasets of paired apo and holo proteins binding soft metals were created. The Zn-ion training set described in Fig. 7 was used for training. Search results for several soft metal binding test sets are shown (color coded, with the number of chains given in parenthesis) following application of the mild and combined stringent filters. Sensitivity indicates the fraction of experimental binding sites predicted; Selectivity indicates the fraction of all predicted sites that are correct. For all metal types, the average stringent selectivity for apo and holo structures is 96% and 95%, respectively.

following stringent filtration, about half could be supported by laborious sequence, structure or literature analyses. For the other half, the CHED predictions represent an entrance point to functional annotation.

The search procedure for metal binding sites was applied to a high resolution, non-redundant set of crystallographic structures from the full PDB. The stringent filter faithfully found 88% of the holo protein sites while predicting that 16% of the apo proteins are soft metal binders (Fig. 9). This percentage is clearly distinguished from the low level (4%) of false positives found for apo proteins.[12]

The Future Belongs to Modeling

A major goal of the current proteomics initiatives is to populate protein sequence space as much as possible. We envision a point in the not too distant future when most translated DNA coding sequences will have a structural template at some usable degree of homology available

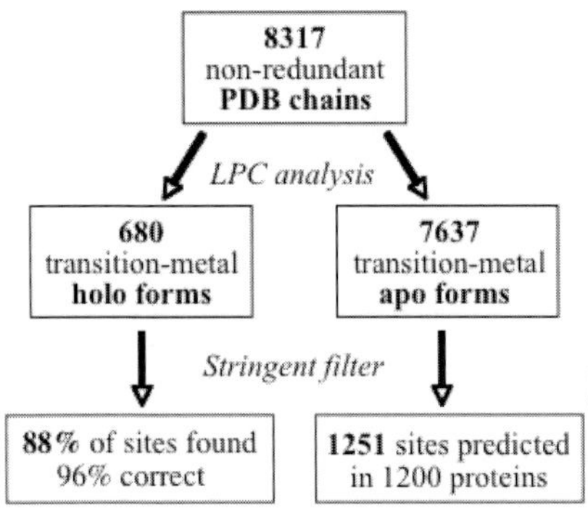

Fig. 9 Prediction of soft metal sites from pre-bound structures in the PDB (from Ref. 12). The CHED algorithm was applied to the full PDB, using a precompiled non-redundant list of 8317 polypeptide chains with resolution better than 2.5 Å and sequence identity less than 90%.[43] In this list, there are 680 proteins with 900 binding sites ligating soft metals (Zn, Co, Ni, Fe, Cu, Mn). The stringent filter found 88% of these sites with a selectivity of 96%. The remaining 7637 polypeptide chains were apo with respect to soft metals. The search procedure found 1251 putative binding sites distributed among 1200 proteins for this apo subset.

in the PDB. A 3D model for almost all proteins could be generated by homology modeling at that point using the existing tools.[44–47] The question, of course, is how good the models are for predicting protein functions such as metal binding. Our preliminary analysis (Levy *et al.*, unpublished) indicates that side chain modeling only moderately reduces sensitivity and selectivity. Further reduction can ensue if the backbone structures for the target sequence and template structure differ. However, we find that minimal reduction occurs if the template structure contains a metal, even if homology between the template and target sequences is very low (20% identity).

Acknowledgments

The research finding on sequence conservation by RL were supported in part by the Avron-Wilstatter Minerva Center for Research in Photosynthesis.

References

1. Friedberg I, Jambon M, Godzik A. (2006) "New avenues in protein function prediction." *Prot Sci* **15**: 1527–29.
2. Ibers JA, Holm RH. (1980) "Modeling coordination sites in metallobiomolecules." *Science* **209**: 223–35.
3. Tainer JA, Roberts VA, Getzoff ED. (1992) "Protein metal-binding sites." *Curr Opin Biotechnol* **3**: 378–87.
4. Williams RJ. (1985) 16th Sir Hans Krebs lecture. "The symbiosis of metal and protein functions." *Eur J Biochem* **150**: 231–48.
5. Bernstein FC, Koetzle TF, Williams GJB, *et al.* (1977) "Protein data bank — computer-based archival file for macromolecular structures." *J Mol Biol* **112**: 535–42.
6. Yamashita MM, Wesson L, Eisenman G, Eisenberg D. (1990) "Where Metal-Ions Bind in Proteins." *Proc Natl Acad Sci USA* **87**: 5648–52.
7. Gregory DS, Martin AC, Cheetham JC, Rees AR. (1993). "The prediction and characterization of metal binding sites in proteins." *Prot Eng* **6**: 29–35.
8. Andreini C, Bertini I, Rosato A. (2004) "A hint to search for metalloproteins in gene banks." *Bioinformatics* **20**: 1373–80.
9. Sodhi JS, Bryson K, McGuffin LJ, *et al.* (2004) "Predicting metal-binding site residues in low-resolution structural models." *J Mol Biol* **342**: 307–20.

10. Lin CT, Lin KL, Yang CH, *et al.* (2005) "Protein metal binding residue prediction based on neural networks." *Int J Neur Sys* **15**: 71–84.
11. Schymkowitz JW, Rousseau F, Martins IC, *et al.* (2005) "Prediction of water and metal binding sites and their affinities by using the Fold-X force field." *Proc Natl Acad Sci USA* **102**: 10147–52.
12. Babor M, Gerzon S, Raveh B, *et al.* (2007) "Prediction of transition metal-binding sites from apo protein structures." *Proteins*, in press.
13. Lin HH, Han LY, Zhang HL, *et al.* (2006) "Prediction of the functional class of metal-binding proteins from sequence derived physicochemical properties by support vector machine approach." *BMC Bioinform* 7(5); S13.
14. Babor M, Greenblatt HM, Edelman M, Sobolev V. (2005). "Flexibility of metal binding sites in proteins on a database scale." *Proteins* **59**: 221–30.
15. Alberts IL, Nadassy K, Wodak SJ. (1998) "Analysis of zinc binding sites in protein crystal structures." *Prot Sci* 7: 1700–16.
16. Auld DS. (2001) "Zinc coordination sphere in biochemical zinc sites." *Biometals* **14**: 271–313.
17. Dudev T, Lin YL, Dudev M, Lim C. (2003) "First-second shell interactions in metal binding sites in proteins: a PDB survey and DFT/CDM calculations." *J Amer Chem Soc* **125**: 3168–80.
18. Tamames B, Sousa SF, Tamames J, *et al.* (2007) "Analysis of zinc-ligand bond lemgths in metalloproteins: trends and patterns." *Proteins* (in press).
19. Bock CW, Katz AK, Markham GD, Glusker JP. (1999) "Manganese as a replacement for magnesium and zinc: functional comparison of the divalent ions." *J Amer Chem Soc* **121**: 7360–72.
20. Glusker JP. (1991) "Structural aspects of metal liganding to functional-groups in proteins." *Adv Prot Chem* **42**: 1–76.
21. Katz AK, Glusker JP, Beebe SA, Bock CW. (1996) "Calcium ion coordination: a comparison with that of beryllium, magnesium, and zinc." *J Amer Chem Soc* **118**: 5752–63.
22. Chakrabarti P. (1990) "Interaction of metal-ions with carboxylic and carboxamide groups in protein structures." *Prot Eng* **4**: 49–56.
23. Harding MM. (2001) "Geometry of metal-ligand interactions in proteins." *Acta Cryst Sec D* **57**: 401–11.
24. Chakrabarti P. (1990) "Geometry of interaction of metal-ions with histidine-residues in protein structures." *Prot Eng* **4**: 57–63.
25. Ouzounis C, Pérez-Irratxeta C, Sander C, Valencia A. (1998) "Are binding residues conserved?" *Pac Symp Biocomput* **3**: 401–12.
26. Janin J, Wodak SJ. (1983) "Structural domains in proteins and their role in the dynamics of protein function." *Progr Biophys Mol Biol* **42**: 21–78.
27. Feinberg H, Greenblatt HM, Shoham G. (1993) "Structural studies of the active-site metal in metalloenzymes." *J Chem Inf Comp Sci* **33**: 501–16.

28. Bruns CM, Anderson DS, Vaughan KG, *et al.* (2001) "Crystallographic and biochemical analyses of the metal-free Haemophilus influenzae Fe3+ –binding protein." *Biochemistry* **40**: 15631–37.
29. Chakrabarti P. (1989) "Geometry of interaction of metal-ions with sulfur-containing ligands in protein structures." *Biochemistry* **28**: 6081–85.
30. Sobolev V, Sorokine A, Prilusky J, *et al.* (1999). "Automated analysis of interatomic contacts in proteins." *Bioinformatics* **15**: 327–32.
31. Maret W. (2005) "Zinc coordination environments in proteins determine zinc functions." *J Trace Elem Med Biol* **19**: 7–12.
32. Murzin AG, Brener SE, Hubbard T, Chothia C. (1995) "SCOP: a structural classification of protein database for the investigation of sequences and structures." *J Mol Biol* **247**: 536–40.
33. Guerois R, Nielsen JE, Serrano L. (2002) "Predicting changes in the stability of proteins and protein complexes: a study of more than 1000 mutations." *J Mol Biol* **320**: 369–87.
34. Andreini C, Banci L, Bertini I, Rosato A. (2006) "Counting the zinc-proteins encoded in the human genome." *J Proteome Res* **5**: 196–201.
35. Dunbrack RL, Cohen FE. (1997) "Bayesian statistical analysis of protein side-chain rotamer preferences." *Prot Sci* **6**: 1661–81.
36. Bondi A. (1964) "Van der Waals volumes and radii." *J Phys Chem* **68**: 441–51.
37. Breiman L, Friedman J, Stone CJ, Olshen RA. (1998) "Classification and Regression Trees." CRC Press, Boca Raton.
38. Chapelle O, Haffner P, Vapnik VN. (1999) "Support vector machines for histogram-based image classification." *IEEE Trans Neural Net* **10**: 1055–64.
39. Scholkopf B, Smola AJ, Williamson RC, Barlett PL. (2000) "New support vector algorithms." *Neural Comput* **12**: 1207–45.
40. Sander C, Schneider R. (1991) "Database of homology-derived protein structures and the structural meaning of sequence alignment." *Proteins* **9**: 56–68.
41. Mirny L, Shakhnovich E. (2001) "Evolutionary conservation of the folding nucleus." *J Mol Biol* **308**: 123–29.
42. Sobolev V, Eyal E, Gerzon S, *et al.* (2005) "SPACE: a suite of tools for protein structure prediction and analysis based on complementarity and environment." *Nucl Acids Res* **33**: W39–43.
43. Wang G, Dunbrack RL. (2003) "PISCES: a protein sequence culling server." *Bioinformatics* **19**: 1589–91.
44. Marti-Renom MA, Stuart A, Fiser A, *et al.* (2000) "Comparative protein structure modeling of genes and genomes." *Ann Rev Biophys Biomol Struct* **29**: 291–325.
45. Schwede T, Kopp J, Guex N, Peitsch MC. (2003) "SWISS-MODEL: an automated protein homology-modeling server." *Nucl Ac Res* **31**: 3381–85.

46. Eyal E, Najmanovich R, McConkey BJ, *et al.* (2004) "Importance of solvent accessibility and contact surfaces in modeling side-chain conformation in proteins." *J Comp Chem* **25**: 712–24.
47. Canutescu AA, Dunbrack RL. (2005) "MollDE: a homology modeling framework you can click with." *Bioinformatics* **21**: 2914–16.

Chapter 9

The Impact of Protein Expression Methodologies on Structural Proteomics

A. Chesneau, H. Yumerefendi and D. J. Hart*

Introduction

Structural proteomics and genomics efforts have greatly increased the number of protein structures in the protein database, accounting for 44% of the total number of novel structures reported in 2005.[1] Key to this success are developments in molecular biology methods for DNA cloning and protein expression that permit the generation and testing of genetic constructs in a high-throughput manner. In contrast to traditional efforts, these experiments are performed in a parallel format, generally that of 96-well plates, and frequently employ automation.

Many classes of targets have been processed through structural proteomics workflows and the well-known observation that some types of protein are more difficult to express than others has become quantifiable. For example, an analysis of the Structural Proteomics in Europe (SPINE) target database after three years of efforts reveals that the efficiencies of soluble protein expression from cloned DNA inserts were: 52% from the bacterial genus *Bacillus*; 30% for human

*Corresponding author: hart@embl.fr

EMBL Grenoble Outstation, 6 rue Jules Horowitz, BP181, 38042 Grenoble Cedex 9, France.

targets; and 11% for those from herpesviruses. These figures are averaged over many processed targets but clearly illustrate that: (i) some target classes are more difficult to express than others; and (ii) that a major problem is obtaining solubly expressing constructs. It is of interest to consider why such apparent differences exist and whether they reflect a difference in the proteins themselves. Bacterial proteins are often smaller than those found in eucaryotes and viruses. When this is the case, construct design is simpler as they can be cloned full-length for initial expression testing and it is not necessary to predict domain boundaries. Should it be necessary to make subconstructs, there are often abundant homologues in the sequence databases, or even solved structures, permitting accurate domain boundary definition. Folding of bacterial proteins in recombinant *E. coli* expression systems is also likely to be more efficient than eucaryotic targets since the chaperones and the cellular environment are similar to that of the original organism. However, an additional, often overlooked explanation for the apparently high efficiency of bacterial protein expression is that these targets are often selected for technology development phases of structural proteomics projects where the aim is to process large numbers of targets through a workflow. In addition to providing good test material, stringent target selection procedures are used to eliminate potentially difficult proteins from the start list. Once a structure pipeline is established using "easy" proteins, it is often applied to difficult targets such as those implicated in health and disease. Here it is the importance of the targets rather than the probability of success that matters and cloning of full-length genes seldom yields well-behaving soluble protein directly. In these cases, it is usually necessary to design and test multiple subconstructs of the target via iterative cycles of domain boundary hypothesis, cloning and testing (Fig. 1). The high-throughput molecular biology and automation platforms are therefore focused intensively on these difficult targets, but the basic strategy of making lots of constructs and testing them for expression remains the same as for projects seeking large scale coverage of diverse targets.

In this chapter, we will review the methods developed within cloning and expression "pipelines," primarily those using *Escherichia coli*

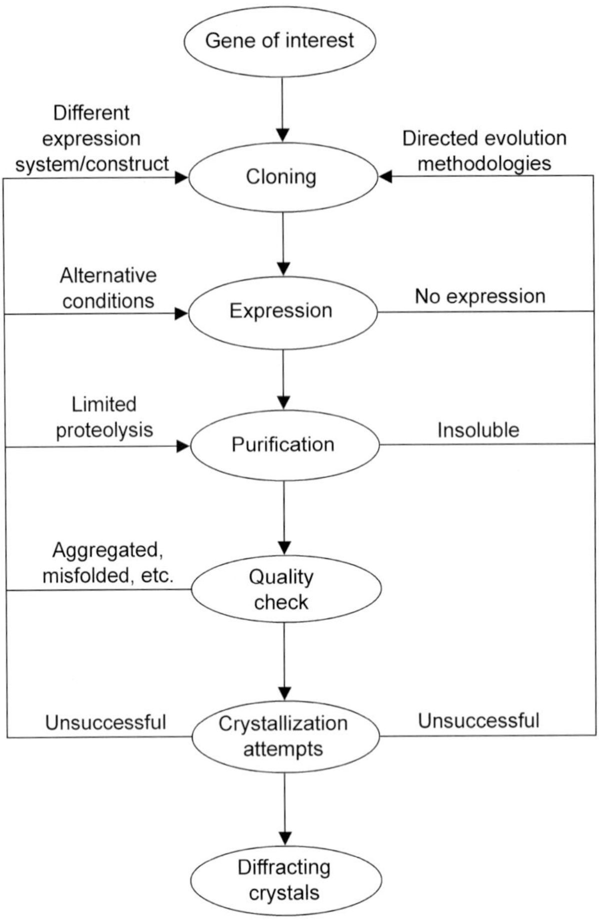

Fig. 1 Schematic of the basic structural proteomics process showing the linear nature of the workflow and approaches for addressing difficult targets. On the left are classical approaches and on the right, directed evolution methodologies such as library-based construct screening and point mutangenesis.

as the expression host, since this remains the main workhorse for most efforts. These comprise robust, high-throughput approaches for cloning target genes and protein expression testing. Firstly, there are genetically encoded parameters in which the DNA sequence of the expression system is altered and are either vector specific (e.g. fusion tags,

promoter type), or target gene specific (e.g. truncated constructs or point mutations). These require cloning steps and have benefited from the development of faster DNA manipulation methods. Environmental variables such as strain, growth temperature and media composition may further improve the yield, solubility and stability of the protein, and are faster and cheaper to test than those that require cloning steps. However, it is generally the case that these will not rescue poorly designed constructs. Therefore, most labs look for a compromise between the two types of variables, aiming to test as many plasmid constructs as possible in a limited matrix of strains and induction conditions. Whilst no universal workflow has yet emerged, protocols from different labs are broadly similar and some benchmarking studies have attempted to draw conclusions about optimal protocols.[2] Finally, new library-based methods drawing from the field of directed evolution will be discussed. These have expanded the multiconstruct and mutational approaches to very large clone numbers and introduced the concepts of random library construction and screening to the structural pipeline. Early successes have demonstrated that this approach can sometimes rescue targets that resist soluble expression using the standard methods.

Genetic Parameters

Vector Properties

As with any cloning experiment, one of the initial decisions to be made is the choice of vector. The repertoire of commercially available and homemade expression vectors is enormous and a comprehensive description is beyond the scope of this review. However, the parameters to consider are similar to those for any expression project and include fusion tags, origin of replication for compatibility with additional vectors, antibiotic resistance marker and cloning strategy.[3]

The high-throughput nature of structural proteomics imposes additional requirements on vector systems, beyond those of a standard laboratory. In some labs, the number of vectors employed is minimal, with only one or two vectors used for all constructs. The plasmid

pSGC-LIC1 (Fig. 2a), developed by the Structural Genomics Consortium (SGC), is a fairly typical example of a high-throughput protein expression vector and contains a number of elements that are commonly employed for efficient parallelization of expression experiments.

a) The T7 expression system[4] utilizes a strong T7 phage promoter to drive transcription via a chromosomally-encoded T7 RNA polymerase. A wide range of BL21 *E. coli* strains are available with this feature and screening of an expression vector in multiple strains can improve yield and soluble expression (discussed below). The use of an IPTG controllable promoter to control the expression of the chromosomal T7 RNA polymerase, and *lac* operator in the vector T7 promoter, ensures that transcription is tightly regulated, and adds the additional option of using auto-induction media.[5] This carefully formulated growth media exploits the effect of catabolite repression of gene expression in the presence of glucose that is gradually metabolized by the replicating cells and ensures that when multiple clones are grown in parallel in multiwell plates, induction of protein expression occurs at similar points in the growth curve of each culture, even if there is significant difference in the growth rates.
b) Like many such vectors, pLIC-SGC1 employs a system for efficient, directional cloning of gene inserts without use of restriction enzymes. Here it is ligation-independent cloning (LIC)[6] that, in common with other sequence-independent cloning strategies (discussed below), ensures that the cloning strategy is universal for all targets. By contrast, classical cloning strategies using restriction enzymes require that additional restriction sites within the gene be silenced by site-directed mutagenesis, limiting their utility in parallel experiments on different open reading frames. These systems have the additional advantage that they require fewer handling steps than traditional manual cloning, e.g. gel purification and buffer exchange steps. They can be configured as serial steps of liquid additions and incubations with few changes of reaction container; thus, they are relatively easy to automate using a liquid-handling robot.

(A)

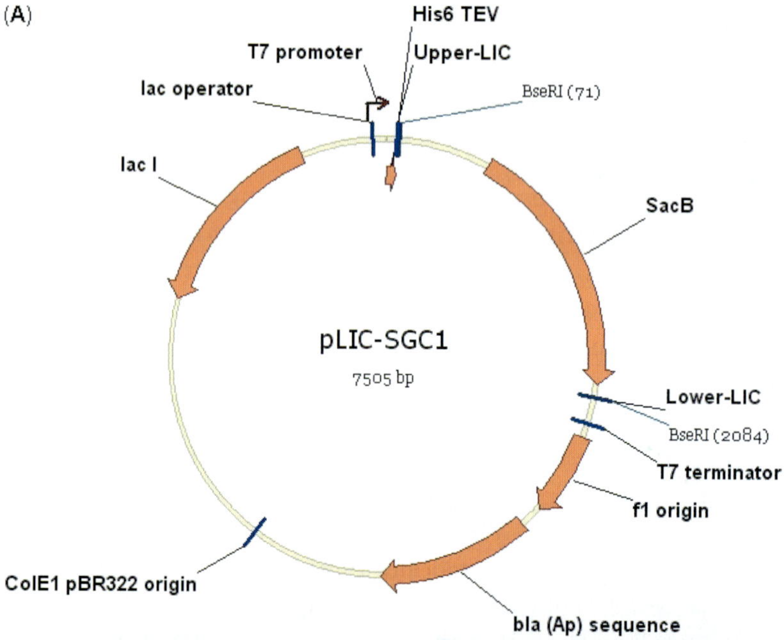

(B)

Fig. 2 (**A**) *Above:* The high-throughput procaryotic cloning vector pLIC-SGCI developed by the Structural Genomics Consortium for ligation-independent cloning (LIC) of target genes. Specific vector features discussed in text; (**B**) *Below:* Theonyx liquid-handling robot configured for automated molecular biology. The PCR blocks are visible in the foreground, temperature-controlled shaker for 96-well plates behind, then sample trays and liquid-dispensing fixed tips.

(C)

(D)

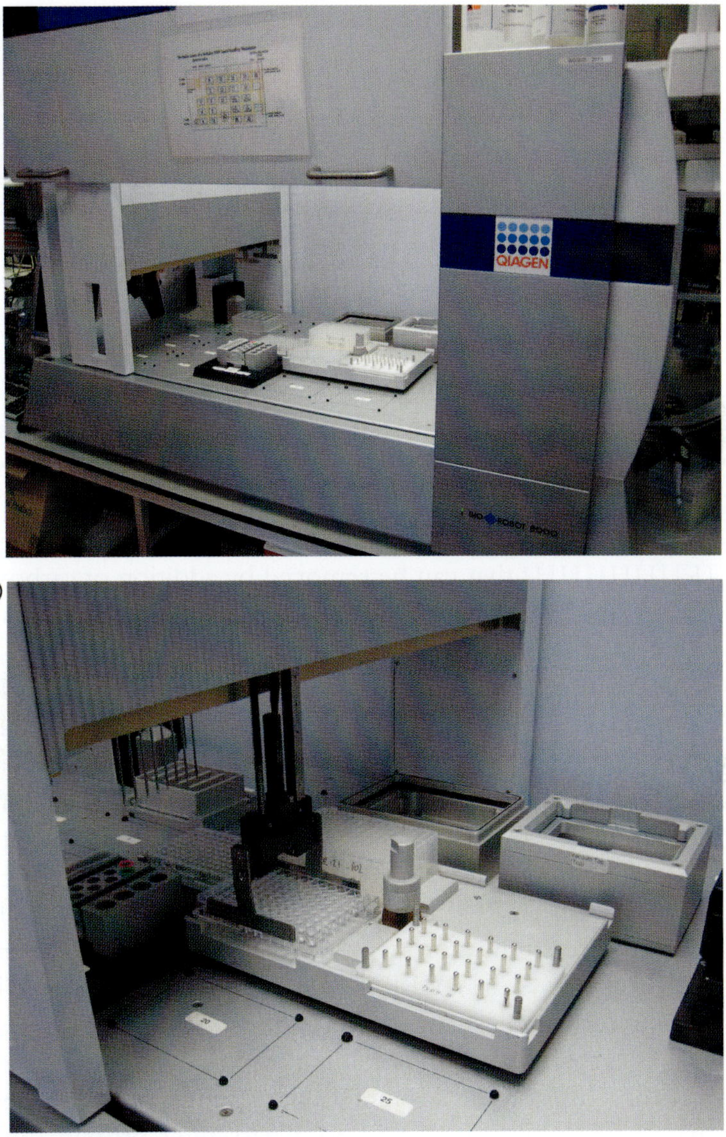

Fig. 2 (*Continued*) (**C**) *Above:* Qiagen Biorobot 8000 commonly used for plasmid minipreps and DNA cleanup steps. Here, it is configured for Ni^{2+} NTA purification of hexahistidine-tagged proteins using magnetic beads; (**D**) *Below:* Details of Biorobot showing magnetic rods for separation of magnetic beads from liquids in 96-well format, and plate-handling arm. (Pictures of robotics courtesy of the Oxford Protein Production Facility.)

c) An affinity purification tag is present to permit rapid purification testing and protein production after scale-up. Here it is an N-terminal hexahistidine tag allowing immobilized metal affinity chromatography (IMAC). One advantage of N-terminal tags is that they can be cleaved from the purified protein using specific proteases due to the C-terminal location of protease cleavage sites relative to the protease recognition sequence. In this vector, the tobacco etch virus (TEV) cleavage site is engineered between the hexahistidine tag and the target, leaving only a single glycine or serine residue after removal. TEV is a popular choice as it is easy to make in the laboratory and therefore inexpensive to use in large amounts. Affinity tags on the C-terminus of a construct are useful in specific cases, but due to the nature of protease cleavage, the recognition site remains fused to the protein afterwards. A solution to this is the use of the C-terminal affinity tag (KHHHHHH) that can be removed by addition of carboxypeptidase Y to hydrolyze the histidines back to the lysine.[7] In addition to facilitating purification, some tags appear to have strong solubilizing effects on the passenger protein. Commonly used are thioredoxin (Trx), hexahistidine (His6), decahistidine (His10), green fluorescent protein (GFP), NusA, glutathione-S-transferase (GST) and others. In a study of N-terminally tagged mammalian protein targets expressed in *E. coli*, enhanced solubility appeared in the following order: His10-Trx = His10-MBP >His10 >His10-GST >His10-GFP.[8] In another study on a set of small and medium sized human proteins, the enhancement of solubility was ranked as Trx >Gb1/MBP >ZZ >NusA >GST >His6.[9] Whilst a comparison of these two studies and others may suggest, for example, that Trx exhibits solubilization advantages, there is no real consensus on the advantages of tagging. It is often observed that the solubility imparted to a passenger protein does not persist after tag removal or that tag-solubilized constructs resist cleavage, suggestive of misfolding. Often these tagged fusion proteins express well, but form large, soluble aggregates. Therefore, in some labs, the strategy is to keep the tagging simple (e.g. TEV-cleavable His6, both N-ter and C-ter variants), and if the construct is not soluble, to change the boundaries of the insert.

d) This vector employs β-lactamase as the resistance marker, conferring resistance to ampicillin and carbenicillin. Often, kanamycin is preferred over ampicillin as it reduces the need to streak out frequently clones on selective agar. This is because β-lactamase (conferring resistance to ampicillin) is a secreted enzyme and allows growth of the neighboring cells that have lost their plasmid (e.g. forming satellite colonies), whilst the kanamycin-inactivating aminoglycoside 3′-phosphotransferase is cytoplasmic, so only those cells containing the vector will replicate.

e) It contains a negative selectable marker that can be used to eliminate parental vectors that contaminate the product of the cloning reaction. Here, *sacB* encoding levansucrase ensures lethality to cells containing the parent vector when the transformation mix is plated out on agar containing sucrose, due to the toxicity of fructose released upon sucrose hydrolysis.[10] The *ccdB* protein, an *E. coli* DNA gyrase inhibitor (11) is also used for the same purpose in Gateway vectors (Invitrogen).

Whilst many labs have built their own vector systems, the common theme is that they facilitate a simple, robust cloning procedure for PCR products.

High-throughput Cloning Strategies

Constructs are generally PCR-amplified using proofreading polymerases such as *Pfu*, *Pwo* or KOD in 96-well plate format. In the next step, it is necessary to clone the insert into the expression vector as reliably and economically as possible. There are a number of possible approaches that are commonly employed:

Restriction enzyme cloning

Some labs, especially those with a medium throughput, still prefer to use restriction enzyme cutting and ligation with T4 DNA ligase to make clones. The use of rare cutting enzymes, where possible, reduces the chance of internal sites cutting the target ORF. Using this traditional approach, a flexible system can be created to generate an array of combinations of insert and vector for rapid expression screening. The Flexi cloning system, based on a directional cloning using two rare-cutting endonucleases, permits shuttling of the protein coding region between vectors. It was compared with the Gateway recombinational system,[12] showing that it performed as well or better and that it could easily be adapted to a high-throughput cloning platform. A similar strategy is used by the Northeast Structural genomics consortium using restriction sites engineered into a collection of pET vectors compatible with automated cloning.[13] Efficient cloning with *Sfi*I has also been described, permitting directional cloning with a single enzyme that varies in the sequence of the overhangs in a user-defined way.[14]

Site-specific recombinational cloning

As an alternative to restriction enzymes, a number of systems have been developed to exploit site-specific recombination, including Gateway (Invitrogen), In-Fusion (BD Clontech) and Cre-lox.[15] Gateway is an efficient cloning technology derived from the natural bacteriophage λ DNA integration system.[16] Gateway vectors contain, between the recombinational *att* sites, the *ccdB* gene that is lethal to bacterial strains (except DB3.1) and provides positive selection after recombination.[11] The recombination reaction enables easy transfer of a sequence-verified master clone to a series of destination vectors. The Gateway system is very efficient and relatively insensitive to DNA concentration, making it easily automatable. A potential disadvantage of Gateway is that it usually adds extra amino acids between the ORF and fused tags due to translation of recombination sites which may reduce the solubility of the target protein. There is also a marked decrease in efficiency for inserts longer than 3 kb. Moreover, this cloning approach is costly and dependent on one supplier for reagents and enzymes. Anecdotally, it is possible to dilute the enzymes significantly with little loss in efficiency and this goes some way towards reducing the cost. Many Gateway-compatible vectors have been custom-engineered by users for particular applications.[17] In one impressive example of use, an automated Gateway cloning pipeline was developed to clone and express 10 167 *C. elegans* genes, corresponding to half the predicted ORFeome.[18] They used a single expression vector, pET15G with a hexahistidine tag and one *E. coli* strain. Protein expression was detected for almost 50% of the genes of which 1536 showed some solubility (15%).

The In-Fusion™ cloning method was originally developed by Clontech to produce a common entry vector for a multi-vector cloning system based on Cre-lox recombination. The recombination reaction is mediated via 15 bp sequences flanking the insert that are also present at the termini of the linearized cloning vector. In common with the Gateway system, at least one of these recombination sites must be translated, although a major advantage of In-Fusion is that the site can actually encode the affinity tag or protease cleavage site, i.e. no additional amino acids are added. This method was adapted by workers at the Oxford Protein Production Facility (OPPF)[19] and elsewhere[20] for direct cloning of PCR products into protein expression vectors. The cloning efficiency observed at the OPPF (94% of 703 PCR products) was higher than that of Gateway (79%).[19]

Ligation Independent Cloning

Ligation-independent cloning (LIC) was developed for the directional cloning of PCR products without restriction enzyme and ligase reactions.[6] The LIC method takes advantage of the 3′- to 5′-exonuclease activity of T4 DNA polymerase to generate, in the presence of a single dNTP, precise 12- to 15-base single-stranded overhangs. Both vector and insert are treated in this way and the resulting complementary ends spontaneously hybridize to generate a nicked circular plasmid that is repaired *in vivo* following transformation. The minimal number of experimental steps and low cost make this method appealing and it is becoming one of the most common systems, employed by the SGC (see vector pLIC-SGC1; Fig. 2a) and several SPINE laboratories, where success rates of 85% or higher are observed.[17]

Enzyme free cloning

Using two pairs of primers per target gene in two parallel PCR reactions, one pair of which adds additional flanking sequence, complementary staggered overhangs can be generated in 25% of molecules via a post-PCR denaturation–hybridization reaction.[21] The low cost of primers means that this approach has become economically feasible and has been used to generate an efficient cloning pipeline.[22] Although 75% of the insert molecules cannot be cloned, this method is still highly efficient and sufficient numbers of transformants are easily obtained.

Environmental Factors and Expression Strains

Once expression plasmids have been constructed and *E. coli* strains transformed, a test can be performed to determine whether, under an initial set of parameters, protein can be expressed. It is necessary to design such an expression matrix carefully as the number of samples can rapidly expand to impractical numbers. For example, if four different constructs were assayed (e.g. different boundaries or solubilizing tags) in three strains at two different temperatures, with both ITPG and autoinduction media, this would generate 48 separate cultures to process and analyze. Most expression trials are performed in 24- or 96-well plates, so this is feasible, but adding a greater range of temperatures or more strains could prove overwhelming if multiple targets are being studied. It is also debatable whether expanding the expression matrix beyond a set of basic conditions really helps, or whether it is case of diminishing returns. However, in some cases, a change in environmental factors can enhance the yield or alter the solubility of a target.

When targets refuse to express solubly, the options are either to design new boundaries, screen constructs extensively using a library strategy (discussed below), or change the expression host. A number of eukaryotic systems are available, such as insect and mammalian cells, and yeast (*Saccharomyces cerevisiae* and *Pichia pastoris*). These are reviewed elsewhere[23] and will not be discussed in detail here; it is worth noting that they provide the additional advantages of posttranslational modification (e.g. glycosylation), secretion, cellular compartmentalization and eukaryotic chaperones. Due to the lower throughput achievable with these systems, they are typically used to rescue valuable targets, and not as a primary expression host unless a particular target class is known *a priori* to conform well, e.g. human cell surface receptors or secreted proteins.[23] Cell-free protein synthesis is another option that is particularly suitable for stable-isotope labeling of proteins for NMR analysis and has been extensively employed by the RIKEN Genomic Sciences Center.[24]

Temperature and Media

Reduction of temperature is a well-known technique to improve the expression of recombinant proteins as it can improve solubility and decrease aggregation. It is therefore a common parameter included in the screening matrices of structural proteomics platforms, e.g. those of the Berlin Protein Structure Factory or the Marseille Structural Genomics platform.[2] Use of promoters for expression at low temperatures and cold induced chaperones can be useful for producing toxic and proteolytically sensitive gene products.[25]

In certain cases, when screening different constructs of a particular protein, it may also be worth investigating the addition of compounds or cofactors to the growth medium, either those with a general effect, e.g. sorbitol, ethanol, NaCl, or specific effect, e.g. staurosporin with some kinases. Many high-throughput platforms use enriched and buffered media such as TB, 2YT or GS96 to ensure the maximum biomass.[17] This compensates for the small size of the expression cultures and permits optical densities of 5–10 as compared with 2–3 for the standard LB medium. Another well-favored broth is the autoinduction medium described by Studier[5] that ensures the cells in different wells of a multiwell plate are induced at the same point in their growth phase (and not at a defined time point as would be the case when using IPTG). Independent of the media used, high-density cultures require high-speed agitation of the deep-well plates to ensure good aeration. A number of commercial incubators are available that achieve this through employing a rotational pitch with the same diameter as the wells of the multiwell plate, e.g. the HiGro by GeneMachines.

Escherichia coli Strains

E. coli is the most widely used organism for recombinant protein production and is well-suited to high-throughput production due to its fast growth, well-established recombinant genetics and low costs. The disadvantages are limits on the size of the expressed proteins, codon bias and absence of posttranslational modifications. The most common strains used in structural proteomics platforms are the protease deficient BL21 (DE3) strain and derivatives such as BL21 (DE3) pLysS, C41 (DE3) for toxic proteins, and Rosetta.[2] These are compatible with the strong T7 promoter system and autoinduction medium (discussed above).

Low expression levels can be due to rare codons within the target ORF and a common approach is to supplement the strain with a plasmid expressing genes for rare tRNAs. BL21 Codon-Plus(DE3)-RIL (Stratagene) is one such strain that compensates for the rare codons AGA, AGG (Arg), AUA (Ile) and CUA (Leu). BL21(DE3)-Rosetta (Novagen) is another commonly used strain that supplements additional tRNAs. Another effective way of addressing this issue is to synthesize the target gene fully from overlapping oligonucleotides employing only codons preferred by *E. coli*.[26]

Other useful strains to consider for use in expression screens include: B834, a methionine-auxotrophic version of BL21(DE3) that permits efficient selenomethionine labeling of proteins[27]; strains that overexpress molecular chaperones such as those available commercially from Takara; the Origami strain (Novagen) for folding of cysteine-rich proteins[28]; and Tuner (Novagen) for regulation of expression levels by IPTG concentration.

Protein Expression Testing

Typically, small-scale expression screens are performed in 96- or 24-deep-well plates in volumes of 1–4 ml of rich media to ensure a maximal biomass. Expression is induced with ITPG, arabinose or autoinducible medium depending on the strain and promoter system being used.[5] A number of methods to lyze the pelleted cells are available, including sonication, freeze-thaw cycles with lysozyme or chemical lysis. Sonication is used successfully by some labs but requires dedicated instrumentation, often homemade, to handle the multiwell format. Many commercially available chemical lysis reagents are available, including B-Per (Pierce) and Bugbuster (Novagen). These are convenient to use, but are occasionally observed to perturb the solubility of some proteins. A common, inexpensive and effective high-throughput protocol uses a classical freeze-thaw alternation in addition to lysozyme and DNase for lysis.

In the next step, the lysates are tested for the presence of soluble protein. The simplest methods involve lysate fractionation, either by centrifugation or filtration plate[29] and observation of total and soluble fractions by SDS-PAGE. Where a hexahistidine tag is present, solubility and purifiability can be determined using Ni^{2+} NTA plates available from several commercial suppliers, or with small amounts (25 μl) of Ni^{2+} NTA agarose dispensed into support plates.[30] Imidazole-eluted, purified fractions can be analyzed by SDS-PAGE where a common success criterion is visibility by Coomassie blue stain. SDS-PAGE of hundreds of samples and even blotting for western blot analysis is simple with the availability of precast gels from many suppliers, or systems such as the Hoefer SE600 Ruby System that permit pouring of multi-lane gels of user-defined composition.

Some groups have dispensed with SDS-PAGE analysis for initial screening and have combined filtration protocols with dot blot.[31] This is highly automatable and capable of a very high throughput, although the information is less detailed in terms of size and condition of the target protein.

Use of Automation

Many labs have embarked upon structural proteomics without the use of robotics and it is often argued that they are not necessary for the majority of "focused" and "medium-throughput" projects. Indeed, a great deal of progress can be made by simply working in multiwell plate format with multichannel pipettes and a vacuum manifold or plate centrifuge. However, there are clear advantages when processing very large numbers of samples and, while not essential in smaller labs, access to simple robotic protocols makes experiments less tedious and reduces human error. The standard instruments used are liquid-handling robots that are available from a number of manufacturers. They generally comprise a flat bed upon which are mounted various functional units such as heater blocks, sample holders, vacuum manifolds and shakers. Over the bed move one or more arms that carry pipetting heads and plate grippers between the various units on the bed (Figs. 2b–d). Syringe-driven pipetting units aspirate and dispense in the 1 to 1000 microliter range and comprise fixed steel washable pins or ejectable plastic tips, usually with eight separate channels. For accurate aspiration and dispensing, conductive tips and electronics permit sensing of the meniscus (i.e. liquid height) of a sample, and monitoring of drop dispensation. Many molecular biology protocols can be automated on these units, although it is advantageous to minimize the number of steps and replace centrifugation (difficult on a robot) with filtration protocols.

A number of manufacturers supply kits for various applications such as DNA or protein purification and are often willing to supply prewritten programs for common robots. Some, like Qiagen, manufacture dedicated robots that are compatible only with their kits (Figs. 2c–d). These are quite inflexible as they are not intended to be programmed by users, and therefore have limited use for other applications. On the positive side, teams of application specialists will have spent much time writing and testing the programs to ensure good reliability and compatibility with their kits. There is also the advantage of reduced setup time as it is not necessary to learn how to write the programs. If a lab decides that the "plug-and-play" solution is too limited, or the kits are too expensive, there are numerous generic robot protocols that work effectively. For example, purification of hexahistidine tagged proteins can be performed equally well on robots with inexpensive Ni^{2+}NTA agarose dispensed into reusable filter plates[30] as with kits costing five times the price.

Ultimately the choice between a bespoke and an off-the-peg robot will depend on the aim of the lab and especially on whether the robot can be dedicated to a single function or is expected to multitask. Note though, that whilst multitasking sounds attractive, it often involves substantial rearrangement of the functional units on the robot bed which is time-consuming and carries the risk of damage due to collision if an error is made.

Library-based Strategies for Soluble Protein Expression

The design of expression constructs is usually performed using sequence alignments, bioinformatic tools such as order and disorder predictors[32] or limited proteolysis data from full-length proteins. Whilst these approaches are quite effective, many targets remain refractory. Sometimes this can be explained by the low quality of the starting information used in the construct design; for example, if there are no homologs in the sequence databases, alignments are not possible. Many viral proteins fall into this category due to their rapid evolution and uncertain ancestry. However, even when good quality information is available, constructs often do not express soluble protein. Faced with such difficulties, several academic and commercial groups have developed approaches that remove the requirement for rational definition and introduce the concept of random screening into construct design (Fig. 1).[33] Using the protein engineering strategy of directed evolution,[34] a library of random genetic constructs is first synthesized. High-throughput screening, often employing robotics, is used to identify the rare clones expressing the desired phenotype. Removing the process of rational design and emulating the natural process of evolution by genetic mutation and selection means that alternative constructs or mutants are tested as compared to those that would have been designed rationally, often leading to solutions that were unpredictable *a priori*.[35]

In structural biology, the desired protein phenotypes are high yield, solubility, protease resistance and crystallizability. To date, no direct genetic screen has been described for the latter, although in one paper it was described how, by selecting for maintenance of enzymatic activity under mutational pressure, surface residue substitutions can be generated that may influence crystallization properties.[36] Several general screening strategies for soluble protein expression have been described,[37] the underlying assumption being that, when a protein is soluble, it is folded correctly. Additionally, where there is an enzymatic activity that can easily be measured, this can be used to screen for solubility and native folding.[38] Such an example is the directed evolution of soluble forms of serum paroxonases (PONS)[39] via family shuffling[40] in which three different homologues were recombined into soluble chimeric variants exhibiting high levels of activity. The soluble construct was then crystallized, allowing structure solution.[41]

When planning directed evolution approaches, it is necessary to consider the type of genetic mutation strategy most likely to yield the desired phenotype. Possible methods include mutagenesis by point mutation or shuffling methods, or modification of the construct ends, either by truncation (where one end of the construct is varied) or fragmentation (both ends simultaneously).[42]

Improvement of soluble expression of difficult targets has been achieved by point mutation. An example was the evolution of highly soluble green fluorescent protein (GFP)[43] by DNA shuffling, an efficient process by which random point mutagenesis is combined with *in vitro* recombination.[44] GFP provides a simple screen for improved soluble expression as its phenotype can be monitored by observation of bacterial colonies, although few proteins are this simple to analyze. This concept was extended further to the monitoring of fusion proteins of a randomized target fusion with GFP,[45] in which several proteins have had their solubility improved by random point mutagenesis.[46] A further application of this approach was the evolution of more soluble variants of TEV protease,[47] a commonly used enzyme for cleavage of affinity tags from recombinant proteins. This was mutated by error-prone PCR and gene shuffling to yield a 5-fold more soluble variant containing 3 mutated residues. An interesting aspect of this work was the use of flow cytometry to select cells with improved green fluorescence (and therefore, solubility). Another example is the solubilization of the kinase domain from the human EphB2 receptor.[48]

Here the target was fused to the reef coral fluorescent protein, ZsGreen, and a fluorescent colony picker used to isolate clones exhibiting increased solubility. Each round of mutation, screening and analysis of 20 000 potential soluble clones was performed in eight weeks, and more soluble kinase variants were identified.

In contrast with the point mutation strategies described above, the more conventional method of improving expression is to reclone alternative constructs. This is most effective when working on multidomain proteins, or when proteins contain predicted disordered regions leading to aggregation and misfolding. The mutation strategies that most closely conform to cloning of subconstructs are gene truncation or fragmentation as these lead to sub full-length gene fragments. Unlike parallel PCR cloning, constructs are generated "all-in-one-pot" from the degraded DNA and then clonally separated by colony picking post transformation. The degradation of the target gene can be enzymatic, including the use of exonuclease III,[49] DNAseI[50] or uracil-DNA glycosylase.[51] A PCR method using random priming[52] has been described, and physical breakage of the target DNA employing methods such as sonication[53] or point sink shearing[54] is also effective. These various methods are all relatively sequence-independent in their cutting positions and fragment sizes can be selected by electrophoresis to fall within a desired range.

A feature of screening randomly is that the vast majority of genetic clones are junk; therefore, a powerful system is required to find the needle in the haystack. Thus, a directed evolution experiment comprises two components: a genetic mutation strategy coupled to a screening or selection process. A number of screens and selections have been demonstrated for the production of soluble constructs. Probably the best known are those that employ C-terminal fluorescent protein fusions (described above). A recent improvement of this system has been reported to overcome false-positives due to passenger solubilization by the highly soluble GFP domain. Here, the mutated target protein was fused to a single short beta-strand of GFP and expression of a soluble target-GFP peptide fusion complements a coexpressed, non-fluorescent, truncated form of the full protein, resulting in strong fluorescence.[55] Two membrane-based methods have been described for assaying soluble protein expression in colony format after plating or robotic gridding of libraries. The CoFi blot method[56] relies on a filter sandwich of a porous Durapore membrane over a nitrocellulose. Following transfer of colonies and lysis under native conditions, soluble members of the library diffuse through the Durapore membrane onto the underlying nitrocellulose membrane, where detection is performed using an antibody against the hexahistidine tag. This has been demonstrated as effective on a series of previously intractable human targets[57] that were randomly 5′-truncated using exonuclease III. The ESPRIT method[58] relies on *in vivo* biotinylation of soluble target proteins by the endogenous biotin ligase enzyme of *E. coli* during colony growth and protein expression. This was used successfully to express and characterize structurally a previously unsuspected domain from influenza polymerase, employing exonuclease III to generate the construct library. Both the CoFi and ESPRIT methods benefit from using only small terminal epitopes for detection, thereby avoiding passenger solubilization artefacts inherent in folded reporter proteins such as GFP.

Library-based methods are limited in their throughput due to the manually intensive nature of the experiment. However, they have been effective in expressing proteins that fail in standard structural proteomics workflows. Thus, they provide an important complementary "rescue" strategy for high value targets and early successes suggest they will contribute significantly as structural proteomics is applied in a focused manner to more difficult targets, such as those implicated in human health and disease.

Conclusion

Structural proteomics and genomics have made much progress over the last decade, and a fairly well defined set of protein expression technologies have emerged that can be variously assembled to form a so-called "pipeline." These systems comprise rapid and efficient methods of plasmid construction, often avoiding traditional cloning with restriction enzymes, combined with strategies for screening various environmental factors for improvements in yield and solubility. *E. coli* has emerged as the primary expression host due to factors of cost and ease of genetic manipulation. Both genetic and environmental factors are tested in order to define the permutation of construct, strain and expression condition that yield the milligram quantities of protein normally required for crystallization trials. Parallelization of work in multiwell plates is a universal approach and this lends itself well to automation using liquid handling robotics. Targets may yield rapidly when screened through an expression matrix, but when they prove refractory, a series of rescue strategies has been developed, such as eukaryotic expression hosts and directed evolution methodologies. Although a recent development, the latter has yielded several impressive results demonstrating that point mutagenesis or DNA fragmentation, coupled with high-throughput screening, can permit soluble expression of otherwise intractable targets.

Acknowledgments

A.C. was funded by the European Commission Framework 6 Integrated Contract "3D Repertoire" (LSHG-CT-2005-512028) and H.Y. by an AstraZeneca fellowship.

References

1. Chandonia JM, Brenner SE. (2006) "The impact of structural genomics: expectations and outcomes." *Science* **311**(5759): 347–51.
2. Berrow NS, Bussow K, Coutard B, *et al.* (2006) "Recombinant protein expression and solubility screening in *Escherichia coli*: a comparative study." *Acta Crystallogr D Biol Crystallogr* **62**(Pt 10): 1218–26.
3. Hartley JL. (2006) "Cloning technologies for protein expression and purification." *Curr Opin Biotechnol* **17**(4): 359–66.
4. Moffatt BA, Studier FW. (1987) "T7 lysozyme inhibits transcription by T7 RNA polymerase." *Cell* **49**(2): 221–7.
5. Studier FW. (2005) "Protein production by auto-induction in high density shaking cultures." *Protein Expr Purif* **41**(1): 207–34.
6. Aslanidis C, de Jong PJ. (1990) "Ligation-independent cloning of PCR products (LIC-PCR)." *Nucl Acids Res* **18**(20): 6069–74.

7. Hayashi R, Moore S, Stein WH. (1973) "Carboxypeptidase from yeast. Large scale preparation and the application to COOH-terminal analysis of peptides and proteins." *J Biol Chem* **248**(7): 2296–302.
8. Dyson MR, Shadbolt SP, Vincent KJ, *et al.* (2004) "Production of soluble mammalian proteins in *Escherichia coli*: identification of protein features that correlate with successful expression." *BMC Biotechnol* **4**: 32.
9. Hammarstrom M, Hellgren N, van Den Berg S, *et al.* (2002) "Rapid screening for improved solubility of small human proteins produced as fusion proteins in *Escherichia coli.*" *Protein Sci* **11**(2): 313–21.
10. Bramucci MG, Nagarajan V. (1996) "Direct selection of cloned DNA in Bacillus subtilis based on sucrose-induced lethality." *Appl Environ Microbiol* **62**(11): 3948–53.
11. Bernard P. (1996) "Positive selection of recombinant DNA by CcdB." *Biotechniques* **21**(2): 320–3.
12. Blommel PG, Martin PA, Wrobel RL, *et al.* (2006) "High efficiency single step production of expression plasmids from cDNA clones using the Flexi Vector cloning system." *Protein Expr Purif* **47**(2): 562–70.
13. Acton TB, Gunsalus KC, Xiao R, *et al.* (2005) "Robotic cloning and protein production platform of the northeast structural genomics consortium." *Methods Enzymol* **394**: 210–43.
14. Pengelley SC, Chapman DC, Abbott MW, *et al.* (2006) "A suite of parallel vectors for baculovirus expression." *Protein Expr Purif* **48**(2): 173–81.
15. Liu Q, Li MZ, Leibham D, *et al.* (1998) "The univector plasmid-fusion system, a method for rapid construction of recombinant DNA without restriction enzymes." *Curr Biol* **8**(24): 1300–9.
16. Walhout AJ, Temple GF, Brasch MA, *et al.* (2000) "GATEWAY recombinational cloning: application to the cloning of large numbers of open reading frames or ORFeomes." *Methods Enzymol* **328**: 575–92.
17. Alzari PM, Berglund H, Berrow NS, *et al.* (2006) "Implementation of semi-automated cloning and prokaryotic expression screening: the impact of SPINE." *Acta Crystallogr D Biol Crystallogr* **62**(Pt 10): 1103–13.
18. Luan CH, Qiu S, Finley JB, *et al.* (2004) "High-throughput expression of C. elegans proteins." *Genome Res* **14**(10B): 2102–10.
19. Berrow NS, Alderton D, Sainsbury S, *et al.* (2007) "A versatile ligation-independent cloning method suitable for high-throughput expression screening applications." *Nucl Acids Res* **35**(6): e45.
20. Benoit RM, Wilhelm RN, Scherer-Becker D, Ostermeier C. (2006) "An improved method for fast, robust, and seamless integration of DNA fragments into multiple plasmids." *Protein Expr Purif* **45**(1): 66–71.
21. Tillett D, Neilan BA. (1999) "Enzyme-free cloning: a rapid method to clone PCR products independent of vector restriction enzyme sites." *Nucl Acids Res* **27**(19): e26.

22. de Jong RN, Daniels MA, Kaptein R, Folkers GE. (2006) "Enzyme free cloning for high throughput gene cloning and expression." *J Struct Funct Genomics* 7(3–4): 109–18.
23. Aricescu AR, Lu W, Jones EY. (2006) "A time- and cost-efficient system for high-level protein production in mammalian cells." *Acta Crystallogr D Biol Crystallogr* **62**(Pt 10): 1243–50.
24. Yokoyama S. (2003) "Protein expression systems for structural genomics and proteomics." *Curr Opin Chem Biol* 7(1): 39–43.
25. Baneyx F. (1999) "Recombinant protein expression in *Escherichia coli*." *Curr Opin Biotechnol* **10**(5): 411–21.
26. Gustafsson C, Govindarajan S, Minshull J. (2004) "Codon bias and heterologous protein expression." *Trends Biotechnol* **22**(7): 346–53.
27. Hendrickson WA, Horton JR, LeMaster DM. (1990) "Selenomethionyl proteins produced for analysis by multiwavelength anomalous diffraction (MAD): a vehicle for direct determination of three-dimensional structure." *Embo J* **9**(5): 1665–72.
28. Lauber T, Marx UC, Schulz A, *et al.* (2001) "Accurate disulfide formation in *Escherichia coli*: overexpression and characterization of the first domain (HF6478) of the multiple Kazal-type inhibitor LEKTI." Protein *Expr Purif* **22**(1): 108–12.
29. Knaust RKC, Nordlund P. (2001) "Screening for soluble expression of recombinant proteins in a 96-well format." *Anal Biochem* **297**: 79–85.
30. Scheich C, Sievert V, Bussow K. (2003) "An automated method for high-throughput protein purification applied to a comparison of His-tag and GST-tag affinity chromatography." *BMC Biotechnol* **3**: 12.
31. Vincentelli R, Canaan S, Offant J, *et al.* (2005) "Automated expression and solubility screening of His-tagged proteins in 96-well format." *Anal Biochem* **346**(1): 77–84.
32. Esnouf RM, Hamer R, Sussman JL, *et al.* (2006) "Honing the in silico toolkit for detecting protein disorder." *Acta Crystallogr D Biol Crystallogr* **62**(Pt 10): 1260–6.
33. Hart DJ, Tarendeau F. (2006) "Combinatorial library approaches for improving soluble protein expression in *Escherichia coli*." *Acta Crystallogr D Biol Crystallogr* **62**(Pt 1): 19–26.
34. Tobin MB, Gustafsson C, Huisman GW. (2000) "Directed evolution: the 'rational' basis for 'irrational' design." *Curr Opin Struct Biol* **10**: 421–7.
35. Arnold FH, Wintrode PL, Miyazaki K, Gershenson A. (2001) "How enzymes adapt: lessons from directed evolution." *Trends Biochem Sci* **26**(2): 100–6.
36. Keenan RJ, Siehl DL, Gorton R, Castle LA. (2005) "DNA shuffling as a tool for protein crystallization." *Proc Natl Acad Sci USA* **102**(25): 8887–92.
37. Roodveldt C, Aharoni A, Tawfik DS. (2005) "Directed evolution of proteins for heterologous expression and stability." *Curr Opin Struct Biol* **15**: 50–56.

38. Aharoni A, Griffiths AD, Tawfik DS. (2005) "High-throughput screens and selections of enzyme-encoding genes." *Curr Opin Chem Biol* **9**(2): 210–6.
39. Aharoni A, Gaidukov L, Yagur S, *et al.* (2004) "Directed evolution of mammalian paraoxonases PON1 and PON3 for bacterial expression and catalytic specialization." *Proc Natl Acad Sci USA* **101**(2): 482–7.
40. Crameri A, Raillard SA, Bermudez E, Stemmer WP. (1998) "DNA shuffling of a family of genes from diverse species accelerates directed evolution." *Nature* **391**: 288–91.
41. Harel M, Aharoni A, Gaidukov L, *et al.* (2004) "Structure and evolution of the serum paraoxonase family of detoxifying and anti-atherosclerotic enzymes." *Nat Struct Mol Biol* **11**(5): 412–9.
42. Neylon C. (2004) "Chemical and biochemical strategies for the randomization of protein encoding DNA sequences: library construction methods for directed evolution." *Nucl Acids Res* **32**(4): 1448–59.
43. Crameri A, Whitehorn EA, Stemmer WP. (1996) "Improved green fluorescent protein by molecular evolution using DNA shuffling." *Nat Biotechnol* **14**: 315–9.
44. Stemmer WP. (1994) "Rapid evolution of a protein *in vitro* by DNA shuffling." *Nature* **370**(6488): 389–91.
45. Waldo GS, Standish BM, Berendzen J, Terwilliger TC. (1999) "Rapid protein-folding assay using green fluorescent protein." *Nat Biotechnol* **17**: 691–5.
46. Pedelacq JD, Piltch E, Liong EC, *et al.* (2002) "Engineering soluble proteins for structural genomics." *Nat Biotechnol* **20**: 927–932.
47. van den Berg S, Lofdahl PA, Hard T, Berglund H. (2006) "Improved solubility of TEV protease by directed evolution." *J Biotechnol* **121**(3): 291–8.
48. Heddle C, Mazaleyrat SL. (2007) "Development of a screening platform for directed evolution using the reef coral fluorescent protein ZsGreen as a solubility reporter." *Protein Eng Des Sel* **20**(7): 327–37.
49. Ostermeier M, Lutz S. (2003) "The creation of ITCHY hybrid protein libraries." *Methods Mol Biol* **231**: 129–41.
50. Anderson S. (1981) "Shotgun DNA sequencing using cloned DNase I-generated fragments." *Nucl Acids Res* **9**: 3015–27.
51. Reich S, Puckey LH, Cheetham CL, *et al.* (2006) "Combinatorial Domain Hunting: an effective approach for the identification of soluble protein domains adaptable to high-throughput applications." *Protein Sci* **15**(10): 2356–65.
52. Kawasaki M, Inagaki F. (2001) "Random PCR-based screening for soluble domains using green fluorescent protein." *Biochem Biophys Res Commun* **280**: 842–4.
53. Deininger PL. (1983) "Random subcloning of sonicated DNA: application to shotgun DNA sequence analysis." *Anal Biochem* **129**: 216–23.

54. Oefner PJ, Hunicke-Smith SP, Chiang L, *et al.* (1996) "Efficient random subcloning of DNA sheared in a recirculating point-sink flow system." *Nucl Acids Res* **24**: 3879–86.
55. Cabantous S, Terwilliger TC, Waldo GS. (2004) "Protein tagging and detection with engineered self-assembling fragments of green fluorescent protein." *Nat Biotechnol* **23**: 102–7.
56. Dahlroth SL, Nordlund P, Cornvik T. (2006) "Colony filtration blotting for screening soluble expression in *Escherichia coli.*" *Nat Protoc* **1**(1): 253–8.
57. Cornvik T, Dahlroth SL, Magnusdottir A, *et al.* (2006) "An efficient and generic strategy for producing soluble human proteins and domains in *E. coli* by screening construct libraries." *Proteins* **65**(2): 266–73.
58. Tarendeau F, Boudet J, Guilligay D, *et al.* (2007) "Structure and nuclear import function of the C-terminal domain of influenza virus polymerase PB2 subunit." *Nat Struct Mol Biol* **14**(3): 229–33.

Chapter 10

Protein Complexes Assembly by Multi-Expression in Bacterial and Eukaryotic Hosts

Christophe Romier

Introduction

Several high-throughput (HTP) studies in the recent years have highlighted the importance of macromolecular complexes in forming a large part of the functional entities within eukaryotic cells.[1] This in turn is increasing the need to purify sufficient quantities of these complexes for biochemical and structural studies to understand their structure-function relationships. Purification of macromolecular complexes is rendered more difficult than for single proteins as additional aspects have to be considered. For instance, obtaining a proper stoichiometry of the different subunits is crucial to prevent the use of nonhomogeneous samples. In the case of structural studies, notably at high resolution, another main concern is the yields of purified complexes that may be lower than for single proteins.

Different techniques are available for forming and purifying complexes. These fall into three major groups: *in vitro* reconstitution, co-expression and endogenous complex purification. Some of these have been used for many years, while others have been developed more

---

IGBMC, 1 rue Laurent Fries, B.P. 10142, 67404 Illkirch Cedex. Tel: (+33) (0)3-88-65-57-98; Fax: (+33) (0)3-88-65-32-76; romier@igbmc.u-strasbg.fr

recently. However, most of them are currently being further refined for the study of larger complexes and are being made compatible with HTP approaches. Indeed, in general the larger the complexes, the more complicated the formation and purification schemes. All these techniques show both different positive and negative aspects that make them more suitable for certain studies than others. Thus, it is important when starting a new analysis to consider the pros and the cons of the different methods and to choose the most appropriate one.

In vitro reconstitution, by mixing of subunits purified independently, offers the advantage of allowing the use of constructs of different sizes, thereby helping to remove unstructured regions or independent structural domains that may hamper crystallization or the recording of good quality NMR data. However, the major risks associated with this technique are: 1) production of insoluble proteins that will require refolding procedures; 2) the formation of non productive complex due to soluble but misfolded proteins; and 3) use of the wrong order in mixing the different subunits, which sometimes prevents reconstitution of the full complex. Altogether, these problems render this technique less amenable to HTP automation.

Endogenous complex purification has recently sparked a huge interest, notably through the use of TAP-tagging.[2] This technique clearly provides the most straightforward approach to complex purification, but other difficulties are associated with it. These include: 1) the generation of stable cell lines expressing the tagged protein; 2) selection the right subunit to be tagged to prevent a failure in complex formation; 3) the presence of sub-stoichiometric subunits that create sample heterogeneity; and 4) low abundance of many complexes that will allow structural studies by electron microscopy but not by other methods such as crystallography.

The technique of co-expression may appear as a hybrid between reconstitution and endogenous complex purification as it combines the major advantages of the other two methods. Notably, complexes can be assembled *in vivo* but, if required, from truncated subunits, thus facilitating further structural studies. In the following sections, a detailed overview of the advantages of this technique is presented, followed by a description of the various pitfalls; the pitfalls can be avoided by careful planning of the experiments. Second, different ways of performing multi-expression are described, especially that in *Escherichia coli* (*E. coli*) which remains a host of choice for initial complex formation and purification. Further, co-expression with the baculovirus system and in mammalian cells will also be presented as they provide interesting or even necessary alternatives to *E. coli*. Finally, host-independent strategies for complex formation and purification that can be automated for HTP studies will be discussed.

Advantages of Co-expression

The co-expression method consists in expressing together several macromolecules within a cell and then to co-purify them, generally with only one of them bearing a tag for affinity purification. The rationale behind this kind of experiment is to form a complex between the different subunits of a complex *in vivo* and thus to avoid two major problems associated with the *in vitro* reconstitution technique: the production of insoluble proteins and the formation of unproductive complexes. Insolubility of subunits can be related to the protein/protein interaction surfaces (e.g. with hydrophobic character) that, if unsatisfactory, can cause aggregation and precipitation, although the protein is properly folded. Another major cause of insolubility is the requirement of a partner for proper folding (e.g. by forming a β-sheet), that may lead once again to the formation of inclusion bodies. By co-expressing the right partners together, these problems can be avoided and a co-solubilization effect can be observed as shown in Fig. 1.[3] This effect is a landmark of the co-expression technique and, even where the interacting partners are themselves soluble, can dramatically increase the yields of some or all the subunits.

In vitro, unproductive complexes can also be formed with soluble proteins that are properly folded but have unsatisfactory protein/protein interaction surfaces, or with soluble proteins that are not correctly folded. In this case, mixing these proteins with other macromolecules that are not bona fide partners but have identical properties may lead to the formation of complexes which represent false positives. One would expect that the same phenomenon be observed when these proteins are co-expressed. Actually, studies have shown that the co-expression technique is robust against false positives,[4] which shows

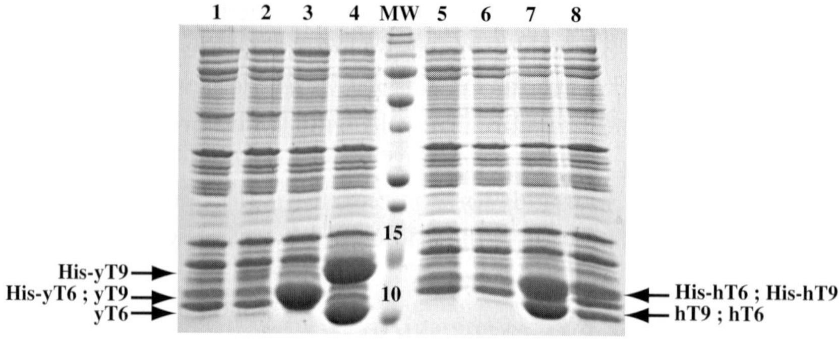

Fig. 1 Co-solubilization and influence of the tag. Coomassie blue stained SDS-PAGE of soluble proteins/complexes obtained after expression and co-expression of yeast (y) and human (h) TAF$_{II}$6 (T6) and TAF$_{II}$9 (T9). All samples represent proteins retained on cobalt beads. Lanes 1, 2, 5, 6: single expression of His-tagged yTAF$_{II}$6, yTAF$_{II}$9, hTAF$_{II}$6, hTAF$_{II}$9, respectively. No soluble protein is observed. Lanes 3, 4, 7, 8: co-expression of His-yTAF$_{II}$6/yTAF$_{II}$9, His-yTAF$_{II}$9/yTAF$_{II}$6, His-hTAF$_{II}$6/hTAF$_{II}$9, His-hTAF$_{II}$9/hTAF$_{II}$6, respectively. A co-solubilization effect is observed upon co-expression, the yield of complex being lower when hTAF$_{II}$9 bears the his-tag. (Reproduced with permission of the International Union of Crystallography. *Crystallog J Online* (http://journals.iucr.org/).[3])

that *in vivo* expression allows a better distinction between productive and non productive complexes. In our hands, although the formation of non productive binary complexes could sometimes be observed by co-expression, the yields were low, or even very low, and having the right partners once again strongly increased the quantities obtained. Clearly, using an even larger number of proteins to be co-expressed simultaneously will further reduce the risk of forming unproductive complexes. In the worst case scenario, these should be easily detected during the setting up of the purification schemes.

Based on these advantages, it is clear that the co-expression technique provides a strong potential, not only to form soluble complexes whose interactions between subunits are known, but also to decipher protein/protein interactions in complexes whose quaternary structures are unknown or poorly characterized. Actually, this is often a first step towards the determination of structure/function relationships of a macromolecular assembly. Once again, the robustness of this technique against false positives makes it a method of choice for this kind of approach in comparison to other well-known methods such as GST-pulldown, yeast two-hybrid or immuno-precipitation which are more prone to generating false interaction data.[4]

Once interactions between the various proteins have been characterized, co-expression can be used for mapping precisely their interaction domains. To do so, constructs of various lengths can be used for the co-expression experiments aimed at finding the precise boundaries by checking whether or not the complex can still be formed (Fig. 2).[4] This might prove extremely useful when crystallization attempts or recording of NMR spectra have to be carried out following purification, especially to avoid crystallization failure due to unstructured regions and/or problems of heterogeneity upon degradation of these regions. It is clear that this approach can quickly lead to a large number of co-expression experiments with increasing number of proteins and constructs used, and the HTP techniques might be well adapted for this kind of studies.

Further, one major advantage of co-expression, as for endogenous complexes, is that purification is carried out on the whole macromolecular assembly rather than on every subunit independently, thus saving a lot of time. Importantly, once a complex has been characterized, it might be useful to perform lysis with different buffers to increase the yields for the complex. As all the subunits are co-expressed, this is not more difficult than for single proteins; in addition, a lot of time is saved during the design of purification protocols.

Finally, co-expression is not restricted to the expression of proteins. Helper vectors such as the pRare vector (Novagen) that encodes *E. coli* rare tRNAs can also be used to increase the yields in protein and complex expression. Other plasmids have also been made to co-express chaperones or enzymes involved in post-translational modifications (e.g. kinases).

Avoiding Pitfalls

As for any other techniques, co-expression is not devoid of drawbacks that could restrict its use. However, careful planning of the experiments is normally sufficient to avoid such drawbacks. The main problem encountered with this approach is the risk of getting false

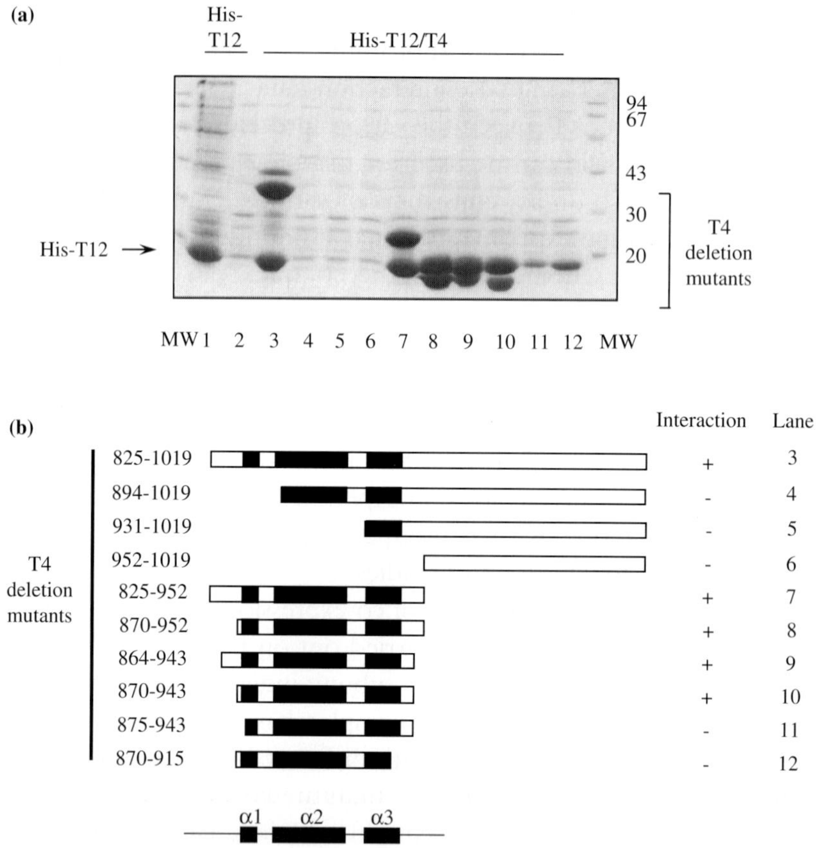

Fig. 2 Deciphering protein interaction domains. (**a**) Coomassie blue stained SDS-PAGE of soluble proteins/complexes obtained after expression and co-expression of human $TAF_{II}12$ (T12) and various constructs of $TAF_{II}4$ (T4). Except for lane 1, that represents the crude extract, and lane 2, that represents the soluble protein retained on cobalt beads upon single expression of His-T12, all samples represent complexes retained on cobalt beads with T12 being his-tagged and T4 constructs untagged. (**b**) T4 constructs used for the co-expression experiments. The gray bars indicate the three α-helices of the histone fold motif that is necessary for interaction with T12. (Reproduced with permission of *J Biol* (http://www.elsevier.com/).[4])

negatives. In this case, although several proteins can form a complex, no interaction data is observed. Several effects can lead to this result: the constructs are too long or too short, the purification tag interferes with the formation of the complex; the purification buffer is inappropriate; one or more proteins are not expressed; and/or post-translational modifications are missing.

The size of the constructs is of great importance in carrying out co-expression experiments, and for *in vitro* reconstitution. Indeed, a construct that is too short and that lacks part of its interaction domain with other proteins will most probably not form a complex, or a rather unstable one. It is important to note that yeast two-hybrid experiments are less amenable to partial truncation.[4] Therefore, one should be careful when trying to co-express partners whose interaction domains have been characterized by yeast two-hybrid. A construct that is too long might also cause problems if it encodes regions that may cause aggregation and precipitation. Even if a complex is formed, it might still be insoluble due to these additional regions. To overcome these problems, it is important to design several constructs that will span different structural domains of the proteins to be studied, unless it is known that there is a single domain. Further, it might be useful to vary the boundaries slightly when some uncertainty exists for a domain, i.e. not to fall too short on one or both sides.

As for the endogenous purification technique, another parameter of paramount importance in co-expressing proteins is the choice and the position of the tag. When expressing single proteins, the tag may sometimes interfere with the proper folding of the protein. In the case of complexes, it can also interfere with the assembly of the complex and-numerous examples are available, including dimers (see Fig. 1). In this respect, small tags (histidine, strep, flag) might be more suitable. For this reason, it appears essential, when performing co-expression, to vary the protein bearing the tag. Ideally, all combinations should be tried (for instance, protein A tagged and proteins B and C untagged; protein B tagged and proteins A and C untagged; protein C tagged and proteins A and B untagged). Clearly the position of the tag on the protein (N- or C-terminus), as well as the kind of tag or fusion used can or should also be varied.

In the case where several proteins can bear a tag without altering the formation or stability of the complex, it might be interesting for the large scale purification to select the tagged protein carefully. Indeed, it might be necessary during the purification to separate the full complex from sub-complexes or from the tagged protein. Choosing the protein that bears the tag as the one which gives the best stoichiometry for the complex (e.g. the least expressed) or as the one that can easily be separated from the full or sub-complex (e.g. a small subunit) will definitely facilitate the large scale purification.

Finally, false negatives might also occur due to the composition of the lysis buffer if it is not adapted to the complex studied, because the proteins are not expressed or because of the absence of post-translational modifications that prevent formation of the complex. In the former case, it can be useful, when performing the initial co-expression tests, to carry out the lysis in different buffers (e.g. low and high salt). In the latter two cases, changing the host might be the best solution. It should also be noted that most of the problems described apply to the *in vitro* reconstitution technique, as well as to a lesser extent to the endogenous complex purification technique, and that similar solutions might also be undertaken in these cases.

Most of the solutions suggested for overcoming the potential drawbacks of the co-expression technique (several constructs, several boundaries, different positions of the tag, different lysis buffers) can sometimes increase the number of experiments to be performed dramatically. Clearly, if information has already been gained on the various domains of the proteins, especially on their interactions with one another, the number of experiments can be reduced. Although one could use a sparse matrix approach or try an "educated guess" to reduce this number, it appears extremely useful to perform as many experiments as possible. HTP platforms (e.g. structural genomics) now offer an easy way of performing rapidly many different small-scale co-expression tests. In our hands, using a bacterial host (e.g. *E. coli*), up to 400 tests can be performed within a week. These include transformation of the bacteria with the plasmids and platting on 24 well-culture plates, cultures in 24 well-deepwells in auto-induction medium, harvesting and lysis per sonication in different buffers, incubation with affinity resin, and washes and analysis on SDS-PAGE gels. Since many of these steps are automated, enough time is gained for having the whole process running correctly and preparing for new experiments. Even if no HTP platform is available, by using the same procedure and a multi-channel pipette, a large number of co-expression tests can be performed. In the case of the baculovirus system or mammalian cells, although the whole process might take a bit longer, a careful set up of the experiment should allow a consequent number of tests to be carried out in parallel.

Multi-expression in *E. Coli*

Co-expression experiments can be performed in a wide variety of hosts. Bacterial hosts, especially *Escherichia coli*, are however, hosts of choice as they grow fast and allow the parallel set-up of a large number of tests, as discussed in the previous section. Besides, multiple technologies permit the rapid generation of expression vectors for these cells. Several ways of performing co-expression within *E. coli* are possible and are described in the following sub-sections. The differences between these different modes are discussed as choosing one over the other can affect the results obtained.

Several Vectors

The easiest way of performing co-expression in *E. coli* is to use several vectors. In the simplest case, each vector bears a single gene. To allow for the proper selection of the cells, every vector has to code for a different antibiotic resistance. This approach is straightforward as many different technologies nowadays allow rapid cloning of a single gene in expression vectors. This also allows the use of different kind of promoters. The next step is then to transform the cells and perform the co-expression experiments.

This method is particularly well adapted for co-expressing of two, three or sometimes four proteins. For a larger number of proteins, some problems might occur. First, the efficiency of the transformation decreases with the increasing number of plasmids used. It may happen that no colonies are observed, even if the quantity of antibiotics in the media is reduced. Although electroporation can be used, chemical transformation appears more convenient when carrying out a large series of tests.

Second, the levels of expression of the different proteins might differ. Normally, the vectors used for co-expression should have similar copy numbers within the cells. As the number of plasmids to be maintained is increased, it is unclear whether there might be a variation in the copy number, resulting in the formation of sub-complexes rather than of the complex itself. To prevent such a problem from occurring, every vector should ideally have a different origin of replication to avoid competition during replication. However, when using a small set of vectors this criterion does not appear so stringent as vectors with identical origin of replications can be used conjointly without problems.[3] Actually, it should be noted that this effect of having different levels of expression is much more likely to occur if some proteins are poorly expressed in *E. coli* or are toxic to the cell.

Finally, searching for *E. coli* expression vectors having different antibiotics resistance, similar copy number and, if possible, different origins of replication strongly restricts the number of potential plasmids to be used for co-expression. Despite these restrictions, the large number of dimeric or trimeric complexes which have already been expressed in large amounts using the multiple vectors strategy clearly shows that this approach is extremely well adapted for this kind of complexes. Furthermore, as the cloning process and the co-expression tests can easily be automated, it represents an important strategy in the field of structural genomics.

Single Vector — Single vs. Multiple Transcripts

The alternative to the multiple vector strategy is the single vector strategy where all the genes are cloned into a single vector. Two main possibilities exist: 1) the genes are under the control of a single promoter, each one being preceded by a ribosome binding site; or 2) the genes are under the control of several promoters. This choice has important implications as in the former case, a single transcript is produced, whereas in the latter case, several transcripts are produced. This raises a certain number of questions. First, the size of the transcripts is clearly different. In the single transcript strategy, this size can be relatively important and it is unclear whether *E. coli* can maintain or synthesize such large mRNAs. However, the presence of operons in the *E. coli* genome suggests that this might not be a problem. Further, the question remains whether the ribosomes preferentially bind to the first ribosome binding sites on the mRNA or whether they do not show any preference. Although various studies have not shown any apparent effect of the order of the genes (Ref. 5; C. Romier, manuscript in preparation), this aspect still remains to be further investigated.

In the case of multiple transcripts, if smaller mRNAs are produced, it is questionable whether downregulation can occur for some of them, especially if some toxicity is associated with the protein that they encode. Another aspect to deal with is the overall size of the plasmid, which is bigger than for a single transcript strategy. This could cause a problem for transformation and plasmid maintenance. On the other hand, this approach allows the use of different promoters for the different genes, which can eventually have positive effects on the expression levels of the proteins, and the promoters can be positioned in different orientations on the plasmids as well.

Finally, it is perfectly feasible to use vectors that harbor several promoters, some of them controlling the expression of a single gene, whereas others control the expression of several genes. This approach might be very useful to obtain complexes with a good stoechiometry, some partner proteins being produced in larger amount when their genes are under the control of a single promoter, whereas others require their own promoter.

Combined Approaches and Existing Systems

Co-expression in bacteria is clearly not restricted to a multiple vector or a single vector strategy. Both approaches can very easily be combined to extend the number of proteins co-expressed, using the simplicity of the multiple vector method together with the larger number of proteins that can be expressed with the single vector method. This approach might be extremely powerful if a complex or a sub-complex has already been characterized and all its genes cloned onto a single vector, with either one promoter or several promoters. This vector can then be used to co-express this complex with potential partners, either single proteins or other multi-protein complexes encoded by one or several plasmids, in order to form higher order assemblies.

Several systems are already available that allow one to perform co-expression experiments. These include notably the pET-DuET system (Novagen; four compatible vectors harboring two promoters); the pST44 system (Ref. 5; one vector allowing the cloning of four genes under the control of a single promoter); and the pET-MCN/pET-MCP systems. (Ref. 3; C. Romier, manuscript in preparation; four compatible vectors for multiple vector strategy that are designed to be transformed for single vector strategy with either a single promoter or multiple promoters, or a combination of both). It is important to note that tests carried out with the last series of vectors have shown that, depending on the strategy used (multiple vector, single vector or a combination of both, as well as position of the tag), the yields and the subunit stoichiometry of a complex can be drastically different, proving that using different approaches in parallel might be extremely useful (C. Romier, manuscript in preparation).

Multi-expression in the Baculovirus System

Co-expression with the baculovirus system has already been extensively used (for review see Ref. 6). However, this has been mostly restricted to multiple infections with several viruses, with one gene incorporated in the genome of each virus. The major drawback of this approach is that there is no possibility of selecting the cells being infected by all viruses. This leads to cells producing only sub-complexes and an overall loss in quantity of the full complex produced, especially when a large number of proteins are co-expressed. In this respect, careful quantification of the virus titer is necessary and may require several attempts before a good ratio between all the viruses is obtained. Recent advances with the pFastBacDual (Invitrogen) and especially the MultiBac[7,8] now allow the incorporation of several genes in the genome of a single virus. It is still unclear whether the large size of the vectors produced by the latter system can easily be accommodated by

E. coli prior to generation of the bacmids. However, the actual *in vitro* plasmid fusion procedure used by the MultiBac system can be automated,[8] allowing its use in HTP procedures.

Still, co-expression of proteins using the baculovirus system is more time consuming than with *E. coli* due to the time spent for the generation of the viruses, although this time has already been considerably reduced. This restricts the number of tests that can be carried out, although most pitfalls described previously apply to this approach as well. Thus, it is more likely that the risk of "missing" an interaction or a complex might occur. It should be noted however that using interaction data already obtained with bacterial co-expression might ease the work to be done. On the other hand, working with this system can be extremely useful when problems are occurring with *E. coli*, for eukaryotic proteins or when particular modifications are required.

Multi-expression in Mammalian Cells

Expression in mammalian cells has raised a lot of interest in the recent past and several laboratories are trying to implement this technique which appears well suited for expressing proteins, especially when they are secreted.[9] In this case, transient transfections are performed with dedicated plasmids and cells are grown for a few days before they are harvested. This approach shares the same advantages as those mentioned for the baculovirus system, with the additional interest of expressing human proteins in human cells (e.g. HEK293). This approach provides further positive aspects when compared with the baculovirus system, notably the possibility of performing small expression tests easily by transient transfection from a small amount of DNA (i.e. miniprep). This implies that the HTP approaches can be set-up with this technique.

Some problems remain, however, that needs to be dealt with. First, even if this approach is extremely useful for secreted proteins, the levels of intracellular expression remain below those obtained with the baculovirus system. A solution to this problem is currently under investigation. Furthermore, for large cultures the quantities of DNA required for the transfection are in the mg range, which implies that maxipreps have to be carried out. Alternatively, the capability of using stable cell lines that can be made using the same vectors as for transient transfection is also being investigated, as they would provide the easiest way for expressing proteins and protein complexes that would have been characterized by small expression tests carried out by transient transfections.

It should be noted that the use of large quantities of plasmid for transfection is not an absolutely negative aspect retaining to this technique. Indeed, since no selection is applied during transient transfection on the different plasmids used for co-expressing complexes, using large quantities of vectors ensure that the vast majority of the cells will contain all the plasmids. Thus, the full complex and not sub-complexes should be expressed. Still, if this technique is to become useful for producing large quantities of complexes, it will be necessary to develop new vectors that could harbor several genes to ease the work to be done and increase the overall yields of complexes.

Strategies for the Assembly of Protein Complexes

Co-expression techniques are offering a large variety of approaches for the assembly of protein complexes. Selecting the right approach is often a matter of scientific project. For instance, if one knows that some subunits forming a complex of interest cannot be expressed in *E. coli*, trials should begin with the baculovirus system or with mammalian cells. Still, due to the fact that every approach might lead to different results, trying several of them might prove to be more successful than restricting oneself to a single approach.

Furthermore, in most cases, e.g. when starting the study of a new complex, little might be known of this complex and a general strategy could be applied. Clearly, due to the ease of its use, *E. coli* appears as a host of choice in a first round of trials. The approach to be used depends then on the size of the complex and whether or not information is available on the subunit/subunit interactions. For small complexes

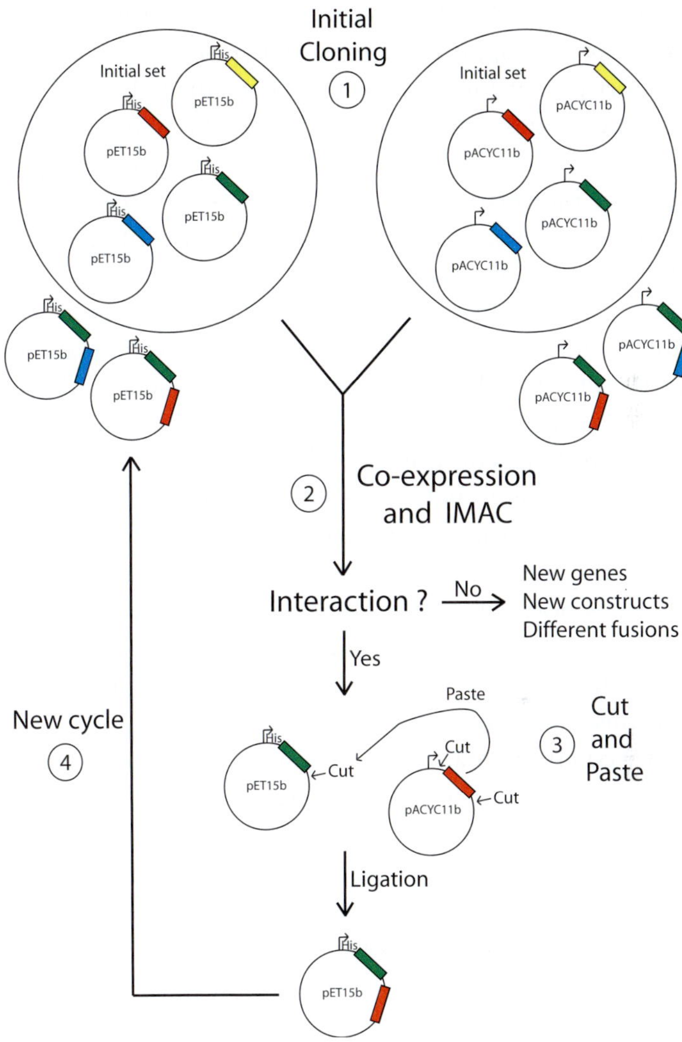

Fig. 3 Strategy for protein complexes reconstitution by co-expression. Flowchart for the reconstitution of protein complexes by iterative co-expression experiments. Initial co-expression tests using a multiple vector approach are carried out to find sub-complexes. The genes encoding the interacting subunits are transferred onto a single vector that can then be used for characterizing larger sub-complexes using again a multiple vector approach. This strategy is carried out until the full complex can be reconstituted. (Reproduced with permission of the International Union of Crystallography. *Crystallogr J Online* (http://journals.iucr.org/).[3])

where information is already known, a multiple vectors- and/or a single vector approach appear the most convenient. In the latter case, sets of vectors permitting easy cloning (e.g. through a LIC strategy) of two or three genes within one vector should be used to avoid any waste of time.

If no prior information is known or if the number of subunits is large, an iterative strategy as described in Fig. 3[3] might be more appropriate and should allow deciphering of subunit/subunit interactions as well as reconstitution of the complex. The use of more than two compatible vectors (three, possibly four) should speed up and improve this strategy, allowing the characterization of trimeric and tetrameric interactions that might be necessary for assembly of sub-complexes, especially when dimers do not form.

It is also important to note that some subunits may not be expressed in *E. coli*, requiring their expression in eukaryotic cells. However, deciphering the interactions between some subunits of a complex using *E. coli* as host, should speedily provide initial interaction data that can then be used for further studies using eukaryotic cells. Thus, both approaches should be seen as highly complementary when seeking the reconstitution of complexes.

Another complementarity can be found with the *in vitro* reconstitution technique. One problem that might be encountered when co-expressing proteins in a host is that the overall yield of the complexes decreases with an increasing number of subunits. In some cases, it will make more sense to co-express sub-complexes that can then be reassembled *in vitro*. This is clearly the case if some sub-complexes can be formed in *E. coli*, while others can only be produced in eukaryotic cells, e.g. insect cells. This approach has recently been used in the structural study of the exosome.[10]

Concluding Remarks

The co-expressions technique represents a major approach for the assembly and purification of large amounts of macromolecular complexes, especially for structural studies at high resolution. Multi-expression systems compatible with HTP technologies have emerged in the recent years that allow a large number of tests that are required for assembling complexes by co-expression to be carried out. Further developments are clearly required that would allow a change from bacterial to eukaryotic hosts without necessitating numerous cloning steps.

The study of complexes appears more complicated than the study of single proteins as many more parameters are present that influence the formation of these complexes. Still, most complexes studied so far have been stable macromolecular assemblies. The next challenge is certainly to assemble and study transient interactions between complexes. It remains to be proven whether the co-expression technique can be used for these studies. But by providing a good technological means by which stable complexes can be assembled, this technique clearly contributes to the study of the transient interactions between macromolecular complexes.

References

1. Devos D, Russell RB. (2007) "A more complete, complexed and structured interactome." *Cur Op Struct Biol* **17**: 370–77.
2. Rigaut G, Shevchenko A, Rutz B, *et al.* (1999) "A generic protein purification method for protein complex characterization and proteome exploration." *Nature Biotech* **17**: 1030–32.
3. Romier C, Ben Jelloul M, Albeck S, *et al.* (2006) "Co-expression of protein complexes in prokaryotic and eukaryotic hosts: experimental procedures, database tracking and case studies." *Acta Cryst* **D62**: 1232–42.
4. Fribourg S, Romier C, Werten S, *et al.* (2001) "Dissecting the interaction network of multiprotein complexes by pairwise coexpression of subunits in *E. coli. J Mol Biol* **306**: 363–73.
5. Tan S, Kern RC, Selleck W. (2005) "The pST44 polycistronic expression system for producing protein complexes in *Escherichia coli.*" *Prot Expr Purif* **40**: 385–95.
6. Kost TA, Condreay JP, Jarvis DL. (2005) "Baculovirus as versatile vectors for protein expression in insect and mammalian cells." *Nat Biotechnol* **23**: 567–75.
7. Berger I, Fitzgerald DJ, Richmond TJ. (2004) "Baculovirus expression system for heterologous multiprotein complexes." *Nature Biotech* **22**: 1583–87.

8. Fitzgerald DJ, Schaffitzel C, Berger P, *et al.* (2007) "Multiprotein expression strategy for structural biology of eukaryotic complexes." *Structure* **15**: 275–79.
9. Aricescu AR, Assenberg R, Bill RM, *et al.* (2006) "Eukaryotic expression: developments for structural proteomics." *Acta Crystallogr* **D62**: 1114–24.
10. Liu Q, Greimann JC, Lima CD. (2006) "Reconstitution, activities, and structure of the eukaryotic RNA exosome." *Cell* **127**: 1223–37.

Chapter 11

The Impact of Structural Proteomics on the Prediction of Protein–Protein Interactions

Christina Kiel* and Luis Serrano

Introduction

Three-dimensional protein structures are crucial for understanding biological processes and they play a central role in the discovery of new drugs. After the complete sequencing of several genomes, the completion of three-dimensional structures of all possible domain folds and interaction types opens up the possibility to fully model and understand biological systems.[1–3] With structural proteomics efforts, some 700–800 different folds[1] of the predicted 1000 domain folds in nature[4] and approximately 2000 of the predicted 10 000 domain–domain interaction types have been found.[1] It is estimated that it will take more than 20 years to obtain a representative structure for each interaction type.[1] However, scientists need not wait until all interaction types are discovered and all possible cell complexes are structurally determined. Large-scale structure-based interaction predictions can be done: a large amount of three-dimensional information is already available for interesting and important protein families, just waiting to

*Corresponding author: christina.kiel@crg.es.
EMBL-CRG Systems Biology partnership Unit, Centre de Regulacio Genomica (CRG), Dr Aiguader 88, 08003 Barcelona, Spain. Tel: +34-93316-0259

be used. Sometimes a simple single *in silico* screen for a binding partner can reveal the missing link in a new signal transduction pathway, as previously shown in the Dosophila gastrulation, where the PDZ domain containing RhoGEF2 was found in a computer screen with a PDZ domain interacting motif.[5]

Assuming that similar sequences have a similar fold and that domains with a similar fold interact in a similar way[6] (Fig. 1), much progress has been made in predicting protein–protein interactions based on structural information.[6–8] One method of structure-based

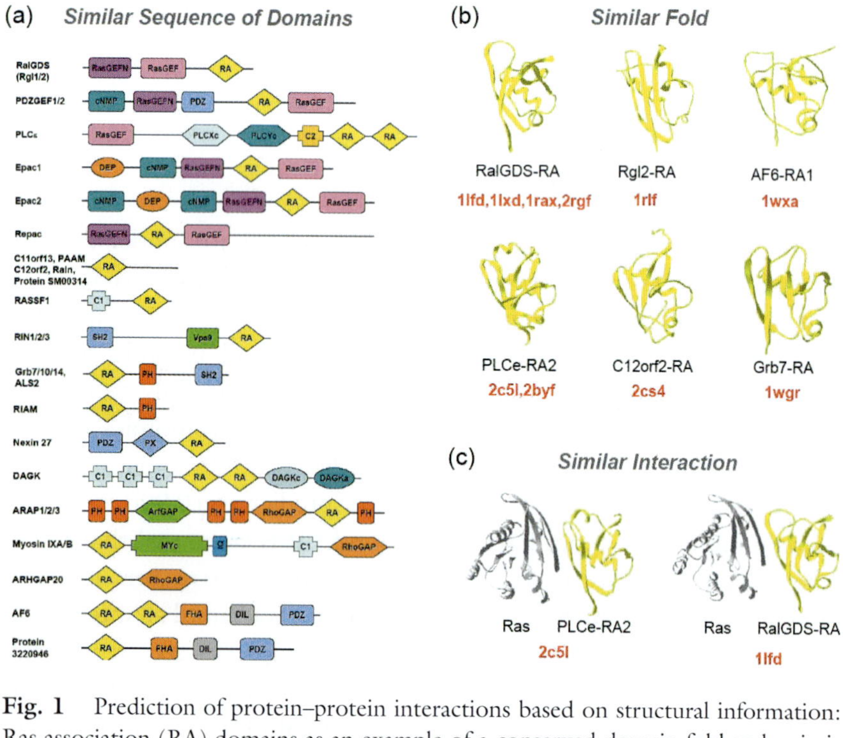

Fig. 1 Prediction of protein–protein interactions based on structural information: Ras association (RA) domains as an example of a conserved domain fold and a similar way of interaction. (**a**) Domain organization of some proteins containing RA domains. (**b**) RA domains have a similar fold, the ubiquitin-like topology, as shown by X-ray (pdb entries 1lfd, 1lxd, and 2c5l) and NMR spectroscopy (pdb entries 1rax, 2rgf, 1rlf, 1wxa, 2byf, 2cs4, and 1wgr). (**c**) They interact in a similar way as Ras proteins, via the interaction of an intermolecular b-sheet, as shown by X-ray (pdb entries 2c5l and 1lfd).

prediction is to generate homology models of proteins which have a similar sequence, and then calculate the interaction energy of the modeled complex using protein design algorithms.[7] To generate a homology model, the amino acid side-chains of a protein complex are replaced by the corresponding amino acid side-chains of two other proteins which belong to the same family, but where no structural information is available.

The prediction accuracy of this method can sometimes be impressive.[8] However, success is not automatic and depends on the availability of several good quality three-dimensional structures, correct (structure-based) sequence alignments, and careful structural inspection of domains and sequences. The latter point is crucial in deciding the sequences that can be modeled reliably based on available three-dimensional complex structures. Other important problems that can be addressed by the careful structural inspection of domains and sequences are the following: Which are the important residues stabilizing the fold? What are the crucial residues stabilizing important loops and secondary structural elements involved in the interaction? Are they conserved in the sequence to be modeled?

The objectives of this review are to provide an overview of the recent developments in structure-based prediction of protein interaction and to discuss the impact of structural proteomics on the success and completeness of the predictions. The main focus will be on the prediction of protein–protein interactions using homology modeling, where we dissect factors that influence the success of this prediction method.

Structure-based Prediction of Protein Interactions

The use of structural information to understand and fully model biological systems has become a valuable complementing tool in systems biology. Structures can be used to predict protein–protein interactions, calculate quantitative data, predict the function of proteins, and help integrate bioinformatics data, and are important for the prediction of SNP effects on proteins involved in diseases (recently reviewed in Ref. 9).

Prediction of Binding Energies and Hot Spot Residues

Progress has been made in the prediction of binding energies using design algorithms and protein complexes. The relative interaction energy based on complex protein structures can be rapidly and accurately predicted using existing protein design algorithms. Successful predictions of binding affinities for wild-type and mutant complexes have been carried out using the protein design algorithms FoldX[10–12] and Rosetta.[13–15] Examples are the prediction of Ras-effector interactions[7,8,16,17] and interactions of PDZ and SH3 domains with their target peptides.[5,18,19] However, the success depends on the quality of the structure: the higher the resolution of the structure, the better the prediction. Ideally structures below 2.2 Å resolution should be used. However, in some concrete cases where information from other members of the family is available, lower resolution in specific cases can be considered. One way to assess the quality of the structure of a complex is to use a protein design algorithm to calculate the interaction energy after refinement. If the $\Delta\Delta G$ binding energy is not negative and is equal to or higher than experimental data, it means that the quality of the structure is not satisfactory and a problem exists.

The structure can be further validated by performing an *in silico* alanine scanning mutagenesis with the original X-ray complex structure and comparing the results with experimental alanine scanning mutagenesis data. If the prediction of single alanine mutants is successful, it indicates that the X-ray structure is of high quality. In the case of Ras effector interactions, alanine mutants have been predicted with a correlation of 0.7/0.8 for the complexes of Ras in complex with RalGDS-RA and Raf-RBD.[16] For the prediction of a member of the TNF family, TRAIL, in complex with its receptor, DR5, a correlation of 0.8 was found for a series of alanine mutants.[20] The prediction of ubiquitin with ubiquitin interacting motifs also gives the qualitative correct trend.[17] In all cases, a regularization of the structure should be done before eliminating small clashes, flip Asn, Gln and His residues, to detect incorrect assignments and problems in the structure.

Protein Design and Homology Modeling as Tools to Modify Protein Interactions

Protein design is a useful tool in systems biology as it allows for rationally changing of the biophysical properties (affinities, or association and dissociation rate constants) of protein complexes, followed by the analysis of the consequence for the system, e.g. a particular pathway. Approaches to predict association rate constants make use of the principle of electrostatic steering.[21] The concept enables the association rate constant of a protein complex to be enhanced by increasing the electrostatic charge complementary in the interface and at the edge of the interface. The protein design algorithm PARE[22] was successfully developed to specifically enhance the rate of association, while not affecting the dissociation rate in various protein systems.[22–24]

Further, protein design exhibits great potential in drug design and in dissecting cellular interaction pathways. In many cases a particular protein in a cell could and will interact with different related and unrelated partners. Thus, for example, Rasp21 interacts with different RBD/RD domains through the same surface, or TRAIL, a member of the TNF family binds to five different cellular receptors. Using protein design tools and homology modeling, one can aim at designing protein/peptide variants that bind selectively to only one of the partner protein/receptors. In this way one can selectively deactivate/activate a certain pathway and study its effect independently of the other interactions. This has been successfully carried out in the TRAIL receptor system, where DR5-selective TRAIL variants, which do not induce apoptosis in DR4-responsive cell lines but show a large increase in biological activity in DR5-responsive cancer cell lines were generated.[20]

Prediction of Protein Interactions as Tools for Analyzing Protein Networks

Considering the large number of protein–protein interaction domains it is important to develop theoretical tools for predicting the protein domains that can interact, before investigating the possible physiological role of the interaction. Based on the assumption that structurally similar domains usually interact in a similar way,[6,25] much progress has been made to predict protein interactions based on sequence and structural information.[6–8,25] One way to predict protein interactions is to use homology modeling. The accuracy of predicting protein–protein interactions based on homology modeling and energy calculations can be impressive and the same methodology can, in principle, be applied to predict binding partners between domains of other families, if enough structural information is available.

Aside from homology modeling, which quantitatively describes the interaction on an atomic detail level, non-atomic detail methods have been successfully used to predict the interaction of proteins.[26,27] These methods, like INTERPReTS[28] and MULTIPROSPECTOR[28] use empirical pair potentials, which describe how well homologous pair of sequences fit into a complex structure. It has been successfully applied to predict the specificities of large domain families, e.g. the complex between fibroblast growth factors and their receptors.[27]

Structural information on domain–domain interactions can also be used to predict *in silico,* the domains in a multi-domain protein that can mediate in the interaction.[29–31] These domain interaction signatures can be used to either predict interactions or validate interactions that have been determined in large scale experimental pull-down assays. This has been first applied on a large scale in *S. serivisiae*, where all known protein–protein interactions were mapped to a possible binding interface, and a curated network was thus generated.[32] The curated network was used to establish that protein properties, such as essentiality, negative selection and co-expression of binding partners are dependent on the number of binding interfaces.

Prediction of Protein Interactions Using Homology Modeling: Factors Influencing Its Success

The accuracy of predicting protein–protein interactions depends on the availability of good quality three-dimensional structures, and correct (structure-based) sequence alignments. Further, a careful structural inspection of domains and sequences is important, in order to weed out the unreliable sequences in modeling from the available template structures. Usually, all available template complex structures are used, to account for small backbone changes in the interface. Homology models can be generated when the templates are ready, the sequences are aligned and the incompatible sequences are removed. The procedure to select the best model of a complex that has been generated using different template structures is to discard models with high intra- or intermolecular van der Waals' clashes and select the one with the lowest interaction energy.

The Interaction Type and the Interface

Around 2000 protein–protein interaction types have already been recorded.[1] They are accessible in 3-DID (3D interacting domains, http://3did.embl.de)[33] and other databases.[34–38] Is an ideal interaction type for the prediction of protein–protein interactions available? Figure 2 shows a selection of interaction types which have already been analyzed using protein design tools. Ras proteins mediate their binding to effector domains using β-sheet interactions and loops. The structural flexibility in these interfaces is low, because of the backbone H-bond interactions. In fact, the main structural changes occur in the helix $\alpha 1$ in the Ras binding domain, which can have significant different conformations, depending on the complex. Thus, the predictions are hightly accurate, although only six different template structures have been used.[8] In contrast, the flexibility of SH3-peptide and PH-peptide complexes is much higher. In addition to previously described motifs important for binding,[39] new motifs in the RT loop and the loop lengths are important for binding (Gregorio Fernandez-Ballester & Luis Serrano personal communication) (ADAN database). Therefore, cases where the loop lengths of sequences to be modeled are similar and share the same motifs can be modeled using available template complex structures. In other cases, where sequences are different, chimera are needed for successful predictions (Fernandez-Ballester, in preparation). The same principle applies to PDZ domains in complex with peptides. Databases for PDZ-peptide and domain-peptide interactions, which collect information about binding motifs

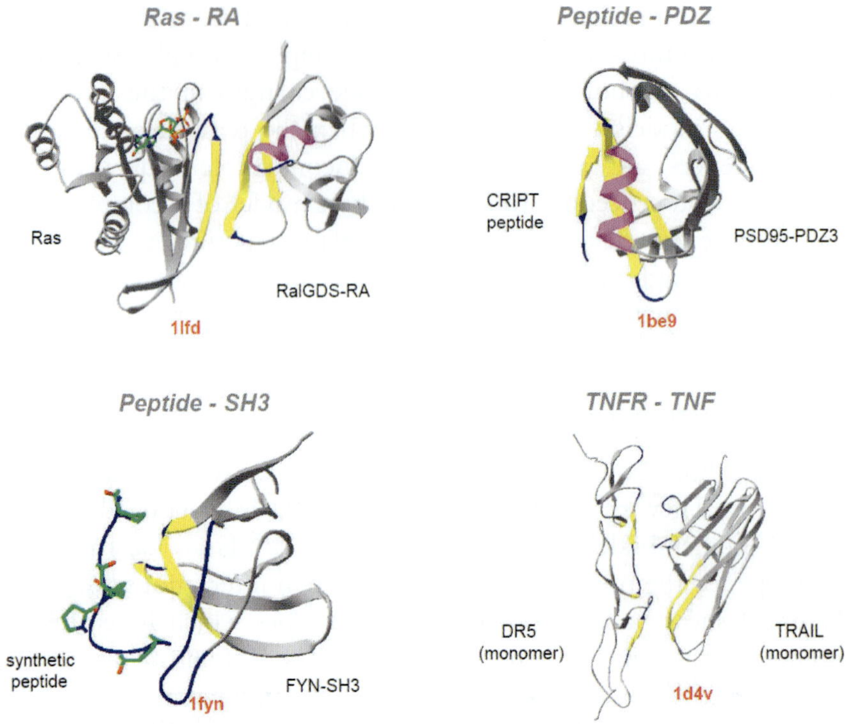

Fig. 2 Selection of interaction types, which have been analyzed using protein design tools. Structures of Ras proteins in complex with RA domains (example Ras-RalGDS-RA, pdb-entry 1lfd), PDZ and SH3 in complex with peptides (pdb entrees 1be9 and 1fyn), and TNF and TNFR domains (pdb entry 1d4v). Secondary structure elements and loops involved in the interface formation are colored (yellow: β-strand, pink: α-helix, blue: loop).

and structural are ADAN (www*****), PDZBase[40] and DOMINO.[41] The tumor necrosis (TNF) family ligand-receptor binding seems to be ideal for structure-based design, since TNF family members assume a very similar tertiary structure, while the diverse feature of surface residues mediate the specificity between different TNF family members.[42,43]

Interactions through the formation of a β-sheet seem to be especially superior, probably because the backbone-backbone H-bonds restrict the possibilities of slightly different conformations in different complexes. Surfaces that involve little or no main-chain H-bonds are more problematic for the simple reason that side chain mutations could change the interaction geometry between molecules A and B slightly and therefore the number of templates required for successful prediction increases enormously.

Structure-based Sequence Alignments

A correct sequence alignment of the domains involved is crucial in predicting protein–protein interactions based on structural information and sequence homology. If the level of sequence identity is very low, a correct sequence alignment is difficult to achieve using conventional multi-sequence alignment programs.[44] One way to improve sequence alignments is to use 3D structural information, as employed in the 3D-Coffee program.[45] However, if the sequences and structures are very different, especially if the loop length varies, automatic sequence alignment programs that take 3D information into account sometimes fail.[17] In these cases, an option is to manually correct sequence alignments using 3D information, by determining positions that must be fully buried and to take secondary structure elements into consideration. Structural proteomics makes it possible to generate alignments of high quality, even if sequence homology is low and automatically generated sequence alignments fail. This has been successfully completed for the ubiquitin fold superfamily, which consists of five sub-families, the RA, RBD, PI3K_rbd, B41 and UBQ domain.[17] Figure 3 shows the RBD domain family as an example of how structural information reveals mistakes in automatically generated multiple sequence alignments: The sequences for RGS12_RBD2 and RGS14_RBD2 would have been aligned wrongly without the structural information on the RGS14_RBD2 (from mouse).

Structural Inspection of Domains and Sequences

In general, similar sequences have a similar fold and domains with a similar fold interact in a similar way. However, single amino acid substitutions can change the fold (backbone) slightly and thus homology

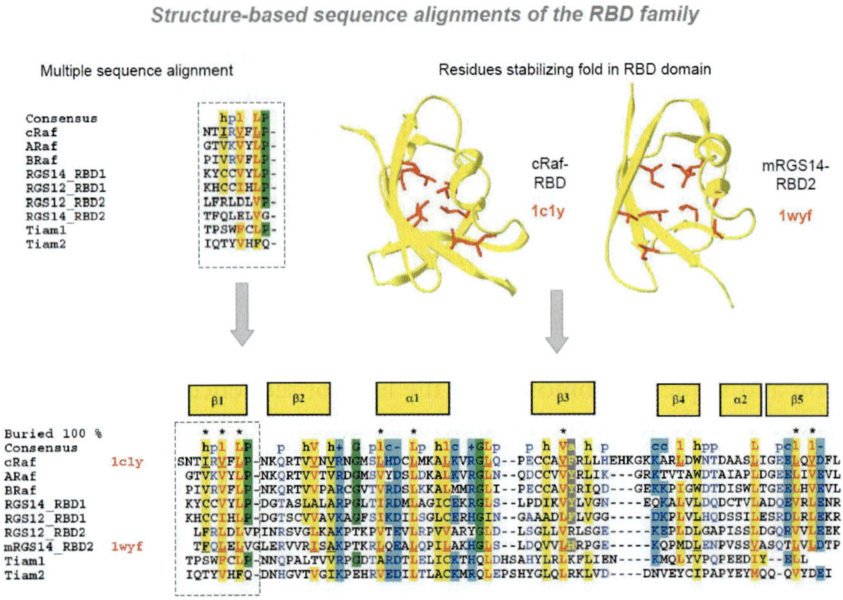

Fig. 3 Structure-based alignments of the Ras binding domain (RBD) family. Residues stabilizing the hydrophobic core in RBD structures should be conserved in all sequences of the RBD family.

modeling (that does not allow backbone changes) with a given template structure would be incorrect. Therefore, it is important to perform a detailed structural inspection of the domains and sequences and identify the crucial residues that stabilize the important loops and secondary structure elements involved in the interaction. Only sequences which have these crucial residues can be modeled reliably using the available template structure. In Fig. 4 we show two examples: Using the Ras-RalGDS complex structure as a homology modeling template, an RA domain sequence with a proline residue in a central position of a β-strand or an α-helix would break (backbone changes) these secondary structure elements. Thus, RA domain sequences containing proline residues at these positions cannot be modeled using the Ras-RalGDS structure as a template.[8] The other important example is the stabilization of a long loop, which is involved in the interface, by a large bulky amino acid (Fig. 4).[46] A tryptophan residue in the E3 RING domain stabilizes the

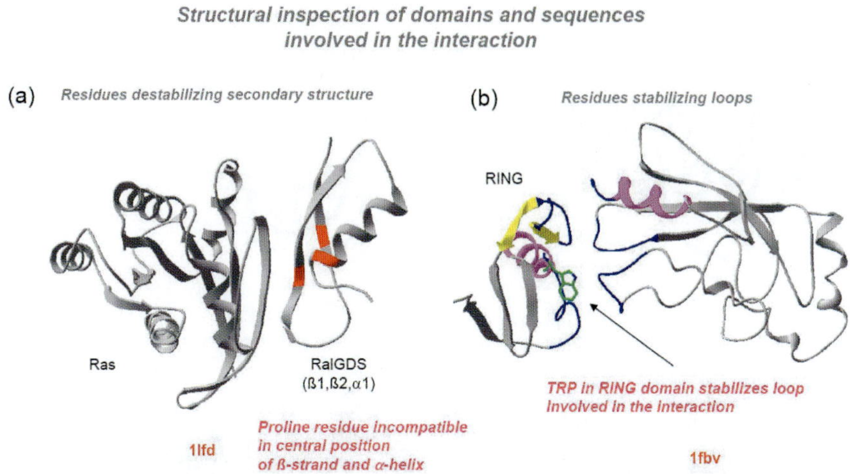

Fig. 4 Structural inspection of domains and sequences involved in the interaction. (**a**) Sequences with residues destabilizing secondary structure elements, like proline residues in a central position of a β-strand, cannot be modeled correctly, as backbone changes will occur. (**b**) Sequences, which do not contain important amino acid residues, involved in stabilizing loops involved in the interaction, cannot be modeled reliably.

conformation of a loop which is involved in complex formation with its E2 partner, UBCH7. A RING sequence with a small residue at this position cannot be modeled reliably, as the loop is expected to have a different conformation.

Template Structures

Based on the original X-ray complex structure, template structures to be used in homology modeling can be generated. The modification of the original template structure depends on the application and the sequence similarity of the domain to be modeled with the original X-ray complex structure (Fig. 5): (a) If the sequences are very similar, and the loops between the secondary structure elements have identical lengths, the complete X-ray complex structure can be assumed to be template structure.[16] The predictions will be very accurate, as they take into consideration regions at the edge of the interface, which

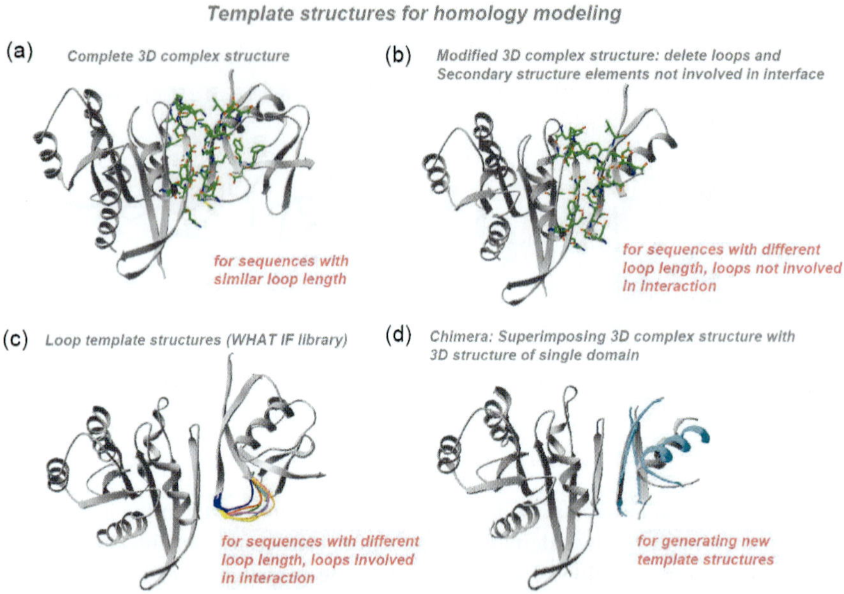

Fig. 5 Template structures used for homology modeling. (**a**) Complete 3D complex structures can be used if the sequences to be modeled have similar loop lengths. (**b**) In order to model sequences with different loops lengths, which are not involved in the interaction, 3D complex structures can be modified by deleting loops and secondary structure elements not involved in the interaction. (**c**) In order to model sequences with different loops length, and the loops are involved in the interaction, loop template structures can be generated using the WHAT IF library (see text). (**d**) If sequences to be modeled are expected to have significant backbone changes, new chimera-template structures can be generated by superimposing 3D complex structures with a 3D structure of a single domain with a similar sequence.

contribute to the overall binding energy. (b) For sequences which have different loop lengths from the original X-ray complex structure, and which are not involved in the interaction, 3D structures can be modified by deleting the structural parts that are not involved in the interaction.[7,8] (c) For sequences which have different loops lengths where the loops contribute significantly to binding, loops can be generated using the WHAT IF library,[47] which contains loops of different lengths from a database of X-ray loop fragments. This has been successfully implemented for the prediction of Ras-effector interaction.[8]

(d) To generate new template structures, if the sequence to be modeled is not compatible with an available template complex structure, one can generate chimera by superimposing the 3D complex structure with the 3D structure of a single domain. This has been successfully done for the prediction of SH3 domains with their target peptides.[19]

Obtaining the Right Reference for Deciding if Two Proteins Interact

After the generation of template structures and the careful structural alignment of sequences to be modeled, homology models can be generated and energy calculations carried out. The question of how the interaction energies can be analyzed and how two proteins interact based on their interaction energies then arises.

As sequences are modeled using different template structures, a successful approach has been to select the model with the lowest interaction energy.[7,8] However, if the total energy of a sequence modeled with a particular template structure is very low, due to high van-der Waals clashes, and indicating that the sequence and the template structure are not compatible, we discard this model. The result of homology modeling will not be reliable in the case, although the interaction energies might be favorable.

After the best model generated using different template structures has been selected, one needs to calculate the criteria/energy thresholds in order to decide whether this sequence will bind or not.

Energy thresholds can be defined by calibrating energy sum values using experimental information.[8] Using threshold information, new binding and non-binding domains can be successfully predicted; however, the rest of the predictions reside in the twilight zone.[8]

Conclusion

The impact of structural proteomics on the prediction of protein–protein interactions exists on many levels: Complex structures of domain–domain interactions are important as template structure to model sequences which are expected to have a similar fold.

Further, complex structures and structures of single domains of a family are essential to generate structure-based alignments for sequences with a low sequence homology. In addition, structural inspection of domains and sequences is crucial to decide the sequences that can be modeled reliably using available template structures.

Acknowledgements

We thank the EU for financial support (INTERACTION PROTEOME, grant-No. LSHG-CT-2003-505520.

References

1. Aloy P, Russell RB. (2004) "Ten thousand interactions for the molecular biologist." *Nat Biotechnol* **22**: 1317–21.
2. Bork P, Serrano L. (2005) "Towards cellular systems in 4D." *Cell* **121**: 507–9.
3. Banci L, Baumeister W, Enfedaque J, *et al.* (2007) "Structural proteomics: from the molecule to the system." *Nat Struct Mol Biol* **14**: 3–4.
4. Chothia C. (1192) "One thousand families for the molecular biologist." *Nature* **357**: 543–4.
5. Kolsch V, Seher T, Fernandez-Ballester GJ, *et al.* (2007) "Control of Drosophila gastrulation by apical localisation of adherens junctions and RhoGEF2." *Science* **315**: 384–6.
6. Aloy P, Russell RB. (2006) "Structural systems biology: modelling protein interactions." *Nat Rev Mol Cell Biol* **3**: 188–97.
7. Kiel C, Wohlgemuth S, Rousseau F, *et al.* (2005) "Recognizing and defining true Ras binding domains II: *In silico* prediction based on homology modelling and energy calculations." *J Mol Biol* **348**: 759–75.
8. Kiel C, Foglierini M, Kuemmerer, N, *et al.* (2007) "A genome-wide Ras-effector interaction network." *J Mol Biol*, in press.
9. Beltrao P, Kiel C, Serrano L. (2007) "Structures in systems biology." *Curr Op Struct Biol*, in press.
10. Guerois R, Nielsen JE, Serrano L. (2002) "Predicting changes in the stability of proteins and protein complexes: a study of more than 1000 mutations." *J Mol Biol* **320**: 369–87.
11. Schymkowitz J, Borg J, Stricher F, *et al.* (2005) "The FoldX web server: an online force field." *Nucleic Acids Res* 33: W382–8.
12. Schymkowitz JWH, Rousseau F, Martins IC, *et al.* (2005) "Prediction of water and metal binding sites and their affinities by using the Fold-X force field." *Proc Natl Acad Sci USA* **102**: 10147–52.

13. Simons KT, Kooperberg C, Huang E, Baker D. (1997) "Assembly of protein tertiary structures from fragments with similar local sequences using simulated annealing and Bayesian scoring functions." *J Mol Biol* **268**: 209–25.
14. Simons KT, Ruczinski I, Kooperberg C, *et al.* (1999) "Improved recognition of native-like protein structures using a combination of sequence-dependent and sequence-independent features of proteins." *Proteins* **34**: 82–95.
15. Kortemme T, Baker D. (2002) "A simple physical model for binding energy hotspots in protein–protein complexes." *Proc Natl Acad Sci USA* **99**: 14116–21.
16. Kiel C, Serrano L, Herrmann C. (2004) "A detailed thermodynamic analysis of Ras/effector complex interfaces." *J Mol Biol* **340**: 1039–58.
17. Kiel C, Serrano L. (2006) "The ubiquitin domain superfold: structure-based sequence alignments and characterization of binding epitopes." *J Mol Biol* **355**: 821–44.
18. Kempkens O, Médina E, Fernandez-Ballester G, *et al.* (2006) "Computer modelling in combination with *in vitro* studies reveals similar binding affinities of *Drosophila* Crumbs for the PDZ domains of Stardust and *Dm*Par-6." *Eur J Cell Biol* **8**: 753–67.
19. Musi V, Birdsall B, Fernandez-Ballester G, *et al.* (2006) "New approaches to high-throughput structure characterization of SH3 complexes: the example of Myosin-3 and Myosin-5 SH3 domains from S. cerevisiae." *Protein Sci* **4**: 795–807.
20. Van der Sloot AM, Tur V, Szegezdi E, *et al.* (2006) "Designed tumor necrosis factor-related apoptosis-inducing ligand variants initiating apoptosis exclusively via the DR5 receptor." *Proc Natl Acad Sci USA* **103**: 8634–9.
21. Sheinerman FB, Norel R, Honig B. (2000) "Electrostatic aspects of protein–protein interactions." *Curr Op Struct Biol* **10**: 153–9.
22. Selzer T, Albeck S, Schreiber G. (2000) "Rational design of faster associating and tighter binding protein complexes." *Nat Struct Biol* 7: 537–41.
23. Kiel C, Selzer T, Shaul Y, *et al.* (2004) "Electrostatically optimized Ras-binding Ral guanine dissociation stimulator mutants increase the rate of association by stabilizing the encounter complex." *Proc Nat Acad Sci USA* **101**: 9223–8.
24. Peleg-Shulman T, Roisman LC, Zupkovitz G, Schreiber G. (2004) "Optimizing the binding affinity of a carrier protein: a case study on the interaction between soluble ifnar2 and interferon beta." *J Biol Chem* **279**: 18046–53.
25. Aloy P, Pichaud M, Russell RB. (2005) "Protein complexes: structure prediction challenges for the 21st century." *Curr Opin Struct Biol* **1**: 15–22.
26. Lu L, Arakaki AK, Lu H, Skolnick J. (2003) "Multimeric threading-based prediction of protein–protein interactions on a genomic scale: application to the Saccharomyces cerevisiae proteome." *Genome Res* **13**: 1146–54.
27. Aloy P, Russell R. (2002) "Interrogating protein interaction networks through structural biology." *Pro Natl Acad Sci USA* **99**: 5896–901.

28. Lu L, Lu H, Skolnick J. (2002) "MULTIPROSPECTOR: an algorithm for the prediction of protein–protein interactions by multimeric threading." *Proteins* **49**: 350–64.
29. Sprinzak E, Margalit H. (2001) "Correlated sequence-signatures as markers of protein–protein interaction." *J Mol Biol* **311**: 681–92.
30. Wojcik J, Schachter V. (2001) "Protein–protein interaction map inference using interacting profile pairs." *Bioinformatics* **17** (Suppl. 1): 296–305.
31. Deng M, Mehta S, Sun F, Chen T. (2002) "Inferring domain–domain interactions from protein–protein interactions." *Genome Res* **12**, 1540–8.
32. Kim PM, Lu LJ, Xia Y, Gerstein MB. (2006) "Relating three-dimensional structures to protein networks provides evolutionary insights." *Science* **314**: 1938–41.
33. Stein A, Russell RB, Aloy P (2005) "3did: interacting protein domains of known three-dimensional structure." *Nucleic Acids Res* **33**: D413–7.
34. Finn RD, Marshall M, Bateman A. (2005) "iPfam: visualization of protein–protein interactions in PDB at domain and amino acid resolutions." *Bioinformatics* **21**: 410–2.
35. Winter C, Henschel A, Kim WK, Schroeder M. (2006) "SCOPPI: a structural classification of protein–protein interfaces." *Nucleic Acids Res* **34**: D310–4.
36. Ogmen U, Keskin O, Aytuna AS, Nussinov R, Gursoy A. (2005) "PRISM: protein interactions by structural matching." *Nucleic Acids Res* **1**: 33.
37. Jefferson ER, Walsh TP, Roberts TJ, Barton GJ. (2007) "SNAPPI-DB: a database and API of structures, interfaces and alignments for protein–protein interactions." *Nucleic Acids Res* **35**: D580–9.
38. Davis FP, Sali A. (2005) "PIBASE: a comprehensive database of structurally defined protein interfaces." *Bioinformatics* **21**: 1901–7.
39. Larson SM, Di Nardo AA, Davidson AR. (2000) "Analysis of covariation in an SH3 domain sequence alignment: applications in tertiary contact prediction and the design of compensating hydrophobic core substitutions." *J Mol Biol* **303**: 433–46.
40. Beuming T, Skrabanek L, Niv MY, *et al.* (2005) "PDZBase: a protein–protein interaction database for PDZ domains." *Bioinformatics* **21**, 827–8.
41. Ceol A, Chatr-aryamontri A, Santonico E, *et al.* (2007) *Nucleic Acids Research* **35**: D557–60.
42. Bodmer J-L, Schneider P, Tschopp J. (2002) "The molecular architecture of the TNF superfamily." *TIBS* **27**: 19–26.
43. Zhang G. (2004) "Tumor necrosis factor family ligand-receptor binding." *Curr Op Struct Biol* **14**: 1–7.
44. Wallace IM, Blackshields G, Higgins DG. (2005) "Multiple sequence alignments." *Curr Opin Struct Biol* **15**: 261–6.
45. O'Sullivan O, Suhre K, Abergel C, *et al.* (2004) "3DCoffee: combining protein sequences and structures within multiple sequence alignments." *J Mol Biol* **340**: 385–95.

46. Zheng N, Wang P, Jeffrey PD, Pavletich NP. (2000) "Structure of a C-Cbl-UBCH7 complex: RING domain function in ubiquitin-protein ligases." *Cell* **102**: 533–9.
47. Vriend G. (1990) "WHAT IF: a molecular modelling and drug design program." *J Mol Graph* **8**: 52–6.

Chapter 12

Cryo-Electron Microscopy in the Era of Structural Proteomics

Alasdair C. Steven*,† and David M. Belnap‡

Summary

Realization of structural proteomics projects must necessarily draw on a comprehensive range of structure determination techniques. Among these, cryo-electron microscopy plays a key role on account of its suitability for imaging large multi-subunit complexes. As currently practiced, three branches of cryo-EM may be distinguished: single particle analysis (SPA); electron crystallography; and cryo-electron tomography. We review the ranges of applicability of these approaches as well as currently accessible resolutions. Focussing on SPA, we consider the progress that has been made towards implementing automated "high throughput" methods to expedite various steps of the pipeline. Strategies for mapping the components of large complexes are also surveyed, as are "hybrid methods" in which cryo-EM is integrated with other experimental approaches. In particular, crystal structures

*Corresponding author: stevena@mail.nih.gov

†Laboratory of Structural Biology, National Institute of Arthritis, Musculoskeletal, and Skin Diseases, National Institutes of Health, Bethesda, MD 20892, USA.

‡Department of Chemistry and Biochemistry, Brigham Young University, Provo, UT 84602, USA.

of individual subunits may be fitted into cryo-EM reconstructions of assembled complexes to define interactions and to leverage the resolution of the cryo-EM data, allowing the formulation of pseudo-atomic models. We estimate the precision of such fittings to be within 1 Å RMSD in favorable circumstances. Summarized in an Appendix are the numerous fitting programs that are now available. We also discuss situations in which fitting is not straightforward as well as the basic compatibility of SPA density maps, which represent solution structures, with crystallographic data, which depict solid-state structures. Intrinsic variability of macromolecular complexes portrayed in cryo-electron micrographs appears to be a resolution-limiting factor in many SPA analyses but has the potential to illuminate multiple conformations (by multiple particle analysis — MPA) and the dynamic properties of the complexes (by variance analysis).

Background

Structural proteomics seeks to determine the three-dimensional structures of all proteins expressed in a given cell or organism. As such, it goes beyond the more narrowly defined objective of structural genomics that aims to determine the folds of all polypeptide chains corresponding to open reading frames in a given genome and, in aggregate, all possible folds. However, most biological functions are performed, not by single domains or subunits but by multi-component macromolecular assemblies.[1] For such an assembly, knowledge of the structure of one domain or subunit *per se*, even at atomic resolution, generally affords little insight into which other subunits it may interact with or how the assembly functions. The large majority of protein structures solved by X-ray crystallography are of single domains or subunits. This approach requires pure samples in large quantities — at a minimum, 10–20 mg — and is limited by crystallizability and the phasing problem. As long as these limitations persist, the scope of unaided X-ray crystallography — and therefore of structural genomics — to give a functionally insightful account of an organism's population of protein molecules is quite restricted. The structures of higher-order assemblies

must be determined directly and other methodologies should be involved. Thus, structural proteomics embraces the capabilities of an extensive range of methodologies that yield structural information. Among them, three-dimensional cryogenic electron microscopy (cryo-EM) plays a central role.

Cryo-EM Today

In the current repertoire, one may distinguish three branches of cryo-EM[2]: electron crystallography (EC) of regular two-dimensional arrays[3]; "single particle analysis" (SPA) of non-crystalline, freestanding, particles[4]; and electron tomography (ET).[5] The term SPA, although in widespread use, is a misnomer because it refers, in fact, to the analysis of very many particles — typically, thousands — whose images are combined to calculate a single three-dimensional density map. Electron tomography, on the other hand, renders three-dimensional density maps of *bona fide* individual particles.

Transmission electron micrographs represent two-dimensional projections of three-dimensional objects. To determine a 3D structure, a specimen must be viewed from multiple directions, and these projections are combined to give a "reconstruction" of the complex. For EC and ET, the specimen is tilted through multiple angles and an image recorded at each position. For SPA, untilted specimens are imaged (in some strategies, a single tilted image is added to the untilted one) and provide the needed range of viewing angles because the complexes are randomly oriented in the solution; however, these orientations must be determined.

Although EC is not restricted to membrane proteins,[6] it is particularly apposite for their study as these molecules may thus be visualized in their natural environment, a lipid bilayer. The resolution obtained in EC density maps is anisotropic, being lower in the dimension perpendicular to the plane of the specimen, both because the specimen may only be tilted through a limited range and because the data quality tends to deteriorate towards higher tilt angles. The highest in-plane resolution achieved to date is 1.9 Å[7] but lower resolutions

tend to be the norm, although they are often sufficient to depict the configurations of transmembrane α-helices and, in a few cases, the protein fold, as first accomplished in 1990.[8]

Resolution in SPA maps, on the other hand, is isotropic unless the specimen fails to assume a full range of viewing orientations. SPA density maps convey the solution structure of the molecule in question, unfettered by crystal contacts. Because SPA requires identifying the orientation of each particle and aligning it with high precision, the images (which are — inescapably — noisy), must be sufficiently well defined for this to be possible. This requirement imposes a lower size limit for which SPA is feasible. The exact limit is debatable but we suggest a value of about 300 kDa for specimens embedded in vitreous ice, although smaller specimens may be analyzed if heavy metal stains are used to enhance contrast — at the risk of compromizing native structure.[9] The largest specimens analyzed by SPA so far have masses of several hundred MDa (e.g. Ref. 10). Attainable resolution depends on a number of factors — in particular, specimen tractability — but the highest resolutions achieved to date are at the 4–6 Å level[11,12]; and resolutions of 7–10 Å, first achieved in 1997,[13–15] are increasingly common.

In SPA, it is assumed that all particles are alike, and therefore it is valid to merge their images in a reconstruction. ET, on the other hand, may be used to reconstruct individual members of heterogeneous populations of particles, as well as suitably thin cells.[16] As in EC, the resolution of a tomogram is anisotropic because the tilt range covered is incomplete. Resolution is limited by signal-to-noise ratio and noise levels in cryo-tomograms are, inevitably, quite high: also, resolution is dependent on particle size. Recently, quantitative empirical criteria have been developed for the resolution of tomograms.[17–19] They, indicate that in-plane resolutions of 5–6 nm are currently achieved in tomograms of large particles (diameter >100 nm, or so), and somewhat higher for smaller particles. However, averaging multiple subtomograms, each containing a representation of the same feature, suppresses noise and overcomes anisotropy and has allowed an approximate doubling of the resolution stated above.[20] Moreover, some repetitive features with spacings finer than 5–6 nm may be

discerned in individual tomograms because their detection implicitly involves averaging.

Suitability of Cryo-EM for The Analysis of Multi-Subunit Protein Complexes

The size range of macromolecular particles eligible for cryo-EM characterization by SPA and/or ET, which spans at least four orders of magnitude in molecular mass starting at ~300 kDa, renders it ideal for investigating macromolecular machines. There are also practical advantages. First, cryo-EM is relatively parsimonious in terms of the amount of material required. As little as 10–50 μg may suffice for analysis to the highest resolutions currently accessible and there is potential for further reduction, as the number of molecules that ultimately contribute to a density map — say, 10^4–10^5 copies — amount to only a very small fraction, ~0.00002%, of the molecules present in a 10 μg sample of a 1MDa complex. Parsimony at this level is a major advantage in that although a few large assemblies, e.g. ribosomes, are abundant and robust, many others are sparse and fragile. Because the technology does not yet exist for bulk co-expression of the >20 protein subunits that are needed for some assemblies — except for viruses which perform this feat routinely in the course of an infection — these specimens must be obtained from natural sources. Second, the fragility of many machines mandates that they be isolated via protocols that are as brief, simple, and respectful of native structure as possible, in which case, their purity may be less than pristine. However, as explained below, this is not a prohibitive obstacle.

Particle Heterogeneity: *in silico* Fractionation and Implications for Dynamics

While homogeneity is desirable in a cryo-EM specimen, it is not essential in that discreet, (relatively) homogeneous, populations of molecules may be distinguished by using classification techniques to sort the images.[21] This practice is sometimes referred to as *in silico*

fractionation. While some particle species may be distinguished visually, most classifications require quantitative methods. Variability among molecular images may arise from several sources: (i) viewing geometry, the images representing many different projections of a three-dimensional object; (ii) compositional heterogeneity: i.e. in addition to the complex of interest, a specimen may contain degraded, aggregated, or otherwise modified variants of the same, as well as other nondescript contaminants; and (iii) intrinsic conformational variability of a given complex. The development of objective classification procedures for sifting out homogeneous populations of particles in the presence of noise and other sources of variability is a focus of intense current activity.[22–27]

A compositionally homogeneous class of particles may exhibit structural variability, potentially arising from several sources: the presence of multiple discreet conformers; concerted movements in molecular "breathing"; or the fluctuations of peripheral mobile elements. Although these phenomena complicate SPA and limit the resolution of unitary density maps, they offer an avenue to investigate the specimen's dynamic properties in terms of the range of conformations that may be populated during its functional cycle. In principle, structural variability may be discreet, involving two or more distinct states, or continuous, involving the fluctuations of mobile elements. These eventualities have to be handled separately. Discreet variability may be confronted by generalizing the SPA approach to include multiple models — one for each distinct conformer; hence the term, multiple particle analysis (MPA).[23] In such cases, the total number of particles required to complete the analysis to a targeted resolution increases in proportion to the number of conformers. Continuous variability poses more of a challenge and is expressed in a unitary density map by local blurring of densities associated with mobile elements. In this context, quantitative analysis of image variance — either in two dimensions[28] or three dimensions[29,30] — offers a way to identify mobile elements (see Fig. 1). Once these elements are identified, if information is available on their sizes and shapes, computational simulation may be used to estimate the amplitudes of their fluctuations.[28]

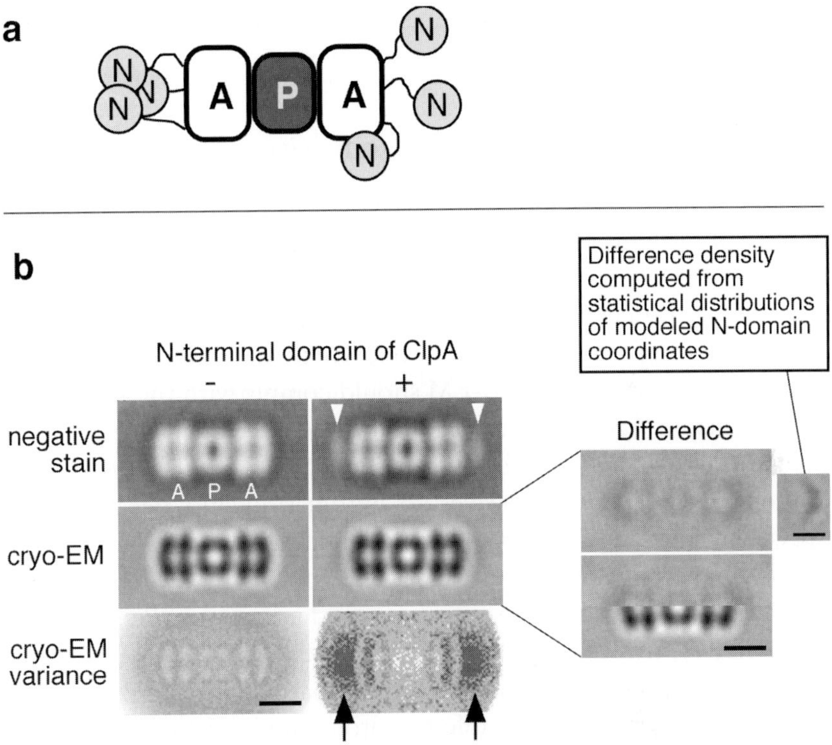

Fig. 1 Highly mobile domains are invisible in an averaged image but are detected in a variance map. (**a**) The ClpAP complex comprises two hexamers of the ClpA ATPase (subunit, 82 kDa) stacked axially on either end of the ClpP peptidase, a double heptamer of 21 kDa subunits. The ClpA hexamer has two rings of AAA+ domains, D1 and D2, and a 16 kDa N-domain. (**b**) Averaged sideviews of a mutant ClpA lacking the N-domains (–; *left*) and wildtype (+; *right*). Additional axial density is seen faintly in the negatively stained image (*top*, *right*, white arrowheads) but not in the vitrified sample (*middle*, *right*). However, the region within which the N-domains fluctuate is defined in the variance map (arrowheads; *bottom*, *right*; high variances are dark). The cryo-EM difference map between the wild-type complex and the deletion mutant shows very faint density in this region, which was reproduced by modeling appropriate spatial distributions of this domain (see Ref. 28; whence the panels of this figure were adapted). Bar = 100 Å.

The current picture, then, is that — for many particles — conformational variability poses an impediment to achieving the resolution in cryo-EM reconstructions that the resolving power of current electron microscopes suggests to be possible. The positive aspect of this situation is that the same phenomenon potentially offers insight into their dynamic properties.

"Less Low Throughput" Techniques in Cryo-EM

The fact of life that the number of ORFs in a cellular genome and the number of proteins expressed in a cell or organism are both very large has mandated, for structural proteomics as well as structural genomics, that analyses proceed along pipelines that are as efficient and as fully automated as possible, i.e. the development of "high throughput" techniques. In principle, cryo-EM should comply with the same operating principle. However, cryo-EM may be less amenable to "high throughput" methods, both because there are rate-limiting steps that do not readily lend themselves to automation and because some aspects of imaging and computational analysis are not yet fully mature. In some ways at least, cryo-EM appears more conducive to hypothesis-driven research than to mass production (see "Perspective"). Nevertheless, there is ample scope for accelerating cryo-EM analyses beyond the tempo presently practiced, particularly in SPA and EC, in those steps that are amenable to automation and by linking them cohesively. The sequence of operations is outlined in Fig. 2. In fact, some degree of automation has already been accomplished for most steps. ET embodies an advanced degree of automation in data acquisition and tomogram reconstruction, while awaiting similar developments in tomogram interpretation (e.g. segmentation) and the averaging of subtomograms. In cryo-EM, therefore, "less low throughput" may be a more realistic goal than "high throughput."

We proceed from the (nontrivial) assumption that biochemical procedures to isolate a specimen of interest have been optimized and it is available in, or may be switched to, a sample volume of 10–50 μl at 1–3 μg/μl protein in buffer without significant content of components that are viscous (e.g. glycerol or sucrose) or of high density (e.g. 1 M ionic strength). Typically, 3 μl are used to make one cryo-EM grid and concentrations of 1–3 μg/µl in bulk solution yield satisfactory distributions of some particles in vitrified thin films. From this point on, there are four stages to an analysis (Fig. 2).

(i) *Specimen preparation for cryo-EM, aka "getting it to work."* Here the goal is to identify conditions that reproducibly yield, in thin vitrified films suspended over holey carbon films, monolayer fields of particles that are not aggregated, i.e. in immediate contact with neighbors but separated by, ideally, 0.5–2 particle diameters for SPA and somewhat more for ET to avoid particle overlap in high-tilt projections. To meet this condition, the particles have to be mutually repelling, e.g. via electrostatic forces. In practice, the time needed to achieve this goal is highly variable, depending on the idiosyncracies of a given specimen and the skill and tenacity of the operator. In refractory cases, it may be necessary to settle for suboptimal distributions and extracting smaller numbers of particles from larger numbers of micrographs. Some prototypic methods for automated grid preparation and screening have been explored,[31,32] but have yet to enter widespread use.

(ii) *Recording and digitizing the data.* Given well optimized specimen preparation conditions, recording an adequate set of micrographs should be an efficient process, requiring no more than a few days, with either manual or automatic[33] data collection. Then, particles must be picked from the digitized micrographs, each roughly centered in a box and surrounded by a small margin of background. A variety of automated approaches have been developed for performing this otherwise laborious task.[34]

A key consideration to completing a targeted analysis is the amount of data required. Let N_p be the number of particles picked, and N_{av}, the number of averaging operations implicitly involved in the final reconstruction. The relationship between these parameters depends on several factors — notably, internal symmetries which, if present, yield N_{sym} equivalent views from a single image. The redundancy factor is specific for a given symmetry, e.g. $N_{sym} = 60$ in the case of icosahedral virus capsids with their 5-3-2 point group symmetry, or $N_{sym} = 6$ in the case of a hexameric ring (C6 symmetry). It is generally the case that not all images picked are of the same quality. Eventually they are ranked in terms of correlation coefficients calculated *versus* the current reconstruction, and an operator may impose

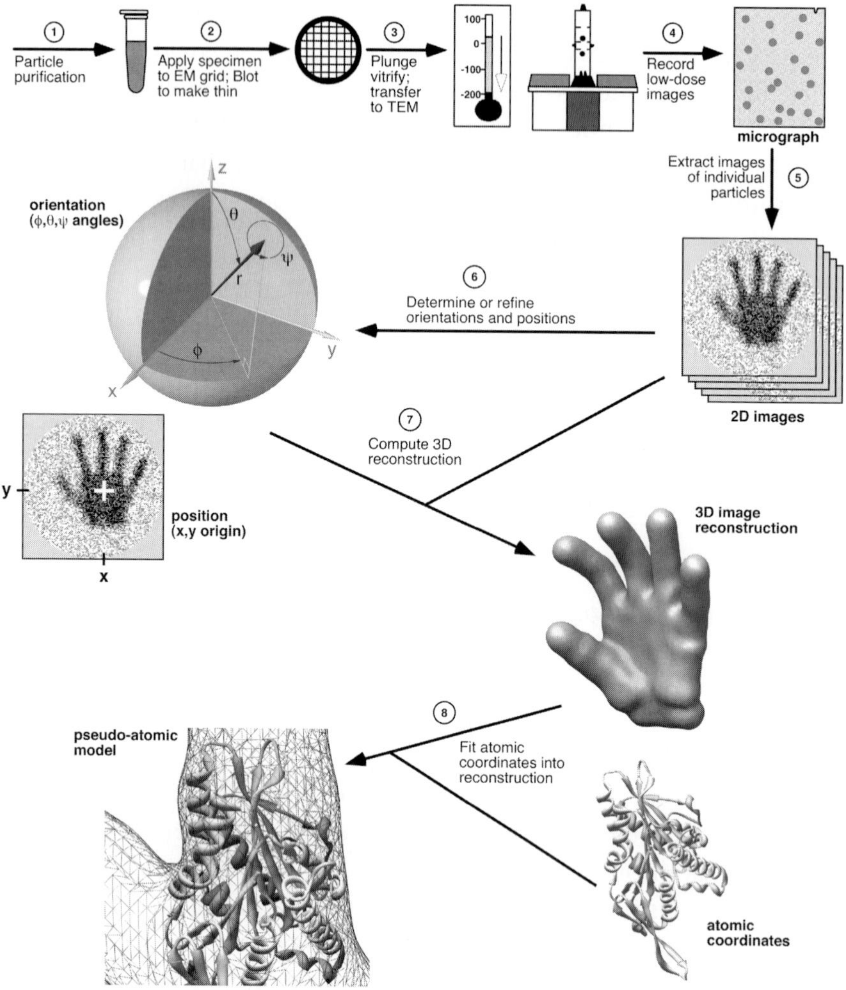

Fig. 2 Pipeline of operations involved in the structural analysis of a macromolecular complex by cryo-EM. Automated or semi-automated procedures are available for all steps, except for step 1, which can be rate-limiting. Steps 1 to 3 are sample preparation; 4, microscopy; 5 to 7, image processing; and 8, interpretation, including fitting:

(1) Sample purification: the sample must be pure enough that the particles of interest can be distinguished in the images.
(2) A droplet of sample (2 to 5 μl) is applied to an EM grid bearing a carbon film (perforated or continuous), then reduced to a thin film by blotting with filter paper.
(3) The grid is plunged into a cryogen, typically liquid ethane. Because the film is thin, it freezes rapidly enough to produce vitreous ice.

an acceptability threshold corresponding to a given fraction, f_{acc}, of the data.

Thus $$N_{av} \sim f_{acc} \cdot N_p \cdot N_{sym}$$

On the important point of what value of N_{av} is required to achieve a given resolution, consensus is still lacking but we note that 4–5 Å resolution has been accomplished with $N_{av} \sim 40{,}000$ for bacterial flagellin.[11] In general, N_{av} will depend both on the quality (i.e. resolution, signal-to-noise ratio, and contrast) of the original images and their number. All other factors being equal, N_{av} is roughly proportional to the particle diameter.[35]

(iii) *Performing the reconstruction(s).* A comprehensive account of the various strategies that may be pursued to obtain a SPA reconstruction

---

Fig. 2 (*Continued*) (4) The grid is kept cold, via liquid nitrogen, until images are recorded. Once the grid is in the microscope, the operator searches for suitable areas (thin ice, appropriate density of particles). Focus and other microscope settings are adjusted by observing an area near, but not at, the area of interest. The beam is then switched to the area of interest only for the recording of the image, so that the specimen is exposed to as few electrons as possible, minimizing radiation damage.
(5) Recorded images are checked for suitability for image processing. Good images have low astigmatism (variation of focus with direction perpendicular to the optical axis) and drift (movement of sample during exposure). If the field-of-view is larger than the desired area to be reconstructed, the area-of-interest is extracted. This may not be necessary for a tomography experiment. For data sets to be used in SPA reconstructions, each particle is extracted as a separate sub-image, as shown. For cryo-EM, the microscope is under-focused to generate image contrast at low resolution. As a result, information in alternating, higher-frequency, resolution bands is conveyed with inverted and unevenly weighted contrasts. This distortion can be largely corrected computationally, a procedure known as contrast transfer function (CTF) correction.
(6) Translational and rotational alignments. The positions $[x, y]$ and orientation angles $[\phi, \theta, \psi]$ are determined. For definition of these parameters, see Ref. 62.
(7) Computation of 3D reconstruction from the 2D images with their assigned orientations and positions. Usually steps 6 and 7 are repeated iteratively until no further improvement in the resolution of the reconstruction is observed.
(8) Atomic coordinates may be fitted into the finished reconstruction either manually or by a computer-aided method.

goes beyond the scope of this chapter (see Ref. 4). In brief, three interconnected operations are involved: determining the particles' orientations and origins; determining the phase contrast transfer function for each micrograph and correcting for its distortions; and calculating the density map. Most studies today employ some form of projection matching.[36] This is an iterative procedure in which the current density map is used to generate a set of projections, and the orientation and translational setting of each particle are determined by pattern recognition techniques that match them to the projection that they most closely resemble. A next-generation density map is then calculated and the cycle repeated until no further improvement is registered. This scenario works smoothly and readily lends itself to automation provided that a robust and suitably detailed starting model is available.[37] Indeed, such specimens may now be processed rapidly and efficiently, potentially within a few days. A more complete level of automation has been accomplished in which the reconstruction is integrated with data collection and particle-picking that is particularly helpful for studies that require very large numbers of particles.[33] However, if a well conditioned starting model is not available, obtaining one may be rate-limiting for the whole analysis.[38] In this context, ET, as enhanced by tomographic averaging, offers a promising source for unbiased starting models.

(a) *Difference mapping*. The principal utility of a cryo-EM map of a protein complex lies in what it is able to disclose about the interactions among its components. For disclosure to be forthcoming, it is necessary to identify the location of each subunit and to delineate the boundaries between them. Interpretation of a density map along these lines may be a complicated process, depending, again, on multiple factors — notably, the number of distinct components and their sizes, shapes, and secondary structures. Within the cryo-EM repertoire, there are several approaches collectively known as "difference mapping." In each case, the goal is to perturb the complex locally in some chemically defined way and then to localize the affected component by comparing density maps of the perturbed and original structures. The perturbation may involve tagging with metal clusters, e.g. Ref. 39; tagging with antibodies (preferably Fab fragments), e.g. Ref. 40; appending or deleting a domain or peptide to serve as a marker for the insertion site, e.g. Ref. 41; selectively extracting loosely bound components, e.g. Ref. 42; excising proteolytically sensitive moieties, e.g. Ref. 43; or comparison with mutants or homologs that happen to lack some component. These procedures are somewhat laborious and it is hard to generalize about timelines for completion. They are, at least, expedited by the consideration that the original density map may be used to initiate projection-matching reconstructions of perturbed structures. Another related approach is possible if high resolution structures have been determined by X-ray crystallography or NMR spectroscopy for one or more constituent subunit (or a close homolog), and the cryo-EM map is sufficiently detailed for its position to be identified. In this way, binding sites and modes of interaction may be specified in greater detail than the nominal resolution of the cryo-EM map would suggest to be likely[40,44]; and conversely, densities recognized as candidates for other components of the complex.[45]

(b) *Automated interpretation.* Even at the relatively modest resolution of 10 Å, a density map of a large macromolecular complex exhibits great complexity. Steps towards automation-assisted interpretation have been taken with the development of programs that seek to perform shape-based identification of secondary structures — i.e. α-helices and β-sheets[46–49] and to pick out density motifs characteristic of certain domains.[46]

The Hybrid Approach to Leveraging Resolution

Systematic integration of data from cryo-EM and X-ray crystallography is the cornerstone of "hybrid methods" in which multiple experimental approaches are simultaneously brought to bear.[50] It offers a means of overcoming, on the one hand, the limited resolution of cryo-EM density maps of complete macromolecular particles, and, on the other hand, the limited insight to be gained from crystal structures of single subunits. In such an experiment, a high resolution structure is fitted as precisely as possible into a cryo-EM density map. (The term "docking" is also used to describe this operation). The first fittings were done by hand, with the operator using molecular graphics to maneuver the fitted molecule into the position and orientation that best fits the cryo-EM map, usually portrayed as a translucent surface (e.g. Fig. 3). This operation has been put on a more quantitative and objective basis by the development of automated and semi-automated fitting programs, of which there is now a substantial number (see Appendix).

Ideally, the fitted structure fits precisely into the molecular envelope defined by the density map. However, in some situations, the procedure must be generalized. In one of them — called "flexible fitting" — the molecule to be fitted is considered not as a single rigid body but as consisting of plastic components (Fig. 3). Another situation that requires procedural adjustment is when multiple, mutually occluding, binding sites exist for an interaction partner on the surface of a complex.

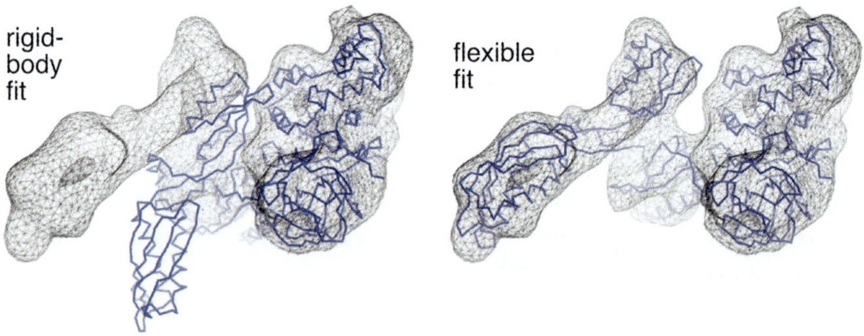

Fig. 3 Flexible fitting of elongation factor EF-G into cryo-EM density (adapted from Ref. 63). Cryo-EM density is represented by the transparent wireframe surface. Molecular models are depicted as blue backbone traces. Here, the crystallographically determined atomic coordinates of EF-G were fitted first as a rigid body into the cryo-EM density (*left* panel). Only the right most part of the density was used for fitting. Here, the fit is good but on the left side, the cryo-EM density appears to have the same shape as the corresponding domains but are differently positioned. It follows that the two conformations are markedly different. Next, the coordinates of these domains were allowed to move flexibly with respect to the remainder of the model. A much better fit was obtained (*right*). The UCSF Chimera package[64] was used to produce this figure as well as parts of Figs. 4 and 5.

If the partner binds randomly to available sites, the corresponding density in a cryo-EM map will not directly match its shape, but rather, represents a merged distribution whose strongest feature is the overlap zone. This contingency may be addressed by performing the fitting to match grayscale sections, not an isodensity contour (Figs. 4 and 5).

In general, the higher the resolution of the cryo-EM reconstruction, the more precise the fitting will be (Fig. 6). α-helices and β-sheets are observable at 10 Å resolution or better. At these resolutions, coordinates can be assigned very precisely, i.e. to within a RMSD of ~1 Å for backbone atoms of the polypeptide chain(s). At lower resolution (e.g. 15–20 Å or so), it may still be possible to achieve high precision if the fitted molecule has a distinctively a symmetric shape that is captured in the cryo-EM density map. At still lower resolutions, ambiguities mount although it may well still be possible to pinpoint the center of mass of the fitted molecule and its point of contact with the rest of the complex.

In any event, the outcome of such an experiment is a "pseudo-atomic model" in which atomic coordinates are assigned for some or all of the amino acids in the complex of interest. Ideally, such a model should emulate an experimentally determined atomic model. In practice, some discrepancies are generally encountered between the fitted structure and the cryo-EM map.

Discrepancies Between Crystal Structures and Cryo-EM Structures

Under what circumstances may observed discrepancies be considered small enough to be dismissed as "rounding errors"? Conversely, by what thresholding criteria should they be deemed significant and requiring explanation? At present, there is no rigorous criterion for resolving this issue. Discrepancies may originate, prosaically, in technical error of some sort in one or the other structure. A more interesting source of discrepancy lies in the consideration that different structures are being observed in the respective experiments; thus, they may both be right but different. For instance, the domains of the ribosomal elongation factor EF-G are configured differently when

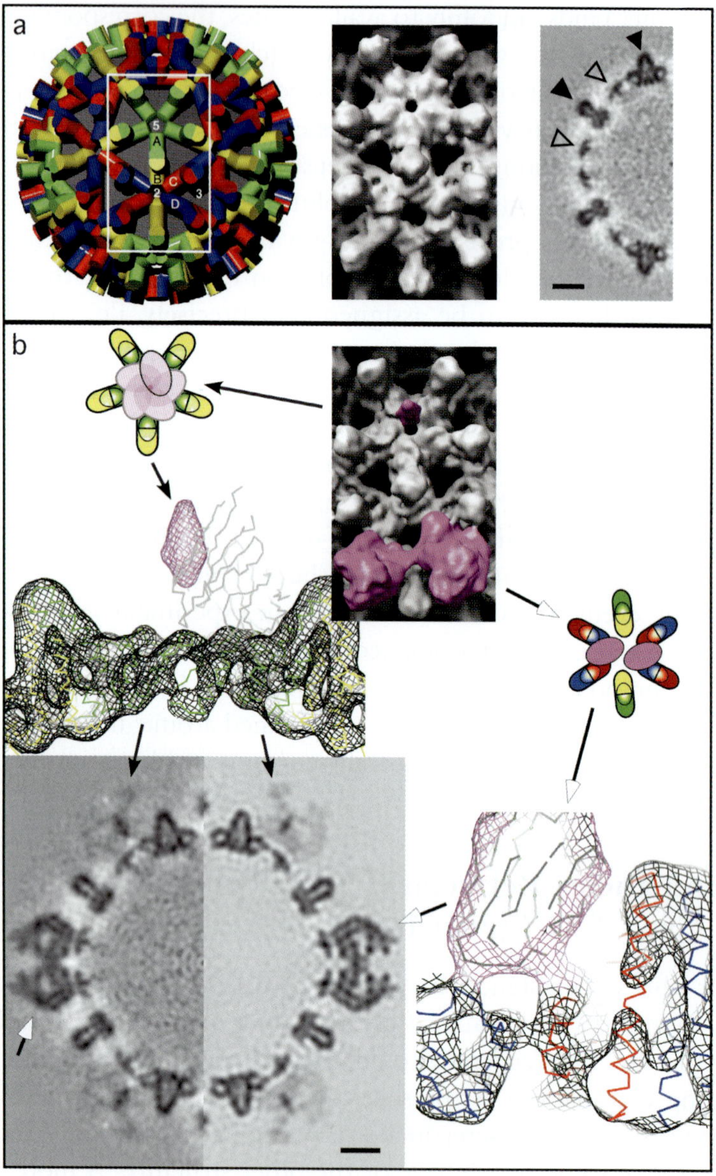

Fig. 4 A case in which the shape of a component in a cryo-EM density map, as represented by isodensity contouring, does not match that of the fitted molecule. These capsids of hepatitis B virus (HBV) are icosahedrally symmetric (triangulation number $T = 4$) and consist of 240 subunits configured as 120 dimers. There are 60 copies

visualized by cryo-EM bound to the ribosome[51] or separate from it, in crystals[52] — Fig. 3. The cryo-EM map represents a solution structure of a functional complex, whereas a crystal-derived map — usually pertaining to molecules experiencing markedly different solvent conditions — is a solid-state structure that may be affected by crystal contacts. A frequently occurring situation with multiple conformational states involves the binding of different nucleotides to ATPase molecules. Recent experience with the motor protein

---

Fig. 4 (*Continued*) each of four quasi-equivalent subunits, labeled A–D on the model (**a** — *left*), (i.e. they are chemically identical but have slightly different conformations and bonding environments), colored green, yellow, red, and blue, respectively. 5-fold, 2-fold, and 3-fold symmetry axes are labeled. The boxed area is conveyed in a cryo-EM density map at 9 Å resolution; Ref. 13 (*middle*). Its most pronounced feature is the protruding spikes. At right is shown a half-plane gray-scale section through the density map. Longitudinal and transverse sections through spikes are marked with solid arrowheads. Open arrowheads designate densities at the spike bases. Bar (*right*) = 50 Å.
(b) Labeling experiment with Fabs of the monoclonal antibody 3120; Ref. 65. Two Fab-related features were observed in the resulting density map (in magenta, *top/middle* panel). One feature has the exact shape of a Fab (two copies are seen at the bottom of this panel). The second feature, much smaller and cigar-shaped, on the 5-fold axis, is also shown in sideview in the wireframe diagram (*top left*). Fitting of the PDB coordinates of a Fab of the same IgG subtype into the first feature gave an excellent fit, as illustrated in the grayscale section at bottom left where the left-hand half shows cryo-EM density and the right-hand half, the corresponding pseudo-atomic model, band-limited to the same resolution (10 Å). Pairs of Fabs, so close as to be essentially in contact, are marked with open-headed arrows. The cigar-shaped density could be reproduced as the overlap volume of Fabs at the symmetry-related sites clustered around the 5-fold axis. In this case, the capsid-binding geometry of the Fabs is such that one Fab, once bound, occludes the other four sites. Overall, the five sites are randomly occupied, each at an average occupancy of 20%, and the only density seen at 100% occupancy is the cigar-shaped occlusion zone. There are, in fact, six potential binding sites — two copies each of three quasi-equivalent sites — around the 2-fold axis where the two complete Fabs are visualized. Serendipitously, one of these three epitopes has markedly higher affinity for the Fab than the other two, giving occupancies of (~100%, ~0%, ~0%), so that two complete Fabs were visualized, not a merging of six Fabs with 16% or 33% occupancy. The schematic diagrams at top left and middle right convey the occupancy/occlusion patterns at the 5-fold and 2-fold (quasi-6-fold) sites. Bar = 50 Å (for grayscale sections).

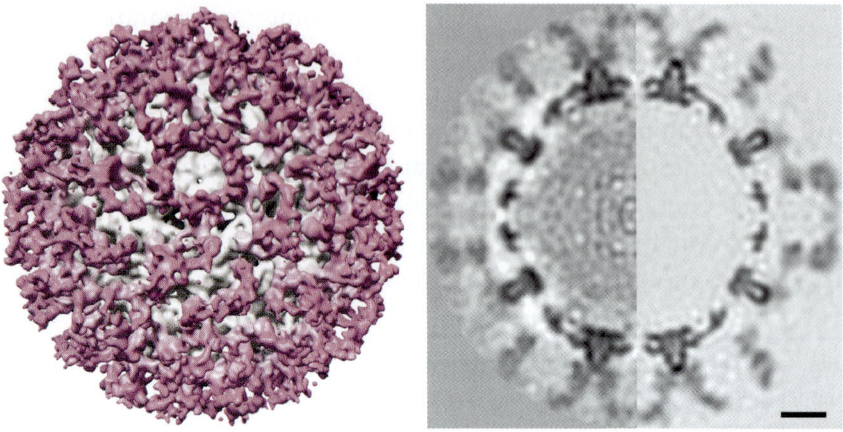

Fig. 5 Density distribution of a substoichiometrically bound Fab on the surface of the HBV $T = 4$ capsid. The capsid architecture is diagrammed in Fig. 4(a). Left, surface rendering: Fab density, magenta; capsid white. As the isodensity contour does not delineate individual Fab molecules, the fitting was done manually, optimizing the match with grayscale sections. At right, the left-hand half shows the density of the cryo-EM map and the right-hand half, the modeled structure. The Fabs bind to the tips of the capsid spikes. Adapted from Ref. 66. Bar = 50 Å.

ncd visualized by cryo-EM bound to microtubules[53,54] or in crystals[55,56] cautions that the correlation between conformational states and nucleotide binding may not be straightforward.

In particular, mobile elements may be visible in one representation but invisible in the other. To illustrate this eventuality, we cite a few examples. (i) The 8-residue C-terminal peptide of the hepatitis B virus capsid protein was not seen in a crystal structure at 3.4 Å resolution[57] but was detected in a cryo-EM map at 9 Å resolution.[41] Presumably, it was seen in the lower resolution data because, although adopting various conformations, it occupied much the same space in each instance. (ii) Conversely, the 154-residue N-terminal domain of the ClpA ATPase was visualized in 2.6 Å crystal structure[58] but essentially invisible in a cryo-EM structure at 11.5 Å[28,a]. It appears that the domain was locked into a fixed position by crystal

[a] The crystal structure was of the monomer in the ADP state, and the cryo-EM structure, of the hexamer in the ATPγS state.

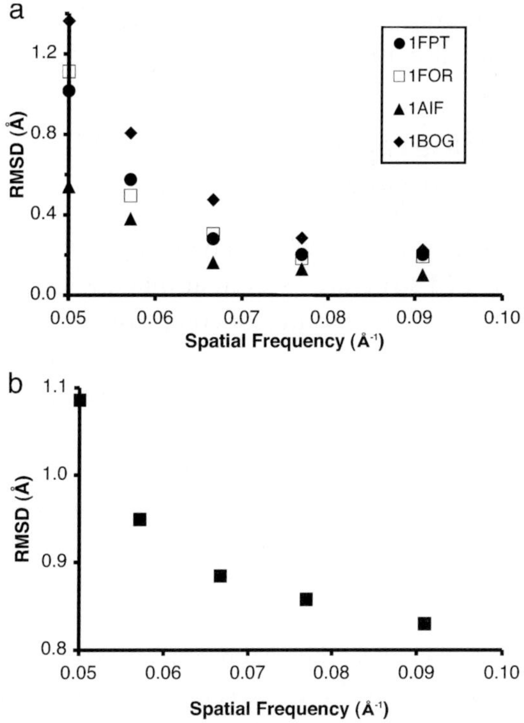

Fig. 6 The precision of atomic coordinates determined by fitting molecules of known structure into cryo-EM maps of complexes improves progressively at higher resolutions. Four Fab coordinates (antibody-binding domain) were fitted into the Fab-related density in a cryo-EM density map at 10 Å resolution[65] via the program CoLoRes (Appendix). Here, prior fittings were repeated but the resolution of the EM map was limited to 11.0, 13.0, 15.0, 17.5, and 20.0 Å, respectively. (**a**) Comparison of repeated fits to the initial fit reported.[65] Protein Data Bank identification codes for the fitted coordinates are listed. (**b**) Root-mean-square deviation (equivalent here to the standard deviation) of all four fits when only $C\alpha$ coordinates for the residues within β-strands were compared. Note how in both (a) and (b), the precision of the fits falls off with decreasing resolution. This example was an exceptionally favorable case in that the cryo-EM density for the Fab was strong and well defined, as was its correlation to the atomic coordinates.

contacts whereas, in solution — although folded — it is connected to the main body of the molecule by a flexible linker, allowing it to undergo fluctuations of 35–40 Å that rendered it invisible after averaging.[28] (iii) The ribosomal protein L7/L12, consisting of two dimers of 12 kDa subunits, was not seen in a 2.4 Å resolution X-ray map of the 50S subunit[59] but was perceived in part in a 13.8 Å cryo-EM map, a discrepancy also attributed to flexible inter-domain linkers.[60]

X-ray Crystallography and Cryo-EM are Mutual Complements

Even for complete particles for which a high resolution crystal structure is feasible, a cryo-EM structure — even at much lower resolution — is not redundant but, rather, a source of complementary information. In the sense that the crystal structure represents a "ground state," the cryo-EM map relates to the solution structure or structures and systematic comparison of the two representations may yield insights into the dynamic properties of the complex. Moreover, cryo-EM affords an avenue via which the particle's interactions with substrates, cofactors, other functional partners, antibodies, and regulatory small molecules may be addressed. In general, it is unlikely that this spectrum of related conformational states will all be susceptible to crystallographic analysis at high resolution. On the other hand, once a structure (cryo-EM or crystal) has been determined for one state, it may be used as a starting model for projection-matching analyses of altered states. As the iterations proceed, the data should impose themselves, ultimately rendering an unbiased representation of the state in question. The logic of this approach is very similar to that of molecular replacement in X-ray crystallography. Once a cryo-EM map of an alternative state has been determined, if a crystal structure is available for the ground state, it may be perturbed as appropriate to fit the altered state and thus to ascertain the molecular movements that accompany this transition.

Perspective

As cryo-EM settles into its role in the structural proteomics enterprise, we may consider future directions. High on the list of desirable methodological advances are digital cameras with improved sensitivity (modulation transfer function) and larger format. Significant enhancement of these features would move the present generation of electronic cameras past the performance level of cut film as a recording medium, and allow full exploitation of their other advantages for high throughput work. Second, there is much to be gained from innovations in methods to isolate low-abundance macromolecular complexes from appropriate source cells or tissues and to co-express the large numbers of protein subunits required for many complexes, with due attention to the mechanisms that are required for their correct assembly.

Further progress may be anticipated in hybrid methods. With the emerging trend of cryo-EM structures being deposited in generally accessible data bases (e.g. www.ebi.ac.uk/msd), the basis for this activity will broaden. Another appealing generalization of fitting studies involves the systematic incorporation of other kinds of data — e.g. FRET, SAXS, crosslinking, molecular dynamics, etc. To be sure, it is not clear at this early juncture how best to combine disparate kinds of data, e.g. in weighting various kinds of observation, or in defining penalty functions for failure to match or Lagrangian functions for optimization. What is clear is that an interdisciplinary crossfire is needed to elucidate the complexities, static and dynamic, of large macromolecular assemblies and the computer is the best place — perhaps the only place — where diverse data sets may be combined. Integrative modeling studies of this kind have the potential to develop into a major branch of computational biology.

Finally, we may anticipate a unification of *in vitro* and *in situ* observations. Cryo-ET of vitrified whole cells and tissue sections offers a way of mapping populations of macromolecules in natively preserved cells.[61] After individual macromolecules are identified as densities within a tomogram, the resolution may be enhanced by fitting in higher resolution structures from SPA, X-ray crystallography, or NMR spectroscopy. In this context, a critical question is whether sufficiently favorable resolution and signal-to-noise ratios can be achieved in the crowded intracellular milieu depicted in cryo-tomograms to allow unambiguous identification of the majority of complexes and to recognize distinct functional states of the same complex. Again, other methodologies — notably, innovative light microscopies — will surely contribute. Looking ahead, this hybrid approach has the potential to bridge cell biology and structural biology. While much remains to be done in structural proteomics, "structural cytomics" may be the next frontier.

Acknowledgments

We thank Drs J. Conway, J. Frank, T. Ishikawa, and M. Valle for contributing figure components, many of the developers of fitting algorithms for their help in compiling the Appendix, and Dr J. Conway for a critical reading of the manuscript. This work was supported, in part, by the Intramural Research Program of the National Institute of Arthritis, Musculoskeletal, and Skin Diseases.

References

1. Alberts B. (1998) "The cell as a collection of protein machines: preparing the next generation of molecular biologists." *Cell* **92**(3): 291–4.
2. Baumeister W, Steven AC. (2000) "Macromolecular electron microscopy in the era of structural genomics." *Trends Biochem Sci* **25**(12): 624–31.
3. Stahlberg H, Fotiadis D, Scheuring S, *et al.* (2001) "Two-dimensional crystals: a powerful approach to assess structure, function and dynamics of membrane proteins." *FEBS Lett* **504**(3): 166–72.
4. Frank J. (2006) Three-dimensional Electron Microscopy of Macromolecular Assemblies. Oxford University Press, New York.
5. Lucic V, Forster F, Baumeister W. (2005) "Structural studies by electron tomography: from cells to molecules." *Annu Rev Biochem* **74**: 833–65.
6. Downing KH, Nogales E. (1999) "Crystallographic structure of tubulin: implications for dynamics and drug binding." *Cell Struct Funct* **24**(5): 269–75.
7. Gonen T, Cheng Y, Sliz P, *et al.* (2005) "Lipid-protein interactions in double-layered two-dimensional AQP0 crystals ." *Nature* **438**: 633–638, corrigendum **441**: 248.
8. Henderson R, Baldwin JM, Ceska TA, *et al.* (1990) "Model for the structure of bacteriorhodopsin based on high-resolution electron cryo-microscopy." *J Mol Biol* **213**(4): 899–929.
9. Adrian M, Dubochet J, Fuller SD, Harris JR. (1998) "Cryo-negative staining." *Micron* **29**(2–3): 145–60.
10. Yan X, Olson NH, Van Etten JL, *et al.* (2000) "Structure and assembly of large lipid-containing dsDNA viruses." *Nat Struct Biol* **7**(2): 101–3.
11. Yonekura K, Maki-Yonekura S, Namba K. (2003) "Complete atomic model of the bacterial flagellar filament by electron cryomicroscopy." *Nature* **424**(6949): 643–50.

12. Ludtke SJ, Chen DH, Song JL, *et al.* (2004) "Seeing GroEL at 6 A resolution by single particle electron cryomicroscopy." *Structure* **12**(7): 1129–36.
13. Conway JF, Cheng N, Zlotnick A, *et al.* (1997) "Visualization of a 4-helix bundle in the hepatitis B virus capsid by cryo-electron microscopy." *Nature* **386**(6620): 91–4.
14. Böttcher B, Wynne SA, Crowther RA. (1997) "Determination of the fold of the core protein of hepatitis B virus by electron cryomicroscopy." *Nature* **386**(6620): 88–91.
15. Trus BL, Roden RB, Greenstone HL, *et al.* (1997) "Novel structural features of bovine papillomavirus capsid revealed by a three-dimensional reconstruction to 9 Å resolution." *Nat Struct Biol* **4**(5): 413–20.
16. Medalia O, Weber I, Frangakis AS, *et al.* (2002) "Macromolecular architecture in eukaryotic cells visualized by cryoelectron tomography." *Science* **298**(5596): 1209–13.
17. Penczek PA. (2002) "Three-dimensional spectral signal-to-noise ratio for a class of reconstruction algorithms." *J Struct Biol* **138**(1–2): 34–46.
18. Cardone G, Grünewald K, Steven AC. (2005) "A resolution criterion for electron tomography based on cross-validation." *J Struct Biol* **151**(2): 117–29.
19. Unser M, Sorzano CO, Thevenaz P, *et al.* (2005) "Spectral signal-to-noise ratio and resolution assessment of 3D reconstructions." *J Struct Biol* **149**(3): 243–55.
20. Forster F, Medalia O, Zauberman N, *et al.* (2005) "Retrovirus envelope protein complex structure *in situ* studied by cryo-electron tomography." *Proc Nat'l Acad Sci USA* **102**(13): 4729–34.
21. Lata R, Conway JF, Cheng N, *et al.* (2000) "Maturation dynamics of a viral capsid: visualization of transitional intermediate states." *Cell* **100**(2): 253–63.
22. Heymann JB, Cheng N, Newcomb WW, *et al.* (2003) "Dynamics of herpes simplex virus capsid maturation visualized by time-lapse cryo-electron microscopy." *Nat Struct Biol* **10**(5): 334–41.
23. Heymann JB, Conway JF, Steven AC. (2004) "Molecular dynamics of protein complexes from four-dimensional cryo-electron microscopy." *J Struct Biol* **147**(3): 291–301.
24. Gao H, Valle M, Ehrenberg M, Frank J. (2004) "Dynamics of EF-G interaction with the ribosome explored by classification of a heterogeneous cryo-EM dataset." *J Struct Biol* **147**(3): 283–90.
25. Scheres SH, Marabini R, Lanzavecchia S, *et al.* (2005) "Classification of single-projection reconstructions for cryo-electron microscopy data of icosahedral viruses." *J Struct Biol* **151**(1): 79–91.
26. Fu J, Gao H, Frank J. (2007) "Unsupervised classification of single particles by cluster tracking in multi-dimensional space." *J Struct Biol* **157**(1): 226–39.
27. Scheres SH, Gao H, Valle M, *et al.* (2007) "Disentangling conformational states of macromolecules in 3D-EM through likelihood optimization." *Nat Methods* **4**(1): 27–9.

28. Ishikawa T, Maurizi MR, Steven AC. (2004) "The N-terminal substrate-binding domain of ClpA unfoldase is highly mobile and extends axially from the distal surface of ClpAP protease." *J Struct Biol* **146**(1–2): 180–8.
29. Penczek PA, Yang C, Frank J, Spahn CM. (2006) "Estimation of variance in single-particle reconstruction using the bootstrap technique." *J Struct Biol* **154**(2): 168–83.
30. Leschziner AE, Nogales E. (2007) "Visualizing flexibility at molecular resolution: analysis of heterogeneity in single-particle electron microscopy reconstructions." *Annu Rev Biophys Biomol Struct* **36**: 43–62.
31. Potter CS, Pulokas J, Smith P, *et al.* (2004) "Robotic grid loading system for a transmission electron microscope." *J Struct Biol* **146**(3): 431–40.
32. Lefman J, Morrison R, Subramaniam S. (2006) "Automated 100-position specimen loader and image acquisition system for transmission electron microscopy." *J Struct Biol* **158**(3): 318–26.
33. Stagg SM, Lander GC, Pulokas J, *et al.* (2006) "Automated cryo-EM data acquisition and analysis of 284742 particles of GroEL." *J Struct Biol* **155**(3): 470–81.
34. Zhu Y, Carragher B, Glaeser RM, *et al.* (2004) "Automatic particle selection: results of a comparative study." *J Struct Biol* **145**(1–2): 3–14.
35. Conway JF, Steven AC. (1999) "Methods for reconstructing density maps of "single particles" from cryoelectron micrographs to subnanometer resolution." *J Struct Biol* **128**: 106–18.
36. Baker TS, Cheng RH. (1996) "A model-based approach for determining orientations of biological macromolecules imaged by cryoelectron microscopy." *J Struct Biol* **116**: 120–30.
37. Mouche F, Zhu Y, Pulokas J, *et al.* (2003) "Automated three-dimensional reconstruction of keyhole limpet hemocyanin type 1." *J Struct Biol* **144**(3): 301–12.
38. Leschziner AE, Nogales E. (2006) "The orthogonal tilt reconstruction method: an approach to generating single-class volumes with no missing cone for *ab initio* reconstruction of asymmetric particles." *J Struct Biol* **153**(3): 284–99.
39. Milligan RA, Whittaker M, Safer D. (1990) "Molecular structure of F-actin and location of surface binding sites." *Nature* **348**(6298): 217–21.
40. Wang GJ, Porta C, Chen ZG, *et al.* (1992) "Identification of a Fab interaction footprint site on an icosahedral virus by cryoelectron microscopy and X-ray crystallography." *Nature* **355**: 275–8.
41. Watts NR, Conway JF, Cheng N, *et al.* (2002) "The morphogenic linker peptide of HBV capsid protein forms a mobile array on the interior surface." *EMBO J* **21**(5): 876–84.
42. Aebi U, van Driel R, Bijlenga RK, *et al.* (1977) "Capsid fine structure of T-even bacteriophages. Binding and localization of two dispensable capsid proteins into the P23* surface lattice." *J Mol Biol* **110**: 687–98.
43. Carrascosa JL, Steven AC. (1978) "A procedure for evaluation of significant structural differences between related arrays of protein molecules." *Micron* **9**: 199–206.

44. Rayment I, Holden HM, Whittaker M, *et al.* (1993) "Structure of the actin-myosin complex and its implications for muscle contraction." *Science* **261**(5117): 58–65.
45. Stewart PL, Fuller SD, Burnett RM. (1993) "Difference imaging of adenovirus: bridging the resolution gap between X-ray crystallography and electron microscopy." *Embo J* **12**(7): 2589–99.
46. Jiang W, Baker ML, Ludtke SJ, Chiu W. (2001) "Bridging the information gap: computational tools for intermediate resolution structure interpretation." *J Mol Biol* **308**(5): 1033–44.
47. Kong Y, Ma J. (2003) "A structural-informatics approach for mining beta-sheets: locating sheets in intermediate-resolution density maps." *J Mol Biol* **332**(2): 399–413.
48. Dal Palu A, He J, Pontelli E, Lu Y. (2006) "Identification of alpha-helices from low resolution protein density maps." *Comput Syst Bioinformatics Conf*, 89–98.
49. Baker ML, Ju T, Chiu W. (2007) "Identification of secondary structure elements in intermediate-resolution density maps." *Structure* **15**(1): 7–19.
50. Hanein D (ed). (2007) "Hybrid methods for the structural analysis of macromolecular assemblies." (Special Issue) *J Struct Biol* **158**(2): 132–254.
51. Allen GS, Zavialov A, Gursky R, *et al.* (2005) "The cryo-EM structure of a translation initiation complex from *Escherichia coli.*" *Cell* **121**(5): 703–12.
52. Roll-Mecak A, Cao C, Dever TE, Burley SK. (2000) "X-ray structures of the universal translation initiation factor IF2/eIF5B: conformational changes on GDP and GTP binding." *Cell* **103**(5): 781–92.
53. Wendt TG, Volkmann N, Skiniotis G, *et al.* (2002) "Microscopic evidence for a minus-end-directed power stroke in the kinesin motor ncd." *Embo J* **21**(22): 5969–78.
54. Endres NF, Yoshioka C, Milligan RA, Vale RD. (2006) "A lever-arm rotation drives motility of the minus-end-directed kinesin ncd." *Nature* **439**(7078): 875–8.
55. Sablin EP, Case RB, Dai SC, *et al.* (1998) "Direction determination in the minus-end-directed kinesin motor ncd." *Nature* **395**(6704): 813–6.
56. Yun M, Bronner CE, Park CG, *et al.* (2003) "Rotation of the stalk/neck and one head in a new crystal structure of the kinesin motor protein, ncd." *Embo J* **22**(20): 5382–9.
57. Wynne SA, Crowther RA, Leslie AG. (1999) "The crystal structure of the human hepatitis B virus capsid." *Mol Cell* **3**(6): 771–80.
58. Guo F, Maurizi MR, Esser L, Xia D. (2002) "Crystal structure of ClpA, an Hsp100 chaperone and regulator of ClpAP protease." *J Biol Chem* **277**(48): 46743–52.
59. Ban N, Nissen P, Hansen J, *et al.* (2000) "The complete atomic structure of the large ribosomal subunit at 2.4 Å resolution." *Science* **289**(5481): 905–20.
60. Bocharov EV, Sobol AG, Pavlov KV, *et al.* (2004) "From structure and dynamics of protein L7/L12 to molecular switching in ribosome." *J Biol Chem* **279**(17): 17697–706.

61. Nickell S, Kofler C, Leis AP, Baumeister W. (2006) "A visual approach to proteomics." *Nat Rev Mol Cell Biol* 7(3): 225–30.
62. Heymann JB, Chagoyen M, Belnap DM. (2005) "Common conventions for interchange and archiving of three-dimensional electron microscopy information in structural biology." *J Struct Biol* **151**(2): 196–207. Corrigendum **153**: 312.
63. Valle M, Zavialov A, Sengupta J, *et al.* (2003) "Locking and unlocking of ribosomal motions." *Cell* **114**(1): 123–34.
64. Pettersen EF, Goddard TD, Huang CC, *et al.* (2004) "UCSF Chimera — a visualization system for exploratory research and analysis." *J Comput Chem* **25**(13): 1605–12.
65. Conway JF, Watts NR, Belnap DM, *et al.* (2003) "Characterization of a conformational epitope on hepatitis B virus core antigen and quasi-equivalent variations in antibody binding." *J Virol* 77: 6466–73.
66. Belnap DM, Watts NR, Conway JF, *et al.* (2003) "Diversity of core antigen epitopes of hepatitis B virus." *Proc Nat'l Acad Sci USA* **100**(19): 10884–9.
67. Noda K, Nakamura R, Nishida Y, *et al.* (2006) "Atomic model construction of protein complexes from electron micrographs and vizualization of their 3D structure using a virtual reality system." *J Plasma Phys* **72**: 1037–40.

Appendix

Table 1. Published Algorithms for Computer-aided Fitting of Atomic-Resolution Coordinates into Lower-Resolution EM Density Maps

Year*	Algorithm, Program, or Package	Principle of Operation	Features of Algorithm	Reference	Availability
1993	Adenovirus-hexon fitting	Density computed from coordinates, step-search, cross-correlation	Manual fitting followed by rigid-body, automated refinement of fit	[1]	contact P. Stewart, phoebe.stewart@vanderbilt.edu
1994	Actin-myosin fitting	Reciprocal-space fitting, R-factor	Rigid-body fitting	[2, 3]	see references for contact information
1994	X-PLOR, etc.	Reciprocal-space, rigid-body refinement; least squares; correlation coefficient; R-factor	Manual fitting followed by rigid-body, automated refinement of fit	[4–7]	see references for contact information
1997	GAP	Correlation coefficient, steepest-ascent refinement	Rigid-body fitting	[8]	contact D. Stuart, equires@strubi.ox.ac.uk
1998	Situs-CoLoRes, FRM	Correlation of topology-representing neural networks, Laplacian correlation, vector	Rigid-body and flexible fitting (Situs, CoLoRes), fast rotational matching (FRM), interactive	[9–20]	situs.biomachina.org (Situs, CoLoRes, FilaSitus, Sculptor); contact W. Wriggers for FRM, wriggers@biomachina.org

Many, if not all, of the methods have grown out of X-ray crystallography and molecular modeling algorithms. The earliest methods employed only rigid-body fitting and the later methods are more complex, generally including algorithms for flexibly fitting coordinates to the maps.

(*Continued*)

Table 1 (*Continued*)

Year*	Algorithm, Program, or Package	Principle of Operation	Features of Algorithm	Reference	Availability
		quantization, anchor-point matching, spherical harmonics, normal-mode analysis	graphical user interface (Sculpter), filament-specific application (FilaSitus)		
1999	COAN	Global real-space correlation, biochemical information function (optional)	Rigid-body fitting, a set of potential fits is given, optional biochemical or biophysical data can be included in assessment of fits	[21–23]	contact N. Volkmann, neils@burnham.org
2000	DOCKEM	Local real-space density correlation: computed only within overlap of coordinates and density	Rigid-body fitting	[24]	contact A. Roseman, roseman@mrc-lmb.cam.ac.uk
2000	EMfit	Real-space density correlation	Iterative, rigid-body fitting	[25, 26]	contact M. Rossmann, mr@purdue.edu
2000	INSOUT	Optimizes a residual based on the sum of Fourier phase errors and sum of Fourier amplitudes squared	Rigid-body fitting, adjusts for neighboring coordinates, including symmetry-related subunits	[27, 28]	contact D. Filman, david_filman@hms.harvard.edu

(*Continued*)

Table 1 (*Continued*)

Year*	Algorithm, Program, or Package	Principle of Operation	Features of Algorithm	Reference	Availability
2000	Eos	Real-space density correlation, spatially restricted	Rigid-body fitting (pdbRhoFit), flexible fitting (mrc image to NAMD Constant Forces)	[29, 67]	www.yasunaga-lab.bse.kyutech.ac.jp/Eos
2001	AIRS (EMAN)	Cross-correlation, assessment of neighbors (SSEhunter); cross-correlation (foldhunter)	Searches for α-helices and β-sheets (SSEhunter); fitting of coordinates into density (foldhunter); segmentation of molecular subunits (AIRS-segment)	[30–35]	ncmi.bcm.tmc.edu/software/AIRS
2001	RSRef	Least-squares or energy minimization, density correlation, molecular dynamics	Flexible fitting, allows inter- and intra-molecular conformational changes in coordinates, observes stereochemical constraints, coordinates may be entire complexes or individual components	[36–39]	xtal.ohsu.edu
2002	NORMA-URO	Reciprocal-space matching, normal-mode analysis	Symmetry accounted for during optimization, independent bodies can	[40, 41]	www.elnemo.org/NORMA

(*Continued*)

Table 1 (*Continued*)

Year*	Algorithm, Program, or Package	Principle of Operation	Features of Algorithm	Reference	Availability
			be simultaneously fitted, both rigid-body (URO) and flexible fitting (NORMA)		
2002	OPUS	Vector quantization to set distinct, finite "cells" within molecule, elastic network analysis to determine elastic deformations of the molecule (QEDM); shape recognition, principal-components analysis, deconvolution, image restoration (sheetminer, sheettracer)	Models conformational flexibility of a protein without sequence or atomic structure, makes models of multiple conformations within a population of particles (QEDM); identifies β-sheets (sheetminer), builds pseudo-C^{α} model of β-sheets (sheettracer), enhances secondary structural elements.	[42–45]	contact J. Ma, jpma@bcm.tmc.edu
2003	EMAP	Core-weighted density correlation, Monte-Carlo sampling search	Rigid-body fitting, extension of CHARRM modeling tools to macromolecules	[46]	part of CHARRM package (www.charmm.org)

(*Continued*)

Table 1 (*Continued*)

Year*	Algorithm, Program, or Package	Principle of Operation	Features of Algorithm	Reference	Availability
2003	RAMOS (SPIDER)	Local, normalized cross-correlation	Allows rigid-body fitting of coordinates or density into density maps	[47]	www.wadsworth.org/spider_doc/spider/docs/techniques
2004	3SOM	Represents coordinates as iso-surface, maximizes overlap of probe and map iso-surfaces; re-ranks best solutions via normalized cross-correlation with all densities	Rigid-body fitting, fast algorithm	[48]	www.russell.embl.de/3SOM
2004	NMFF-EM	Normal-mode analysis: correlation coefficient, steepest-descent gradient optimization	Flexible fitting, coordinate model allowed to flex, user can specify which parts of model flex or stay rigid	[49, 50]	mmtsb.scripps.edu
2005	DensityFit	Rigid-body alignment of model coordinates and density, deform model	Flexible fitting	[51]	dirac.cnrs-orleans.fr/plone/software/densityfit

(*Continued*)

Table 1 (*Continued*)

Year*	Algorithm, Program, or Package	Principle of Operation	Features of Algorithm	Reference	Availability
		via normal-mode analysis, energy minimization			
2005	Hex	Spherical polar Fourier correlation	Rigid-body fitting, economical in computer time and memory	[52]	www.csd.abdn.ac.uk/hex
2005	MODELLER	cross-correlation, Monte Carlo or exhaustive search in real space	Rigid-body fitting (flexible fitting available**), can fit alternative models calculated from different sequence alignments	[33, 53]	salilab.org/modeller
2005	S-flexfit	Variability in evolutionarily related proteins (i.e. protein superfamilies) is model for how protein structure may vary, find best fit via local cross-correlation	Flexible fitting, relies on three outside packages in addition to its own software	[54–56]	biocomp.cnb.uam.es/Biocomp/public/Software/S_flexfit_web
2007	ADP_EM	Spherical harmonics, translational scanning	Rigid-body fitting, fast algorithm	[57]	sbg.cib.csic.es/Software/ADP_EM

(*Continued*)

Table 1 (*Continued*)

Year*	Algorithm, Program, or Package	Principle of Operation	Features of Algorithm	Reference	Availability
2007	Bsoft	Monte Carlo Metropolis algorithm (bmonte), Newtonian mechanics (bmd, bwater)	Rigid-body refinement of fitted structures (bmonte); flexible fitting via simple molecular dynamics (bmd) or molecular dynamics in water (bwater)	[58]	bsoft.ws
2007	Chimera	Local optimization, density correlation	Interactive and automated rigid-body fitting	[59]	www.cgl.ucsf.edu/chimera
2007	EMatch	Detect helices, match helices to atomic-resolution structures via secondary-structure alignment, build pseudo-atomic model from matched structural homologs	"Flexible fitting," protein structure deduced from density, then matched to protein structure database	[60, 61]	bioinfo3d.cs.tau.ac.il/EMatch

(*Continued*)

Table 1 (*Continued*)

Year*	Algorithm, Program, or Package	Principle of Operation	Features of Algorithm	Reference	Availability
2007	FRODA (FIRST)	Constrained geometric simulations: identifies rigid and flexible regions, Monte Carlo simulation, real-space correlation coefficient, Metropolis criterion	Flexible fitting, rigid portions (e.g. secondary structural elements) and stereochemistry are maintained throughout simulation	[62]	flexweb.asu.edu/software

*Year of first publication for the fitting procedure.
**M. Topf, personal communication.

References (for Appendix)

1. Stewart PL, Fuller SD, Burnett RM. (1993) "Difference imaging of adenovirus: bridging the resolution gap between X-ray crystallography and electron microscopy." EMBO *Journal* **12**: 2589–99.
2. Mendelson RA, Morris E. (1994) "The structure of F-actin: results of global searches using data from electron microscopy and X-ray crystallography." *J Mol Biol* **240**: 138–54.
3. Mendelson R, Morris EP. (1997) "The structure of the acto-myosin subfragment 1 complex: results of searches using data from electron microscopy and X-ray crystallography." *Proc Natl Acad Sci USA* **94**: 8533–8.
4. Cheng RH, Reddy VS, Olson NH, *et al.* (1994) "Functional implications of quasi-equivalence in a T = 3 icosahedral animal virus established by cryo-electron microscopy and X-ray crystallography." *Structure* **2**: 271–82.
5. Che Z, Olson NH, Leippe D, *et al.* (1998) "Antibody-mediated neutralization of human rhinovirus 14 explored by means of cryoelectron microscopy and X-ray crystallography of virus-Fab complexes." *J Virol* **72**: 4610–22.
6. Wikoff WR, Wang G, Parrish CR, *et al.* (1994) "The structure of a neutralized virus: canine parvovirus complexed with neutralizing antibody fragment." *Structure* **2**: 595-607.
7. Cheng RH, Kuhn RJ, Olson NH, *et al.* (1995) "Nucleocapsid and glycoprotein organization in an enveloped virus." *Cell* **80**: 621–30.
8. Grimes JM, Jakana J, Ghosh M, *et al.* (1997) "An atomic model of the outer layer of the bluetongue virus core derived from X-ray crystallography and electron cryomicroscopy." *Structure* **5**: 885–93.
9. Chacón P, Wriggers W. (2002) "Multi-resolution contour-based fitting of macromolecular structures." *J Mol Biol* **317**: 375–84.
10. Wriggers W, Milligan RA, McCammon JA. (1999) "Situs: a package for docking crystal structures into low-resolution maps from electron microscopy." *J Struct Biol* **125**: 185–95.
11. Wriggers W, Milligan RA, Schulten K, McCammon JA. (1998) "Self-organizing neural networks bridge the biomolecular resolution gap." *J Mol Biol* **284**: 1247–54.
12. Birmanns S, Wriggers W. (2003) "Interactive fitting augmented by force-feedback and virtual reality." *J Struct Biol* **144**: 123–31.
13. Wriggers W, Chacón P, Kovacs JA, *et al.* (2004) "Topology representing neural networks reconcile biomolecular shape, structure, and dynamics." *Neurocomputing* **56**: 365–79.
14. Birmanns S, Wriggers W. (2007) "Multi-resolution anchor-point registration of biomolecular assemblies and their components." *J Struct Biol* **157**: 271–80.
15. Kovacs JA, Wriggers W. (2002) "Fast rotational matching." *Acta Cryst D* **58**: 1282–6.

16. Kovacs JA, Chacón P, Cong Y, *et al.* (2003) "Fast rotation matching of rigid bodies by fast Fourier transform acceleration of five degrees of freedom." *Acta Cryst D* **59**: 1371–6.
17. Wriggers W, Birmanns S. (2001) "Using Situs for flexible and rigid-body fitting of multiresolution single-molecule data." *J Struct Biol* **133**: 193–202.
18. Wriggers W, Agrawal RK, Drew DL, *et al.* (2000) "Domain motions of EF-G bound to the 70S ribosome: insights from a hand shaking between multi-resolution structures." *Biophys J* **79**: 1670–8.
19. Tama F, Wriggers W, Brooks CL, III. (2002) "Exploring global distortions of biological macromolecules and assemblies from low-resolution structural information and elastic network theory." *J Mol Biol* **321**: 297–305.
20. Chacón P, Tama F, Wriggers W. (2003) "Mega-dalton biomolecular motion captured from electron microscopy reconstructions." *J Mol Biol* **326**: 485–92.
21. Volkmann N, Hanein D. (1999) "Quantitative fitting of atomic models into observed densities derived by electron microscopy." *J Struct Biol* **125**: 176–84, Erratum 28: 223.
22. Shacham E, Sheehan B, Volkmann N. (2007) "Density-based score for selecting near-atomic models of unknown structures." *J Struct Biol* **158**: 188–95.
23. Volkmann N, Hanein D. (2003) "Docking of atomic models into reconstructions from electron microscopy." *Methods in Enzymology* **374**: 204–25.
24. Roseman AM. (2000) "Docking structures of domains into maps from cryo-electron microscopy using local correlation." *Acta Cryst D* **56**: 1332–40.
25. Rossmann MG. (2000) "Fitting atomic models into electron-microscopy maps." *Acta Cryst D* **56**: 1341–9.
26. Rossmann MG, Bernal R, Pletnev SV. (2001) "Combining electron microscopic with X-ray crystallographic structures." *J Struct Biol* **136**: 190–200.
27. Belnap DM, Filman DJ, Trus BL, *et al.* (2000) "Molecular tectonic model of virus structural transitions: the putative cell entry states of poliovirus." *J Virol* **74**: 1342–54.
28. Bubeck D, Filman DJ, Cheng N, *et al.* (2005) "The structure of the poliovirus 135S cell entry intermediate at 10-Angstrom resolution reveals the location of an externalized polypeptide that binds to membranes." *J Virol* **79**: 7745–55.
29. Kikkawa M, Okada Y, Hirokawa N. (2000) "15 Å resolution model of the monomeric kinesin motor, KIF1A." *Cell* **100**: 241–52.
30. Jiang W, Baker ML, Ludtke SJ, Chiu W. (2001) "Bridging the information gap: computational tools for intermediate resolution structure interpretation." *J Mol Biol* **308**: 1033–44.
31. Baker ML, Yu Z, Chiu W, Bajaj C. (2006) "Automated segmentation of molecular subunits in electron cryomicroscopy density maps." *J Struct Biol* **156**: 432–41.
32. Baker ML, Ju T, Chiu W. (2007) "Identification of secondary structure elements in intermediate-resolution density maps." *Structure* **15**: 7–19.

33. Topf M, Baker ML, John B, *et al.* (2005) "Structural characterization of components of protein assemblies by comparative modeling and electron cryo-microscopy." *J Struct Biol* **149**: 191–203.
34. Baker ML, Jiang W, Wedemeyer WJ, *et al.* (2006) "*Ab initio* modeling of the herpesvirus VP26 core domain assessed by cryoEM density." *PLoS Computational Biology.* **2**: 1313–24.
35. Ju T, Baker M, Chiu W. (2007) "Computing a family of skeletons of volumetric models for shape description." *Computer-AIDED Design* **39**: 352–60.
36. Chen JZ, Fürst J, Chapman MS, Grigorieff N. (2003) "Low-resolution structure refinement in electron microscopy." *J Struct Biol* **144**: 144–51.
37. Chen LF, Blanc E, Chapman MS, Taylor KA. (2001) "Real space refinement of acto-myosin structures from sectioned muscle." *J Struct Biol* **133**: 221–32.
38. Fabiola F, Chapman MS. (2005) "Fitting of high-resolution structures into electron microscopy reconstruction images." *Structure* **13**: 389–400.
39. Sachse C, Chen JZ, Coureux P-D, *et al.* (2007) "High-resolution electron microscopy of helical specimens: a fresh look at tobacco mosaic virus." *J Mol Biol* **311**: 812–35.
40. Navaza J, Lepault J, Rey FA, *et al.* (2002) "On the fitting of model electron densities into EM reconstructions: a reciprocal-space formulation." *Acta Cryst D* **58**: 1820–5.
41. Suhre K, Navaza J, Sanejouand Y-H. (2006) "NORMA: a tool for flexible fitting of high-resolution protein structures into low-resolution electron-microscopy-derived density maps." *Acta Cryst D* **62**: 1098–100.
42. Ming D, Kong Y, Lambert MA, *et al.* (2002) "How to describe protein motion without amino acid sequence and atomic coordinates." *Proc Natl Acad Sci USA* **99**: 8620–5.
43. Kong Y, Ma J. (2003) "A structural-informatics approach for mining β-sheets: locating sheets in intermediate-resolution density maps." *J Mol Biol* **332**: 399–413.
44. Kong Y, Zhang X, Baker TS, Ma J. (2004) "A structural-informatics approach for mining β-sheets: building pseudo-Cα traces for β-strands in intermediate-resolution density maps." *J Mol Biol* **339**: 117–30.
45. Brink J, Ludtke SJ, Kong Y, *et al.* (2004) "Experimental verification of conformational variation of human fatty acid synthase as predicted by normal mode analysis." *Structure* **12**: 185–91.
46. Wu X, Milne JLS, Borgnia MJ, *et al.* (2003) "A core-weighted fitting method for docking atomic structures into low-resolution maps: application to cryo-electron microscopy." *J Struct Biol* **141**: 63–76.
47. Rath BK, Hegerl R, Leith A, *et al.* (2003) "Fast 3D motif search of EM density maps using a locally normalized cross-correlation function." *J Struct Biol* **144**: 95–103.

48. Ceulemans H, Russell RB. (2004) "Fast fitting of atomic structures to low-resolution electron density maps by surface overlap maximization." *J Mol Biol* **338**: 783–93.
49. Tama F, Miyashita O, Brooks CL, III. (2004) "Normal mode based flexible fitting of high-resolution structure into low-resolution experimental data from cryo-EM." *J Struct Biol* **147**: 315–26.
50. Tama F, Miyashita O, Brooks CL, III. (2004) "Flexible multi-scale fitting of atomic structures into low-resolution electron density maps with elastic network normal mode analysis." *J Mol Biol* **337**: 985–99.
51. Hinsen K, Reuter N, Navazza J, *et al.* (2005) "Normal mode-based fitting of atomic structure into electron density maps: application to sarcoplasmic reticulum Ca-ATPase." *Biophys J* **88**: 818–27.
52. Ritchie DW. (2005) "High-order analytic translation matrix elements for real-space six-dimensional polar Fourier correlations." *J Appl Crystallog* **38**: 808–18.
53. Topf M, Baker ML, Marti-Renom MA, *et al.* (2006) "Refinement of protein structures by iterative comparative modeling and cryo-EM density fitting." *J Mol Biol* **357**: 1655–68.
54. Velázquez-Muriel JA, Sorzano COS, Scheres SHW, *et al.* (2005) "SPI-EM: towards a tool for predicting CATH superfamilies in 3D-EM maps." *J Mol Biol* **345**: 759–71.
55. Velazquez-Muriel J-Á, Valle M, Santamaría-Pang A, *et al.* (2006) "Flexible fitting in 3D-EM guided by the structural variability of protein superfamilies." *Structure* **14**: 1115–26.
56. Velazquez-Muriel JA, Carazo J-M. (2007) "Flexible fitting in 3D-EM with incomplete data on superfamily variability." *J Struct Biol* **158**: 165–81.
57. Garzón JI, Kovacs J, Abagyan R, *et al.* (2007) "ADP_EM: fast exhaustive multi-resolution docking for high-throughput coverage." *Bioinformatics* **23**: 427–33.
58. Heymann JB, Belnap DM. (2007) "Bsoft: image processing and molecular modeling for electron microscopy." *J Struct Biol* **157**: 3–18.
59. Goddard TD, Huang CC, Ferrin TE. (2007) "Visualizing density maps with UCSF Chimera." *J Struct Biol* **157**: 281–7.
60. Dror O, Lasker K, Nussinov R, *et al.* (2007) "EMatch: an efficient method for aligning atomic resolution subunits into intermediate-resolution cryo-EM maps of large macromolecular assemblies." *Acta Cryst D* **63**: 42–9.
61. Lasker K, Dror O, Shatsky M, *et al.* (2007) "EMatch: discovery of high resolution structural homologues of protein domains in intermediate resolution cryo-EM maps." *IEEE/ACM Transactions on Computational Biology and Bioinformatics* **4**: 28–39.
62. Jolley C, Wells SA, Fromme P, Thorpe MF. (2008) "Fitting low-resolution cryo-EM electron density maps of proteins using constrained geometric simulations." *Biophys J* **94**: 1613–21.

Chapter 13

On NMR-based Structural Proteomics

Thomas Szyperski

Summary

NMR spectroscopy plays an important role in the derivation of the atomic resolution structures of soluble proteins in structural proteomics projects because NMR neatly complements X-ray crystallography: 1) many proteins do not form diffraction quality crystals and their structures are thus amenable to determination solely by NMR; 2) the success of structural determination by NMR and X-ray, i.e., the quality of NMR spectra and crystallization success, are hardly correlated; 3) the cost-effectiveness of NMR and X-ray crystallographic structure production is nowadays comparable; and 4) NMR is about equally successful for pro- and eukaryotic proteins, while eukaryotic proteins crystallize less frequently than prokaryotic proteins. Since the inception of structural proteomics around the year of 2000, all components required for high-throughput NMR structure production of soluble proteins were established or developed further. Currently, about 10% of the soluble protein structures solved in the framework of the US Protein Structure Initiative are obtained by using NMR. Apart from contributing novel structures to accomplish "structural

---

816 Natural Sciences Complex, Chemistry Department, State University of New York at Buffalo, Buffalo, NY 14260, USA. szypersk@chem.buffalo.edu

coverage" of families of sequence homologs, many of these NMR structures are of immediate importance for biomedical and biological research. More recently, projects focusing on methodology development for NMR-based structural proteomics of (integral) membrane proteins were also established. In general, the methodology developed for structural proteomics projects nowadays broadly impacts on the scientific infrastructure of NMR-based structural biology.

Background

In the United States, "structural proteomics" (also referred to as "structural genomics") is pursued within the framework of the *Protein Structure Initiative* (PSI) of the National Institutes of Health (NIH). The mission of PSI (http://www.nigms.nih.gov/Initiatives/PSI/Background/MissionStatement) is defined as the "*long-range goal to make the three-dimensional atomic-level structures of most proteins easily obtainable from knowledge of their corresponding DNA sequences.*" This implies that the PSI does not aim at finding new protein "folds" as much as determining the structural coverage of the families of sequence homologues. Achievement the mission evidently requires both high-throughput atomic-resolution protein structural determination and computational tools to calculate the structures ("models") of proteins for which the atomic resolution structure of a sequence homologue is known.

As far as nuclear magnetic resonance (NMR) spectroscopy is concerned, both requirements appeared quite ambitious when NMR-based structural proteomics projects were first initiated in North America, Japan and Europe around the year of 2000 (e.g. Refs. 1–4): NMR data acquisition, processing and analysis was slow, typically requiring several months *per* structure, and it was not clear if high-quality NMR structures suitable for homology modeling could be obtained in high-throughput (HTP). As a result, some NMR researchers predicted that NMR will not be able to compete with X-ray crystallography in structural proteomics. It can thus be considered to be a significant success, primarily accomplished during the years 2000–2005, that the importance of NMR for structural genomics projects is nowadays widely acknowledged.

Role of NMR for Structural Proteomics

As of September 2007, NMR contributed ~10% of the ~2200 protein structures solved within PSI (Fig. 1A; for updated numbers, see http://www.mcsg.anl.gov under "SG progress"). Among the four PSI large-scale structure production centers (http://www.nigms.nih.gov/Initiatives/PSI/Centers/), only the Northeast Structural Genomics Consortium (NESG, http://www.nesg.org) established X-ray crystallography and NMR HTP structure production pipelines with comparable capacity. As a result, NESG contributed ~76% of all the PSI NMR structures (Fig. 1B), and it is expected that in the future, NESG may also function as the "NMR branch" of the other centers. Outside of the United States, the Riken Genomic Science Center in Japan dominates structure production, having contributed ~93% of the ~2200 non-PSI structural proteomics structures solved since 2000 (97% and 81% of all the NMR and X-ray structures, respectively).

NMR is important for structural proteomics due to several reasons. First, many proteins do not form diffraction-quality crystals so that

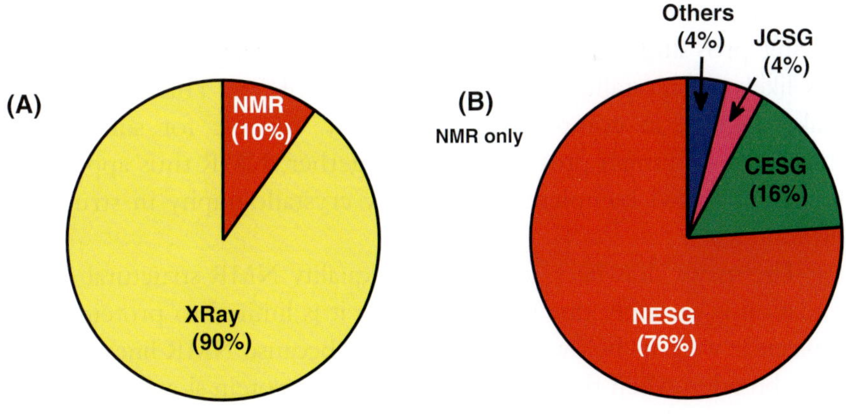

Fig. 1 (**A**) Protein structure production by X-ray or NMR within the framework of the United States Protein Structure Initiative (PSI; see text). (**B**) Breakdown of PSI NMR structure production by center: NESG (Northeast Structural Genomics Consortium; www.nesg.org); CESG (Center for Eukaryotic Structural Genomics; www.uwstructuralgenomics.org); JCSG (Joint Center for Structural Genomics; www.jcsg.org). "Others" represent other specialized PSI centers.

their structures can solely be obtained by NMR. Moreover, three independent studies showed that a protein's "crystallizability" does not correlate with the quality of its 2D [^{15}N,^{1}H] HSQC NMR spectrum, which shows one signal for each amino acid residue and reliably predicts the success of an NMR structural determination.[5–7] Hence, many proteins giving rise to high-quality NMR spectra do not form diffraction-quality crystals and, vice versa, many proteins forming crystals diffraction to high resolution give rise only to low-quality NMR spectra. This can be at least partially rationalized by considering that flexibly disordered polypeptide segments may efficiently prevent crystallization while they hardly affect the quality of the NMR spectra. On the other hand, transient formation of oligomeric aggregates or unfolded states usually has a devastating impact on the quality of the NMR spectra,[1] while a high-quality crystal may still form out of the equilibrium of the species present in solution.[5–7] Second, the cost-effectiveness of NMR and X-ray structure production is nowadays comparable. This is because in recent years: 1) efficient production of ^{13}C/^{15}N stable isotope labeled protein samples was established[8–10]; 2) the NMR measurement time *per* structure was dramatically reduced[11]; and 3) (semi-) automated data processing and analysis protocols were devised.[12] Third, NMR is about equally successful for solving pro- and eukaryotic proteins, while eukaryotic proteins are less likely to crystallize than prokaryotic proteins. This observation makes NMR, in comparison, particularly valuable for solving the structures of human proteins. Taken together, NMR thus appears to be a well suited complement for X-ray crystallography in structural proteomics.

The major drawback of HTP high-quality NMR structural determination is certainly due to the fact that it is limited to proteins with molecular weights below ~25 kDa. This is because NMR lines increasingly broaden when the overall tumbling of a protein slows down with increasing molecular weight, while the number of resonance lines concomitantly increases with size.[13] Although NMR methodology has been developed in the last decade for approaching proteins with significantly higher molecular weights, these approaches either do not yield high-quality structures suitable for homology modeling, or are currently too costly for high-throughput applications (see "Perspectives" section concluding this chapter "on NMR for structural proteomics").

High-Throughput NMR Structure Production

HTP NMR Structural determination requires: 1) preparation of $^{13}C/^{15}N$-labeled protein samples; 2) rapid acquisition of NMR data for resonance assignment and derivation of ^{1}H–^{1}H upper distance limit constraints; 3) (semi-)automated data analysis and structure calculation; and 4) comprehensive structure validation. These components of a structural determination pipeline are surveyed in the following.

Protein Expression and Preparation of $^{13}C/^{15}N$-labeled Samples

Rapid structural determination is based on heteronuclear, $^{13}C/^{15}N$-resolved multidimensional NMR spectroscopy.[13] The generation of the required $^{13}C/^{15}N$-labeled protein samples in HTP represents a major challenge of NMR-based structural proteomics. Two approaches are nowadays available, i.e., 1) heterologous expression in (mostly) *Escherichia coli* cells[4,9,14,15]; and 2) wheat-germ based cell-free expression.[16–18]

When employed for structural proteomics in HTP, the first approach benefits from both our exceptional knowledge and experience with *E. coli* for protein expression and recent innovations to improve expression yield, throughput and rapid purification. For example, the use of auto-inducible media for parallel expression,[19] ligation independent cloning[20] and combinatorial (robotic) cloning systems[21] have greatly increased the success rates and throughput. However, for many structural proteomics targets, the protein expression levels are too low, or the protein turns out to be insoluble. Hence, an alternative protein synthesis methodology is required.

A recently established cell-free protein synthesis complements *E. coli* based protein expression and promises to significantly increase the fraction of genes for which soluble and folded protein samples are obtained.[16,17] Cell-free expression is characterized by a high success rate, in particular also for eukaryotic proteins, and the costs per stable isotope labeled protein sample are nowadays comparable to those of cell-based approaches.[18] Furthermore, cell-free expression greatly facilitates protein purification and enables one to devise selective labeling schemes that are not feasible in cells.

Rapid NMR Data Acquisition

Rapid acquisition of NMR data is of central importance for structural proteomics as the NMR measurement time is expensive. Moreover, proteins often precipitate slowly and the ability to acquire the required data rapidly may in these cases become a prerequisite to solving the structure. The methodology for rapid data acquisition[11] aims at avoiding the "sampling limited data acquisition regime"[22] in which instrument time is invested to sample the indirect dimensions of a multidimensional NMR experiment. Instead, HTP structural determination is preferably pursued in the "sensitivity limited data acquisition regime"[22] in which instrument time is invested only to the extent that the signal-to-noise ratios of (most) peaks reach the threshold for reliable peak identification. Conventional acquisition of multidimensional NMR spectra[13] is not suited to avoid sampling limited data acquisition because the minimal measurement time of an N-dimensional spectrum scales with the *product* of the number of points sampled in the N-1 indirect dimensions. This key limitation of conventional data acquisition has been named the "NMR sampling problem".[11] The introduction of cryogenic NMR probes, which increase the spectrometer sensitivity about 3-fold and thus reduce the measurement times by about an order of magnitude, created additional demand for rapid NMR sampling methods to tackle the NMR sampling problem.

GFT Projection NMR

A straightforward and attractive solution of the NMR sampling problem is to simply avoid the independent sampling of many indirect dimensions by implementing the joint sampling of several shift evolution periods. The major problem associated with this concept is due to the fact that only phase-sensitively detected pure absorption mode spectra are usable in practice.[13] The invention of G-matrix Fourier Transform (GFT) NMR spectroscopy,[23] a generalization of reduced-dimensionality (RD) NMR,[11,22] has solved this problem and enables one to derive arbitrary *N*-dimensional spectral information from projected sub-spectra of lower dimensionality (Fig. 2). As a result, the minimal measurement times for GFT NMR spectra scale with the *sum* of the number of points that had to be sampled in the indirect dimensions of the conventional high-dimensional NMR experiment.[22] A set of GFT NMR spectra for complete resonance assignment (Ref. 24; Fig. 3) contributed to establishing an NMR data

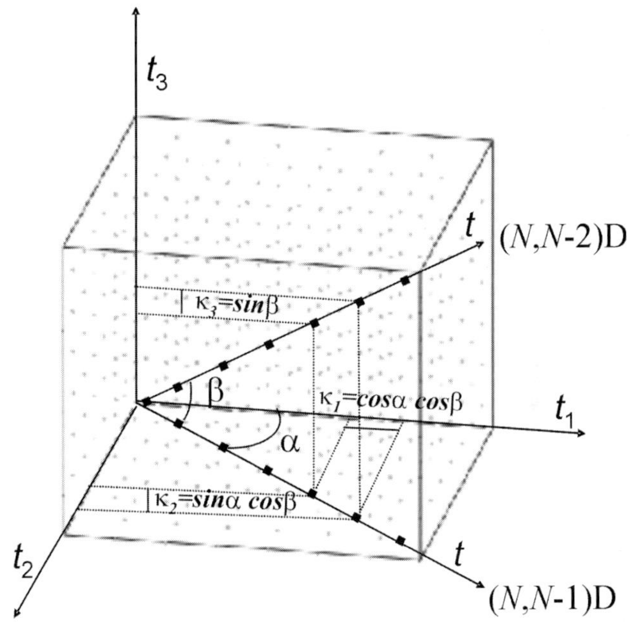

Fig. 2 Spectral parameters defining time domain data sampling in GFT projection NMR spectroscopy. The three-dimensional time domain sub-space of an *N*-dimensional NMR experiment is shown. Gray dots indicate data points which need to be sampled in a conventional FT-NMR experiment,[13] and black squares represent data points acquired 2^K times in a projected GFT NMR experiment with varying sine and cosine modulations of the jointly sampled shifts.[24–26,28] The projection tilt angles, which are adjusted by setting the scaling factors (κ) of the individual chemical shift evolution periods,[27] are also indicated. (Reproduced with permission from Ref. 27.)

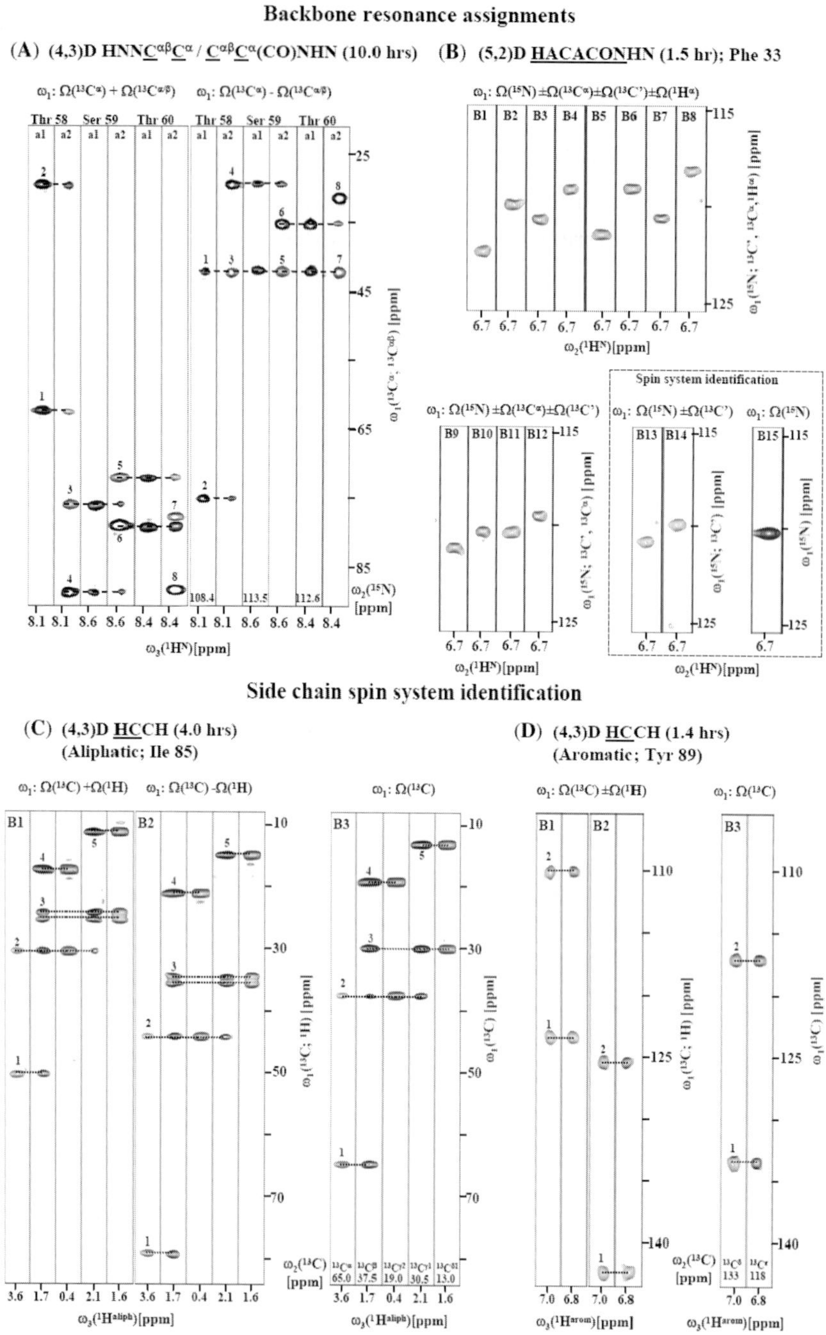

Fig. 3 (see next page for caption)

collection and analysis protocol for high-throughput protein structural determination (Ref. 25; Fig. 4). This protocol, or variants thereof, has thus far been used to solve the structures of about 30 proteins with molecular weights ranging from 10 to 23 kDa, including several structures of homo-dimeric proteins.

It was subsequently proposed[26] to use GFT projection NMR sub-spectra for the reconstruction of the multidimensional parent spectrum, thereby taking advantage of the option to scale the jointly sampled chemical shifts in the GFT spectra.[11,23] For that purpose, the GFT NMR formalism was translated into a form which is related to the terminology used in the projection-reconstruction (PR) theory. It can, however, be readily shown that the projection ("P") part of PR NMR is strictly equivalent to recording several GFT NMR spectra with different scaling of shift evolution periods.[27] Several algorithms have been proposed for reconstructing the parent multidimensional spectrum ("R" part of PR NMR), but the extent to which the reconstruction introduces spectral artefacts remains a matter of debate.[28] The resulting limitations appear

---

Fig. 3 The resonance assignment based on GFT NMR experiments recorded in 16.9 hours is exemplified for 14 kDa protein YqfB.[33] The individual measurement times as well as sequence specific resonance assignments are indicated above each panel. (**A**) [$\omega_1(^{13}C^{\alpha};^{13}C^{\alpha\beta})$,$\omega_3(^1H^N)$]-strips taken from (4,3)D $\underline{C}^{\alpha\beta}\underline{C}^{\alpha}$(CO)NHN ("a1") and (4,3)D HNN$\underline{C}^{\alpha\beta}\underline{C}^{\alpha}$ ("a2") which provide four-dimensional spectral information. The strips were taken at $\omega_2(^{15}N)$ (the corresponding ^{15}N chemical shifts are indicated at the bottom of the strips) and are centered along $\omega_3(^1H^N)$ about their backbone $^1H^N$ shifts. Along $\omega_1(^{13}C^{\alpha};^{13}C^{\alpha\beta})$, peaks are observed at $\Omega\ (^{13}C^{\alpha})\pm\Omega(^{13}C^{\alpha/\beta})$ of residue *i*-1 in "a1" and of residue *i* in "a2" (the type of linear combination of chemical shifts that is measured is indicated above the plots). The combined use of (4,3)D $\underline{C}^{\alpha\beta}\underline{C}^{\alpha}$(CO)NHN/HNN$\underline{C}^{\alpha\beta}\underline{C}^{\alpha}$ thus yields three sequential "walks" along the polypeptide backbone which are indicated by dashed lines. Panel (**B**) shows plots from (5,2)D $\underline{\text{HACACON}}$HN[24] yielding also the $^1H^{\alpha}$ shift assignments. Panel (**C**) demonstrates assignment of aliphatic side chains. For aliphatic spin system identification, sums and differences of shifts of covalently attached ^{13}C and ^1H nuclei are delineated in B1 and B2, while ^{13}C shifts are matched in B3 (indicated by dashed lines). Panel (**D**) describes the identification of aromatic spin systems which was accomplished in a manner similar to that described in the legend of panel (c) for aliphatic spin systems. (Reproduced with permission from Ref. 32.)

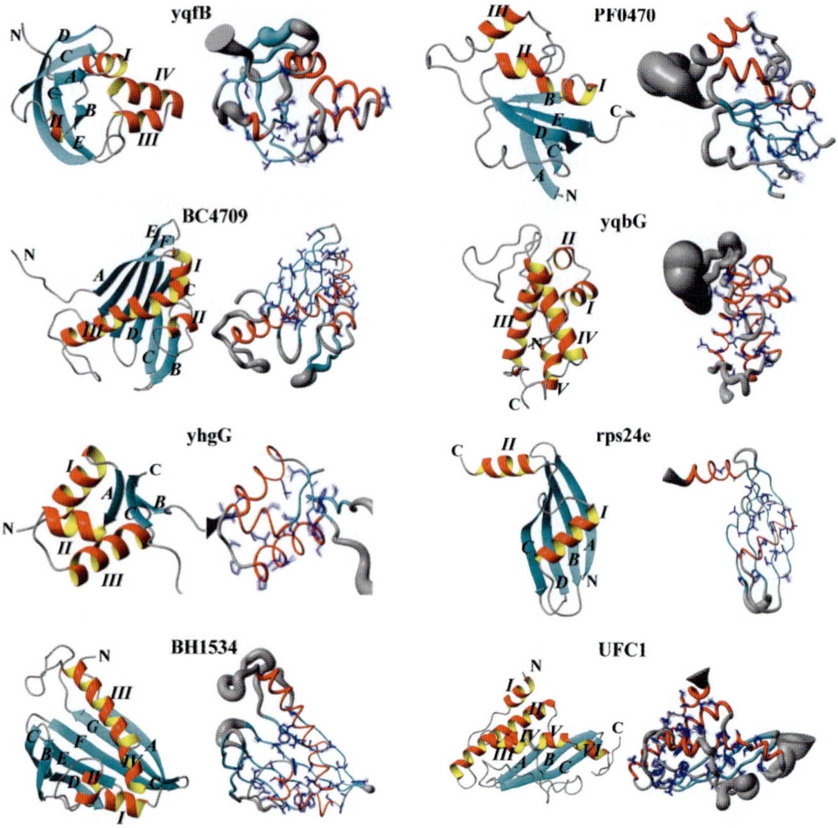

Fig. 4 Gallery of representative high-quality NMR solution structures of proteins with molecular weights ranging from 10 to 20 kDa. The structures were obtained using a GFT projection NMR based protocol,[26] and total NMR measurement times were between 1 and 9 days per structure. For each structure, a ribbon drawing is shown on the left. On the right, a "sausage" representation of the backbone is shown, where the thickness of the cylindrical rod reflects the precision achieved for the determination of the polypeptide backbone conformation. Superposition of the best defined side-chains are also shown in order to indicate the precision of the determination of side-chain conformations. Helices are shown in red; the β-stands are depicted in cyan; other polypeptide segments are displayed in gray; and the side-chains of the molecular core are shown in blue. The gene name is indicated above each structure. (Reproduced with permission from Ref. 25.)

to be significant, and, to the best of the author's knowledge, no protein structure has thus far been solved using solely reconstructed multidimensional NMR spectra. Other valuable GFT projection NMR based approaches include the "Automated projection NMR" (Ref. 29; APSY), which considers features of pattern recognition based peak picking in GFT spectra,[30] as well as Hi-Fi NMR,[31] which aims at adjusting the data acquisition times based on completeness of the resonance assignment obtained while the data acquisition is in progress. As for PR NMR, the projection NMR data acquisition is identical to recording multiple GFT NMR spectra with varying scaling of shift evolution periods.

The replacement of the GFT NMR nomenclature within the framework of PR NMR has led to a situation in which every research group developing the GFT NMR experiments has adopted a different nomenclature. This evidently impedes comparison of the experiments. To enable the researchers to readily compare implementations, we provide on our GFT NMR web page (http://www.nsm.buffalo.edu/Research/GFT/) a table listing all the published projection NMR experiments thus far (under "Reference Table: Projection NMR nomenclature").

Simultaneously acquired $^{13}C/^{15}N$ resolved NOESY

Nuclear overhauser enhancement spectroscopy ("NOESY") provides ^{1}H–^{1}H distance constraints for structural determination. For $^{13}C/^{15}N$-resolved labeled proteins, these are routinely derived from three 3D experiments, i.e., 3D $^{13}C^{aliphatic}$-, $^{13}C^{aromatic}$- and ^{15}N-resolved [^{1}H,^{1}H] NOESY.[13] For HTP, NOESY data collection times can be reduced by about a factor of 2 when collecting these three experiments simultaneously in 3D [H]-NOESY-[CH^{ali}/CH^{aro}/NH].[32]

L-optimization

Longitudinal relaxation optimization[33] can accelerate data acquisition for experiments based on initial excitation and detection of polypeptide backbone amide proton[11,24,34,35] or aromatic proton magnetization.[37]

This is because such L-optimization uses aliphatic proton polarization to enhance the longitudinal relaxation of amide and/or aromatic protons. As a result, very short relaxation delays between scans can be employed, shortening the measurement time by a factor 2–3 without significant loss of intrinsic sensitivity.[24]

Sparse sampling

"Sparse" sampling (e.g. Refs. 11, 37–39) of the time domain of a conventional or a GFT projection NMR experiment combined with alternative data processing techniques such as maximum entropy reconstruction[37] or multidimensional decomposition,[38,39] promises to become an important option for rapid HTP NMR data acquisition. The protocols best suited for routine HTP applications are currently being explored (Ref. 15, Vladislav Orekhov, Jeffrey Hoch, personal communication).

(Semi-)Automated Data Analysis and Structure Calculation

(Semi-)automated NMR data processing and analysis is critical for HTP Structural determination.[1,12,15,25,29,31,40–43] Currently, a suite of programs is available,[12] allowing researchers to efficiently handle all the steps from processing the raw time domain data to calculating a high-quality NMR structure. For a medium-sized protein of about 15 kDa, data processing and assignment of the resonances of the polypeptide backbone using the program AUTOASSIGN[12] typically takes a day or two. Assignment of side-chain resonances is preferably pursued in conjunction with the assignment of intra-residue, sequential and medium-range NOEs, and usually takes less than about a week. Notably, the ABACUS strategy[15] primarily relies on obtaining sequential connectivities from NOEs,[13] thereby avoiding acquisition of comparably insensitive "intraresidue" $^{13}C/^{15}N$-tripel resonance spectra. The subsequent semi-automated assignment of long-range NOEs and the calculation of the structure profit from using two programs in parallel which rely on different algorithms.[25] Specifically, the top-down operating program CYANA[41] and the bottom-up operating program AUTOSTRUCTURE[40] can be effectively "coupled." As a result, about two thirds of the long range NOEs are typically assigned nearly error-free.[25] This provides a suitable basis to finish the structure within a few days, followed by MD refinement in explicit solvent.

Structure Validation

Prior to submission of the coordinates to the Protein Data Base (PDB),[44] a newly solved protein structure needs to be validated. This is particularly important for a "factory" type structural proteomics pipeline where researchers have to deal with vast amount of information. The recently developed protein structure validation software suite (PSVS)[45] relies on using several programs to provide: 1) constraint analyses; 2) statistics on goodness-of-fit between structures and experimental data; and 3) knowledge-based structure quality scores. PSVS is available on-line (Fig. 5) and provides both global and site-specific measures of protein structure quality, and the global quality measures are reported as z-scores, based on calibration with a set of high-resolution

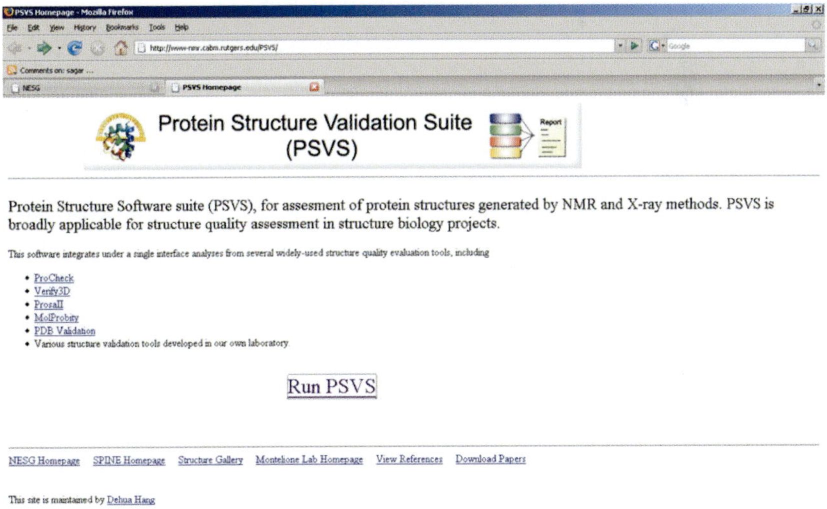

Fig. 5 Interface of the Protein Structure Validation Suite (PSVS; http://www-nmr.cabm.rutgers.edu/PSVS/).[46]

X-ray crystal structures. An alternative route for structure validation is to experimentally measure residual dipolar couplings (RDCs) and check on the consistency of these RDC data with the structure.[46] This route is currently routinely pursued in the NESG and is particularly valuable for validating structures of homo-dimeric complexes.

Benchmarking NMR Data Acquisition

In view of the central importance of NMR data acquisition speed and the dimensionality of the spectral information obtained, additional practical aspects are discussed in this section. To the best of the author's knowledge, the shortest NMR measurement time reported thus far for obtaining a high-quality NMR structure of a medium-sized $^{13}C/^{15}N$-protein amounts to 16.9 hours for GFT NMR based resonance assignment and 9 hours for acquiring 3D [H]-NOESY-[$CH^{ali}/CH^{aro}/NH$]: the structure of 14 kDa protein YqfB (PDB ID 1TE7) was solved[32] as a pilot project to set-up the HTP pipeline described in Ref. 25. The highest dimensional spectral information obtained thus far for a protein is six dimensional:* (6,3)D $\underline{H}^{\alpha\beta}\underline{C}^{\alpha\beta}\underline{C}^{\alpha}\underline{CO}$NHN for sequentially correlating the chemical shifts of C'–C^{α}H–C^{β}H moieties of residue *i*–1 and the NH group of residue *I* was described in Ref. 24.

* Comparable to or even of higher dimensional spectral information has been claimed for APSY (7D information in Ref. 47) and for what has been called "hyperdimensional NMR spectroscopy" (10D information in Ref. 48). However, analysis of tilt angles and jointly sampled evolution times reveals that these studies rely exclusively on recording a large number of "basic" spectra of (3,2)D and (4,2)D GFT NMR experiments, i.e., in no case are more than 4 chemical shifts directly correlated (see "Reference Table: Projection NMR nomenclature" at http://www.nsm.buffalo.edu/Research/GFT/). Given the facts that: (i) an equally large number of 3D and 4D conventional experiments does not yield 7D or 10D spectral information; and that (ii) the authors do not provide evidence that the 7D or 10D information can be extracted from their set of recorded GFT NMR experiments, these claims of highest dimensional spectral information appear unsubstantiated currently. Importantly, 7D and 10D experiments are supposed to break, respectively, 6-fold and 9-fold chemical shift degeneracy. Recording of *N*-dimensional spectral information requires that the *N* chemical shifts are directly correlated in a single experiment; one way to prove that the claimed *N*-dimensional information is indeed obtained is through recursive central peak detection,[23] or different scaling of shift evolution periods[11] in (*N*,*K*)D GFT NMR.

Microgram Scale NMR-based Structural Proteomics

Microcoil NMR probes[49] allow one to assign resonances[50] and solve[51] NMR structures on a microgram scale. This approach is attractive for structural proteomics whenever the expression yield is low for a given target protein with a molecular weight smaller than ~15 kDa, while its solubility is high (> 1 mM). It has been estimated that these constraints are met for ~25% of the targets of an NMR HTP structure production pipeline.[51] Moreover, microcoil NMR probes also allow protein production groups to establish routine screening of foldedness and optimization of the buffer condition by acquiring 2D [^{15}N,^{1}H] HSQC NMR spectra with small amounts of protein before up-scaling of protein expression to milligram quantities.[51]

NMR Structures for Homology Modeling

Nowadays, the average accuracy of NMR structures is lower than the average accuracy of X-ray crystal structures. This fact has been named the "NMR-X-ray structure quality gap." Structural proteomics is based on homology modeling of protein structures for families of sequence homologs from experimentally determined atomic resolution structures.[52] Hence, it was evident early on that NMR-based structural proteomics could be successful only if high-quality NMR structures could be generated which at least match the accuracy of medium-quality 1.5–2.5 Å resolution X-ray crystal structures.[1] It was thus of central importance that it could be shown that high-quality NMR structures emerging from the NESG NMR pipeline were as valuable as X-ray structures for homology modeling.[52,53] Ongoing improvement of the quality of structural proteomics NMR structures, which is already higher than the average quality of non-structural proteomics NMR structures,[45] promises to further increase the leverage value of NMR structures.

Sructural Proteomics of Membrane Proteins

Even though: 1) 20–30% of a typical proteome are membrane proteins; 2) membrane proteins are pivotal for living cellular systems;

and 3) 60–70% of current drug targets are membrane proteins,[14] not even 0.5% of all atomic resolution structures in the PDB are from membrane proteins.[54] These facts reflect the challenges encountered for membrane protein structural determination. Efficient recombinant overexpression and purification systems became available only recently for prokaryotic membrane proteins (e.g. Ref. 55). In contrast, the creation of systems for eukaryotic membrane proteins remains a major bottleneck for structural proteomics projects. This is of key importance because only 15% of the human membrane proteins have prokaryotic homologs.[54] In the future, cell-free approaches promise to complement cell-based expression.[56] Furthermore, a methodology for accurate homology modeling needs to be developed and the recently accomplished global topology analysis of the *Escherichia coli* inner membrane proteome[55] has provided invaluable experimental insights which can be used to constrain homology modeling calculations.

For NMR, preparation of $^{13}C/^{15}N$-labeled protein samples is required along with suitable protocols for embedding the purified membrane protein into a membrane mimic. Subsequently, either solution or solid-state NMR can contribute to obtaining a high-quality NMR structure (e.g. Refs. 57–59). For both approaches, however, chemical shift degeneracy is a major obstacle: the shift dispersion in membrane proteins is narrow due to: 1) their limited amino acid composition; and 2) regular α-helical secondary structure. Hence, the use of GFT projection NMR[11,23,24,27,36] for obtaining precise highest-dimensional NMR spectral information appears to be quite promising for membrane proteins.[60] To resolve the bottlenecks for structural proteomics of membrane proteins, "specialized centers" for methodology development, such as the New York Consortium of Membrane Protein Structure (NYCOMPS) (http://www.nycomps.org/), were founded within the framework of the Unites States PSI in order to complement the research efforts on a large scale, such as the NESG.

Perspectives

HTP NMR structural determination of soluble proteins is currently limited to proteins with a molecular weight of less than about 25 kDa, and comparable size limitations exist for structural determination of membrane proteins embedded in membrane mimics. Transverse relaxation optimized spectroscopy ("TROSY", Ref. 13) allows recording of amide proton or aromatic proton detected multidimensional experiments for large (partially) deuterated proteins. However, resonance assignment of the aliphatic side-chains,[13] for which TROSY-type experiments cannot be devised, is pivotal for obtaining the high-quality NMR structures yielding high-quality homology models.[52,53] Very recently, the stereo-array isotope (^2H) labeling (SAIL) approach was developed,[61] which enables one to obtain high-quality NMR structures for proteins up to about 50–75 kDa. This approach is based on cell-free protein expression[16–18,56] with chemically synthesized stable isotope labeled amino acids and is thus currently too costly for HTP applications. However, provided that the costs for SAIL-sample preparation are significantly reduced in the future, SAIL will greatly impact on NMR-based structural proteomics. Likewise, single protein production (SPP) in *E. coli* represents an important innovation for the synthesis of labeled protein, and allows one to selectively overexpress the desired target protein while the ^{13}C-labeled carbon source is supplied to the medium.[62] SPP also promises to facilitate in-cell NMR,[63] and the mapping of structural interactions in cell NMR (STINT-NMR; Ref. 64).

It has recently become clear that structural models derived from chemical shifts alone[65–67] will greatly impact on the speed and reliability of NMR structure production. The shift derived models indicate structural novelty in an early phase of the structural determination, and can be considered to derive and validate NOE distance constraint assignments. Similarly, statistical analysis of RDCs,[68] in particular when several types of RDCs are measured in a correlated manner,[68,69] promises to provide insights into structural novelty at the outset of the NMR structural determination. In turn, this should allow one to pursue a structural determination protocol tailored to the expected degree of structural novelty. Structure validation, especially of homodimeric protein complexes, might in the future also benefit from the use of solution small-angle X-ray scattering,[70] and for smaller proteins, NOESY in supercooled water promises to yield additional distance constraints for validation and refinement. Overall, it can be expected that nearly fully automated HTP NMR structural determination and validation will become feasible within the next couple of years.

One goal of structural proteomics is to rely on atomic resolution structures to support functional annotation through the use of bioinformatics tools.[72] In many cases, however, the structure combined with sequence alignments alone does not reveal function. This is manifested by the fact that about one third of the structures which have so far been solved by structural genomics initiatives remain without functional annotation.[73] For these proteins, additional experimental data are currently required to elucidate function, and NMR-based screening of compound libraries combined with computational docking approaches offers the opportunity to identify metabolites interacting with the protein. The functional annotations of the metabolites can then be considered in a straightforward manner to annotate protein function.

NMR is best suited to identify flexibly disordered polypeptide segments in solution.[13] These segments may increase the functional complexity of the proteome,[74] and bioinformatics tools to identify those segments based on sequence alone are desired. Primarily based on NMR derived information, such tools have recently been developed.[75] Hence, it can be expected that NMR-based structural proteomics will continue to greatly increase our empiric knowledge of natively unstructured polypeptide segments, thereby further contributing to the functional annotation of proteomes.

Acknowledgments

The author is indebted to all the colleagues and co-workers in the Northeast Structural Genomics Consortium and the New York Consortium on Membrane Protein Structure for stimulating discussions, and to Dr. Arindam Ghosh and Mr. Bharathwaj Sathyamoorthy for help in preparing the manuscript.

References

1. Montelione GT, Zheng D, Huang YJ, *et al.* (2000) "Protein NMR spectroscopy in structural genomics." *Nat Struct Biol* 7: 982–85.

2. Yokoyama S, Hirota H, Kigawa T, *et al.* (2000) "Structural genomics projects in Japan." *Nat Struct Biol* 7: 943–45.
3. Heinemann U. (2000) "Structural genomics in Europe: slow start, strong finish?" *Nat Struct Biol* 7: 940–42.
4. Yee A, Chang X, Pineda-Lucena A, *et al.* (2002) "An NMR approach to structural proteomics." *Proc Natl Acad Sci USA* **99**: 1825–30.
5. Tyler RC, Aceti DJ, Bingman CA, *et al.* (2005) "Comparison of cell-based and cell-free protocols for producing target proteins from the *Arabidopsis thaliana* genome for structural studies." *Proteins* **59**: 633–43.
6. Snyder DA, Chen Y, Denissova NG, *et al.* (2005) "Comparisons of NMR spectral quality and success in crystallization demonstrate that NMR and X-ray crystallography are complementary methods for small protein structure determination." *J Am Chem Soc* **127**: 16505–11.
7. Yee AA, Savchenko A, Ignachenko A, *et al.* (2005) "NMR and X-ray crystallography, complementary tools in structural proteomics of small proteins." *J Am Chem Soc* **127**: 16512–17.
8. Yokoyama S. (2003) "Protein expression systems for structural genomics and proteomics." *Curr Opin Chem Biol* 7: 39–43.
9. Acton TB, Gunsalus KC, Xiao R, *et al.* (2005) "Robotic cloning and protein production platform of the Northeast Structural Genomics Consortium." *Methods Enzymol* **394**: 210–43.
10. Vinarov DA, Loushin Newman CL, Markley JL. (2006) "Wheat germ cell-free platform for eukaryotic protein production." *FEBS J* **273**: 4160–69.
11. Atreya HS, Szyperski T. (2005) "Rapid NMR data collection." *Methods Enzymol* **394**: 78–108.
12. Huang YJ, Moseley HN, Baran MC, *et al.* (2005) "An integrated platform for automated analysis of protein NMR structures." *Methods Enzymol* **394**: 111–41.
13. Cavanagh J, Fairbrother W, Palmer AG, *et al.* (2007) "*Protein NMR Spectroscopy.*" Academic Press, New York.
14. Lundstrom K. (2007) "Structural genomics and drug discovery." *J Cell Mol Med* **11**: 224–38.
15. Yee A, Gutmanas A, Arrowsmith CH. (2006) "Solution NMR in structural genomics." *Curr Opin Struct Biol* **16**: 611–17.
16. Kigawa T, Yabuki T, Matsuda N, *et al.* (2004) "Preparation of *Escherichia coli* cell extract for highly productive cell-free protein expression." *J. Struct Funct Genom* **5**: 63–68.
17. Sawasaki T, Ogasawara T, Morishita R, Endo Y. (2002) "A cell-free protein synthesis system for high-throughput proteomics." *Proc Natl Acad Sci USA* **99**: 14652–57.
18. Vinarov DA, Loushin Newman CL, Markley JL. (2006) "Wheat germ cell-free platform for eukaryotic protein production." *FEBS J* **273**: 4160–69.

19. Studier FW. (2005) "Protein production by auto-induction in high density shaking cultures." *Protein Expr Purif* **41**: 207–34.
20. Marsischky G, LaBaer J. (2004) "Many paths to many clones: a comparative look at high-throughput cloning methods." *Genome Res* **14**: 2020–28.
21. Hunt I. (2005) "From gene to protein: a review of new and enabling technologies for multi-parallel protein expression." *Protein Expr Purif* **40**: 1–22.
22. Szyperski T, Yeh DC, Sukumaran DK, *et al.* (2002) "Reduced-dimensionality NMR spectroscopy for high-throughput protein resonance assignment." *Proc Natl Acad Sci USA* **99**: 8009–14.
23. Kim S, Szyperski T. (2003) "GFT NMR, a new approach to rapidly obtain precise high dimensional NMR spectral information." *J Am Chem Soc* **125**: 1385–93.
24. Atreya HS, Szyperski T. (2004) "G-matrix Fourier Transform NMR spectroscopy for complete protein resonance assignment." *Proc Natl Acad Sci USA* **101**: 9642–47.
25. Liu G, Shen Y, Atreya HS, *et al.* (2005) "NMR data collection and analysis protocol for high-throughput protein structure determination." *Proc Natl Acad Sci USA* **102**: 10487–92.
26. Kupce E, Freeman R. (2004) "Projection-reconstruction technique for speeding up multidimensional NMR spectroscopy." *J Am Chem Soc* **126**: 6429–40.
27. Szyperski T, Atreya HS. (2006) "Principles and applications of GFT projection NMR spectroscopy." *Magn Reson Chem* **44**: S51–60.
28. Venters RA, Coggins BE, Kojetin D, *et al.* (2005) "(4,2)D Projection — reconstruction experiments for protein backbone assignment: application to human carbonic anhydrase II and calbindin D(28K)." *J Am Chem Soc* **127**: 8785–95.
29. Hiller S, Fiorito F, Wüthrich K, Wider G. (2005) "Automated projection spectroscopy (APSY)." *Proc Natl Acad Sci USA* **102**: 10876–81.
30. Moseley HN, Riaz N, Aramini JM, *et al.* (2004) "A generalized approach to automated NMR peak list editing: Application to reduced dimensionality triple resonance spectra." *J Magn Reson* **170**: 263–77.
31. Eghbalnia HR, Bahrami A, Tonelli M, *et al.* (2005) "High-resolution iterative frequency identification for NMR as a general strategy for multidimensional data collection." *J Am Chem Soc* **127**: 12528–36.
32. Shen Y, Atreya HS, Liu G, Szyperski T. (2005) "G-matrix fourier transform NOESY based protocol for high-quality protein structure determination." *J Am Chem Soc* **127**: 9085–99.
33. Pervushin K, Vögeli B, Eletsky A. (2002) "Longitudinal ^{1}H relaxation optimization in TROSY NMR spectroscopy." *J Am Chem Soc* **124**: 12898–902.
34. Diercks T, Daniels M, Kaptein R. (2005) "Extended flip-back schemes for sensitivity enhancement in multidimensional HSQC-type out-and-back experiments." *J Biomol NMR* **33**: 243–59.

35. Schanda P, Brutscher B. (2005) "Very fast two-dimensional NMR spectroscopy for real-time investigation of dynamic events in proteins on the time scale of seconds." *J Am Chem Soc* **127**: 8014–15.
36. Eletsky A, Atreya HS, Liu G, Szyperski T. (2005) "Probing structure and functional dynamics of (large) proteins with aromatic rings: L-GFT-TROSY (4,3)D HCCH NMR spectroscopy." *J Am Chem Soc* **127**: 14578–79.
37. Stern AS, Li KB, Hoch JC. (2002) "Modern spectrum analysis in multidimensional NMR spectroscopy: comparison of linear prediction, extrapolation and maximum-entropy reconstruction." *J Am Chem Soc* **124**: 1982–93.
38. Jaravine V, Ibraghimov I, Orekhov VY. (2006) "Removal of a time barrier for high-resolution multidimensional NMR spectroscopy." *Nat Methods* **3**: 605–07.
39. Malmodin D, Billeter M. (2005) "Multiway decomposition of NMR spectra with coupled evolution periods." *J Am Chem Soc* **127**: 13486–87.
40. Huang YJ, Tejero R, Powers R, Montelione GT. (2006) "A topology-constrained distance network algorithm for protein structure determinations from NOESY data." *Proteins* **62**: 587–603.
41. Güntert P. (2003) "Automated NMR protein structure calculation." *Prog Nucl Magn Reson Spectrosc* **43**: 105–25.
42. Habeck M, Rieping W, Linge JP, Nilges M. (2004) "NOE assignment with ARIA 2.0: the nuts and bolts." *Methods Mol Biol* **278**: 379–402.
43. Ab E, Atkinson AR, Banci L, *et al.* (2006) "NMR in the SPINE structural proteomics project." *Acta Crystallogr D Biol Crystallogr* **D62**: 1150–61.
44. Berman HM, Westbrook J, Feng Z, *et al.* (2000) "The Protein Data Bank." *Nucleic Acids Res* **28**: 235–42.
45. Bhattacharya A, Tejero R, Montelione GT. (2007) "Evaluating protein structures determined by structural genomics consortia." *Proteins* **66**: 778–95.
46. Prestegard JH, Valafar H, Glushka J, Tian F. (2001) "Nuclear magnetic resonance in the era of structural genomics." *Biochemistry* **40**: 8677–85.
47. Hiller S, Wasmer C, Wider G, Wüthrich K. (2007) "Sequence — Specific resonance assignment of soluble nonglobular proteins by 7D APSY-NMR spectroscopy." *J Am Chem Soc* **129**: 10823–28.
48. Kupce E, Freeman R. (2006) "Hyperdimensional NMR spectroscopy." *J Am Chem Soc* **128**: 6020–21.
49. Schroeder FC, Gronquist M. (2006) "Extending the scope of NMR spectroscopy with microcoil probes." *Angew Chem Int Ed Engl* **43**: 7122–31.
50. Peti W, Norcross J, Eldridge G, O'Neil-Johnson M. (2004) "Biomolecular NMR using a microcoil probe-new technique for the chemical shift assignment of aromatic side chains in proteins." *J Am Chem Soc* **126**: 5873–78.
51. Aramini JM, Rossi P, Anklin C, *et al.* (2007) "Microgram-scale protein structure determination by NMR." *Nat Methods* **4**: 491–93.

52. Mirkovic N, Li Z, Parnassa A, Murray D. (2007) "Strategies for high-throughput comparative modeling: applications to leverage analysis in structural genomics and protein family organization." *Proteins* **66**: 766–77.
53. Liu G, Li Z, Chiang Y, *et al.* (2005) "High-quality homology models derived from NMR and X-ray structures of *E. coli* proteins YdgK and Suf E suggest that all members of the YdgK/Suf E protein family are enhancers of cysteine desulfurases." *Protein Sci* **14**: 1597–608.
54. Granseth E, Seppälä S, Rapp M, *et al.* (2007) "Membrane protein structural biology — How far can the bugs take us?" *Mol Membr Biol* **24**: 329–32.
55. Daley DO, Rapp M, Granseth E, *et al.* (2005) "Global topology analysis of the *Escherichia coli* inner membrane proteome." *Science* **308**: 1321–23.
56. Klammt C, Schwarz D, Dötsch V, Berhard F. (2007) "Cell-free production of integral membrane proteins on a preparative scale." *Methods Mol Biol* **375**: 57–78.
57. Gao FP, Cross TA. (2005) "Recent developments in membrane protein structural genomics." *Gemome Biol* **6**: 244.
58. Sorgen PL, Hu Y, Guan L, *et al.* (2002) "An approach to membrane protein structure without crystals." *Proc Natl Acad Sci USA* **99**: 14037–40.
59. Tamm LK, Abildgaard F, Arora A, *et al.* (2003) "Structure, dynamics and function of the outer membrane protein A (OmpA) and influenza hemagglutinin fusion domain in detergent micelles by solution NMR." *FEBS Lett* **555**: 139–43.
60. Atreya HS, Eletsky A, Szyperski T (2005) "Resonance assignment of proteins with high shift degeneracy based on 5D spectral information encoded in G^2FT NMR experiments." *J Am Chem Soc* **127**: 4554–55.
61. Kainosho M, Torizawa T, Iwashita Y, *et al.* (2006) "Optimal isotope labeling for NMR protein structure determinations." *Nature* **440**: 52–57.
62. Suzuki M, Mao L, Inouye M. (2007) "Single protein production (SPP) in *Escherichia coli.*" *Nat Protoc* **2**: 1802–10.
63. Serber Z, Corsini L, Durst F, Dötsch V. (2005) "In-cell NMR spectroscopy." *Methods Enzymol* **394**: 17–41.
64. Burz DS, Dutta K, Cowburn D, Shekhtman A. (2006) "Mapping structural interactions using in-cell NMR spectroscopy (STINT-NMR)." *Nat Methods* **3**: 91–93.
65. Cavalli A, Salvatella X, Dobson CM, Vendruscolo M. (2007) "Protein structure determination from NMR chemical shifts." *Proc Natl Acad Sci USA* **104**: 9615–20.
66. Shen Y, Bax A. (2007) "Protein backbone chemical shifts predicted from searching a database for torsion angle and sequence homology." *J Biomol NMR* **38**: 289–302.
67. Berjanskii MV, Neal S, Wishart DS. (2006) "PREDITOR: a web server for predicting protein torsion angle restraints." *Nucl Acids Res* **34**: W63–69.

68. Valafar H, Prestegard JH. (2003) "Rapid classification of a protein fold family using a statistical analysis of dipolar couplings." *Bioinformatics* **19**: 1549–55.
69. Atreya HS, Garcia E, Shen Y, Szyperski T. (2007) "*J*-GFT NMR for precise measurement of mutually correlated nuclear spin-spin couplings." *J Am Chem Soc* **129**: 680–92.
70. Grishaev A, Wu J, Trewhella J, Bax A. (2005) "Refinement of multidomain protein structures by combination of solution small-angle X-ray scattering and NMR data." *J Am Chem Soc* **127**: 16621–28.
71. Yang S, Szyperski T. (2007) "NMR structure of protein BPTI derived with NOEs measured in supercooled water indicates a new way for validating and refining highest quality protein solution structures." *Angew Chem Int Ed Engl* **46**, in press.
72. Laskowski RA, Watson JD, Thornton JM (2005). "Protein function prediction using local 3D templates." *J Mol Biol* **351**: 614–26.
73. Mercier KA, Baran M, Ramanathan V, *et al.* (2006) "FAST-NMR: functional annotation screening technology using NMR spectroscopy." *J Am Chem Soc* **128**: 15292–99.
74. Dyson HJ, Wright P. (2002) "Coupling of folding and binding for unstructured proteins." *Curr Opin Struct Biol* **12**: 54–60.
75. Schlessinger A, Liu J, Rost B. (2007) "Natively unstructured loops differ from other loops." *PLoS Comput Biol* **3**: e140.

Chapter 14

Structural Proteomics in Relation to Signaling Pathways

Florence Bedez, Arnaud Poterszman and Dino Moras*

Introduction

Signaling pathways allow the regulation of cell activity in response to chemical, physical or cellular stimuli. The ultimate goal is a rapid and sensitive adaptation of unicellular or pluricellular organisms to environmental changes. In the human body, signaling events are very complex processes because a given physiological response involves different stimuli at the same time. In the cell, the complexity stems from multiple interactions between the molecules involved in distinct pathways. In this chapter, we will illustrate how structural proteomics contributes to our knowledge of three specific cellular signaling steps, i.e. stimulus recognition by receptor, signal transduction and the activation of effector systems. We will focus particularly on nuclear receptor signaling pathways which control the expression of complex gene networks in response to a large variety of hormonal or metabolic signals within specific tissue and which play a crucial role in numerous physiological processes such as growth, development, homeostasis and therefore make them obvious drug targets.

*Corresponding author: Dino.MORAS@igbmc.u-strasbg.fr.
Institut de Génétique et de Biologie Moléculaire et Cellulaire, UMR 7104, 1 rue Laurent Fries, BP 10142, 67404 Illkirch Cedex, France.

Nuclear Receptor Signaling Pathways

Nuclear Receptors and Structural Proteomic Data

To date, 48 human nuclear receptors (NR) have been identified, including receptors of the classic ligands such as the steroid hormone receptor (SHR), thyroid receptor (TR), vitamin D receptor (VDR), retinoic acid receptor (RAR) as well as receptors for which no ligand was known (when the receptor was cloned) such as peroxisome proliferator–activated receptor (PPAR) or oestrogen related receptor (ERR). Based on sequence alignments, phylogenetic and structural analyses, NRs have been divided into two groups that reflect their oligomeric behavior.[1] Although the 3-dimensional structure of a complete NR has not yet been solved, two domains have been of particular interest for structural studies: the C domain or DNA binding domain (DBD) and the E domain termed ligand binding domain (LBD). The first structure of a DBD was solved in 1991. Since then, only 20 structures of DBDs have been solved, often in complex with DNA. Due to the enormous potential of LBD as drug targets, more than 200 X-ray structures of LBD have been determined, mainly in the form of monomers or homodimers and in complex with ligands or co-activator peptides. Note that although half of all NRs act as heterodimers, only a few structures of heterodimeric LBDs have been obtained (Fig. 1).

Nuclear Receptor Signaling Pathways

NRs participate in two distinct signaling cascades, generating either a genomic response or a non genomic response, also called rapid response (Fig. 2). Initially, nuclear receptors were thought to be located in the nucleus or in the cytoplasm where they regulate gene expression by binding to specific DNA response elements as monomers, homodimers or heterodimers with RXR. Their action triggers a genomic response that can take between a few hours and several days and that can be stopped by inhibitors of gene expression. In the last few years, some NRs such as VDR have also been shown to be implicated in membrane signaling pathways. These pathways

(**A**)

Nom.	Name	Structures
NR1A1	TR-alpha	+(1)
NR1A2	TR-beta	+(3)
NR1B1	RAR-alpha	+(2)
NR1B2	RAR-beta	+(2)
NR1B3	RAR-gamma	+(2)
NR1C1	PPAR-alpha	+(2)
NR1C2	PPAR-beta	+(8)
NR1C3	PPAR-gamma	+(12)
NR1D1	Rev-erb alpha	-
NR1D2	Rev-erb beta	-
NR1F1	ROR-alpha	+(2)
NR1F2	ROR-beta	+(3)
NR1F3	ROR-gamma	-
NR1H1	LXR-alpha	+(1)
NR1H2	LXR-beta	+(6)
NR1H4	FXR-alpha	+(3)
NR1H5	FXR-beta	-
NR1I1	VDR	+(20)
NR1I2	PXR	+(6)
NR1I3	CAR	+(3)

Nom.	Name	Structures
NR2A1	HNF4-alpha	+(1)
NR2A2	HNF4-gamma	+(1)
NR2B1	RXR-alpha	+(7)
NR2B2	RXR-beta	+(1)
NR2B3	RXR-gamma	+(7)
NR2C1	TR2	-
NR2C2	TR4	-
NR2E1	TLX	-
NR2E3	TLL	-
NR2F1	COUP-TFI	-
NR2F2	COUP-TFII	-
NR2F6	EAR2	-
NR3A1	ER-alpha	+(30)
NR3A2	ER-beta	+(26)
NR3B1	ERR-alpha	+(1)
NR3B2	ERR-beta	-
NR3B3	ERR-gamma	+(9)
NR3C1	GR	+(2)
NR3C2	MR	+(11)
NR3C3	PR	+(5)
NR3C4	AR	+(24)
NR4A1	NGFI-B	+(2)
NR4A2	NURR1	+(1)
NR5A1	SF1	+(3)
NR5A2	LRH1	+(6)
NR6A1	GCNF	-
NROB1	DAX-1	-
NROB2	SHP	-

(**B**)

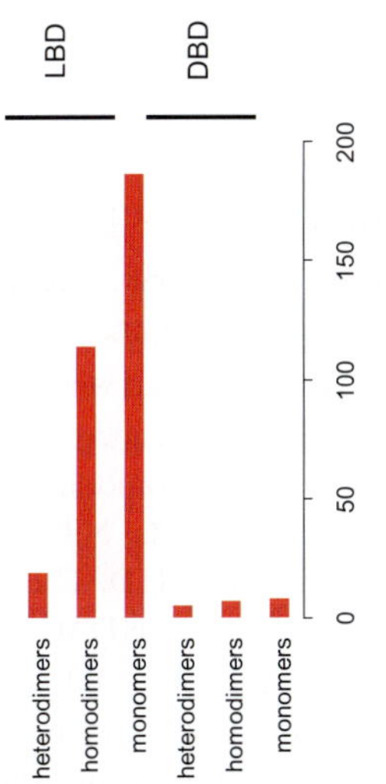

(C)

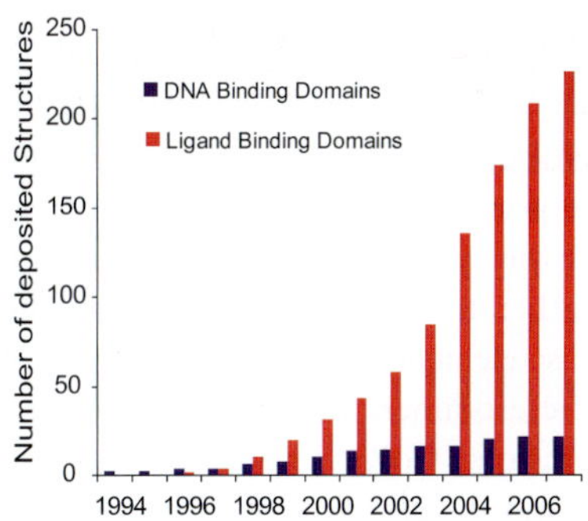

Fig. 1 Analysis of deposited NRs structures (on 22/09/2007). (**A**) Solved nuclear receptor LBD structures. (Representatives of all families have been determined with the exception of the NR6 and NRO superfamilies. "+" indicates a solved structure; in parenthesis, the number of PDB entries. "–" indicates that no structure has been deposited.) (**B**) Distribution of PDB entries of DBDs and LBDs. (**C**) Chart showing growth of PDB entries for DBD and LBDs.

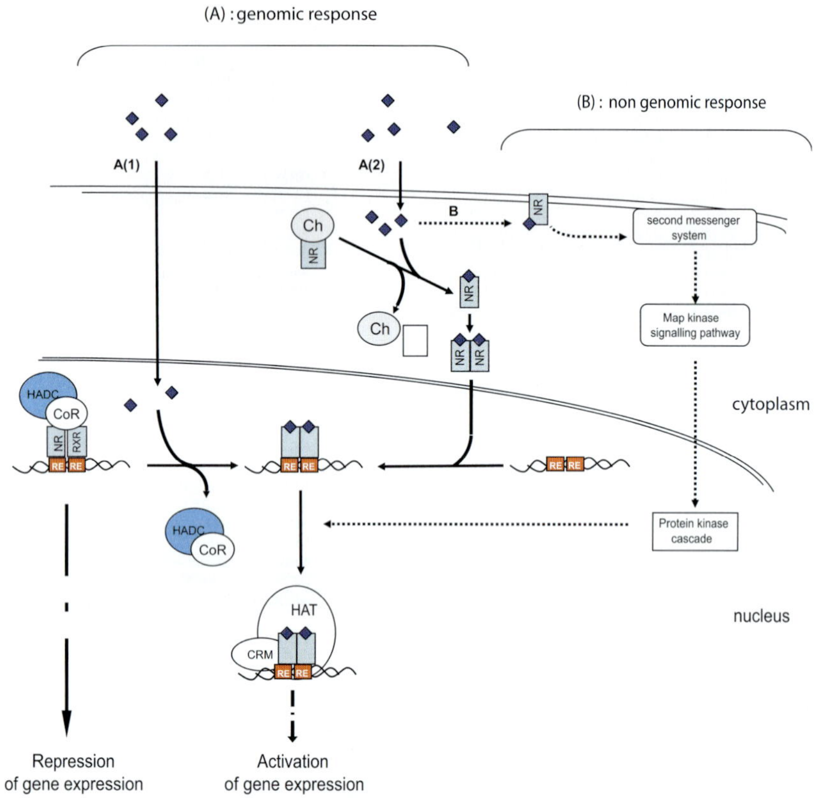

Fig. 2 Nuclear receptors signaling pathways. (**A**) Signaling pathways generating a genomic response. Steroid hormone receptors (SHRs) are associated with chaperone proteins in the cytoplasm. The ligand binds to NRs, releasing chaperone proteins. Holo-NRs then translocate to the nucleus where they bind as a homodimers to specific response elements (RE). In the absence of ligand, non steroidal NRs are bound to RE as heterodimers with the RXR and are associated with HDAC complexes and corepressors (CoR). Ligand binding induces the release of HDAC and CoR complexes and the recruitment of histone acetyltransferase (HAT) and chromatin-remodeling (CRM) complexes. Finally, the transcriptional machinery is recruited, resulting in activation of gene expression. (**B**) Signaling pathways generating a non genomic response (rapid response). After diffusion through the cytoplasmic membrane, the ligand can interact with their cognate receptors located in the caveolae of the plasma membrane to generate rapid response. Ligand binding may result in the activation of one or more second messenger systems, including protein kinase C, G protein-coupled receptors, or phosphatidylinositol-3-kinase, some of which (such as RAF/MAPK) may engage in cross-talk with the nucleus to modulate gene expression.

are mediated by second-messengers and produce a rapid biological response, called non genomic response, varying in time from a few seconds to a few hours.

In genomic responses, two pathways need to be described (Fig. 2A). The first involves steroid NRs which in the absence of the ligand, are located in the cytoplasm and are associated with chaperone proteins. After hormone binding, the receptors release the chaperone protein, dimerize, translocate into the nucleus and bind as homodimers to their specific DNA sequence called response elements (RE). Finally, this binding of DNA with holo-NR enhances transcription by recruitment of the transcription machinery through interaction with co-activators or general transcription factors (i.e. p160, p300/CBP).[2] In the second pathway, nuclear receptors are initially bound as dimers to specific response element and associated with corepressors or histone deacetylase complexes (HDAC). This results in chromatin condensation and silencing of target genes. Ligand binding induces structural changes that trigger the release of corepressors/ HDAC and subsequently the sequential recruitment of coactivators. In both pathways, coactivators associated with chromatin remodeling and modifying factors (i.e. histone acetyltransferase activity (HAT), methyltransferase, SWI/SNF) decompact the chromatin and enhance gene expression.

In contrast to the genomic response, the rapid response pathway (Fig. 2B) involves the binding of the natural ligand to a nuclear receptor located in the cytoplasmic membrane. This response triggers either an instant biological response such as Ca^{2+} channel opening or a crosstalk that modulates the genomic response by MAP kinase signaling activation.

For twenty years, structural determinations by X-ray or magnetic nuclear resonance (MNR) have given us insights into the relationships between the structure and function of signaling proteins which participate in the signal transduction towards the transcription machinery. Collectively, the structural data provide important information about agonist or antagonist ligand specificity, target gene selectivity, and gene expression activation through the recruitment of multiprotein complex.

Specificity of Ligand Recognition

The nuclear receptor LBD structures share a common fold composed of 12 alpha helices (H1 to H12) and a short β sheet, packed in a three layered anti-parallel α helical sandwich. The LBD contains the ligand binding pocket (LBP), the main dimerization interface and the C terminal region in helix 12, which houses the ligand dependent transactivation domain (AF-2) (Fig. 3).[3] Crystallographic studies revealed how NRs are able to discriminate a specific signal among natural ligands as diverse as steroid hormones, fatty acids or retinoic acid, taking into consideration both the chemical similarity of the ligand and the LBP sequence identity.

Discrimination of Cognate Ligand in Genomic Response

The ligand binding pocket is topologically conserved in all LBDs. It is generally located behind helix H3 and in front of helices, H5 and H7 and is lined by hydrophobic residues. The numerous LBD structures show that the selectivity and specificity of ligand binding is determined by the size and shape of the binding pocket and by hydrophobic or hydrophylic interactions.[4]

In the case of steroid receptors, the small volume available within the LBP selects the rigid, steroidal backbone, while the polar side chains establish specific hydrophilic interactions with the natural ligand. In this way, the LBP selects the rigid ligand with high affinity among closely related steroidal structures. For instance, the crystal structure of the estradiol receptor (ER) bound to hormone showed that the binding specificity is caused by the interaction between the glutamate side chain (Glu-353) of the ER-LBP and the hydroxyl groups of estradiol acting as hydrogen bound donors, whereas in the progesterone receptor (PR), the glutamine (Gln-725) takes the place of the glutamic acid to establish a specific binding with the 3-ketone group of progesterone.[5,6]

In contrast, the size of PPAR-LBP is variable and allows the accommodation of various ligands with a lower binding affinity.[4] These observations can be correlated with the functions of NRs. Indeed, the PPARs are receptors for compounds from dietary origin such as

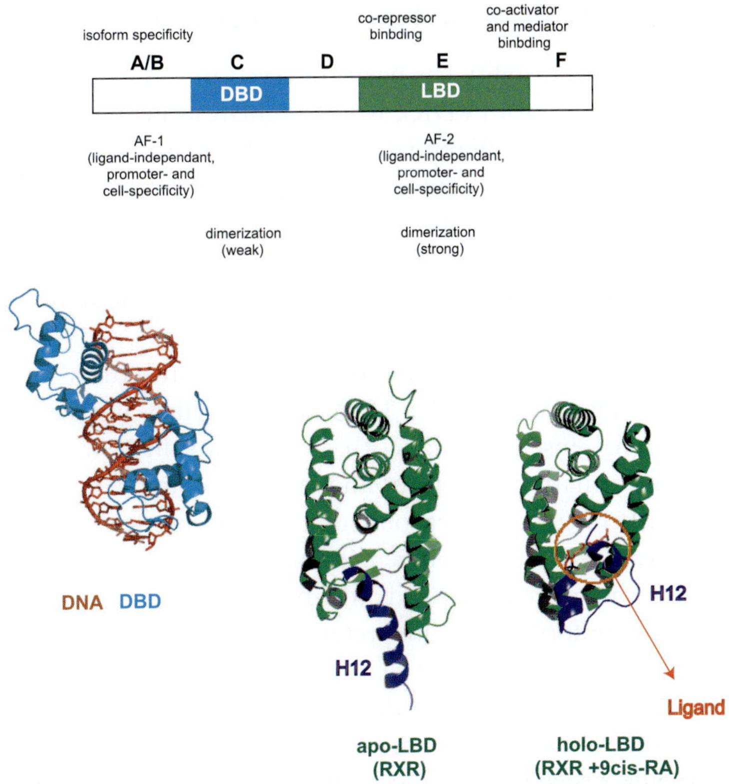

Fig. 3 Schematic representation of structural and functional organization of NRs. The A/B region is of variable length and sequence and contains a transactivation domain AF-1 that is ligand independent and cell specific. The C region or DBD has two zinc-finger motifs that are common to the overall family and recognizes a hexanucleotide sequence motif, within the regulatory region of the target gene. The D region is a flexible linker. The E region or LBD whose overall architecture is conserved in the NRs family, is responsible for selective ligand recognition and for the ligand-induced activation function (known as AF-2). 3D structures of an RAR/RXR heterodimer DBD bound to DNA (1DSZ), of RXR LBD in the absence (Apo-form, 1LBD) or the presence (Holo-form, 1FBY) of 9-*cis* retinoic acid, are represented. Experimental structures of N-terminal domain (A/B), hinge region (D), and C-terminal domain (E) have not been determined.

metabolic intermediates fatty acids, triglycerides present in human body at high concentration (μM range). In contrast, the steroid receptors specifically select the cognate ligand, present in minute quantities as for all endocrine molecules.

Alternative Pocket in the Vitamin D Receptor

Conformational flexibility has been observed in the natural ligand of the vitamin D receptor (VDR), which regulates calcium metabolism, cell proliferation and differentiation. Indeed, the natural VDR ligand, 1 α,25$(OH)_2$-vitamin D_3 can either adopt a 6-s-*cis* or 6-s-*trans* conformation. One conformation generates a genomic response while the other triggers a rapid response, but both responses are mediated by the same VDR receptor, which is located either in the nucleus or in association with the plasma membrane.

Using the atomic coordinates of VDR X-ray structures[7] and computer modeling, an alternative LBP[8] was identified and suggested a conformational ensemble model that could explain how, depending on the conformation of the bound ligand, a single receptor may generate either a genomic response or a rapid response. In this model, apo-VDR can exist in three different conformational states that are in equilibrium with one another and differ by the configuration of helix 12. The 6-s-*cis* conformer would preferentially occupy the alternative pocket and become an agonist to induce a rapid response, whereas the 6-s-*trans* conformer will bind the classical LBP leading to initiation of a genomic response. This model also suggests how this receptor preferentially selects one ligand conformation from a population of flexible conformers.[9,10]

Signal Transduction in the NR Signaling Pathways

Structural Basis of the Transactivation Mechanism

The LBD structures of apo-RXRα and holo-RARγ led to the proposition of a mechanism called the "mousetrap model"[11,12] by which the C-terminal region of LBD transduces the agonist signal to the transcriptional machinery.

In the apo- form, helix H12 is positioned away from the LBD core and is unable to recruit co-activators. Binding of an agonist prompts a conformational change which repositions H12 against the core, sealing the LBP and generating a novel surface which is able to recognize and recruit co-activators such as p160 proteins and p300/CBP associated factor. NR activation is thus mediated through accurate localization of the AF2 helix, which forms a charge clamp pocket to accommodate the co-activators and simultaneously excludes the co-repressors.[13] The LBD interaction surface defines a hydrophobic cleft that recognizes and docks a short α helix within the NR interacting domain (NID) of coactivators.

Specificity of Comodulator Recruitment

The short LxxLL motif of the NID makes contacts with helices H3, H4 and H12, while the recognition specificity seems to involve electrostatic interactions between the LBD and variable residues flanking the conserved LxxLL motif. Unliganded NRs bound to the specific promoter can exhibit a repressive activity similar to that of antagonist bound NRs. This repressive function is mediated by the binding of co-repressors such as the nuclear receptor co-repressor (NCoR) or the silencing mediator for retinoid and thyroid receptor (SMRT). The structure of an antagonist-bound PPARα in complex with a corepressor peptide reveals that the co-repressor peptide is located between H3 and H4, and occupies the same hydrophobic groove as the coactivators. The interaction is mediated through a similar but longer motif, LxxxIxxxL/I, that adopts an additional turn in the N terminus of the α helix. This N-terminal extension seems to require the shift of helix H12 from the active position to accommodate co-repressors.[14,15]

Structural studies also propose a mechanism that explains the specific recognition of the co-regulators to the AF-2 helix but nevertheless, does not show how the choice of binding between the co-activator and co-repressor is determined. In the cellular context, the co-repressors are associated with the histone deacetylase complex, resulting in condensed chromatin to the target DNA, whereas the co-activators are associated with the co-activator complex chromatin modelers and modifiers, in order to decompact the repressive chromatin and thus facilitate the transcription machinery recruitment. These multi-protein co-activator complexes include different proteins such as p160, coactivator-associated arginine methyltransferase-1 (CARM1), p300/CBP or histone acetyl-transferases (HATs).

Histone Acetylase Complexes

HATs acetylate specific lysine residues of histones, resulting in important effects on the gene expression signaling pathways. The accurate mechanism linking acetylation and transcription is not yet fully understood. However, it is accepted that HATs could: 1) participate in the formation of a platform to recruit transcription regulators; 2) lead to an increase of chromatin accessibility for transcriptional complexes; and 3) be implicated in the histone code with other histone-modifying enzymes.

The histone acetyl-transferases are grouped into three families (Gcn5/PCAF, MYST and p300/CBP family) and have been the subject of various structural studies.

The crystallographic data have been obtained mainly with three HATs (Hat1, Gcn5/PCAF and Esa1) in binary or ternary complex with the coenzyme A (CoA) or the acetyl-coenzyme A (acetyl-CoA) and a histone peptide substrate. The structures of the catalytic domain of yeast Gcn5 (yGCN5), *Tetrahymena thermophila* Gcn5 (tGCN5), yeast Esa1 from MYST family (yEsa1) and human PCAF display a structurally conserved catalytic core, including a three stranded anti-parallel β sheet followed by an α helix that is responsible for acetyl-CoA binding in the HATs of the Gcn5/PCAF family. Interestingly, the structure of the tGCN5/CoA/H3 peptide shows that the structurally variable N and C terminal segments flanking the core domain are also implicated in substrate binding and confer lysine specificity.[16]

Catalysis Mechanism of HATs

In agreement with enzymatic studies, the X-ray structures allowed the establishment of a sequential mechanism by which HAT recognizes and acetylates the histone substrate (Fig. 4A).

This process requires acetyl-CoA, subsequent histone binding to form a ternary complex to allow the nucleophilic attack of the ε-amine on the thioester of acetyl-CoA. In this well established model, the

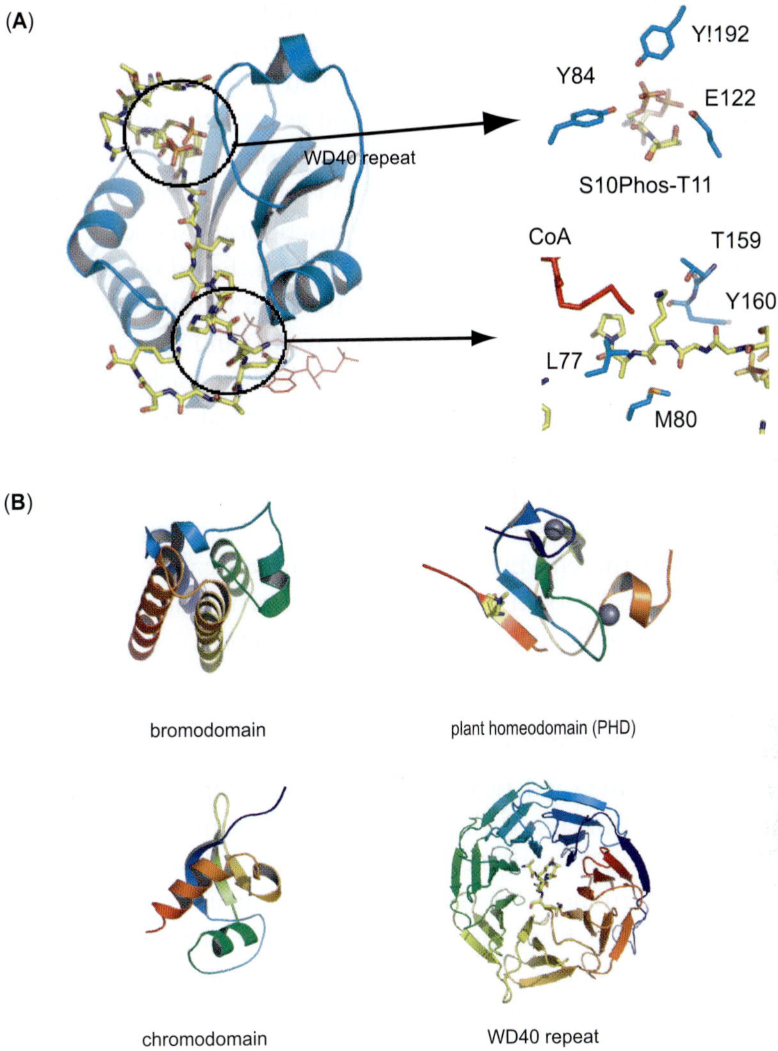

Fig. 4 Structures of histone acetyltransferases. (**A**) Crystal structure of histone acetyl-transferase domain of *Tetrahymena hydrophila* Gcn5 (in cyan) with bound coenzyme A (in red) and H3 peptide (in yellow) (PDB code 1PUA). In the lower insert, key residues involved in lysine (in yellow) substrate recognition (bottom) are shown. In the upper insert, the two conformations of the phosphoserine 10 from the H3 peptide as well as residues from the enzyme involved in the recognition are indicated. (**B**) Examples of chromatin binding domains found in HAT complexes: bromodomain, 1F68; chromodomain, 2DY7; WD40 repeat, 2G9A.

acetyl-coA configures the HAT domain to promote H3 binding, thus demonstrating that both cofactor and enzyme participate in the substrate binding.[17,18] Initially, this process had been described for HATs of the Gcn5/PCAF family that function as co-activators for a subset of transcriptional activators. A distinct and controversial mechanism[19] has been proposed for MYST family HATs that are implicated in more varied biological processes such as in transcriptional silencing and activation, DNA repair, and cell cycle progression. Based on the observation of an acetylated cysteine in the active site, this alternative model suggests that the CoA cofactor has to leave to bind the peptide lysine to the acetyl-enzyme intermediate and to allow the nucleophilic attack. In addition to the catalysis mechanism, structural studies describing the complex tGCN5 bound to CoA and H3 peptide bearing phosphoserine showed how the modification of a single residue from the H3 peptide induces a conformational change and subsequently enhances the affinity of the enzyme for the lysine substrate.[20] It revealed how post-translational modifications bearing on proximal residues of the lysine substrate influence the substrate recognition.

Docking the Hat Complex to Promoter or Gene Target

In addition to the catalytic core domain, HATs exhibit various chromatin binding domains such as the bromo-domain, chromo-domain, PHD fingers, WD40 repeat and Tudor domain, whose 3-D structures have been determined (Fig. 4). The bromo-domain is found at the C-terminus of HATs from the Gcn5 family, while the chromo-domain that is located in the N-terminus, seems to be a feature of the MYST family members. These domains cooperate to recruit the HATs to the appropriate location on the genome by recognition of modified lysines surrounding the substrate lysine. NMR titration revealed that the P/CAF bromodomain binds the acetylation sites on histone H3 or H4 with high specificity and that the strong interaction is dependent upon lysine acetylation.[21] In contrast, the chromo-domain, WD40 repeat domains and PH fingers have been shown to be involved in recognition of methylated lysine tails. Crystal structures of the WD40 domain from WDR5 and of PHD fingers reveal how these domains

specifically bind dimethylated or trimethylated H3 substrates, respectively.[22–24]

The complexity of HAT functions stems from the fact that a single HAT catalytic subunit can reside within different HAT multimeric complexes. In the cellular context, each HAT complex is made up of multiple and various subunits that enable it to carry out specialized functions and to be recruited to a distinct region of the genome, thanks to their different chromatin binding domains.

In yeast, the complexity of HAT function can be illustrated by the existence of two complexes SAGA and Elongator complexes that exhibit similar substrate specificity but that acetylate histones either within the promoter sequences or within gene coding regions, respectively.

Conclusion

Today, a major challenge in structural proteomics is the analysis of multimeric signaling complexes in order to understand how signal transduction influences the composition and conformation of these complexes so that they can modulate effector systems allowing a highly efficient fine tuning of cellular activity. A major bottleneck lies in the preparation of concentrated, fully defined and functional complexes. A second issue is the understanding of the roles of unstructured regions within signaling proteins. Recent studies have demonstrated their implication in various processes such as transcription regulation, signal transduction, or self-assembly of multiprotein complexes. Indeed, these regions fold on binding to their physiological targets or act as flexible linkers. The recent improvement in the sensitivity and resolution of spectroscopic methods now makes it possible to analyze and characterize the structural behavior and dynamics of such intrinsically disordered peptides in solution.[25]

Structural proteomics represents a crucial step towards gaining an in-depth understanding of the structure-function relationships between signaling proteins and their ligands/cofactors, the stability of multimeric complexes, the catalysis mechanisms, and the effects of mutations. Based on structural data, mutant cell lines and animal

models can be generated to better understand the function of isoforms in different cell types and during the different stages of development in relation to human health and disease. Structural proteomics is also crucial in the definition of the principles of rational drug design, since a good knowledge of signaling pathways enables the determination of which one, among the potential drug candidates, might act in a cell-type-specific manner with reduced side-effects. In the case of the signaling pathway mediated by VDR, the challenge will consist in designing an agonist ligand that will specifically activate either the genomic or the non genomic pathway. The aim is to develop an agonist ligand for treatment of psoriasis, cancer or autoimmune diseases without the calcemic side effects of the natural ligand.

References

1. Brelivet Y, Kammerer S, Rochel N, *et al.* (2004) "Signature of the oligomeric behaviour of nuclear receptors at the sequence and structural level." *EMBO Rep* **5**: 423–29.
2. Roeder RG. (1996) "Nuclear RNA polymerases: role of general initiation factors and cofactors in eukaryotic transcription." *Methods Enzymol* **273**: 165–71.
3. Renaud JP, Moras D. (2000) "Structural studies on nuclear receptors." *Cell Mol Life Sci* 57: 1748–69.
4. Li Y, Lambert MH, Xu HE. (2003) "Activation of nuclear receptors: a perspective from structural genomics." *Structure* **11**: 741–46.
5. Williams SP, Sigler PB. (1998) "Atomic structure of progesterone complexed with its receptor." *Nature* **393**: 392–96.
6. Tanenbaum DM, Wang Y, Williams SP, Sigler PB. (1998) "Crystallographic comparison of the estrogen and progesterone receptor's ligand binding domains." *Proc Natl Acad Sci USA* **95**: 5998–6003.
7. Rochel N, Wurtz JM, Mitschler A, *et al.* (2000) "The crystal structure of the nuclear receptor for vitamin D bound to its natural ligand." *Mol Cell* **5**: 173–79.
8. Mizwicki MT, Keidel D, Bula CM, *et al.* (2004) "Identification of an alternative ligand-binding pocket in the nuclear vitamin D receptor and its functional importance in 1alpha,25(OH)2-vitamin D3 signaling." *Proc Natl Acad Sci USA* **101**: 12876–81.
9. Norman AW, Mizwicki MT, Norman DP. (2004) "Steroid-hormone rapid actions, membrane receptors and a conformational ensemble model." *Nat Rev Drug Discov* **3**: 27–41.

10. Rochel N, Moras D. (2006) "Ligand binding domain of vitamin D receptors." *Curr Top Med Chem* **6**: 1229–41.
11. Bourguet W, Ruff M, Chambon P, *et al.* (1995) "Crystal structure of the ligand-binding domain of the human nuclear receptor RXR-alpha." *Nature* **375**: 377–82.
12. Renaud JP, Rochel N, Ruff M, *et al.* (1995) "Crystal structure of the RAR-gamma ligand-binding domain bound to all-trans retinoic acid." *Nature* **378**: 681–89.
13. Bledsoe RK, Montana VG, Stanley TB, *et al.* (2002) "Crystal structure of the glucocorticoid receptor ligand binding domain reveals a novel mode of receptor dimerization and coactivator recognition." *Cell* **110**: 93–105.
14. Xu HE, Stanley TB, Montana VG, *et al.* (2002) "Structural basis for antagonist-mediated recruitment of nuclear co-repressors by PPARalpha." *Nature* **415**: 813–17.
15. Germain P, Staels B, Dacquet C, *et al.* (2006) "Overview of nomenclature of nuclear receptors." *Pharmacol Rev* **58**: 685–704.
16. Marmorstein R. (2001) "Structure and function of histone acetyltransferases." *Cell Mol Life Sci* **58**: 693–703.
17. Rojas JR, Trievel RC, Zhou J, *et al.* (1999) "Structure of Tetrahymena GCN5 bound to coenzyme A and a histone H3 peptide." *Nature* **401**: 93–98.
18. Tanner KG, Langer MR, Denu JM. (2000) "Kinetic mechanism of human histone acetyltransferase P/CAF." *Biochemistry* **39**: 15652.
19. Berndsen CE, Albaugh BN, Tan S, Denu JM. (2007) "Catalytic mechanism of a MYST family histone acetyltransferase." *Biochemistry* **46**: 623–29.
20. Clements A, Poux AN, Lo WS, *et al.* (2003) "Structural basis for histone and phosphohistone binding by the GCN5 histone acetyltransferase." *Mol Cell* **12**: 461–73.
21. Zeng L, Zhou MM. (2002) "Bromodomain: an acetyl-lysine binding domain." *FEBS Lett* **513**: 124–28.
22. Couture JF, Collazo E, Trievel RC. (2006) "Molecular recognition of histone H3 by the WD40 protein WDR5." *Nat Struct Mol Biol* **13**: 698–703.
23. Li H, Ilin S, Wang W, *et al.* (2006) "Molecular basis for site-specific read-out of histone H3K4me3 by the BPTF PHD finger of NURF." *Nature* **442**: 91–95.
24. Wysocka J, Swigut T, Xiao H, *et al.* (2006) "A PHD finger of NURF couples histone H3 lysine 4 trimethylation with chromatin remodelling." *Nature* **442**: 86–90.
25. Dyson HJ, Wright PE. (2005) "Intrinsically unstructured proteins and their functions." *Nat Rev Mol Cell Biol* **6**: 197–208.

Chapter 15

The Impact of Structural Proteomics on Drug Design

Yuan-Ping Pang

Introduction

The completion of the Human Genome Project in 2003 ushered in the age of proteomics, viz., the study of all the gene products of a given organism and, in particular, the study of the structures and functions of its proteins.[1–3] Structural proteomics — also known as structural genomics — is the determination of the three-dimensional (3D) protein structures on a genome-wide scale by using high-throughput X-ray crystallography,[4] solution nuclear magnetic resonance spectroscopy,[5] and computational approaches including *de novo* structure prediction,[6] homology modeling,[7] threading,[8] and multiple molecular dynamics simulations for protein structure refinement.[9] Unlike traditional structural biology, determining a protein structure in such a context often precedes any knowledge of the function and biological role of a given protein. The structural proteomics effort has greatly increased the number of 3D protein structures solved, and has concomitantly advanced the methodologies for cloning, protein expression and production, high-throughput robotic crystallization, imaging, tracking crystallization, and X-ray data collection. Although

Computer-Aided Molecular Design Laboratory, Mayo Clinic, Rochester, Minnesota, United States of America. pang@mayo.edu

the number of 3D protein structures resulting from the structural proteomics effort has, perhaps, not increased at the rate anticipated originally, structural genomics has, nevertheless, substantially impacted drug design as will become apparent below.

Drug design is an approach to finding small-molecule drugs by using the knowledge of their biological targets, which are essential macromolecules involved in disease states. By capitalizing on structural proteomics, modern drug design — also termed structure-based drug design — relies on the iterative use of 3D macromolecular structures and computer simulations to identify and optimize small molecules that can effectively interfere with the functions of the biological targets. This interference is often achieved through blocking or modifying the active sites of their protein targets by the use of small molecules that are designed to have high affinity and selectivity for those sites.

Designing small molecules with high affinity and selectivity requires knowledge of their interactions with their protein targets in various conformational states. This knowledge can be obtained directly or extrapolated, with the aid of computer simulations, from 3D protein structures in the absence of or in complex with small molecules and their analogs by knowledge specialists — also known as molecular architects. Because the success of drug design hinges on whether the designed molecules can be made in practice, the knowledge of intermolecular interactions is best used to guide the design of molecules that not only have high affinity and selectivity but also can be readily synthesized.[10–13] One example of structure-based drug design is the development of Trusopt, a carbonic anhydrase inhibitor designed to lower increased intraocular pressure in open-angle glaucoma and ocular hypertension.[14,15]

In addition to affinity and selectivity properties, successfully moving a drug candidate from bench to bedside depends on other properties such as ADMET (absorption, distribution, metabolism, excretion, and toxicity) that appear to have no connection with structural proteomics. However, the knowledge of intermolecular interactions can offer insights into structural modifications, showing ways to adjust the ADMET properties without compromising affinity and selectivity, thus indirectly expediting the process of optimizing ADMET.

Thus, the relationship between structural proteomics and modern drug design is that structural proteomics generates the knowledge required by modern drug design. Given this relationship, it is plausible that the greatly increased number of 3D protein structures resulting from the structural proteomics effort will permit more diversified applications of modern drug design. It is also likely that high-throughput methodologies that are a spin-off of the structural proteomics effort will expedite more the drug design process because high-throughput determination of co-crystal structures of small molecules with a given target will yield timely insights into structural modifications of small molecules for optimal affinity, selectivity, and ADMET properties. However, it is less obvious but more important that, given 3D structures of more distinct proteins within a genome and more 3D structures of a common protein conserved across species, structural proteomics will revolutionize the strategies adopted in modern drug design. The potential for this transformation is demonstrated by the paradigm described below for developing individualized human-safe pesticides utilizing the 3D structures of insect and mammalian acetylcholinesterases (AChEs).

Serendipitous Discovery of the Present Pesticide Target

AChE is a serine hydrolase vital for regulating the neurotransmitter acetylcholine in mammals and insects. It has a deep and narrow active-site gorge, with the catalytic site near its bottom, and the so-called peripheral anionic site near its entrance (Fig. 1).[16,17] Inhibitors of this enzyme in a pest, which are currently in use as anticholinesterase pesticides such as methamidophos, chemically modify a catalytic serine residue that is crucial to the catalysis of pest AChE, thus blocking the function of the enzyme and incapacitating the pest. Targeting the serine residue of pest AChEs by pesticides was a serendipitous outcome of a World War II research on organophosphate nerve agents. Because the catalytic serine residue is conserved in mammalian AChEs, the use of these pesticides has been severely restricted by their toxicity to mammals. The use of anticholinesterase pesticides has also been

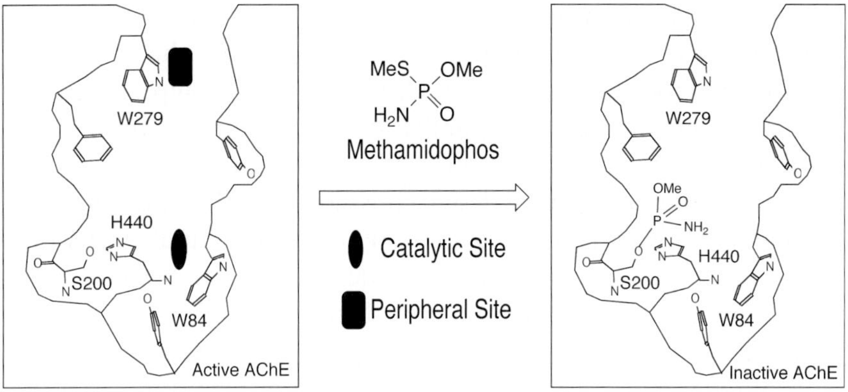

Fig. 1 Cross-section view of acetylcholinesterase (AChE) showing the conserved catalytic serine residue (S200) being chemically modified by the currently employed anticholinesterase pesticide methamidophos.

limited by the appearance of pest strains that are resistant to anticholinesterase pesticides. For example, there are strains of mosquitoes that bear the Gly119Ser mutation and are resistant to many of the pesticides currently employed.[18]

Although it has long been assumed that humans are not harmed by low doses of the anticholinesterase pesticides because insects are more sensitive to them than humans, a recent report by the Office of Inspector General of the U.S. Environmental Protection Agency indicates that some anticholinesterases can enter the brain of fetuses and young children and may destroy cells in the developing nervous system.[19] There is an urgent need for novel pesticides whose pest toxicity is so high that trace amounts will be capable of incapacitating specific pests without having measurable deleterious effects on humans.

Use of Structural Proteomics to Discover New Pesticide Targets

Conceptually, the need for a human-safe pesticide can, in general, be met by developing an irreversible inhibitor that specifically targets a unique and conserved binding site of an essential pest protein at an extremely low concentration of the inhibitor. Sequence analysis of

nine AChEs in humans, lancelets, and insects such as the greenbug (*Schizaphis graminum*) or the African malaria-bearing mosquito (*Anopheles gambiae*) identified a cysteine residue, Cys289 in greenbug AChE, that is absent from human AChE but conserved in insect and lancelet AChEs (Fig. 2).[20] This sequence-based finding was consistent with the reports that aphid AChEs were sensitive to sulfhydryl inhibitors[21,22] and led to the speculation that Cys289 or its equivalent in other aphid AChEs might be located in the active site of aphid AChEs and might thus be a putative target for developing selective aphidicides.[20]

However, the idea of targeting Cys289 of greenbug AChE or its equivalent in other aphid AChEs was mooted without a 3D structure or a model of an aphid AChE, viz., without knowing where the cysteine residue is located in the structures of aphid AChEs and without knowing whether this residue is accessible for covalent bond formation with sulfhydryl agents. The notion of targeting Cys289 became compelling after independent analysis of 112 AChE sequences (Fig. 3)[23] and after the 3D models of AChEs in the greenbug, the English grain aphid (*Sitobion avenae*), and the malaria-bearing mosquito had been created by homology modeling based on eight AChE crystal structures and by using a protein structure refinement technique that had shown success in the 2006 Continuous CASP Model Refinement Experiment (http://predictioncenter.org/caspR/).[23,24] This context underscores the requirement for protein crystal structures to compute

```
                                                     289
                                                      |
Schizaphis graminum (AChE)        LRKANSSVMVEKEWDHVA---ICFFPFVPVVDGAFLDD
Branchiostoma floridae (AChE2)    MRGVPAQTISDNEWVVWG---LCQFPFAPIVDGNFIRE
Anopheles gambiae (AChE1)         LRGKDPHVLVNNEWGTLG---ICEFPFVPVVDGAFLDE
Culex pipiens (AChE1)             LRTKDPNELVDNEWGTLG---ICEFPFVPVVDGAFLDE
Aphis gossypii (AChE2)            LRKVNSSAMVEKEWDHVA---MCFFPFVPVVDGAFLDD
Myzus persicae (AChE)             LRKANSSTMVEKEWDHVA---ICFFPFVPVVDGAFLDD
Apis mellifera (AChE)             LRVKDPVELVKNEWGTLG---ICEFPFVPVIDGAFLDE
Homo sapiens (AChE)               LRTRPSQVLVNHEWHVLPQESVFRFSFVPVVDGDFLSD
Torpedo californica (AChE)        LREKKPQELIDVEWNVLPFDSIFRFSFVPVIDGEFFPT
                                  :*   .  : . **       :  *.*.*::** *:
```

Fig. 2 Sequence alignments of the nine acetylcholinesterases reported in Ref. 20 (note: the annotations for sequences of *B. floridae*, *A. gossypii*, and *S. graminum* are revised according to the records at the U.S. National Center for Biotechnology Information; also the sequence order is changed for clarity.).

```
                                                         289
                                                          |
Schizaphis graminum                  MVEKEWDHVA------ICFFPFVPVVDGAFLDDHPQKSLS
Aedes aegypti                        ISVQQWNSYSG-----ILGFPSAPTIDGVFMTADPMTMLR
Anopheles gambiae (ace1,*)           LVNNEWGTLG------ICEFPFVPVVDGAFLDETPQRSLA
Anopheles gambiae (ace1,**)          LVNNEWGTLG------ICEFPFVPVVDGAFLDETPQRSLA
Anopheles stephensi (ace2)           ISVQQWNSYSG-----ILGFPSAPTIDGVFMTADPMTMLR
Aphis gossypii (ace2)                MVEKEWDHVA------MCFFPFVPVVDGAFLDDHPQKSLS
Apis mellifera (ace1)                LVKNEWGTLG------ICEFPFVPVIDGAFLDETPQRSLA
Bactrocera dorsalis                  ISVQQWNSYSG-----ILSFPSAPTIDGAFLPDHPMKMME
Bactrocera oleae                     ISVQQWNSYSG-----ILSFPSAPTIDGAFLPDHPMKMME
Bemisia tabaci                       VSSQQWSSYFG-----ILGFPSAPTIDGEFLPKHPLELMK
Blattella germanica (ace1)           LVNNEWGTLG------ICEFPFVPIIDGTILDGPPQRSLA
Bombyx mori (ace1)                   LVNNEWGTLG------ICEFPFVPIIDGSFLDEMPVRSLA
Culex pipiens (ace1)                 LVDNEWGTLG------ICEFPFVPVVDGAFLDETPQRSLA
Culex tritaeniorhynchus(ace2)        LVDNEWGTLG------ICEFPFVPVVDGAFLDETPQRSLA
Cydia pomonella (ace1)               LVNNEWGTLG------ICEFPFVPIIDGSFLDEMPIRSLA
Drosophila melanogaster              ISVQQWNSYSG-----ILSFPSAPTIDGAFLPADPMTLMK
Haematobia irritans                  ISVQQWNSYSG-----ILSFPSAPTIDGAFLPADPMTLLK
Helicoverpa armigera (ace1)          LVNNEWGTLG------ICEFPFVPIIDGSFLDELPVRSLV
Helicoverpa assulta                  ISVQQWNSYTG-----ILGFPSAPTVDGVFLPKDPDTMMK
Leptinotarsa decemlineata            ISLQQWNSYSG-----ILGFPSTPTIEGVLLPKHPMDMLA
Lucilia cuprina                      ISVQQWNSYSG-----ILSFPSAPTIDGAFLPADPMTLMK
Musca domestica                      ISVQQWNSYSG-----ILSFPSAPTIDGAFLPADPMTLLK
Myzus persicae (aceM)                MVEKEWDHVA------ICFFPFVPVVDGAFLDDYPQKSLS
Nephotettix cincticeps               ISVQQWNSYFG-----ILGFPSAPTIDGVFLPKHPLDLLK
Nilaparvata lugens                   ISVQQWNSYSG-----ILGLPSAPTIDGIFLPKHPLDLLK
Oulema oryzae (ace2)                 LVNNEAGTLG------ICDFPFVPVIDGAFLDEHPVRALA
Plutella xylostella (ace1)           LVNNEWGTLG------ICEFPFVPIIDGSFLDEMPIRSLA
Rhopalosiphum padi (ace1)            MVEKEWDHVA------ICFFPFVPVVDGAFLDDHPQKSLS
Sitobion avenae (ace1)               MVEKEWDHVA------ICFFPFVPVVDGAFLDDYPQKSLS
Spodoptera exigua (ace1)             LVNNEWGTLG------ICEFPFVPIIDGSFLDELPARSLA
Trialeurodes vaporariorum            VSSQQWSSYFG-----ILGFPSAPTIDGVFLPKHPLELMK
Bos taurus                           LVDHEWRVLPQEH---VFRFSFVPVVDGDFLSDTPEALIN
Canis familiaris                     LVDHEWHVLPQES---VFRFSFVPVVDGDFLSDTPEALIS
Felis catus                          LVDHEWHVLPQES---VFRFSFVPVVDGDFLSDTPEALIN
Homo sapiens                         LVNHEWHVLPQES---VFRFSFVPVVDGDFLSDTPEALIN
Macaca mulatta                       LVNNEWHVLPQES---VFRFSFVPVVDGDFLSDTPEALIN
Mus musculus                         LVDHEWHVLPQES---IFRFSFVPVVDGDFLSDTPEALIN
Oryctolagus cuniculus                LVDHEWRVLPQES---IFRFSFVPVVDGDFLSDTPEALIN
Rattus norvegicus                    LVDHEWHVLPQES---IFRFSFVPVVDGDFLSDTPDALIN
```

Fig. 3 Part of the sequence alignments of acetylcholinesterases in insects (*blue*) and mammals (*red*) reported in Ref. 23 showing that greenbug C289 (*magenta*) is conserved in many other insect species. *wild-type *A. gambiae* AChE (GenBank accession number: BN000066). **mutant *A. gambiae* AChE that is insusceptible to current pesticides (GenBank accession number: AJ515149).

the 3D models of protein homologs, and the need for such protein models to clarify the structure/function relations en route to design of drugs and agrochemicals.

In the 3D models of aphid and mosquito AChEs that have been constructed by homology modeling and protein structure refinement

(Protein Data Bank codes: 2HCP and 2AZG),[23,24] greenbug AChE Cys289 and the equivalent mosquito AChE Cys286 are located at the entrance of the AChE active-site gorge. In greenbug AChE, Cys289 makes a favorable sulfur-aromatic interaction with Tyr336,[25] and Cys289 is fully accessible for covalent interaction with sulfhydryl reagents binding at the active-site gorge (Fig. 4). In the mosquito AChE, the corresponding residue, Cys286, has favorable sulfur-aromatic interactions with both Tyr333 and Trp280, and Cys286 is not fully accessible for covalent bond formation (Fig. 4). In human AChE, the residue corresponding to Cys289 is Val294, which obviously would not react with sulfhydryl agents (Fig. 5).

In general, a native or engineered free cysteine residue near the active site of an enzyme can "hook" (covalently bond to) a small molecule that binds, even loosely, at the active site, as long as the molecule contains an electrophilic group able to react with the thiol group of the cysteine residue.[26] It has been reported that sulfhydryl reagents will react covalently with a cysteine residue inserted at the peripheral site of mammalian AChE, by the His287Cys mutation (Fig. 5), and indeed interfere with substrate binding and, as a consequence, inhibit

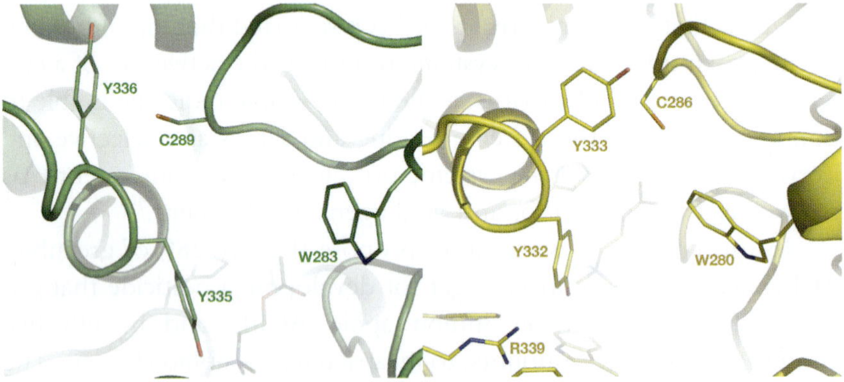

Fig. 4 Close-up view of Cys289 and Cys286 in greenbug (*green*) and malaria-carrying mosquito (*yellow*) acetylcholinesterases, respectively, showing that the two cysteine residues have favorable interactions with aromatic residues and that targeting C286 requires additional interacting sites such as Arg339 in order to disrupt the interaction between Cys286 and Trp280 (see Ref. 23).

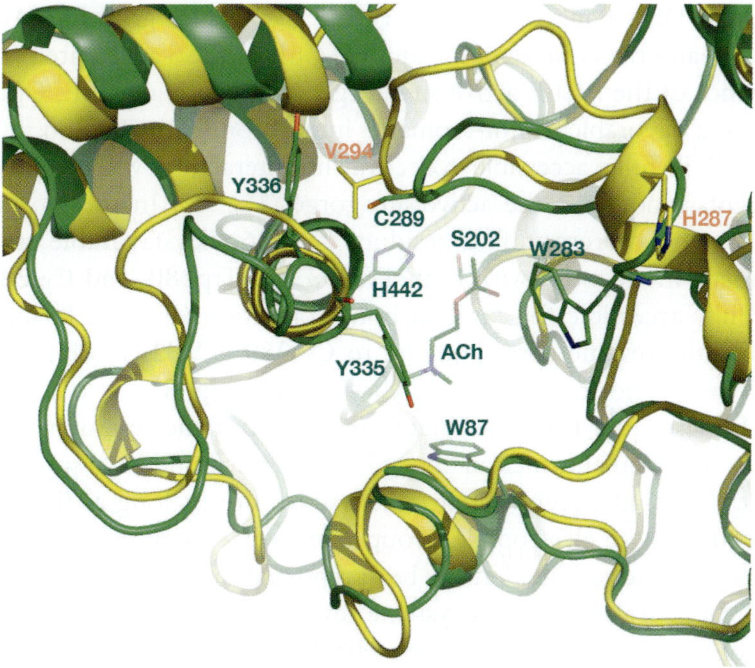

Fig. 5 Overlay of greenbug (*green*) and human (*yellow*) acetylcholinesterases.

catalytic activity.[27,28] Similarly, it has been reported that, upon binding to the proximity of a native cysteine residue at the active site of a cysteine proteinase of adenovirus, a chemically stable molecule is able to bond covalently to the cysteine residue and serves as a selective and irreversible inhibitor of the cysteine proteinase, and, consequently, as a novel lead compound for anti-viral agent development.[29]

In this context, it is logical to propose that Cys289 of greenbug AChE may serve as a novel target for developing a pesticide that can selectively and irreversibly inhibit aphid AChE, and be effective enough, even in trace amounts, to incapacitate the aphid (Fig. 6). The absence of any free cysteine residues in the active sites of mammalian AChEs means that inhibitors targeting Cys289, or its equivalent in other aphids or insects, will be much less toxic to mammals than the current anticholinesterase pesticides that target the catalytic serine residue that is conserved in both insect and mammalian AChEs.

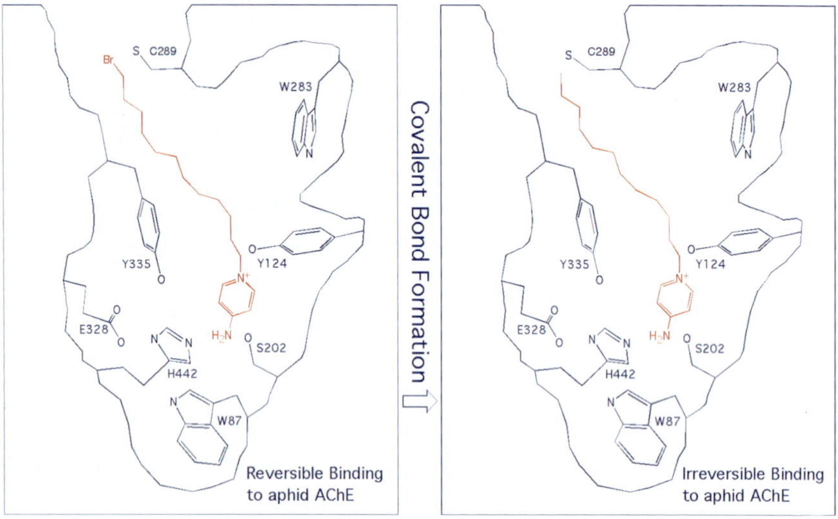

Fig. 6 Diagram representation of a chemically stable compound (*red*) that reacts with greenbug acetylcholinesterase (*blue*) upon its binding at the active-site gorge.

Due to the interaction of Cys289 with Tyr336, Cys289 plays a structural role in stabilizing the conformations of the aromatic residues within the active-site gorge of aphid AChEs; thus, Cys289 is conserved in aphid AChEs as its mutations into any non-aromatic residues destabilize the gorge. Indeed, sequence analysis shows that the AChEs of both green peach and cotton/melon aphids contain Cys289, although these aphids are resistant to most current pesticides.[23,24] Targeting Cys289 would thus encounter less resistance problem than targeting the catalytic serine residue in the case of most of the anticholinesterase pesticides currently in use, such pesticides having been in use for a long time.

The notion of targeting Cys289 of greenbug AChE or its equivalence in other aphid AChEs has now gained experimental support from kinetics studies. Testing of prototypic irreversible inhibitors designed on the basis of the 3D model of greenbug AChE showed that, at an inhibitor concentration of 30 μM, one inhibitor caused 100% reversible inhibition of greenbug AChE and 95% irreversible inhibition of the same enzyme but without irreversibly inhibiting

human AChE at all and with only 10% reversible inhibition of the human enzyme.[30] This inhibitor also selectively and irreversibly inhibited the AChE of the soybean aphid which is prevalent in Minnesota.[30] The species preference of this inhibitor is great enough to justify its classification as "pest-specific." It is worth noting that Minnesota agriculture is a multibillion-dollar enterprise annually, representing the second largest economic sector in the state's economy.[31] Soybean is the state's second largest agricultural commodity. Each year Minnesotan farmers harvest approximately 10% of the total U.S. soybean crop (www.nass.usda.gov). Nationally, combined yield losses and increased production costs caused by the soybean aphid exceed $1 billion annually.[32] The control of aphids currently hinges on the use of anticholinesterase pesticides, which contaminate the air, water, soil, and food, are toxic to other insects and to humans, but are essential for the production of food and fiber to feed an ever-growing world population.

The novel pesticide target discovered using the knowledge generated by structural proteomics holds promise for developing a pest-specific or individualized crop protection chemical that incapacitates soybean aphids yet poses no harm to humans or other mammals. Such an agent could potentially revolutionize the way in which we control aphids, and lead, eventually, to individualized pesticides so safe that they would be harmless if inadvertently consumed, e.g. on unwashed apples.

Concluding Remarks and Outlooks

As apparent from the effort to develop individualized pesticides outlined above, structural proteomics has not only broadened the scope of modern drug design and advanced the techniques used by structure-based drug design but has also begun to transform the strategies employed by modern drug design. As opposed to traditionally working on one protein target within one species, structural proteomics permits a comparative study of protein targets across many species, facilitating the identification of subtle differences in protein structure that contribute to the diversity and complexity of life. Such subtle structural differences can lead to a paradigm shift, like that described

above — a shift from targeting a ubiquitous serine residue to targeting a species-specific cysteine residue, a shift from developing generic pesticides to developing individualized pesticides, and ultimately a shift from curing a disease (reaction) to preventing a disease through improving environmental health (pro-action).

Conceptually, the paradigm shift to targeting the species-specific cysteine residue is relevant to and can serve as an incentive for the design of anti-cancer drugs, as the latter faces the same challenges of toxicity and drug-resistance. It is logical to hope that, as protein structural information resulting from structural proteomics becomes more prevalent, one would be able to perform comparative studies of more 3D protein structures in both normal and cancerous cells, and, in particular, the studies of more 3D protein structures with somatic mutations to be revealed by the ongoing cancer genome project,[33] and ultimately such studies would successfully identify new and better protein targets for developing anticancer drugs with minimal problems of toxicity and drug-resistance.

Perhaps still dimly from the above example, structural proteomics is ushering in a period of transformation in modern drug design: a transformation to be driven by the perception that the vast amount of information on 3D protein structures produced by structural proteomics — so immense that one risks being swamped by the information — should not merely be utilized for doing faster what we have been doing in designing drugs, but for implementing conceptual innovations with the potential to improve the success rate of moving discoveries from the laboratory to the patient and to ultimately shift from curing diseases to preventing diseases.

Acknowledgments

The author is currently supported by the U.S. Army Medical Research Acquisition Activity (W81XWH-04-2-0001), the U.S. National Institutes of Health (5R01GM061300-06), the State of Minnesota of the United States of America, and the Mayo Foundation for Medical Education and Research. The opinions or assertions contained herein belong to the author and are not necessarily the official views of the

U.S. Army, the U.S. National Institutes of Health, or the State of Minnesota of the United States of America.

References

1. Venter JC, Adams MD, Myers EW, *et al.* (2001) "The sequence of the human genome." *Science* **291**(5507): 1304–51.
2. Banks RE, Dunn MJ, Hochstrasser DF, *et al.* (2000) "Proteomics: new perspectives, new biomedical opportunities." *Lancet* **356**(9243): 1749–56.
3. Burley SK, Almo SC, Bonanno JB, *et al.* (1999) "Structural genomics: beyond the Human Genome Project." *Nature Gen* **23**(2): 151–57.
4. Jhoti H. (2001) "High-throughput structural proteomics using X-rays." *Trends Biotechnol* **19**(10): S67–71.
5. Yee A, Gutmanas A, Arrowsmith CH. (2006) "Solution NMR in structural genomics." *Curr Opin Struct Biol* **16**(5): 611–17.
6. Bradley P, Misura KM, Baker D. (2005) "Toward high-resolution de novo structure prediction for small proteins." *Science* **309**(5742): 1868–71.
7. Peitsch MC, Guex N. (2000) "Comparative protein modelling." http://www.expasy.ch/swissmod/course/text/chapter6.htm.
8. Kelley LA, MacCallum RM, Sternberg MJE. (2000) "Enhanced genome annotation using structural profiles in the program 3D-PSSM." *J Molecul Biol* **299**(2): 499–520.
9. Pang Y-P. (2007) "*In silico* drug discovery: solving the 'target-rich and lead-poor imbalance' using the genome-to-drug-lead paradigm." *Clin Pharmacol Ther* **81**(1): 30–34.
10. Pang Y-P, Quiram P, Jelacic T, *et al.* (1996) "Highly potent, selective, and low cost bis-tetrahydroaminacrine inhibitors of acetylcholinesterase: steps toward novel drugs for treating Alzheimer's disease." *J Biol Chem* **271**(39): 23646–49.
11. Pang Y-P, Kollmeyer TM, Hong F, *et al.* (2003) "Rational design of alkylene-linked bis-pyridiniumaldoximes as improved acetylcholinesterase reactivators." *Chem Biol* **10**(6): 491–502.
12. Park JG, Sill PC, Makiyi EF, *et al.* (2006) "Serotype-selective, small-molecule inhibitors of the zinc endopeptidase of botulinum neurotoxin serotype A." *Bioorg Med Chem* **14**(2): 395–408.
13. Tang J, Park JG, Millard CB, *et al.* (2007) "Computer-aided lead optimization: improved small-molecule inhibitor of the zinc endopeptidase of botulinum neurotoxin serotype a." PLoS ONE **2**: e761.
14. Greer J, Erickson JW, Baldwin JJ, Varney MD. (1994) "Application of the three-dimensional structures of protein target molecules in structure-based drug design." *J Med Chem* **37**(8): 1035–54.

15. Grover S, Apushkin MA, Fishman GA. (2006) "Topical dorzolamide for the treatment of cystoid macular edema in patients with retinitis pigmentosa." *Am J Ophthalmol* **141**(5): 850–58.
16. Sussman JL, Harel M, Frolow F, *et al.* (1991) "Atomic structure of acetylcholinesterase from Torpedo californica: a prototypic acetylcholine-binding protein." *Science* **253**(5022): 872–79.
17. Raves ML, Harel M, Pang YP, *et al.* (1997) "Structure of acetylcholinesterase complexed with the nootropic alkaloid, (-)-huperzine A." *Nature Struct Biol* **4**(1): 57–63.
18. Weill M, Lutfalla G, Mogensen K, *et al.* (2003) "Comparative genomics: insecticide resistance in mosquito vectors." *Nature* **423**(6936): 136–37.
19. Fialka JJ. (2006) "EPA scientists cite pressure in pesticide study." *Wall Street J*, May 25, Sect. A4.
20. Pezzementi L, Rowland M, Wolfe M, Tsigelny I. (2006) "Inactivation of an invertebrate acetylcholinesterase by sulfhydryl reagents: the roles of two cysteines in the catalytic gorge of the enzyme." *Invert Neurosci* **6**(2): 47–55.
21. Zahavi M, Tahori AS, Klimer F. (1972) "An acetylcholinesterase sensitive to sulfphydryl inhibitors." *Biochim Biophys Acta* **276**: 577–83.
22. Smissaert HR. (1976) "Reactivity of a critical sufhydryl group of the acetylcholinesterse from aphids (Myzys persicae)." *Pest Biochem Physiol* **6**: 215–22.
23. Pang Y-P. (2006) "Novel acetylcholinesterase target site for malaria mosquito control." PLoS ONE **1**: e58.
24. Pang Y-P. (2007) "Species marker for developing novel and safe pesticides." *Bioorg Med Chem Lett* **17**: 197–99.
25. Zauhar RJ, Colbert CL, Morgan RS, Welsh WJ. (2000) "Evidence for a strong sulfur-aromatic interaction derived from crystallographic data." *Biopolymers* **53**(3): 233–48.
26. Erlanson DA, Braisted AC, Raphael DR, *et al.* (2000) "Site-directed ligand discovery." *Proc Nat Acad Sci USA* **97**(17): 9367–72.
27. Boyd AE, Marnett AB, Wong L, Taylor P. (2000) "Probing the active center gorge of acetylcholinesterase by fluorophores linked to substituted cysteines." *J Biol Chem* **275**(29): 22401–08.
28. Johnson JL, Cusack B, Hughes TF, *et al.* (2003) "Inhibitors tethered near the acetylcholinesterase active site serve as molecular rulers of the peripheral and acylation sites." *J Biol Chem* **278**(40): 38948–55.
29. Pang Y-P, Xu K, Kollmeyer TM, *et al.* (2001) "Discovery of a new inhibitor lead of adenovirus proteinase: steps toward selective, irreversible inhibitors of cysteine proteinases." *FEBS Lett* **502**(3): 93–97.
30. Pang Y-P, Singh SK, Lassiter TL, *et al.* "Selective and irreversible inhibitors of aphid acetylcholinesterases. Submitted."

31. Cheney Q. (2006) "State of Minnesota, 2008–2009 Biennial Budget." In: Department A, editor, 2006.
32. Ragsdale DW, McCornack BP, Venette RC, *et al.* (2007) "Economic threshold for soybean aphid (Homoptera: Aphididae)." *J Econ Entomol* **100**(4): in press.
33. Kaiser J. (2005) "National Institutes of Health. NCI gears up for cancer genome project." *Science* **307**(5713): 1182.

Chapter 16

Structural Proteomics of Emerging Viruses: The Examples of SARS-CoV and Other Coronaviruses

Rolf Hilgenfeld*,†, Jinzhi Tan†, Shuai Chen†,
Xu Shen‡ and Hualiang Jiang‡

Summary

This chapter discusses the contribution that structural proteomics can make to the understanding of the mechanisms through which viral pathogens infect human cells and replicate in these hosts, and how the information obtained can be used in the discovery of antiviral drugs. Given the limitations of space, the focus is on the most prominent outbreak of a new virus in the recent past — that of the coronavirus causing severe acute respiratory syndrome (SARS). The structural proteomics work that has been published on SARS-CoV and related coronaviruses since 2002/2003 is reviewed, and the question of whether structural proteomics can facilitate the rapid identification and characterization of drug targets, and hence the

* Corresponding author. Email: hilgenfield@biochem.uni-luebeck.de

† Institute of Biochemistry, Center for Structural and Cell Biology in Medicine, University of Lübeck, Ratzeburger Allee 160, 23538 Lübeck, Germany.

‡ Drug Discovery and Design Center, State Key Laboratory of Drug Research, Shanghai Institute of Materia Medica, Chinese Academy of Sciences, Zuchongzhi Rd. 555, Shanghai 201203, China.

discovery of new antivirals in case of a similar outbreak in the future, is examined.

Introduction

The number of viral outbreaks has been increasing dramatically recently. In the past 10 years, the world has seen at least one major outbreak of either a new virus or a new variant of a known virus per year. Incidentally, almost all of these outbreaks were caused by RNA rather than DNA viruses. Vaccination is hardly suitable to stop such outbreaks, because even with advances in development, vaccines will not be available for six to 12 months following the first appearance of a new pathogen. Immediate containment of viral outbreaks will therefore depend on quarantine and antiviral drugs. Unfortunately, for most known viral diseases of humans, let alone for newly emerging ones, no drug treatment is available. We believe that in view of this scenario, it is necessary to develop lead compounds with activity against all major families of viruses, both those that infect humans and those that so far have been restricted to animals but may cross the species-barrier by zoonotic transition. Ultimately, we should aim at discovering antiviral compounds with a relatively broad specifity, which may be active against a range of new viruses should they emerge.

In order to discover such leads for antiviral compounds, a sophisticated approach is necessary. Random screening for antivirals may have its merits, but the case of HIV has demonstrated that this approach has not led to a single marketed drug in the past 20 years, while no less than 26 antiretroviral drugs are on the market or in late development phases that have been designed based on insights into the structure and function of viral proteins. Therefore, we believe in the merits of structure-based approaches to discover new compounds with activity against RNA viruses. Methods applied include *de novo* design, virtual screening, and structure-guided medicinal chemistry, all of which require a detailed knowledge of the structure and function of the viral target proteins.

As an example of an emerging RNA virus, we will discuss the newly discovered coronavirus that was responsible for the outbreak of severe acute respiratory syndrome (SARS) in 2003. Structural biology undoubtedly made an important contribution to understanding aspects of SARS coronavirus replication within a few weeks after the start of the outbreak. We will show that this was only possible because of prior structural work on proteins from related coronaviruses. Recognizing this, we advocate the initiation of structural proteomics projects for every family of RNA viruses that has a potential to cause outbreaks of disease in humans. In fact, this is exactly the philosophy of VIZIER ("Comparative structural genomics of viral enzymes involved in replication"; www.vizier-europe.org), an integrated structural virology project funded by the European Commission.[1]

Coronaviruses: A Growing Family of Pathogens

Coronaviruses such as human coronavirus 229E (HCoV-229E) or HCoV-OC43 are responsible for variants of the common cold, while SARS-CoV caused the outbreak of severe acute respiratory syndrome (SARS) in 2003, which affected about 8500 people worldwide and killed more than 800. Increased research activities since then have led to the discovery of two hitherto unknown human coronaviruses. The human coronavirus NL63, first described in March 2004, has been identified as the causative agent of respiratory disease in very young children and in immunocompromised adults.[2–6] Furthermore, HCoV-NL63 infection has been associated with laryngotracheitis (croup) and Kawasaki disease, although the latter is not supported by a recent study.[7] The other recently discovered human coronavirus is HKU1, which also causes respiratory disease.[8,9]

Coronaviruses belong to the order *Nidovirales* and have been classified into three groups. Group 1 includes the human coronaviruses 229E and NL63 as well as the animal viruses, (porcine) transmissible gastroenteritis virus (TGEV) and feline peritonitis virus. Group 2 consists of the human coronaviruses OC43 and HKU-1, as well as mouse hepatitis virus (MHV) and bovine coronavirus (BCoV). Group 3 so far comprises only the avian coronaviruses: infectious bronchitis virus (IBV) and turkey coronavirus. The SARS coronavirus has been classified as an outlier of group 2.[10,11] In addition to the results of the SARS-CoV structural proteomics projects, we will also present here the structural information that is available for the other coronaviruses.

The emergence of new coronaviruses during the last decade and the wide variety of the diseases caused by them, from harmless common cold to potentially lethal SARS, demonstrate the need to be prepared for future outbreaks of coronaviral diseases. This is particularly true in view of the frequent occurrence of coronaviruses highly similar to SARS-CoV in horseshoe bats[12]; the detection of other coronaviruses in almost all kinds of bats (these animals constitute roughly 20% of all mammals on Earth!)[13,14]; and the wide range of other mammals that are susceptible to coronavirus infection. In order to be able to design inhibitors with anti-coronavirus activity, we need to understand the replication and transcription mechanisms of these viruses.

Organization of the Coronavirus Genome and Proteome

The coronaviral genome is the largest RNA genome known (up to 32 kb). The single-stranded genomic RNA has positive polarity and is capped (at the 5′-end) as well as polyadenylated (at the 3′-end). In case of SARS-CoV (29.7 kb), it comprises 14 open reading frames (ORFs) encoding 28 proteins (Fig. 1). Three classes of proteins can be distinguished: non-structural proteins (Nsps), structural proteins, and accessory proteins.[10,15]

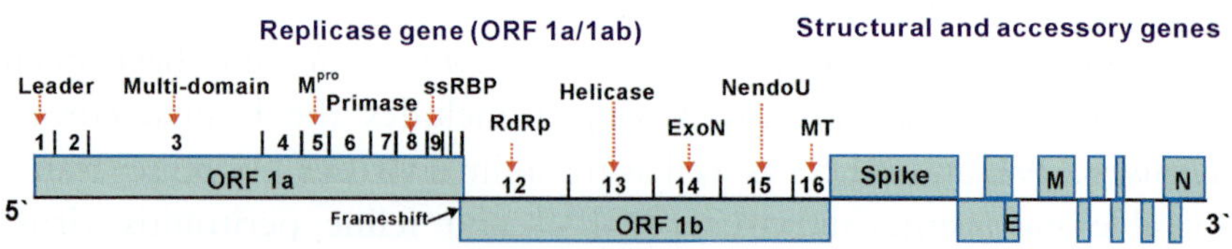

Fig. 1 Organization of the SARS-CoV RNA genome. The individual non-structural proteins (Nsps) are indicated as domains of the replicase, which is encoded by ORFs 1a and 1b. The genes for the major structural proteins (spike, E, M, and N) are also indicated. Open reading frames encoding accessory proteins are interspersed between these genes.

The 16 non-structural proteins of the coronaviruses are required for genomic RNA synthesis (replication) and subgenomic RNA synthesis (transcription; up to eight subgenomic RNAs), and/or for interaction with components of the host cell. They are encoded by the replicase gene, which comprises two open reading frames, ORF1a and ORF1b. ORF1a encodes the replicative polyprotein 1a (pp1a, ≈490 kD). The much larger pp1ab (>750 kD) is encoded by ORFs 1a and 1b; expression of the latter involves a (−1) ribosomal frameshift during translation, just upstream of the ORF1a stop codon. The polyproteins are processed into individual polypeptides by the virus-encoded proteases, the papain-like cysteine proteases (PL^{pro}s; two domains of non-structural protein 3 (Nsp3) in most coronaviruses, but only one such domain in SARS-CoV) and the main protease (Nsp5). Even though the functions of a few non-structural proteins are known,[16,17] e.g. Nsp5 being the main protease, Nsp13 being a helicase and Nsp12 being the RNA-dependent RNA polymerase, the mechanisms of replication and transcription have not been elucidated. It is our aim to tackle this problem through structural and functional studies of individual non-structural proteins and of the complexes they form with one another, with RNA, and possibly with proteins of the host cell.

The 3′ third of the SARS-CoV genome codes for the structural proteins (spike, S; envelope, E; membrane (or matrix), M; and nucleocapsid, N). In addition, this region of the genome contains six or more non-conserved open reading frames (3a, 3b, 6, 7a, 7b, 8a/b, 9b), some of which overlap with one another.[10] Since the "accessory proteins" encoded by these ORFs are absent from other coronaviruses, they may carry out unique functions in SARS-CoV replication or assembly, or they may be related to the extraordinary pathogenicity of the SARS virus. In fact, in other viruses, such accessory proteins are often not vital for virus replication in cell culture, but have been shown to circumvent the host innate and adaptive immunity response.[18] Indeed, accessory proteins 3a and 6, along with the nucleocapsid protein, of SARS-CoV have been implicated in the suppression of the type-I interferon response of the host cell.[19] Some of the accessory proteins, such as 3a and 9b, of the SARS virus also have a structural role and interact with the E, M, and N proteins. Furthermore, protein 3a has recently been shown to specifically interact with the 5′-untranslated region of the viral genome.[20]

For the synthesis of the structural and accessory proteins, an interesting mechanism involving subgenomic messenger RNAs (sgmRNAs) is responsible. Each of these sgmRNAs (there are a total of at least six and probably eight of them in SARS-CoV) contains a 5′-leader sequence corresponding to the 5′-end of the entire genome. This 5′ leader is joined to an mRNA "body", which extends from the 3′-poly(A) tail to a ≈10-nucleotide, AU-rich segment upstream of the 5′-end of each ORF, which is called transcription-regulating sequence (TRS). It has been proposed that the process of discontinuous transcription occurs during the synthesis of minus-strand subgenome-length templates.[21] This could involve the following steps: (i) assembly of a functional replicase/transcriptase complex (RTC) and initiation of minus-strand synthesis at the 3′-end of genomic RNA; (ii) elongation of the nascent minus-strand RNA until encounter of the first TRS is achieved; (iii) at this point, a fixed proportion of the RTCs will disregard the TRS and continue elongation of the nascent strand, while (iv) the remaining RTCs will stop synthesis of the nascent minus strand, relocate and complete its synthesis; (v) this relocation will be guided by the complementarity between the 3′-end of the nascent minus strand and the first TRS motif from the 5′-end (the leader, TRS). As a consequence, the translocated minus-strand will be extended by copying the 5′-end of the genome; (vi) the completed minus-strand RNA will then serve as a template for mRNA synthesis. Unfortunately, for most of these proposed steps, it is not understood how they may work in detail.[21–23]

In what follows, we will first discuss the non-structural proteins, i.e. the components of the replicase/transcriptase complex encoded by ORF1, followed by the structural proteins and, finally, the accessory proteins. We will start the discussion of the non-structural proteins with the main proteinase (M^{pro}, Nsp5), because this was the first structure of any coronaviral protein to be solved, prior to and during the SARS outbreak of 2003, and therefore generated quite some excitement when the new virus emerged. Also, it is still the one major target for the design of anti-coronavirus compounds, as we will see below, and the starting point for the coronavirus structural genomics projects. Following the M^{pro}, we will discuss the other non-structural proteins, from Nsp1 to Nsp16, in the order in which they are encoded by the viral genome, from 5′ to 3′. We will include all non-structural proteins in this discussion, provided some functional and/or structural information is available; because of the lack of any such information to date, the integral membrane proteins Nsp4 and Nsp6 will be omitted. This will be followed by a discussion of the data available for the structural proteins, and finally, we will briefly review the (still very limited) information on the structures of the accessory proteins of the virus.

Non-Structural Proteins

Coronavirus Main Protease: Work Prior to SARS and the Hectic Months After the Discovery of SARS-CoV

In late March, 2003, several parallel reports appeared as electronic prepublications, describing the discovery of a new coronavirus as the etiological agent of severe acute respiratory syndrome (SARS), which had been spreading in the Guangdong province of China, Hong Kong, Singapore, and Vietnam since the end of February, and reached Toronto, Beijing, and Inner Mongolia a little later.[24–27] Within three weeks after this discovery, on April 13, 2003, the genome of the virus had been sequenced and published on the internet (Refs. 28 and 29; see also Ref. 30). Shortly before, the group of one of us (RH) had determined the crystal structure of the main proteinase (M^{pro}) of human coronavirus 229E (HCoV 229E) and also a structure of a complex between the homologous enzyme of the transmissible gastroenteritis virus (TGEV) and a hexapeptidyl chloromethyl ketone inhibitor[31] that had been designed on the basis of the crystal structure of the free TGEV enzyme, published by RH's group in 2002 (Ref. 32; see below). When the genomic sequence of the SARS virus appeared on the Internet, the region coding for the main protease was rapidly identified by RH and his coworkers, and it surfaced from comparison with other coronaviral main proteinase sequences that the new virus was most related to group 2 of the coronavirus family. The group went on to build a preliminary homology model for the SARS-CoV enzyme and conducted a comparison of the binding mode of the hexapeptidyl chloromethyl ketone inhibitor with all the cysteine protease inhibitor complexes contained in the Protein Data Bank. This revealed that AG7088 (ruprintrivir), a peptidomimetic vinylogous ethyl ester developed by Pfizer as a drug against the common cold caused by human rhinovirus, binds to its target enzyme, the rhinovirus 3C proteinase, in a similar way as does the hexapeptidyl inhibitor to TGEV M^{pro}. When docking of AG7088 into the homology model of the SARS-CoV M^{pro} was performed, it appeared that the compound would largely fit into the substrate-binding site, after covalent attachment to the active-site Cys145 of the protease through Michael addition, although there seemed to be some steric hindrance exerted by the p-fluorobenzyl group in the P1 position of AG7088. Consequently, RH's group proposed in their May 13, 2003, publication in *Sciencexpress on-line* that AG7088 could be a good starting point for the design of anti-SARS drugs.[31] Indeed, AG7088 itself was later shown to block the SARS-CoV main protease, albeit with a rather high IC_{50} of $\approx 100\ \mu M$.

This case shows that the use of homology models does have its merits in the absence of an experimental structure of the target protein.[33] In the first few weeks after the identification of the new coronavirus, many researchers used this approach, all of them using the crystal structure of the main proteinase (M^{pro}) of the TGEV as a basis. This structure had been published by RH's group in 2002,[32] i.e. prior to the SARS outbreak. HJ's group used their homology model for the SARS-CoV M^{pro}[34] for virtual screening of large chemical libraries, and within weeks of the discovery of the virus, they found that cinanserin is an inhibitor of the enzyme.[35] This was of considerable interest, since cinanserin is a serotonin antagonist that had undergone clinical trials in the 1960s. Even though the drug is not free of side-effects, this discovery would have offered an opportunity for the therapy of SARS, had the number of infected people increased at the same pace as it did during April and May, 2003. Cinanserin was later shown to be a very efficient inhibitor of the SARS virus in Vero E6 cells, along with low toxicity,[36] although surprisingly, the inhibitory activity against the main protease in the standard *in vitro* assays is rather weak, even though binding has been demonstrated by surface plasmon resonance.[35]

The discovery of AG7088 and cinanserin as potential inhibitors of the SARS virus, or at least as starting points for the design of such inhibitors, demonstrates the importance of prior knowledge on the identity and structures of potential target proteins from the same virus family. At the time of the SARS outbreak, RH's crystal structure of the main proteinase (M^{pro}) of TGEV was the only structure of any coronavirus protein that had been published. The enzyme, just like later the HCoV-229E M^{pro} and the SARS-CoV M^{pro}, was shown to form a homodimer with two protomers oriented almost at right angle to each other.[32] Each protomer is composed of three domains, which include the N-terminal domains I (residues 8–101) and II (residues 102–184), both having an antiparallel β-barrel structure. In contrast, the C-terminal domain III (residues 201–303) contains five α-helices arranged into a large globular cluster, and is connected with domain II through a loop region (residues 185–200). The Cys-His catalytic dyad is located in a cleft between domains I and II. Together, the two N-terminal domains resemble the fold of chymotrypsin. The C-terminal domain is responsible for orienting the N-terminal segment (the "N-finger," residues 1 to 8) for interaction with the substrate-binding site of the other monomer in the dimer. This latter feature is absolutely required for enzymatic activity, because the N-finger helps shape the substrate-binding site of the neighboring active site; in particular, it supports the active conformation of the oxyanion loop (residues 137–144). Anand *et al.*[32] have shown that deletion of residues 1 to 6 leads to a completely inactive M^{pro}, and so does deletion of domain III. Verschueren *et al.*[37] have recently determined the structure of a SARS-CoV M^{pro} that has residue Ser1 deleted ("amputated N-finger"). This enzyme has an activity that is reduced by ≈60%, as compared with the wild-type. The conformation of residue Gly2 is seen to have changed, with this residue now interacting with Cys300 of domain III of the parent protomer, rather than with the oxyanion loop of the opposing monomer. Others have shown that residues 1–3 are not essential for dimerization, but for full enzymatic activity, whereas deletion of residue Arg4 and beyond leads to an inactive monomer, presumably due to the destruction of the Arg4...Glu290 ion pair (SARS-CoV numbering).[38] However, there has been some disagreement concerning this latter observation. HJ's laboratory has proposed that the M^{pro} can dimerize in spite of deletion of the seven residues at the N-terminal. Molecular dynamics simulations suggested that dimerization in this case is mainly mediated by interactions between the domains III of the two monomers.[39] This is conceivable in view of the observation that the isolated domain III of the M^{pro} forms a stable dimer by itself[40] (see also Ref. 41 for the role of the C-terminal domain).

Crystal structures of SARS-CoV M^{pro}[42–44] (Fig. 2) show largely the same features as its homologues from TGEV[32] and HCoV 229E.[31]

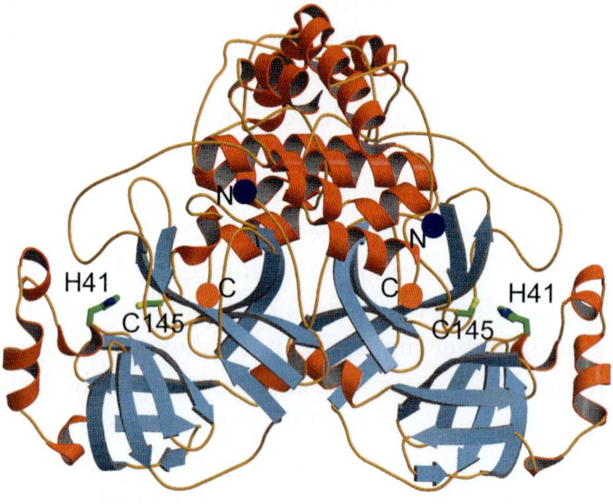

Fig. 2 Structure of the SARS-CoV main proteinase (M^{pro}) dimer[42] (PDB code: 1UJ1). α-Helices are indicated as red ribbons, β-strands as blue ones. The two monomers are oriented at a right angle to one another. Domains I and II are β-barrels and harbor the catalytic site (C145...H41) between them. Domain III is α-helical and partly responsible for the dimerization of the enzyme. To a large extent, dimerization is due to the interaction of the "N-finger" of one subunit with domains II and III of the other subunit. The amino-termini are marked by a dark-blue sphere and the letter N. Carboxy-termini are labeled by a bright red sphere and the letter C.

However, this enzyme was originally crystallized at pH 6.0, where it displays low catalytic activity. In agreement with this, one of the two subunits of the homodimer was found to exist in an inactive conformation.[42] The plausible reason for this was that at the low pH of crystallization, a conserved histidine (His163) at the bottom of the S1 pocket, which is responsible for the enzyme's absolute specificity for cleavage after P1 glutamine residues, is protonated (Fig. 3b). To neutralize this positive charge in a partly hydrophobic environment, residue Glu166 changes conformation and forms a salt bridge with His163, thereby partly occupying the S1 specificity pocket. In the active state of the enzyme (Fig. 3a), Glu166 forms an ion pair with His172 at the rim of the S1 pocket, and, importantly, another one with the N-finger of the other subunit in the dimer. The latter in turn also interacts with the main chain of residue Phe140, within the oxyanion loop, as discussed above for the TGEV M^{pro} (not shown). All of these interactions break down when Glu166 moves into the S1 pocket, towards His163, in the inactive form induced by low pH (Fig. 3b). When the same crystals were incubated at pH 7.6, both subunits were found in the active conformation,[42] whereas at pH 8.0, the ion pair between His172 and Glu166 is partly lost, probably due to deprotonation of His172 at this pH.[43] Thus, the pH-activity profile of the SARS-CoV M^{pro} appears to be governed by the protonation of His163 at low pH and the deprotonation of His172 at high pH. These pH-controlled conformational changes have been nicely reproduced in molecular dynamics simulations, carried out at different pH values, by Tan *et al.*[43] Also, other crystal structures of the SARS-CoV M^{pro} in different space groups and grown at different pH values confirmed the important role of pH in the activity state of the enzyme.[43,45] Depending on pH of crystallization and crystal symmetry, structures of the SARS-CoV M^{pro} have been determined which have both monomers in the inactive or active conformation, or one in the active and the other in the inactive conformation.[46–48] Since an expression construct was used in the original pH-titration studies[42] that yielded an M^{pro} with five extra residues attached to the N-terminus (which were mobile and hence not seen in electron density), RH's group repeated the pH-titration with M^{pro} with authentic chain

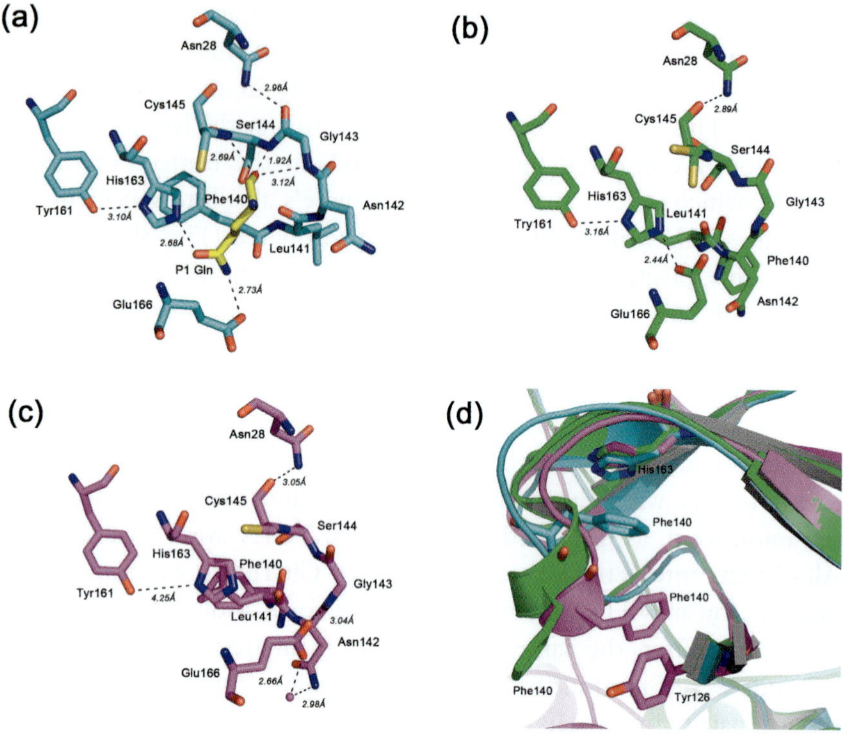

Fig. 3 The S1 specificity subsite of the SARS-CoV M^{pro} (with permission from the *American Society for Biochemistry and Molecular Biology*; Ref. 49). **(a)** The situation in the active conformer of the enzyme[42] (PDB code: 1UJ1). The P1 glutamine side chain (yellow) interacts with Glu166 at the rim of the S1 pocket and His163 at its bottom. The histidine is not protonated in this conformer. Phe140 of the oxyanion loop also interacts with His163; this interaction contributes to a catalytically competent conformation of the oxyanion loop (residues 140–145). **(b)** Protonation of His163 leads to a reorientation of Glu166, which moves into the pocket to neutralize the positive charge on the histidine. As a consequence, the S1 pocket is no longer accessible to substrate and the oxyanion loop collapses[42] (PDB code: 1UJ1). **(c)** The situation in the same region of the monomeric Gly11Ala mutant of SARS-CoV M^{pro}[49] (PDB code: 2PWX). Here, Glu166 interacts with the main-chain amides of residues 142 and 143, also resulting in a catalytically incompetent oxyanion loop. **(d)** Overlay of the three structures. Note the different orientations of the side chain of Phe140.

termini and obtained the same result as previously (Verschueren *et al.*, unpublished).

In fact, the mechanism of proenzyme activation in the case of coronavirus M^{pro} resembles that of chymotrypsin: the oxyanion-binding loop only adopts its catalytically competent conformation after interaction with the amino-terminal NH_3^+ group, i.e. at Ile16 in chymotrypsin (created through trypsin cleavage of the 14–15 and 15–16 bonds), and at Ser1 of the neighboring monomer in the dimer in the coronavirus M^{pro}s. In chymotrypsin, this interaction is *intra*molecular, involving Asp194 of the oxyanion loop, whereas in coronavirus M^{pro}, it is *inter*molecular, involving both Phe140 of the oxyanion loop and Glu166 of the S1 site.

As long as it is part of the coronaviral polyprotein, the M^{pro} proenzyme will be in an inactive, monomeric state. The M^{pro} monomer appears to be inactive, because it is not able to maintain a catalytically competent conformation of its oxyanion loop and of its S1 specificity pocket in the absence of support from the "N finger" of the neighboring monomer in a dimer. The crystal structure of an M^{pro} monomer has long been sought after. Recently, HJ's group has succeeded in generating a monomeric SARS-CoV M^{pro} by replacing residue Gly11 by Ala.[49] In the wild-type protein, Gly11 is in contact with Gly11 of the opposing monomer as well as with Gln14 and there is no space for a side chain on residue 11 in this closely packed interface. As a consequence, the α-helix 10–16 is shortened by two residues in the Gly11Ala mutant and the "N finger," residues 1–7, changes orientation and can no longer interact with Glu166 and the oxyanion hole of the other monomer; hence, the dimer collapses, and so does the oxyanion hole, leading to a complete loss of enzymatic activity. Devoid of the "N finger" as a partner, Glu166 accepts hydrogen bonds from the amide groups of the oxyanion loop, residues 143 to 145 (Figs. 3c,d). Furthermore, domain III of the monomer has an orientation relative to domains I and II that is very different from what has been seen in the dimer.

It is not clear whether autoactivation of the coronaviral M^{pro} proenzyme occurs in *cis* or *trans*, but in view of the requirement for dimerization, the latter is more likely. The crystallographically observed mixed complexes between one active and one inactive M^{pro} molecule.[42,43] (Verschueren *et al.*, unpublished) might be a good model for the initial step of proenzyme activation in *trans*: the subdomains III of two monomeric M^{pro} domains (both of which are in the inactive conformation) may form an initial dimer (the isolated domain III alone can form dimers[40]), followed by the penetration of the "N finger" of one of the monomers into the cleft between its own domain III and domain II of the neighboring monomer.[49] This would lead to a conformational change in the neighboring monomer and its catalytic activation. (A somewhat related model of proenzyme activation has been proposed by Hsu *et al.*[44]).

Still, the resulting mixed active:inactive dimer would have reduced catalytic activity and this may be stabilized by low local pH. As is common for positive-strand RNA viruses replicating in eukaryotic cells, the coronavirus replicase associates with modified intracellular membranes hijacked from the endoplasmic reticulum or from endosomes. These membranes are modified to form special vesicles. For coronaviruses, double-membrane vesicles have been proposed as the locus of viral RNA synthesis.[50] The purpose of such a compartmentalization may be to hide double-stranded RNA intermediates from detection by host-response factors that would trigger the antiviral type-I interferon response of the infected cell. The coronaviral polyproteins will be directed to these membraneous structures by their hydrophobic domains (Nsp4, Nsp6, and parts of Nsp3). In particular, the main proteinase (Nsp5) is thus "framed" by membrane-bound domains, suggesting its localization close to these membrane structures. It may well be that local pH is acidic in such structures, and that the relatively slow activation of the main proteinase (after polyprotein synthesis) occurs only after the occasional, spontaneous liberation of the M^{pro} proenzyme domain from its membrane-attached locus and exposure to neutral pH.

If activation of the low pH form of the M^{pro} (mixed inactive:active dimers) is indeed an important step in polyprotein processing, then inhibitors should be of interest that specifically prevent this step. RH's group has designed one such inhibitor, which when added to the enzyme at pH 5.5, prevents the conformational changes necessary for activation at higher pH values (Pumpor *et al.*, unpublished).

After activation, the coronaviral main protease is responsible for cutting the polyproteins at 11 sites, all with the sequence pattern (small)-X-(Leu/Phe/Met)-Gln-(small) for P4-P3-P2-P1-P1′ ("X" is any amino acid). It is clear that inhibition of this process will prevent maturation of viral proteins; hence, the M^{pro} is a key antiviral target.[51] HPLC- and fluorescence-based assays have been used to characterize the protease and to determine the potency of the inhibitors. Monitoring the increase of fluorescence from the cleavage of a peptide substrate containing an Edans-Dabcyl fluorescence quenching pair at the two ends of the peptidic substrate, the fluorogenic method has enabled the use of high-throughput screening to speed up the drug discovery process. Several groups of inhibitors have been identified through high-throughput screening and rational drug design approaches. The original idea of using Michael acceptor compounds such as AG7088 was refined further; while the parent inhibitor proposed by Anand *et al.*[31] displayed a K_i value of almost 100 μM, rather minor modifications, some of which had already been proposed in the original publication, led to inhibitors with K_i values in the single-digit micromolar range (and relatively large values for k_3, the rate of inactivation by covalent reaction with the active-site cysteine of the enzyme).[52] In addition, α,β-unsaturated peptidomimetics, anilides, epoxides, metal-conjugated compounds, boronic acids, quinolinecarboxylate derivatives, thiophenecarboxylates, phthalhydrazide-substituted ketoglutamine analogues, isatin and natural products have been identified as potent inhibitors of the SARS-CoV main protease. Some of these inhibitors could perhaps be developed into potential drug candidates, which may provide a solution to combat possible reoccurrence of the SARS virus and other life-threatening viruses with proteases of the M^{pro} type.

Due to limitations of space, we are unable to discuss here the structures of SARS-CoV M^{pro} in complexes with inhibitors[37,45,53–58] that are available at this time. We only mention one approach to M^{pro} inhibition because of its methodological novelty: Schmidt *et al.*[59] have made use of the reversible covalent binding of aldehydes to the active-site cysteine of the SARS-CoV M^{pro}.[60] The enzyme-inhibitor complex is then used to screen a library of non-peptidic nucleophiles (mainly amines) which react with the aldehyde component, replacing it from the cysteine. In the next step, the amine moiety of the non-peptidic component is replaced by an aldehyde group which reacts with the active-site cysteine. As before, this complex is then used to again screen the nucleophile library. This way, a non-peptidic inhibitor that inhibits the enzyme through non-covalent binding, with a K_i value in the lower micromolar range, has been created. The entire process is called "Dynamic Ligation Screening."[59]

Nsp1: Divergent between Coronaviruses

The replicase gene (ORFs 1a and 1b) shows relatively low conservation between coronaviruses in its 5′-terminal third, especially in the region coding for non-structural proteins 1 to 3. This is particularly true for Nsp1, which has been suggested as a group-specific marker for coronaviruses.[10] In group 1 coronaviruses (HCoV 229E, HCoV NL63, TGEV), Nsp1 comprises about 110 residues, whereas in group 2a (MHV, BCoV, HCoV OC43), it is more than double this size (245 residues). The Nsp1 of SARS-CoV, a group 2b virus, consists of 180 amino-acid residues with little sequence similarity to that of group-2a and none to group 1. Finally, group 3 coronaviruses (infectious bronchitis virus, IBV; turkey coronavirus), in contrast, do not encode an Nsp1.

For MHV Nsp1, a number of functional studies have been published. Interaction of the protein with Nsp7 and Nsp10 as well as colocalization with other replicase components and nucleocapsid protein in the cytoplasm in the early stages of infection have been demonstrated.[61] Nsp1 appears to be an important pathogenicity factor of MHV. Mutations in the Nsp1/Nsp2 cleavage site that prevented the release of Nsp1 from the polyprotein resulted in slower growth and reduced RNA synthesis.[62] Deletion of its coding region from the MHV genome yielded virus that replicated almost normally in cell culture,[63] but was unable to productively infect mice.[64] However, in mice deficient for the type-I interferon receptor, growth of the ΔNsp1 virus was normal. Thus, Nsp1 appears to be part of the viral system that suppresses the interferon response of the host cell. This has recently also been observed for SARS-CoV Nsp1.[64a] It was also reported that exogeneous expression of a gene coding for Nsp1 in mammalian cells leads to arrest of the cell cycle in the G_0/G_1 phase and inhibition of cell proliferation.[65]

SARS-CoV Nsp1 was shown to enhance the degradation of cellular mRNA, thus reducing the rate of synthesis of host-cell proteins, including those involved in the antiviral response (e.g., type-I interferon).[66] As far as interaction with other coronaviral proteins is concerned, Nsp1 could be immuno-coprecipitated with the E protein and this interaction was also positive in a yeast-two-hybrid approach.[67] The three-dimensional structure of Nsp1 of the SARS virus has been determined by NMR spectroscopy (Fig. 4).[68] The 179-residue protein

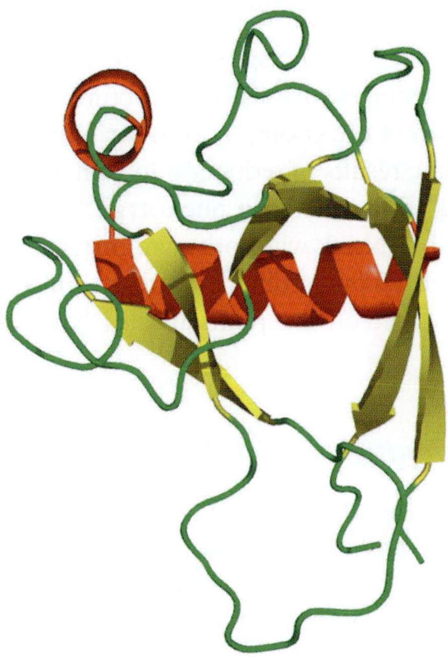

Fig. 4 NMR structure of SARS-CoV Nsp1[68] (PDB code: 2HSX).

contains a core fragment (residues 13–121), which is folded into a mixed (parallel/antiparallel) six-stranded β-barrel. There is an amphiphilic, 14-residue α-helix located across one barrel opening and a short 3_{10}-helix alongside the barrel. Whereas a number of six-stranded β-barrels have been found in various proteins, the connectivity of the strands seen in SARS-CoV Nsp1 constitutes a new fold. The C-terminal third of full-length Nsp1 appears to be intrinsically disordered.[68] At the moment, the three-dimensional structure of the protein does not add much insight to its still unknown function, but it suggests future mutational studies that might answer the quest for Nsp1's function. Obviously, it is difficult to derive potential functions from the three-dimensional structure of SARS-CoV proteins if their fold is new and there is no sequence similarity to other proteins outside the coronaviruses. Similar observations have been made in the other structural genomics projects.

Nsp2: Structure and Function Still Unknown

Along with Nsp1, the 70-kD Nsp2 is the most variable non-structural protein among the coronaviruses. Its structure and function are still unknown. Deletion of the coding region for Nsp2 from the MHV or SARS-CoV genome results in reduction in viral growth rate and RNA synthesis but is otherwise without phenotype.[69] A number of interactions of SARS-CoV Nsp2 with other non-structural proteins have been demonstrated in one study,[67] whereas similar investigations by Imbert *et al.*[70] disagree on this point and list Nsp2 as an "interaction orphan."

Nsp3: A Large Multi-Domain Protein

With a molecular mass of >180 kD, Nsp3 is the largest of the non-structural proteins of coronaviruses. In SARS-CoV, the protein is even larger (213 kD) than in other coronaviruses and contains at least five subdomains: an N-terminal acidic domain (Ac, also called Nsp3a); an X domain (also designated as ADRP, or Nsp3b); a domain that is unique to SARS coronavirus and absent in all the other members of the family sequenced so far, and therefore appropriately called "SARS-unique domain" (SUD, or Nsp3c); a papain-like proteinase, PL^{pro} (also called Nsp3d), and a Y domain (Nsp3e) that includes an N-terminal transmembrane (TM) region. Recently, another non-canonical papain-like protease domain has been proposed to exist between the PL^{pro} and the Y domain,[70] but this idea, which was derived from bioinformatics, has yet to be confirmed experimentally. At present, it is completely unclear whether and how the individual domains of Nsp3 interact with one another or with other components of the coronaviral replicase complex. Also, several of them appear to interact with components of the infected host cell, as will be discussed in the description of the individual domains. However, we have to realize that very little is known about the interactions involved, and even less about structural aspects of these interactions.

The Acidic Domain of Nsp3 ("Nsp3a")

The NMR structure of the N-terminal part of the acidic domain has been determined very recently.[71] Comprising 112 amino-acid residues, this subdomain displays a ubiquitin fold. However, in addition to the four β-strands and two α-helices that are characteristic of this fold, there are two additional short α-helices (Fig. 5). NMR chemical shift analysis suggests that these non-canonical structural elements might bind single-stranded RNA with some specificity for AUA-containing sequences, although the K_d values observed are relatively high ($\approx 20\ \mu$M). Also, the additional helices feature an overall negative electrostatic potential, so that the proposed binding of ssRNA is surprising and the observed chemical shifts may be due to indirect effects. The specificity for the AUA sequence suggests that Nsp3a might bind to the 5′-UTR of the SARS-CoV genome, as there are several repeats of this sequence. Proteins binding to the 5′-UTR may be involved in cap-dependent translation, in genome replication, or in synthesis of subgenomic RNAs.

The C-terminal 70 residue-segment of the "acidic domain" (Nsp3a) is the part that is responsible for the name of the domain; it comprises a large number of glutamates (38% of the residues) and aspartates (12%)

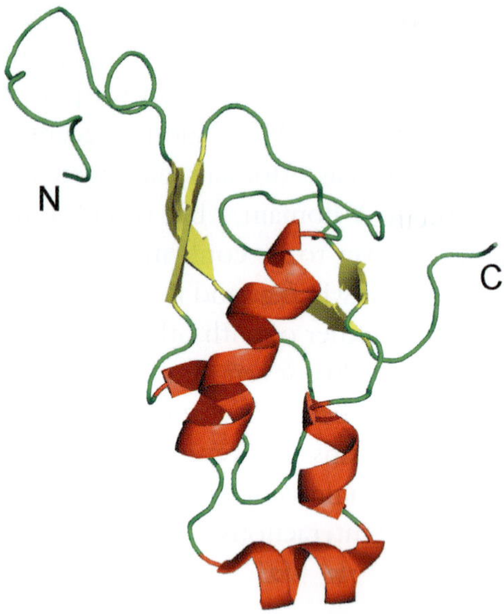

Fig. 5 NMR structure of the ordered N-terminal fragment of SARS-CoV Nsp3a ("acidic domain"), which displays a non-canonical ubiquitin fold[71] (PDB code: 2IDY).

and does not have an ordered structure. Similar disordered acidic regions have been found previously in bacterial single-stranded DNA-binding proteins (SSBs; see e.g. Ref. 72).

Why does Nsp3a contain a ubiquitin-like domain? In fact, as we shall see, a second domain of Nsp3, the papain-like protease (PL^{pro}, Nsp3d), also contains an N-terminal ubiquitin-like subdomain. Possible roles for these domains will be discussed below.

The X Domain of Nsp3 ("Nsp3b"): Fold Suggests Possible Function

Among the subdomains of the Nsp3 multidomain protein, there is the so-called "X domain" (Nsp3b), which shows structural homology to macrodomains. The latter name refers to the non-histone-like domain of histone macro2A. In animal cells, such domains are occasionally physically associated with enzymes involved in ADP ribosylation or ADP-ribose metabolism. Because of this linkage and on the basis of sequence similarity to Poa1p, a yeast protein involved in the removal of the 1″-phosphate group from ADP-ribose 1″-phosphate (a late step in tRNA splicing), it has been proposed that the coronaviral X domains may have the function of ADP-ribose-1″-phosphatases (ADRPs; Ref. 73). The crystal structures of X domains of SARS-CoV[74,75] and HCoV NL63 (Piotrowski *et al.*, unpublished) (Fig. 6) showed that the protein has the three-layer $\alpha/\beta/\alpha$ fold characteristic of the macrodomains. Recent evidence has shown that the X domain is not essential for replication of coronaviruses in cell culture.[73] Also, the ADRP activity measured for the X domains of SARS-CoV, HCoV-229E, and TGEV[75,76] is unusually low for an enzyme supposed to be part of an important RNA metabolic pathway. Furthermore, binding of the product of the ADRP reaction, ADP-ribose, is relatively weak for the SARS-CoV X domain,[75] and even weaker for that of HCoV NL63 (Piotrowski *et al.*, unpublished). Also, unless major

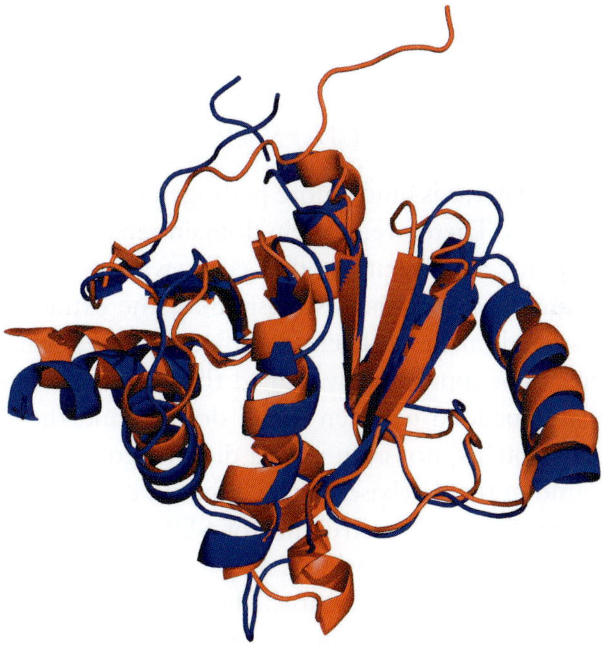

Fig. 6 Superimpositions of the crystal structures of the X domains of SARS-CoV (red; PDB code: 2ACF) and HCoV NL63 (blue) (Ref. 74; Piotrowski *et al.*, unpublished).

structural rearrangements occur, the ADRP substrate, ADP-ribose 1″-phosphate, would not fit into the binding site for ADP-ribose, as revealed by the structural studies.[75] Finally, two of the three catalytic residues proposed for ADRP enzymes of other organisms are not conserved in the coronavirus X domains. Based on these observations, Egloff *et al.*[75] suggested that the X domain of coronaviruses (and alphaviruses as well as hepatitis E virus) are binding modules for poly(ADP-ribose), rather than enzymes involved in the degradation of ADP-ribose 1″-phosphate. Upon viral infection, poly(ADP-ribosylation) of target proteins is induced by the host cell as a stress signal leading to apoptosis and/or necrosis. Poly(ADP-ribose) synthesizing activity (by PARPs) is normally localized to the nucleus, but a cytosolic antiviral zinc-finger protein has been described that belongs to the PARP family.[77] The viral X domain could target this protein in order to prevent its antiviral activity. Thus, the coronavirus X domain is a good example of the usefulness of elucidation of three-dimensional structures for suggesting possible functions of target proteins, and for making alternative ones less likely.

The SARS-unique Domain (SUD, "Nsp3c")

The SARS coronavirus is much more pathogenic for humans than any other coronavirus. Therefore, protein domains encoded by the SARS-CoV genome that are absent in other coronaviruses are of particular interest, because they may be responsible for the extraordinary virulence. One such domain has been identified by bioinformatics as part of the huge Nsp3; appropriately named the "SARS-unique domain" (SUD), it is embedded between the X domain and the PL^{pro}.[10] No homologous proteins, not even with a distant sequence relationship, could be found in the databases. In an effort to characterize the function and three-dimensional structure of SUD, RH's laboratory has produced full-length SUD (residues 349 to 726 of Nsp3) by *in-vitro* protein synthesis, and a more stable, shortened 174-residue version by expression in *E. coli*. It was shown that both full-length SUD and its fragment bind oligo(G) strings in both RNA and DNA with high specificity, with K_d values of <1 μM for $d(G)_{14}$ and G_{14}.[78] Since no component of SARS-CoV Nsp3 has been reported to localize to the nucleus of the infected host cell, viral or host RNA is the likely interaction partner of SUD. Oligo(G) sequences are found in 3′-UTRs of mRNAs for several proteins involved in cellular signaling, such as the pro-apoptotic protein Bbc3, MAP kinase 1, RAB6B (a member of the Ras oncogene family), and TAB3, a component of the NF-κB signaling pathway. These proteins are prime candidates for an interference of the virus with cellular signaling pathways. Changes in the stability and/or translation efficiency of these mRNAs due to the binding of a regulatory factor could result in an altered response of the infected cell to extracellular signals, or it could silence the antiviral response. RH's group has obtained crystals of SUD fragments and structure elucidation is underway. Also, Wüthrich's group has determined (but not yet published) the NMR structure of a small C-terminal fragment of SUD.

Papain-like Proteinases (PL^{pro}): Multifunctional Enzymes

In addition to the main protease, most coronaviral genomes encode two papain-like proteinases (PL^{pro}s) within their nsp3 multidomain segment of the viral genome. SARS-CoV constitutes an exception here, as it has only one such enzyme ($PL2^{pro}$; Nsp3d). The other exception is the avian infectious bronchitis virus (IBV), the genome of which does encode two PL^{pro} domains, one of which has an inactivating mutation in the active site. PL^{pro}s cleave the coronaviral polyproteins after two consecutive Gly residues; thus, SARS-CoV PL^{pro} processes pp1a/pp1ab at the sites ${}_{177}$LNGG↓AVT${}_{183}$, ${}_{815}$LKGG↓API${}_{821}$, and ${}_{2737}$LKGG↓KIV${}_{2743}$, to release Nsp1, Nsp2, and Nsp3, respectively. In addition to its proteolytic activities on the N-terminal third of the polyproteins, the SARS-CoV PL^{pro} has also been shown to be a deubiquitinating enzyme.[79–82] In fact, the P4-P1 pattern of residues cleaved by the PL^{pro} exhibits a striking resemblance to the conserved C-terminus of ubiquitin (LXGG). The amino-acid sequence of the SARS-CoV PL^{pro} displays similarities to ubiquitin C-terminal hydrolase, ubiquitin-specific protease 14 (USP14), and herpes-associated ubiquitin-specific protease (HAUSP, also known as USP7).[83] Thus, it seems that PL^{pro} may have a role in subverting cellular ubiquitination mechanisms to facilitate viral replication. Lindner *et al.*[84] have shown that in addition to its proteolytic and deubiquitinating activity, the SARS-CoV PL^{pro} acts as a de-ISGylating enzyme. Induction of ISG15, which has a C-terminal LRLRGG sequence, and its subsequent conjugation to proteins protects cells from the effects of viral infection.[85,86] Thus, there appears to be an analogy in SARS-CoV to influenza B virus, whose NS1 protein scavenges ISG15 from the host cell, one of the manifold actions of this viral protein suppressing the antiviral response.[87] The crystal structure of SARS-CoV PL^{pro}[88] revealed the presence of four distinct domains, three of which — termed thumb, palm, and fingers — form a right-hand configuration, whereas the N-terminal 66 residues adopt a ubiquitin-like β-grasp fold (Fig. 7).

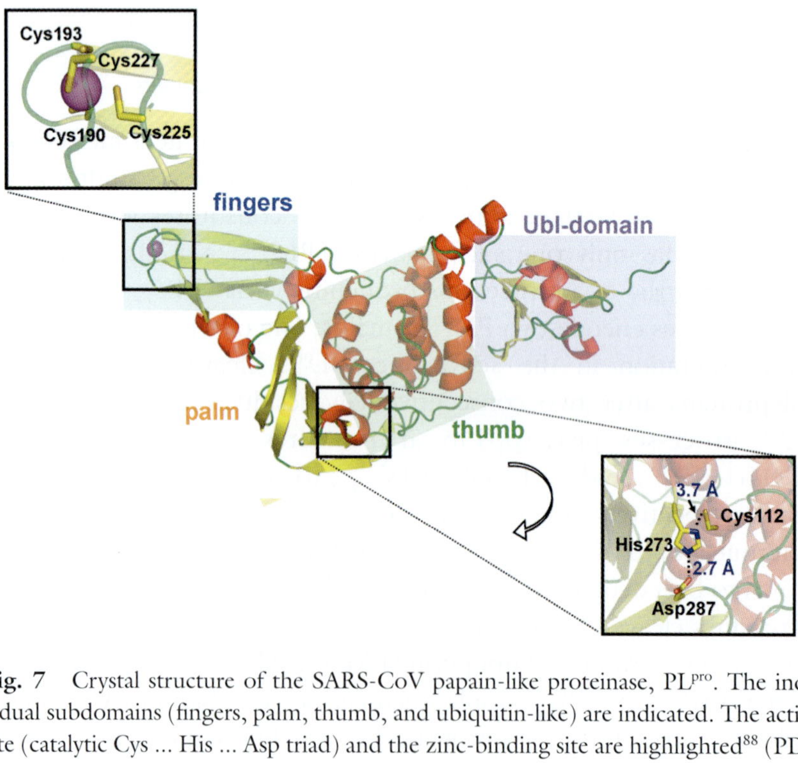

Fig. 7 Crystal structure of the SARS-CoV papain-like proteinase, PL^{pro}. The individual subdomains (fingers, palm, thumb, and ubiquitin-like) are indicated. The active site (catalytic Cys ... His ... Asp triad) and the zinc-binding site are highlighted[88] (PDB code: 2FE8).

In contrast to earlier predictions,[89] the active-site contains a catalytic Cys-His-Asp triad proper and is located in a cleft between the palm and the fingers domain. The tip of the fingers domain contains two β-hairpins which together bind a zinc ion via four cysteine residues.

The role of the ubiquitin-(or ISG15-) like domain of the PL^{pro} is not so clear; it has been speculated that it might act as a decoy to distract cellular ubiquitinating enzymes from their viral targets, or that it may mediate protein-protein interactions in the coronaviral replicase complex.[88] The same can be said of the second subdomain with a ubiquitin fold found in Nsp3, that of Nsp3a (acidic domain, Ref. 71, see above). The ubiquitin-like subdomains of the two Nsp3 domains are similar in structure to the above-mentioned ISG15 and may compete with this immunomodulatory protein for interaction partners in the infected host cell. Since the ISG15 gene is induced by interferon as part of the antiviral response of the innate immune system, the de-ISGylation activity of Nsp3d and the presence of the ISG15-mimicking subdomain of Nsp3d could explain the suppression of the interferon response by the papain-like protease (Nsp3d), which was reported recently.[90] The ISG15-like subdomain of Nsp3a could play an auxiliary role in such a scenario, but clearly, we are just beginning to understand the structural basis of viral suppression of the interferon response.

It is also possible that the ubiquitin-like moieties of Nsp3a and Nsp3d are involved in interactions with host-cell proteins that are not directly related to the innate immune response. In eukaryotes, the ubiquitin fold occurs in several proteins interacting with the cellular GTPase Ras,[91] such as the Ras-binding domains (RBDs) of c-Raf[92] (see Ref. 93 for a review); phosphatidyl-inositol-(3) kinase γ[94]; and RalGDS[95]; as well as AF6.[96] One of the many fundamental cellular processes, Ras is involved in cell cycle progression from the G_0 to the G_1 phase. Some of the proteins interacting with Ras inhibit this process. Interestingly, mouse hepatitis virus (MHV) has been shown to induce cell cycle arrest in the G_0/G_1 phase during the lytic cell cycle.[97] Also, some accessory SARS-CoV proteins such as 7a and 3b are able to induce apoptosis or cell cycle arrest in transfected cells.[98,99]

Nsp7/Nsp8: A Hexadecamer Embracing the Viral RNA

RNA synthesis in coronaviruses is primer-dependent.[100] In many (+)-stranded RNA viruses, the Vpg (viral protein genome-linked), covalently linked to oligonucleotides, serves as a primer for the RdRp.[101] However, the coronaviral genome does not encode a Vpg. Imbert *et al.*[102] have recently shown that short primers (<6 nucleotides) for the SARS-CoV RdRp are synthesized by Nsp8, which acts as a primase and, therefore, second RNA-dependent RNA polymerase of the virus. In a manganese-dependent reaction, primers are synthesized with specificity for internal 5′(G/U)-CC-3′ sites on the template RNA.

The C-terminal domain of SARS-CoV Nsp8 displays some homology to the palm domain of the RdRp and it has therefore been speculated that the two enzymes may share a common origin.[102] The three-dimensional structure of Nsp8 has been determined by Zhai *et al.*,[103] as part of an 8:8 complex between Nsp7 and Nsp8 (Fig. 8a). The inner cylinder formed by this hexadecameric structure is large enough (diameter ≈30 Å) to accommodate double-stranded RNA, leading Zhai *et al.*[103] to propose that the complex might act as a processivity factor for the RdRp, similar to the β_2 clamp in case of bacterial DNA polymerase, or PCNA for the corresponding eukaryotic enzyme. The structure has been described as consisting of an Nsp8 scaffold, the interstices of which are filled by Nsp7, the "mortar" between the "bricks." Two different conformations of Nsp8 are found in this structure: four molecules exist in a conformation which has been termed "golf club"-like because of its appearance in a ribbon plot. The N-terminal ("shaft") domain consists of three α-helices, the third of which is very long and forms the linker to the C-terminal domain (Fig. 8b). In the other four of the eight Nsp8 molecules in the hexadecamer, the long helix connecting the two domains is unfolded in its central part, so that two helices exist that are separated by a number of amino-acid residues (Fig. 8b). In this conformer of Nsp8, the N-terminal region is disordered,[103] and intrinsic disorder has also been suggested for isolated Nsp8 by prediction programs and one-dimensional ^{1}H-NMR spectroscopy.[104] The C-terminal "head" domain displays an α/β fold consisting of a 7-stranded antiparallel β-barrel and three peripheral α-helices.[103] Interestingly, the domain has structural

(a)

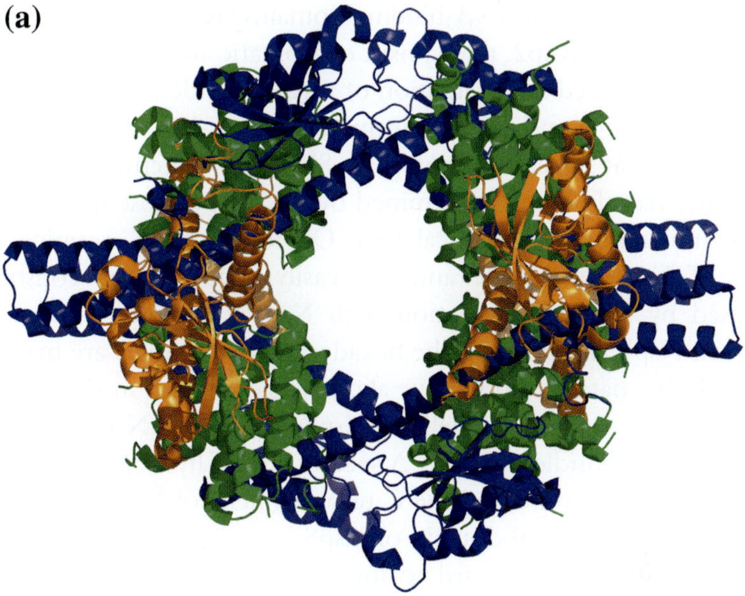

(b)

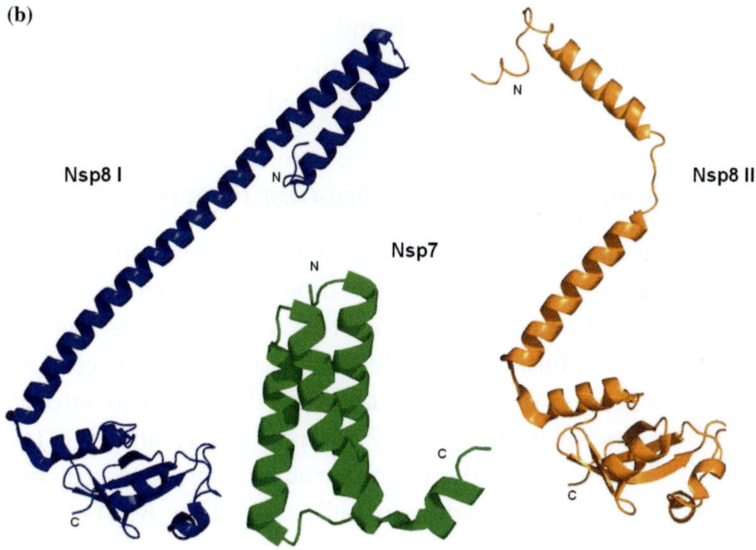

Fig. 8 (a) Crystal structure of the hexadecameric (8:8) supercomplex of Nsp7 (green) and Nsp8 (blue and ochre) (103) (PDB code: 2AHM). **(b)** Structures of the individual components of the Nsp7-Nsp8 supercomplex. Nsp7 (green) is a 3-helix bundle, whereas Nsp8 occurs in two different forms (blue and ochre), depending on the conformation of the N-terminal helices of the polypeptide[103] (PDB code: 2AHM).

similarity to a family of RNA-binding domains (RBDs) comprising two motifs, Rnp1 and Rnp2, that consist of aromatic and basic residues and are usually involved in recognition of single-stranded RNA.

Nsp7 mainly consists of an antiparallel three-helix bundle with an extra α-helix at its C-terminus (Fig. 8b). The structure of an Nsp7 monomer has also been determined by NMR spectroscopy,[105] revealing a similar but not identical fold. Obviously, the C-terminal segment of Nsp7 is flexible and can easily adapt to the constraints imposed by complex formation with Nsp8. The interaction forces between Nsp7 and Nsp8 in the hexadecameric complex are mainly of hydrophobic nature and involve aliphatic side-chains in the C-terminal half of the connecting helix of Nsp8 and on the N-terminal helix of the 3-helix bundle of Nsp7. A second site of hydrophobic interactions involves the third and the C-terminal helix of Nsp7 and the first helix of the C-terminal domain of Nsp8.

It is not straightforward to locate the catalytic primase site on Nsp8; in particular, there are no obvious pairs of aspartate residues that could act as metal-ion ligands and/or general base. Probably, the active site involves an Nsp8 dimer, with Lys58 of one monomer and Arg75 as well as Lys82 of the other involved in phosphoryl transfer.[102]

Nsp9: A Single-stranded RNA-binding Protein

By analytical ultracentrifugation, it has been shown that Nsp8 also interacts with Nsp9,[106] although according to our own measurements using surface plasmon resonance, this interaction is very weak (Ponnusamy, unpublished data). In their yeast-2-hybrid/coimmunoprecipitation study, von Brunn *et al.*[67] also detected this interaction. In MHV, colocalization of Nsp7, Nsp8, Nsp9, and Nsp10 was observed.[107] Very likely, these non-structural proteins are directly involved in the replication complex built around the RNA-dependent RNA polymerase (Nsp12).

Two crystal structures for SARS-CoV Nsp9 were published independently in 2004, by Egloff *et al.*[108] (Fig. 9a). and Sutton *et al.*[106] There are significant differences between the two studies, which are due to the presence of a long N-terminal tag resulting from the cloning

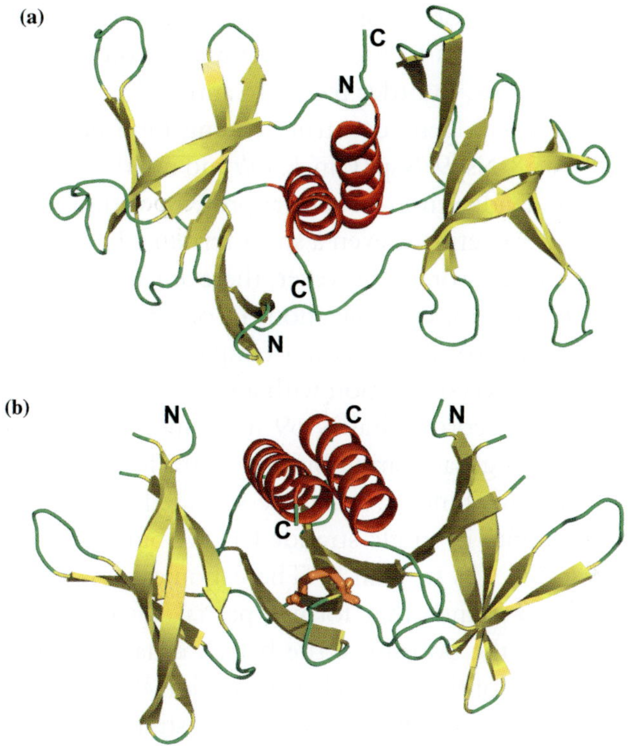

Fig. 9 (a) Crystal structure of the SARS-CoV Nsp9 dimer. Dimerization is largely due to interactions between the C-terminal α-helices (red) which are oriented *parallel* to one another[108] (PDB code: 1QZ8). **(b)** Crystal structure of the HCoV-229E Nsp9 dimer. Dimer formation is governed by a disulfide bridge (orange) between Cys69 of each monomer, and involves an *antiparallel* interaction of the C-terminal α-helices (Ponnusamy *et al.*, unpublished) (PDB code: 2J97).

procedure.[106] This extension of >30 amino-acid residues contributes to the dimerization of the protein by forming an additional two-stranded β-sheet with the same region in the other monomer. Our experience over the years has shown that extensions of the chain termini can lead to artifacts in the resulting structure, most of which are usually minor, though. Therefore, we always make sure that extensions at the chain termini, including $(His)_6$ tags, can be cleaved off after purification. This is particularly important in case of the long N-terminal extensions that often result from recombination cloning using the Gateway™

(Invitrogen) system. For coronavirus proteins, we have developed a procedure that allows us to use the viral main proteinase to perform the cleavage reactions, in order to create authentic chain termini. For example, we express a gene construct coding for a polyprotein comprising Nsp7-Nsp8-Nsp9-Nsp10 in *E. coli* and use the SARS-CoV M^{pro} to release the individual non-structural proteins without any terminal extensions.[109,110] Sometimes, even a short terminal His tag can prevent crystallization of the protein; however, there are also cases where such a tag may be needed for crystallization. In our hands, Nsp9 of SARS-CoV (but not of HCoV 229E) is an example for the latter. Therefore, it is advisable to try crystallization with and without such tags present.

The fold of the coronaviral Nsp9 monomer is a variation of the oligonucleotide/oligosaccharide-binding (OB) fold. This fold is characteristic of proteins binding to single-stranded nucleic acids[111] and occurs, for example, in single-stranded DNA-binding proteins from bacteria (e.g., Ref. 72) to man.[112] The canonical OB fold comprises five antiparallel β-strands that form a partial barrel, and an α-helix that packs against the bottom of the barrel, usually in an orientation along the long axis of the β-barrel cross-section.[111] In the classical OB fold, the α-helix is interspersed between β-strands 3 and 4, but in Nsp9, the helix is appended at the C-terminal of the polypeptide chain (residues 92–108). Also, Nsp9 has two extra β-strands (strands 6 and 7) forming a long hairpin.

Through *parallel* packing of the C-terminal helices of two monomers, SARS-CoV Nsp9 forms a homodimer, both in the presence of an N-terminal His tag[108] (Fig. 9a) and the >30-residue extension.[106] In spite of 45% sequence identity between SARS-CoV and HCoV-229E Nsp9, the wild-type structure of the latter exhibits a mode of homodimerization that is entirely different from what has been observed in the crystal structure of the former (Ponnusamy *et al.*, unpublished). 229E Nsp9 forms a symmetric homodimer linked by a disulfide bond involving the Cys69 residue of each monomer (Fig. 9b), in spite of the continuous presence of 5 mM DTT in the crystallizing sample. Here, the α-helix of each monomer is involved in dimerization through *antiparallel* interaction with its counterpart on the other subunit. Replication and translation of the coronavirus genome occurs in the cytosol of the infected cell,[113] where the overall milieu is reductive and disulfide bonds are rare, although a few cytosolic proteins containing them have been described.[114,115] In order to probe the effect of the disulfide bridge, Cys69 of HCoV-229E Nsp9 was mutated to alanine. Surprisingly, the crystal structure of the mutant displays the same dimerization mode as SARS-CoV Nsp9[106,108] and is thus very different from the 229E wild-type dimer. On the other hand, gel-shift assays and surface plasmon resonance measurements indicate that only the wild-type HCoV-229E Nsp9, not the Cys69Ala mutant, binds strongly to single-stranded RNA (Ponnusamy *et al.*, unpublished).

The cysteine residue involved in the disulfide linkage in HCoV-229E Nsp9 is conserved in SARS-CoV, and yet, no disulfide is observed in this case. The reason may be that the latter protein contains two additional free cysteine residues which might reduce any transiently formed inter-subunit disulfide *in trans*. In fact, apart from plant and mammalian cysteine proteases, where a free cysteine constitutes the active principle, the presence of free cysteines in addition to disulfide bonds in one and the same protein is rare.

We also have to entertain the possibility that in cells infected by HCoV 229E, both the disulfide-linked homodimer and the "normal" homodimer may exist at different states of the viral life cycle. The disulfide-bonded form of HCoV-229E may possibly be formed in response to oxidative stress inside the infected cell. There are several reports suggesting the regulation of DNA/RNA-binding proteins by redox processes. Thus, the redox state has been shown to determine the interaction with DNA of the multifunctional eukaryotic SSB protein RP-A.[116] LEF3 (the SSB of baculovirus) shows a DNA-annealing effect in its oxidized state, whereas in the reduced state, its DNA-unwinding activity is favored.[117] The E2 protein of bovine papilloma virus type 1[118] and ICP8 of herpes simplex virus type 1 also show DNA-binding activity depending on the redox state of the environment.[119–121] In view of these data, it can be hypothesized that the disulfide-mediated dimer of Nsp9 might play a role during oxidative stress in the infected host cell.

In the crystal and (as suggested by Dynamic Light Scattering) in solution, wild-type HCoV-229E Nsp9 forms hexamers, which appear to depend on the presence of sulfate ions. In contrast, the Cys69Ala mutant as well as the SARS-CoV Nsp9 form endless polymeric rods in all the crystal forms described[106,108] (Ponnusamy *et al.*, unpublished). By making use of sulfate ion positions around these rods, single-stranded RNA has been modeled winding around them (Ponnusamy *et al.*, unpublished), in a way reminiscent of the wrapping of single-stranded DNA around *E. coli* SSB.[122] In this crude model, the ssRNA forms a left-handed helix wrapping around the Nsp9 polymer, reminiscent of the model proposed for the nucleocapsid protein of SARS-CoV interacting with ssRNA[123] (see below).

Nsp10: An RNA-binding Protein Containing Two Zinc Fingers

Nsp10 is one of the better conserved non-structural proteins in coronaviruses. The corresponding protein (also termed "p15") of mouse hepatitis virus (MHV) has been shown by immunofluorescence microscopy to be part of the membrane-associated cytoplasmic replicase complex. A temperature-sensitive MHV-A59 Nsp10 mutant in which the highly conserved Gln65 was substituted by Glu appeared to be defective in negative-strand synthesis of the viral RNA[124] and in activation of the main proteinase (Nsp5; Ref. 125). RH's group has shown that MHV Nsp10 binds double-stranded RNA or DNA, as well as single-stranded RNA and contains two zinc ions per monomer, coordinated by two zinc finger-like modules.[126] Also, the group proposed the existence of monomeric and oligomeric forms of the protein. Indeed, structures of a monomeric form and of a dodecameric form of SARS-CoV Nsp10 have been elucidated by X-ray crystallography.[127,128] The fold of the monomer (Fig. 10) was found to be unique, with a helical hairpin at the N-terminus followed by an irregular β-sheet surrounded by additional helices, and a C-terminal loop subdomain. The first zinc finger module is of a novel CCHC type, with three cysteines and one histidine as metal-ion ligands,

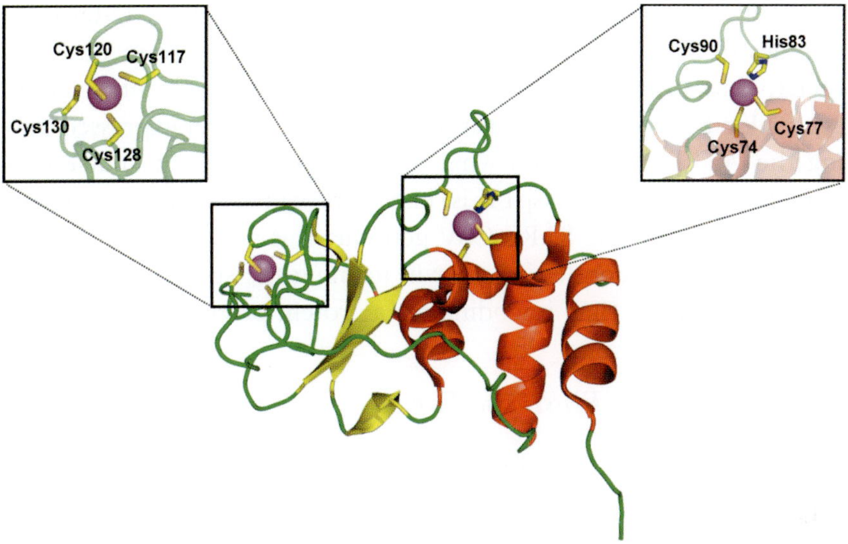

Fig. 10 Crystal structure of the SARS-CoV Nsp10 monomer. The two zinc-binding modules are highlighted. Zn^{2+} is indicated by purple spheres[127] (PDB code: 2FYG).

whereas the second holds the zinc through four cysteines (CCCC) and is reminiscent of the "polymerase gag knuckle fold" which is found, e.g., in yeast RNA polymerase. Together, the two zinc fingers resemble the arrangement of metal-ion ligands in the HIT-type proteins, named after the first such protein from yeast (HIT1), where this feature was found. The Nsp10 dedecamer forms a hollow ring-like structure, with an internal hole of 36 Å diameter. Both the inner and the outer surface of the oligomer display positive electrostatic potential, compatible with binding to RNA.[128] It has been shown that upon addition of nucleic acids, the MHV Nsp10 oligomer dissociates into monomers.[126]

Nsp11: Is It Ever Made?

In the crystallographic study by Su *et al.*,[128] which revealed the structure of the SARS-CoV Nsp10 dodecamer, a fusion protein of Nsp10 and the short Nsp11 (11 amino-acid residues) had been used.

However, it was found that the Nsp11 part as well as C-terminal residues of Nsp10, including the 4th zinc ligand of the second zinc finger, Cys130, had been removed by spontaneous proteolytic cleavage.[128] Up to now, it is not clear whether Nsp11 is ever expressed during the viral life cycle, and what its role may be. As the nucleotide sequence coding for Nsp11 is just 5′-proximal to the ribosomal frameshift site on the SARS-CoV genome, it is well possible that the role of these nucleotides is rather in forming the special structure of the frameshift site than in coding for a protein product.

Nsp12: Structure Elusive so Far

Second in importance only to the coronavirus main protease, the RNA-dependent RNA polymerase is an attractive target for drug design. After all, most antiviral drugs on the market or in clinical trials are directed against either a polymerase or a protease. Unfortunately, the structure of the SARS-CoV RdRp has not yet been determined due to difficulties in expression and purification of the enzyme in an active form. A homology model suggested that the protein adopts the classic "right-hand" fold featuring the characteristic finger, palm and thumb subdomains.[129] The finger domain is composed of three α-helices, a unique helix-loop-helix supersecondary structure, and two β-sheets. The palm domain comprises two helices and a β-hairpin, which harbors two aspartate residues possibly responsible for catalyzing the nucleotide transfer reaction. The thumb domain consists essentially of two α-helices and a large loop connecting it to the finger domain. According to this model, the active site is buried in the center of the protein. Potential nucleoside analogue and non-nucleosidic inhibitors of SARS-CoV RdRp have been proposed by molecular docking.[130]

Nsp13: Helicase with a Binuclear Zinc Finger

As with several of the other coronavirus enzymes, the helicase appears to be significantly different from the corresponding activities of the other (+)-strand RNA viruses. First, it follows the RdRp coding sequence in the genome, while the opposite is true for most viral helicases. Second, it contains an N-terminal binuclear zinc-binding module consisting of as many as 12 conserved cysteine or histidine residues. The presence of this domain is essential for the 5′→3′ unwinding activity of this enzyme. Finally, the enzyme does not only feature NTPase and dsRNA-unwinding activities, but works equally efficiently with deoxynucleotides and dsDNA.[131,132] As in some other (+)-strand RNA viruses, the enzyme also has a 5′-triphosphatase activity which likely plays a role in RNA capping.[132] The Nsp13 helicase belongs to the superfamily 1 of helicases. Although the crystallizability of the SARS-CoV helicase has occasionally been mentioned at conferences, no three-dimensional structure is available yet; however, a homology model constructed on the basis of *E. coli* Rep helicase has been published.[133]

Nsp14: Part of a Proofreading System

Their unusually large RNA genome constitutes a problem for the coronaviruses, as errors would unavoidably accumulate if one assumes that the error rate of the RdRp is similar to that observed for other RNA viruses. Therefore, it has been suspected that these viruses have some kind of proofreading system. Indeed, with the identification of Nsp14 as a 3′→5′ exonuclease (ExoN), a component of such a system has been characterized.[134,135] In MHV, mutation of ExoN active-site residues results in a virus that accumulated 15-fold more mutations than wild-type virus, without changes in growth fitness.[135] From three conserved sequence segments, it has been concluded that the SARS-CoV ExoN belongs to the DEDD family of nucleases[10] and this is supported by mutagenesis of the key residues responsible for coordinating the two metal ions involved in catalysis.[134] The enzyme was shown to hydrolyze both single-stranded RNA (ssRNA) and double-stranded RNA (dsRNA). The minimum product lengths observed with ssRNA is 8–12 nucleotides, whereas dsRNA is degraded to fragments of 4–7 nucleotides, suggesting that dsRNA is the better substrate of the two. Furthermore, addition of dsRNA increases the ribonucleolytic activity of Nsp14 on ssRNA, which is degraded to a predominant 7-nucleotide product in its presence. No hydrolytic activity is observed with ssDNA or dsDNA. On the basis of these observations, it has been concluded that dsRNA is the major substrate of the coronavirus ExoN, although a single-stranded 3′ end is probably required, as is the case with other nucleases.[136] The enzymatic activity requires magnesium or manganese ions, but even with low concentrations of Zn^{2+}, some activity is observed, whereas at higher concentrations, this metal ion is inhibitory.[134,137] Interestingly, Nsp14 contains a putative zinc finger between the conserved DEDD sequence motifs I and II. Together with the zinc fingers associated with the $PL1^{pro}$ (absent in SARS-CoV), $PL2^{pro}$ (also called Nsp3d in SARS-CoV), the dsRNA-binding protein Nsp10, the helicase (Nsp13), this brings the number of these structural modules encoded by the coronavirus genome to at least six.

Minskaia *et al.*[134] constructed HCoV-229E mutants with substitutions in the presumed active site of the ExoN. They found that the amount of subgenomic RNA synthesized and/or accumulated a few days after infection of susceptible cells was significantly reduced as compared with the wild-type virus, suggesting a distinct role of Nsp14 in transcription, in addition to its function in replication. In agreement with this, no infectious virus progeny could be recovered after transfection of full-length genomic RNA encoding a replicase with deficient ExoN.

It should also be mentioned that Nsp14 comprises another, largely uncharacterized domain C-terminal to the ExoN. Interestingly, mutations in this region in MHV strain A59 lead to a reduction in virulence in mice without affecting virus replication in cultured cells.[138] Nsp14 could therefore be another multifunctional element of the coronaviral polyprotein. The additional functions remain to be discovered and structural information on Nsp14 is still elusive.

Nsp15: Yet Another Viral Nuclease

In addition to the 3′→5′ exonuclease, Nsp14, the coronavirus replicase comprises another RNA processing enzyme, the endonuclease Nsp15 (also called NendoU). It has been proposed that its activity might be required in minus-strand RNA synthesis to digest the nascent RNA.[19–21] The presence of an endonuclease is characteristic of nidoviruses; it does not occur in any other RNA virus. Nsp15 is a Mn^{2+}-dependent enzyme, which leaves a 2′,3′-cyclic phosphate 5′ to the cleaved bond.[139,140] Cleavage occurs preferably on the 3′-side of unpaired uridylates, suggesting that single-stranded RNA or single-stranded regions in double-stranded RNA are the primary substrates of the enzyme.[141] In contrast to earlier reports,[139,140] there is no cleavage on the 5′-side of uridylate.[141] Crystal structures of Nsp15 have been determined for both SARS coronavirus[142,143] and MHV[144]; in addition, a structural model derived from cryo-electron microscopy at 8.3 Å resolution is also available for the SARS-CoV protein.[141] The Nsp15 monomer consists of three domains: a small N-terminal domain comprising a three-stranded, antiparallel β-sheet and two α-helices packed against its concave side; a central α/β domain featuring a mixed 3-stranded sheet, two α-helices, and an additional β-hairpin as well as a large portion of irregular secondary structure including one additional α-helix; and finally a large C-terminal domain containing an α-helical core flanked on both sides by two three-stranded antiparallel β-sheets (Fig. 11a). As with so many of the coronaviral proteins, this type of polypeptide fold has not been described so far. Nsp15 monomers assemble into a hexamer, which is a back-to-back dimer of trimers (Fig. 11b), with a central tunnel of 15 Å diameter, i.e. too narrow to accommodate RNA. Instead, cryo-electron microscopy suggested that RNA binds at the interface between the two trimers in the hexamer.[141] Oligomerization seems to be largely driven by interactions between the N-terminal domains of each monomer. Each monomer carries its own active site, which is embedded in the C-terminal domain, at the periphery of the hexamer. The active site is similar to the well-characterized catalytic center of RNase A and consists of two histidine residues and a lysine. The two histidines are proposed to act as general base and acid, respectively (although the negatively charged histidine postulated in Ricagno *et al.*[142] is an unlikely intermediate), whereas the lysine is believed to stabilize the transition state centered on a pentavalent phosphorus atom. Interestingly, mutation of either histidine, and

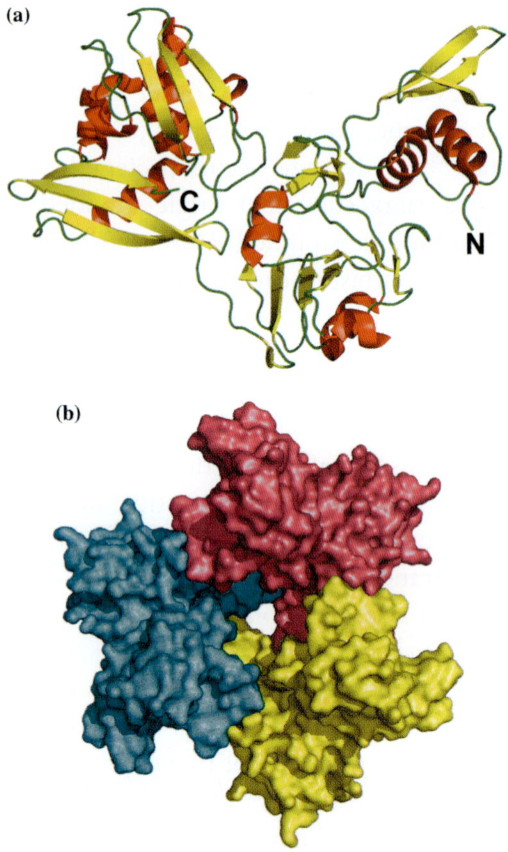

Fig. 11 **(a)** Structure of the Nsp15 (NendoU) monomer within the hexamer of the enzyme[142] (PDB code: 2H85). **(b)** Space-filling illustration of an Nsp15 trimer. Individual monomers are indicated in different colors. The hexamer observed in the crystals and also in solution is formed by back-to-back dimerization of the trimers (the second trimer would be behind the one shown in this figure).

even both histidines together, does not lead to complete loss of NendoU activity of MHV Nsp15 *in vitro*, and the corresponding mutant viruses have a reduced growth rate but are viable.[145] The specificity for uridine is probably due to a serine residue (in SARS-CoV; threonine in MHV) positioned to form two specific hydrogen bonds with this base.[145] No structural basis for the requirement for manganese ions for the NendoU catalytic activity was unveiled, but it has been speculated that a tyrosine residue could be involved in coordinating a Mn^{2+} ion through π-cation interactions supported by one of the two catalytic histidines.[142] Alternatively, it has been proposed that manganese may facilitate RNA binding to Nsp15, rather than being directly involved in catalysis.[141] The crystal structure of a trimmed SARS-CoV Nsp15, which lacks N-terminal residues involved in hexamerization and is therefore monomeric, reveals conformational changes in two loops supporting residues involved in the active site[143] and this explains why only the hexamer is catalytically active.[146] The requirement for hexamerization may be important in protecting the genome of the coronaviruses from untimely cleavage by Nsp15.

How are the 2′,3′-cyclic phosphates produced by Nsp15 hydrolyzed? Is it necessary at all to remove them? It is interesting to observe that some coronaviruses of group 2a, mouse hepatitis virus (MHV), bovine coronavirus (BCoV), and human coronavirus OC43 (HCoV OC43), encode a presumable cyclic phosphodiesterase (sometimes called Ns2) in their open reading frame 2 (ORF2).[138,147] However, this is not observed in the SARS-CoV genome, nor in any other coronavirus sequenced so far.

Nsp16: A Methyltransferase Involved in Capping the Viral mRNA

In June, 2003, i.e. during the ongoing SARS outbreak in China, von Grotthuss *et al.*[148] published a prediction of the three-dimensional structure of Nsp16 (which was erroneously assigned as Nsp13 then). Using the 3D Jury software, these authors predicted that Nsp16 would have the fold of an S-adenosylmethionine (AdoMet)-dependent O-methyltransferase (MTase), the role of which is to form the mRNA cap-1 (mGpppNm). To date, this enzymatic function of Nsp16 has not been demonstrated experimentally, but very likely, it is correct. If so, this protein would be the third coronaviral RNA processing enzyme, in addition to Nsp14 and Nsp15. There is no experimental 3D structure available yet for Nsp16.

If Nsp16 is indeed the cap-1 forming MTase, then one should also expect the presence of a cap-0 enzyme, which would promote the formation of mGpppN. Using similar prediction methods, the same group predicted the SUD subdomain of Nsp3 to perform this task.[149] In our laboratory, we also predicted the fold of a typical AdoMet-dependent MTase for the SUD, with the AviRa MTase, which acts on ribosomal RNA, being the closest relative; however, we failed to show MTase activity for SUD experimentally (Schmidt *et al.*, unpublished). Instead, we discovered the oligo(G)-binding properties of SUD (see the paragraph on Nsp3 above; Ref. 78).

Structural Proteins

Spike Protein: Receptor Binding and Fusion Core

SARS-CoV spike (S) protein, which forms prominent projections from the envelope of the viral particle, is a 1255-residue glycoprotein that possesses a signal peptide at the N-terminus, a putative fusion peptide, a single ectodomain and a transmembrane region followed by a short cytoplasmic tail at the C-terminus. As a class I viral fusion protein, the S protein can directly mediate the infection of host cells by specifically binding to the cellular SARS- virus receptor (angiotensin-converting enzyme 2, ACE2) and subsequently inducing the fusion of viral and cellular membranes.[150] After being translated as a large single-chain precursor, SARS-CoV S protein can be cleaved by cellular proteases to produce two mature functional subunits, named receptor-binding (S1) and membrane fusion (S2) fragments. The S1 subunit binds to ACE2 via its receptor-binding domain (RBD), while the S2 subunit is responsible for driving viral and cellular membrane fusion by forming the "fusion core" structure involving two heptad repeats, HR1 and HR2.

The RBD — ACE2 Complex

The crystal structure of the complex between the receptor-binding domain (RBD) of the S1 fragment and angiotensin-converting enzyme 2 (ACE2)[151] is certainly one of the highlights of structural biology of SARS coronavirus so far (Fig. 12). The RBD was shown to contain two subdomains: a core and an extended loop. The core is a five-stranded anti-parallel β-sheet with three short connecting α helices. An extended loop lies at one edge of the core and presents a gently

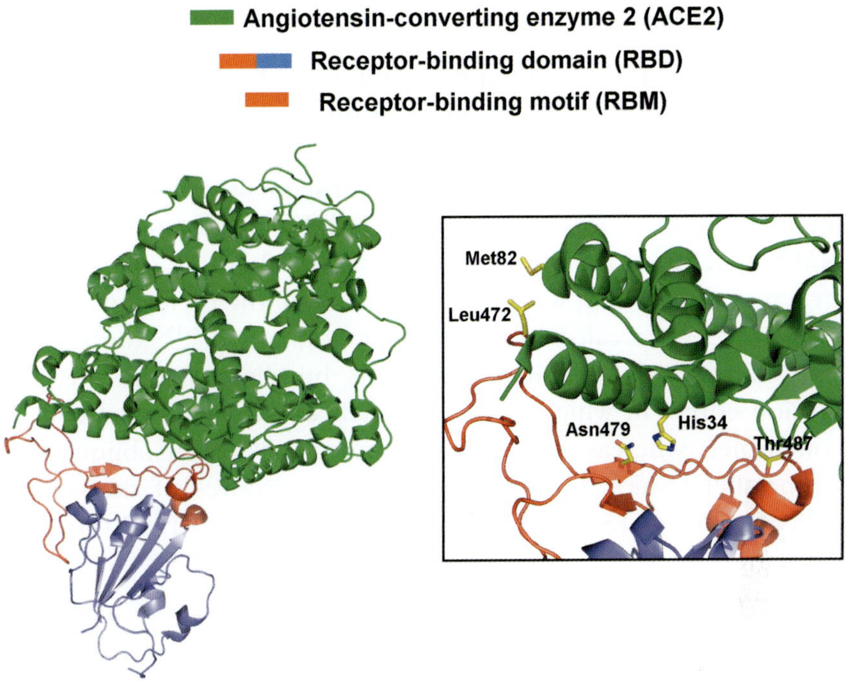

Fig. 12 Structure of the complex between the angiotensin-converting enzyme 2 (ACE2; green) and the receptor-binding domain (RBD; red and blue) of the SARS-CoV spike (S) protein (S1 fragment). The specific interactions at the interface between the two proteins define the host range of the virus[151] (PDB code: 2AJF).

concave surface, which cradles the N-terminal lobe of the peptidase domain of ACE2. The atomic details at the interface between the RBD and ACE2 nicely explain the host range specificity and identify mutations that facilitate efficient cross-species infection and human-to-human transmission of SARS-CoV. Furthermore, the structure of the complex also suggests a possible strategy to prepare truncated disulfide-stabilized RBD variants for the design of coronavirus vaccines. In addition, the structure of a complex between the RBD and a neutralizing monoclonal antibody ("m396") has been reported recently.[152] Although the overall structure of the m396-bound RBD is not significantly different from that of the ACE2-bound RBD, this structure provides a structural rationale for understanding the major determinant of SARS-CoV immunogenicity and neutralization.

The Fusion Core

After binding of the S1 subunit to ACE2, the heptad repeat regions HR1 and HR2 in the S2 fragment of the SARS-CoV spike protein are believed to undergo a series of conformational changes, from prefusion to intermediate state, and subsequently postfusion during the fusion process of viral and cellular membranes.[153–158] By NMR spectroscopy, it was shown that HR2 forms a coiled-coil structure in solution, which is assumed to be the prefusion conformation.[156] As revealed by X-ray crystallography,[153–155] the postfusion state of S2 subunit is a six-helix bundle structure, i.e. the "fusion core," in which three HR1 helices form a parallel coiled-coil trimer, whereas three HR2 peptides pack in an oblique and antiparallel fashion into the hydrophobic grooves of the central coiled coil (Fig. 13). Packing of the helical parts of HR2 onto the

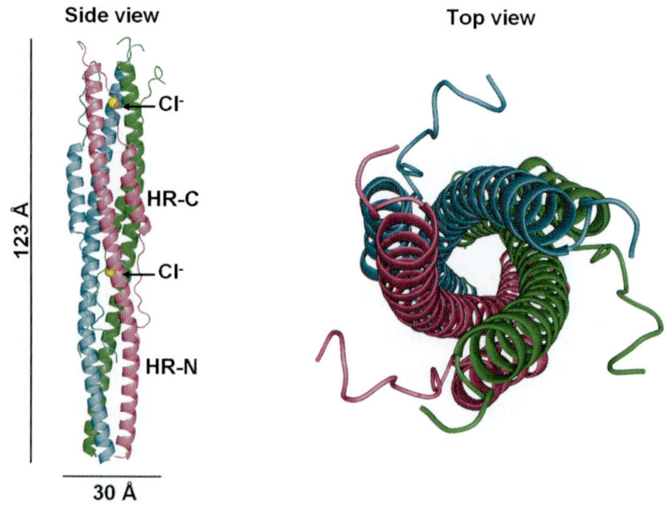

Fig. 13 Structure of the fusion core of the SARS-CoV spike protein (S2 fragment). The three-helix bundle formed by the N-terminal HR regions (HR-N) is complemented by the C-terminal HR regions (HR-C) so that a hexameric arrangement is formed. Chloride ions from the crystallization buffer bind on the quasi-threefold axis of symmetry[154] (PDB code: 1WNC).

HR1 trimer grooves likely plays an important role in stabilizing the postfusion structure. Since the formation of the fusion core is necessary for the fusion of the viral and cellular membranes, blocking the conformational changes of the S protein from the intermediate state to the postfusion state is expected to inhibit viral entry. Peptides designed to perform this task have successfully been established as fusion inhibitors for the treatment of HIV infections in their early stages.

Nucleocapsid Protein: Another Multifunctional Player

As with several other (or perhaps most?) coronavirus proteins, the nucleocapsid protein (NP) is multifunctional. One of its functions is to recognize the packaging signal on the viral RNA and thus initiate the formation of the ribonucleoprotein (RNP) during virus assembly. Formation of the RNP may also be important for timely replication and transmission of the genetic material.[159] Furthermore, it will protect the genomic RNA from the processing activities of the genome-encoded exonuclease (Nsp14) and the endonuclease (Nsp15), as well as from host-cell nucleases. Even though it is a structural protein, NP has also been found to be part of the MHV replicase-transcriptase complex.[107,160] Furthermore, the SARS-CoV NP has been shown to interfere with signal transduction processes in the infected host cell,[161–163] and there have been reports that it may be involved in down-regulation of the type-I interferon response of the host cell.[19,164] Moreover, it has been shown by us to specifically interact with cyclophilin A[165] and human RNP A1.[166] Of the three domains of NP, the N- and the C-terminal domain are involved in RNA binding. The linker between these two domains is flexible, rich in serine and arginine, and the site of extensive phosphorylation; it appears to contribute to the oligomerization of NP.[167] The very carboxy terminus of the C-terminal domain is also involved in multimerization[168] and is the site of interaction between the NP and the membrane (M) protein (also called matrix protein), which thus links the paracrystalline nucleocapsid to the viral envelope.[169]

The N-terminal domain (NTD, residues ≈45–180) of the SARS-CoV NP has been investigated by NMR spectroscopy and X-ray crystallography.[170,171] Also, the same domain of infectious bronchitis virus (IBV) has been studied by X-ray crystallography.[172,173] The structure consists of a five-stranded antiparallel β-sheet with a unique arrangement of long loops connecting the strands. Even though this fold matches no other known 3D structure, it has features typical of RNA-binding proteins. There are similarities in the arrangement of four of the five strands to the β-sheet found in the family of RNA-binding proteins containing the RRM motif and represented by the U1A spliceosomal protein.[174] Also, an extended β-hairpin carrying several basic amino-acid residues is protruding from the sheet in NP, similar to what is seen in RNPs as well as in coronavirus Nsp9 and other nucleic-acid binding proteins with an OB or OB-like fold. It is believed that the role of this hairpin is to clamp the nucleic acid against the β-sheet core of the protein. Binding of RNA to the NTD was demonstrated through analysis of chemical shifts and line-broadening of resonances assigned to residues at the junction between the hairpin and the β-sheet. Also, some initial screening for small-molecule binders was carried out by NMR.[170]

The C-terminal domain (CTD, residues 248–365) of the SARS-CoV NP had its three-dimensional structure determined by X-ray crystallography[123,175] and also by NMR spectroscopy.[176] Furthermore, the same domain of IBV has also bee investigated by X-ray crystallography.[173] The core of this domain (residues 280–365) is responsible for dimerization of the NP, while its N-terminus (residues 248–280) is also involved in RNA binding, with higher affinity than is observed for the NTD. In the crystal structure, an octamer of the CTD is found, which can also be described as a tetramer of a dimeric building block. Dimerization of the CTD occurs, in part, through formation of a 4-stranded β-sheet comprising two strands from each subunit. In addition, three of the eight α-helices of each subunit also contribute to dimerization, via hydrophobic contacts between their side chains. The CTD dimer seems to be the most prominent species in solution, but the dimers of dimers (tetramers) and dimers of tetramers (octamers) in the crystal are likely to exist in the context of the viral particle and in the presence of RNA.

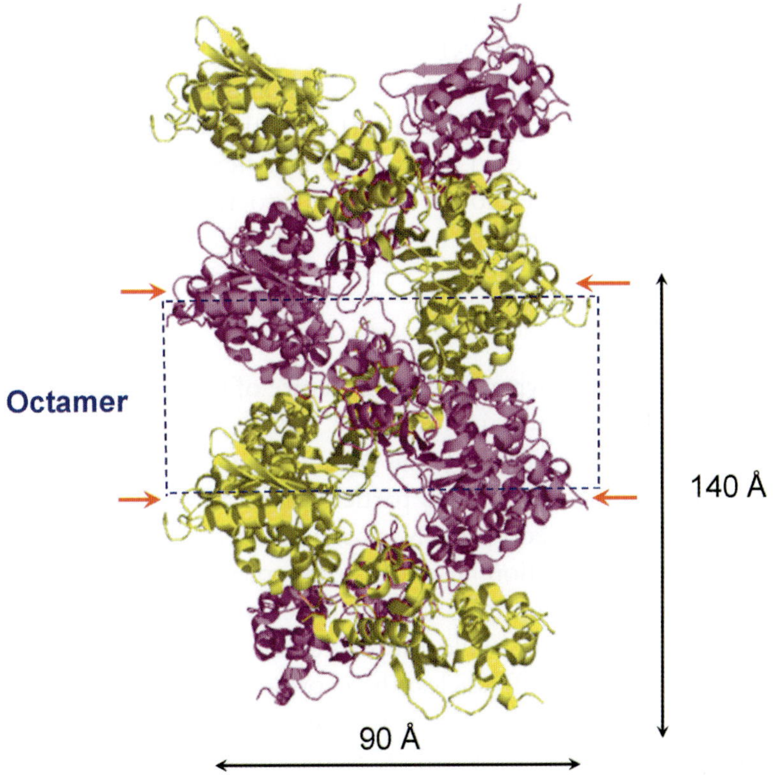

Fig. 14 Through translation stacking of nucleocapsid protein (NP) octamers, a left-handed twin superhelix is formed that features two positively charged grooves wound around it. These can be assumed to be the binding sites for the viral RNA, which would thus wrap around the NP superhelix[123] (PDB code: 2CJR).

Through translation stacking of octamers, a left-handed twin superhelix is formed (Fig. 14) that features two positively charged grooves wound around it. Since these can be assumed to be the binding sites for the viral RNA, it has to be concluded that the latter would wrap around the NP superhelix. How then can the RNA be protected from cleavage by RNases of the virus and the host cell? There is a large number of conserved aromatic residues that could intercalate into the single-stranded RNA, similar to the base stacking by aromates demonstrated for single-stranded DNA-binding proteins such as *E. coli* SSB[122] and proposed for coronavirus Nsp9 (Ponnusamy *et al.*, unpublished; see above). Such intercalation is believed to protect the nucleic acid from destruction. Intriguingly, the N-terminal domain of SARS-CoV NP also comprises several conserved aromatic amino-acid residues in the RNA-binding site and likely makes use of the same principle. In concert with the CTD superhelix (of which the NTD, connected to the CTD by a flexible linker, will form protrusions), the NTD could contribute to the protection of the coronavirus genetic material.

Envelope Protein

Relatively little is known about the structure of the envelope (E) protein. It has been shown in the MHV system that coexpression of genes coding for coronaviral E and M (membrane or matrix) proteins leads to formation of virus-like particles, indicating that neither the spike (S) nor the nucleocapsid (N) proteins are required for viral budding.[177,178] The interaction between the cytoplasmic domains of the E and M proteins is thought to take place in pre-Golgi compartments.[179] Furthermore, while the expression of M alone does not induce the formation of vesicles, the expression of E does, underlining the important role of the latter for the budding process.[180,181] Coronaviral E proteins are small (≈75 amino-acid residues), but carry a rather long (≈25 residues) hydrophobic segment between the hydrophilic terminal regions. While this overall property is well conserved between the various coronaviruses, the amino-acid sequence has significant similarities only within, but not between, the three groups of coronaviruses. Being an outlier to group 2, the SARS-CoV has an E protein that shares just 20% sequence identity with its homologues in other coronaviruses. Using FT-IR spectroscopy, artificial lipid bilayers, and a hydrophobic peptide corresponding to the transmembrane segment of SARS-CoV E protein, it has been suggested that this segment could form a helical hairpin traversing the membrane twice and leading to distortions of membrane order.[182,183] In contrast, early predictions of the E protein topology from HJ's laboratory had proposed a single α-helix formed by a somewhat shorter (≈20 residues) transmembrane segment.[184] The importance of individual amino-acid residues within the transmembrane helix has recently been assessed by alanine scanning.[185] It has also been shown that upon expression in *E. coli*, SARS-CoV E protein can form disulfide-mediated dimers and trimers, leading to an increase in membrane permeability.[186,187] More recently, it has been shown that the coronaviral E protein can also form pentamers and may have the function of a "viroporin" ion channel, similar to the influenza virus M2 protein or the Vpu of HIV-1.[187–190] Of the transmembrane topologies proposed for the E protein so far, only the pentamer would be compatible with the observed ion channel properties.[191]

Membrane Protein

The coronaviral membrane (M) protein, also called matrix protein (221 amino-acid residues in SARS-CoV), spans the membrane bilayer three times, with a rather large carboxy-terminal cytoplasmic domain inside the virion and a small amino-terminal domain outside.[192] In addition to its interactions with the envelope (E) protein (see above), the M protein also interacts with the nucleocapsid (N) protein *in vitro* and *in vivo*.[193,194] The M protein is highly glycosylated, and this glycosylation may be essential for interaction with the host. Recently, it has been shown that SARS-CoV M interacts directly with IKKβ, the major IκB kinase, and inhibits the latter to prevent an inflammatory response to the viral infection.[195] Normally, proinflammatory stimuli, including viruses, will be recognized by Toll-like receptors (in case of viruses, TLR3, TLR7, or TLR9), leading to an activation of the IKK signalosome. In addition to IKKβ, the IKK signalosome comprises another kinase, IKKα, and a regulatory protein, IKKγ. The activated IKK signalosome phosphorylates the inhibitor of NF-κB, IκB, which subsequently undergoes ubiquitination and degradation, exposing a nuclear localization signal on NF-κB and thereby allowing it to translocate into the nucleus. Once in the nucleus, NF-κB activates target genes such as *cox-2*, the protein product of which (cyclooxygenase 2) catalyzes the synthesis of prostaglandins, leading to the promotion of inflammation through various mechanisms. Through its interaction with IKKβ, SARS-CoV M protein appears to suppress this mechanism, allowing the virus to evade innate immunity.

Accessory Proteins

Accessory Protein 7a: A Structural Protein

ORF7a of SARS-CoV encodes a type-I transmembrane protein that comprises an 81-residue luminal domain, a 21-residue transmembrane segment, and a very short (5-residue) cytoplasmic tail. This protein has been identified in all isolates of SARS-CoV from humans and animals, but not in other coronaviruses. The three-dimensional structure of the luminal domain has been determined by X-ray crystallography[196] and by NMR spectroscopy.[197] Surprisingly, it adopts a compact, immunoglobulin(Ig)-like β-sandwich fold, with interesting deviations from the canonical Ig fold (Fig. 15). Thus, strand A does not run antiparallel to strand B, but has switched from the first to the second sheet of this sandwich structure and runs parallel to strand G. This rearrangement could make strand B available for interaction with other proteins. Also, there are two non-canonical disulfide bonds linking the two sheets. The closest structural relatives to SARS-CoV protein 7a turned out to be the N-terminal domain of human intercellular adhesion molecules 1 and 2 (ICAM-1 and -2), and the N-terminal domain of human interleukin-1 receptor (IL1R). Along with the observed (limited) sequence similarity to ICAM-1, this suggested that the ORF7a protein might bind to human lymphocyte function-associated antigen-1 (LFA-1), and this has indeed been shown to be the case for LFA-1 on the surface of Jurkat cells.[198] This suggests that LFA-1 could be an attachment factor or a receptor for SARS-CoV on human leukocytes. An alternative — or additional — function of the ORF7a protein appears to be the induction of apoptosis of the host cell,[199] which was recently shown to depend on the interaction of ORF7a with Bcl-XL.[200]

Interestingly, SARS-CoV protein 7a contains an intracellular targeting signal, which consists of a Lys-Arg-Lys sequence in the cytoplasmic

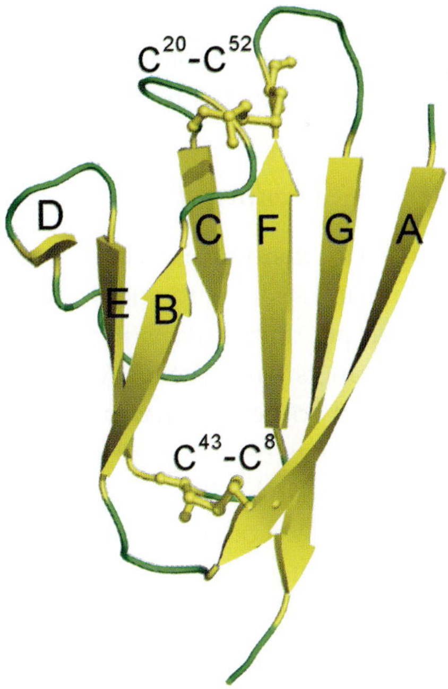

Fig. 15 Crystal structure of the accessory protein 7a of SARS-CoV. The molecule displays a non-canonical immunoglobulin fold[196] (PDB code: 1XAK). One of the deviations is the parallel orientation of strand A with respect to strand G. The disulfide bond between the two sheets is indicated.

tail and may also involve parts of the transmembrane segment. Coronaviruses acquire their envelope by budding into the lumen of an ER-to-Golgi intermediate compartment.[201,202] The proteins budding into this envelope are M (most prominent), E, S, and — in coronaviruses such as MHV that have it — the hemagglutinin esterase (HE). Both S and HE localize to the pre-Golgi by interactions with M.[203] If SARS-CoV follows the same trafficking strategy as other coronaviruses, then it is possible that protein 7a may play a role in viral assembly or budding. The protein has also been shown to interact with the product of ORF3a, which in turn interacts with M, E, and S.[204,205] The "intra-ORFeome interaction" study by von Brunn *et al.*[67] yielded only relatively weak hints of possible interactions between protein 7a and the N-terminal half of Nsp3 as well as Nsp16. These potential interactions were only detected in one direction of the yeast-two-hybrid assay, and were not supported by co-immunoprecipitation.

Accessory Protein 9b — Product of an Alternative ORF

Although the coronavirus genome is relatively large, these viruses make use of a common viral strategy to enrich the variety of their proteomes: the use of multiple start codons *within* a gene that give rise to different protein products. In SARS-CoV, this mechanism is used more frequently than in any other coronaviruses. The phenomenon of alternative reading frames has been observed within ORFs 3, 7, and 8 of SARS-CoV, leading to ORFs 3b, 7b, 8b. In addition, within the N gene of SARS-CoV, ORF9b codes for such a "protein within a protein". Alternative reading frames may be read out if the original AUG of the main gene is overread due to the unfavorable secondary structure of the RNA segment surrounding it and a fraction of the ribosomal scanning complexes continue scanning for start sites ("leaky scanning"). In case of ORF9b, the second start codon is just separated from the first by 7 nucleotides. In addition to "leaky scanning," other possible mechanisms are translational reinitiation after termination, "ribosomal shunting," or internal ribosome entry sites.[206–208] Such alternative ORFs should be of general interest in terms of their evolution, which is certainly restricted due to structural requirements of the protein encoded by the regular ORF that encompasses the alternative ORF. Hallmarks of such limited evolutionary freedom could be less-than-optimal folds and structural disorder, and indeed this is observable for some of the proteins encoded by alternative ORFs in viruses. In any case, it is important to realize that protein 9b is not an artifact without any role in the life cycle of SARS coronavirus. Rather, it is indeed produced during infection of host cells, as antibodies against it have been detected in SARS patients.[209]

The three-dimensional structure of SARS-CoV protein 9b has been determined by X-ray crystallography.[208] The protein forms a 2-fold symmetric dimer consisting of two twisted β-sheets (Fig. 16). Each of these β-sheets is formed by strands contributed from both monomers in the dimer, leading to a highly interlocked arrangement.

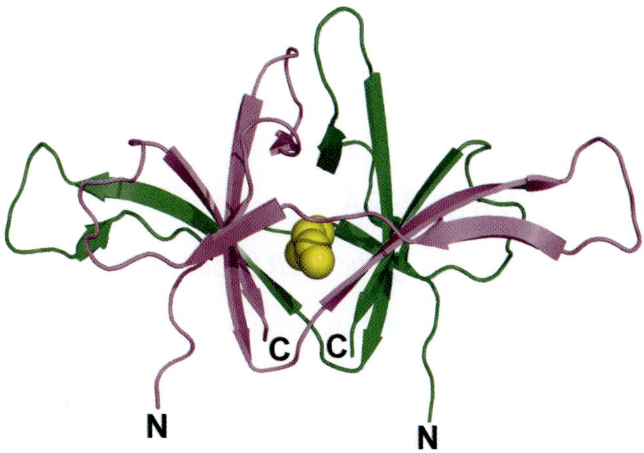

Fig. 16 Crystal structure of the accessory protein 9b of SARS-CoV. The symmetric homodimer embraces a hydrophobic channel, which is believed to bind lipid molecules (yellow spheres)[208] (PDB code: 2CME).

No other protein with a similar fold is known. One side of the dimer features a distinct positive electrostatic potential, while the opposite side is negative. At the center of the dimer, there is a 22 Å long tunnel that is lined with hydrophobic residues. In the crystal structure, this cavity is partly filled with elongated electron density suggesting the presence of a lipid molecule, most probably a fatty acid or fatty acid ester. This component has likely been accommodated from *E. coli* during heterologous expression of ORF9b, but a similar function in virus assembly is conceivable. While the protein is water-soluble when produced in *E. coli*, it has been shown that expression of ORF9b in mammalian cells leads to association with intracellular vesicles.[208] One possible function of dimeric SARS-CoV protein 9b is thus that it may be associated with membranes of the ER-Golgi network through interaction of its positively charged side with the phosphate head groups of the lipids, and also by accommodating one or more lipid molecules into its central tunnel. It could thus function as an accessory protein in virus assembly. In MHV and bovine coronavirus (coronavirus group 2a), the homologue of SARS-CoV protein 9b is found in the virus particle[210] and has been suggested to mediate the interaction between the N protein and the membrane.[211] Thus, in contrast to the product of ORF7a, protein 9b would have to be regarded as a structural protein. This does not exclude interactions with Nsp8 and Nsp14 of the replicase complex, as have been found using yeast-two-hybrid technology and immuno-coprecipitation.[67]

Structures of Viral RNA

3′-Untranslated Region of the SARS-CoV RNA Genome

The structure of a 48-nucleotide fragment of the 3′-untranslated region of SARS coronavirus has been determined by X-ray crystallography.[212] Although the term "structural proteomics" certainly does not apply to these results, "structural genomics" does, in the original meaning of the word. Within the 3′-untranslated region, the first ≈150 nucleotides adjoining the poly(A) stretch contain the stem-loop II motif (s2m) and the sequence 5′-GGAAGAGC-3′ that is present in all coronavirus genomes. The three-dimensional structure comprises two distinct helical regions that are oriented perpendicular to one another (Fig. 17). The smaller of the two helices contains a 5-nucleotide loop that mimics a tetraloop, with one U being "looped out."

This highly structured RNA fold is similar to that of the "530 loop" in 16S ribosomal RNA. The latter loop is known to bind to initiation factors, suggesting that the s2m of the SARS-CoV 3′-UTR might have a similar function. Possibly, it is involved in binding eukaryotic initiation factor 1A (eIF1A) and thereby hijacking the cellular translation machinery for use by the virus.[212]

Further upstream, there are two predicted structures, a bulged stem-loop and a hairpin-type pseudoknot that overlap and are therefore mutually exclusive. It has been proposed that they constitute a "molecular switch" related to different modes of RNA synthesis.[213] It is conceivable that elements such as this control the decision between replication and transcription.[19]

Conclusions

Four and a half years after the SARS outbreak of 2003, three-dimensional structures have been determined for about half of the non-structural proteins of the coronavirus (if we count Nsp3 as five separate

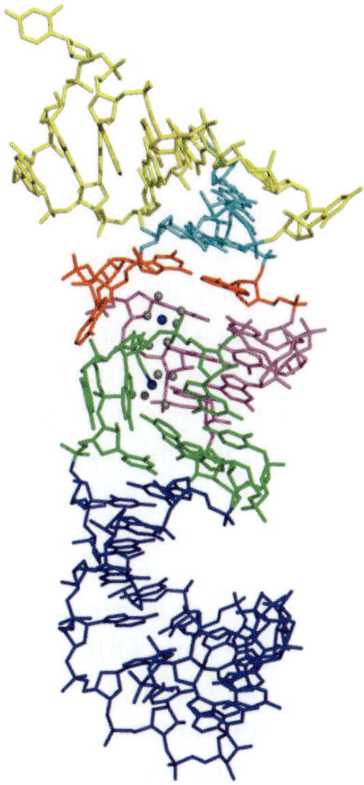

Fig. 17 Crystal structure of the stem-loop II region (s2m) in the 3′-UTR of the SARS-CoV mRNA[212] (PDB code: 1XJR). Note the 90° angle between the shorter helix (yellow) and the longer helical segment. The smaller helical segment ends in a loop that mimics a tetraloop with a looped-out uridine.

domains). In addition, we know structures for fragments of two of the structural proteins, S and N, and have reasonable models for the structure of the E protein. Of the accessory proteins, only the structures of 7a and 9b have been determined so far. Overall, this is a tremendous success that has only been possible due to the technological advances of recent years in macromolecular crystallography and NMR spectroscopy. Much of these advances have been achieved through the various structural genomics projects, as outlined elsewhere in this book.

Of course, the problems remaining are probably greater than those that have already been solved. In a way, it is possible that so far, the lower-hanging fruits have been harvested, although it should not be forgotten that some of these have been hanging quite high already on the "proteome tree"... It is obviously difficult to obtain structural information about the membrane-associated domains of the viral polyproteins (a segment of Nsp3, Nsp4, and Nsp6). With the RNA-dependent RNA polymerase (RdRp, Nsp12) and the helicase (Nsp13), two important potential drug targets have yet to have their structures determined. There have been early conference reports on the crystallization of Nsp13, but this was so far not followed by a structure. One of the difficulties with Nsp13 is the presence of a zinc-binding domain, which tends to cause aggregation of the recombinant sample. In the absence of this domain, the helicase is not active. RH's laboratory has obtained preliminary crystals of the N-terminal domain of the RdRp, but their diffraction properties are poor. Many laboratories are working on the catalytic C-terminal domain, but it is difficult to produce active enzyme in *E. coli*. Therefore, expression trials in the yeast *Pichia pastoris* and in mammalian cells as well as in cell-free systems are being carried out. Other structures that are still missing are those of Nsp14 and Nsp16. In these cases, the problem is not so much the recombinant production in an active form, but crystallization. The very short Nsp11 is probably unstructured and may in fact never be produced in the viral life cycle.

With so many building blocks still missing, and not a single structure of a complex with ssRNA or dsRNA being available, we still have a long way to go to catch a first glimpse of the architecture of the large, membrane-associated replicase complex. Only this structure, or at least significant parts thereof, will pave the way towards a more-than-partial understanding of the mechanisms underlying replication and transcription in coronaviruses. It will be necessary to employ cryo-electron microscopy with single-particle reconstruction in these studies; important steps in this direction have already been made (e.g., Refs. 103 and 214).

Also, much has yet to be learnt in the field of coronavirus-host interaction. The structure of the receptor-binding domain of the spike protein in complex with the SARS-CoV receptor, ACE2, was a highlight in the structural biology of coronaviruses in 2005,[151] but since then, new information was kind of limited. It is still a matter of debate how SARS coronaviruses suppresses the type-I interferon antiviral response of the infected host cell. No less than five individual proteins of the virus have been proposed to be involved in this process: the PL^{pro} domain of Nsp3,[90] the nucleocapsid protein,[19,164] ORF3b protein,[19] ORF6 protein,[19] and (as shown in case of MHV) Nsp1.[64,64a] The SARS-unique domain (Nsp3c) has been speculated to be involved in the regulation of the antiviral response or apoptosis,[78] but very little is known about the signaling steps involved. Also, ORF3b and ORF7a proteins, and the nucleocapsid protein as well, have been shown to enhance apoptosis of SARS-CoV-infected cells.[99,199,162] Clearly, we are only beginning to understand how the virus modulates signal transduction in the host (for a recent review, see Ref. 215).

Several of the early structural genomics projects initiated in the year 2000 and thereafter aimed at determining as many new protein folds as possible. While the coronavirus projects did not have this aim *per se*, they revealed many new folds indeed, e.g. in Nsp1, Nsp3a, the C-terminal domain of Nsp5, Nsp7/8, Nsp10, Nsp15, domains of the nucleocapsid protein, as well as accessory proteins 7a and 9b. This probably reflects the fact that viral proteins are more related to eukaryotic proteins than to prokaryotic ones (most of the eukaryotic homologues of coronaviral non-structural proteins, if at all existent, yet have to be discovered and their 3D structures determined), and also the rather large variability of the coronavirus genome, which is evident from the large distance between members of the individual groups of coronaviruses.

A number of non-structural and accessory proteins of SARS-CoV (e.g., Nsp1, Nsp3a, Nsp3c, Nsp8, proteins 7a and 9b) feature a significant amount of disorder in their structures. In some cases, this could be predicted from the amino-acid sequences, using methods described elsewhere in this book. Flexibility of the chain termini of non-structural proteins may be required to allow access of the proteases to the scissile bonds separating the domains. One-dimensional ^{1}H-NMR spectra can be indicative of the presence of disorder; this has been used, for instance, for Nsp8.[104] Proteins with intrinsic disorder are often difficult to crystallize and may be more amenable for structure determination by NMR spectroscopy; this is certainly one reason for the relatively large number of NMR structures available for SARS-CoV proteins (Nsp1, Nsp3a, Nsp7, protein 7a).

The SARS-CoV structural proteomics project has occasionally been described as "a rapid response to the emergence of a new pathogen." While it is true that great efforts were made to determine as many structures of coronaviral proteins as possible, the vast part of the structural information that is available today was created in 2005 and 2006 (see Fig. 18). In 2007, the numbers of PDB coordinate sets for coronaviral proteins becoming newly accessible in the PDB was lower than in the years before; the likely reason for this is that the structures of the majority of "well-behaved," soluble and properly folded proteins have been determined by now and what remains are the more difficult cases. It can only be hoped that funding agencies will continue to support structural biology of coronaviruses, even though some time has passed since the SARS epidemic. As outlined

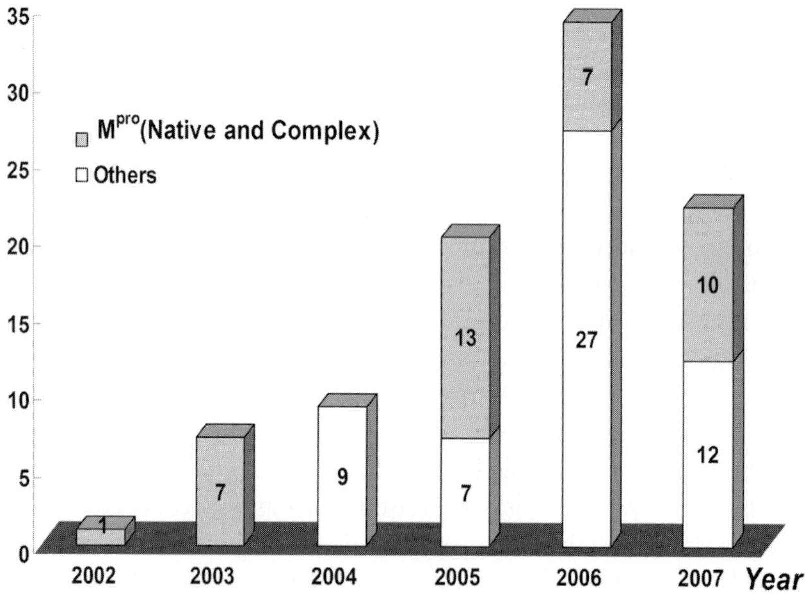

Fig. 18 Number of coordinate sets for coronavirus proteins (and nucleic acids) in the Protein Data Bank (PDB), according to year entered. Structures of the main proteinase and its complexes are indicated in gray.

above, we are just beginning to understand some mosaic pieces of the grand puzzle of the replicase complex and even less of the interaction between virus and host, and it would be disastrous if activities had to be decreased or even stopped because of lack of funding. Funding of research projects has to allow for mid- to long-term development and should not only follow what is just fashionable; otherwise, we will end up with a large amount of very incomplete data on many systems, rather than a deeper insight into a single system or a few related ones.

In any case, Fig. 18 shows that the overall response to the 2003 SARS outbreak *of course took some time*; structural proteomics projects were initiated immediately in some laboratories, but on average, funding did not arrive until late 2004. The rapid determination, within 10 weeks of the discovery of the virus, of the crystal structure of the SARS-CoV main proteinase (M^{pro})[42] relied to a large extent on the availability of the structures of the homologous M^{pro}s of the human coronavirus 229E[31] and transmissible gastroenteritis virus.[32] Even before the SARS-CoV M^{pro} structure was solved, these two related enzymes allowed the construction of homology models for the SARS-CoV protein and, thereby, successful discovery of inhibitors. Thus, the take-home lesson is that structural proteomics, in spite of all the technical advances that came with it, is still not capable of mounting a really rapid response in case of an emerging pathogen, but months and years can be saved through the availability of homologous structures from related viruses. Therefore, "preparedness" is the key word here, rather than "rapid response."[1,216]

Where do we stand in terms of discovery of compounds with anti-SARS activity? The most attractive viral targets are usually the enzymes of the pathogen, not the least because of the relatively easily available *in vitro* assays.[217] With the structures of the RdRp (Nsp12) and the helicase (Nsp13) still unknown, the main proteinase has remained the prime molecular target so far; dozens of inhibitors have been described, but the majority of these is still peptidic and binds with affinities in the lower micromolar range. The best compounds available to date have K_i values in the two- to three-digit nanomolar range. Thus, there is still a lot of room for improvement. One advantage of the M^{pro} target is that it is relatively well conserved across the coronaviruses; therefore, inhibitors can be designed that are active against a wide range of viruses.[31,51] In fact, RH's laboratory has recently designed protease inhibitors that show remarkable antiviral activity against coronaviruses, feline calicivirus (a model for norovirus), and coxsackievirus B3, with no cell toxicity up to and beyond concentrations of 100 μM (Pumpor *et al.*, unpublished). These lead compounds appear to constitute a promising first step towards fulfilling the quest for broad-band antivirals. The three-dimensional structures of the coronavirus M^{pro},[31,32] the norovirus 3C-like protease,[218] and the coxsackievirus B3 3C proteinase (Anand *et al.*, unpublished) were essential for this development. After preclinical and clinical development of these and other antiviral candidate compounds, we will hopefully be in a position to combat the viral outbreaks that lie ahead, both the ones to be expected (yellow fever? pandemic flu?) and the unexpected.

Acknowledgments

A large portion of the work described in this review was supported by the European Commission through the SEPSDA project (Sino-European Project on SARS Diagnostics and Antivirals; www.sepsda.info). Support from the Sino-German Center for the Promotion of Research, Beijing, and the Deutsche Forschungsgemeinschaft (Hi 611/4-1), the Schleswig-Holstein Innovation Fund, and the Fonds der Chemischen Industrie is also acknowledged. We thank the following coworkers, past and present, for their contributions to those of the described research results that were obtained in our Lübeck laboratories: Kanchan Anand, Jeroen R. Mesters, Ralf Moll, Doris Mutschall, Krishna Nagarajan, Yvonne Piotrowski, Rajesh Ponnusamy, Ksenia Pumpor, Christian Schmidt, Silke Schmidtke, and Koen Verschueren. Figures showing molecular structures have been prepared using Pymol (DeLano WL (2002) The PyMOL Molecular Graphics System on World Wide Web; http://www.pymol.org).

References

1. Coutard B, Gorbalenya AE, Snijder EJ, *et al.* (2008) The VIZIER project: Preparedness against pathogenic RNA viruses. *Antiviral Res*, in press (Epub ahead of print: Nov 29, 2007).
2. van der Hoek L, Pyrc K, Jebbink MF, *et al.* (2004) Identification of a new human coronavirus. *Nature Med* **10**: 368–373.
3. van der Hoek L, Pyrc K, Berkhout B. (2006) Human coronavirus NL63, a new respiratory virus. *FEMS Microbiol Reviews* **30**: 760–73.
4. van der Hoek L, Sure K, Ihorst G, *et al.* (2005) Croup is associated with the novel coronavirus NL63. *PLoS Medicine* **2**: 764–70.
5. Fouchier RAM, Hartwig NG, Bestebroer TM, *et al.* (2004) A previously undescribed coronavirus associated with respiratory disease in humans. *Proc Natl Acad Sci USA* **10:** 6212–16.
6. Chiu SS, Chan KH, Hu KW, *et al.* (2005) Human coronavirus NL63 infection and other coronavirus infections in children hospitalized with acute respiratory disease in Hong Kong, China. *Clin Infect Dis* **40**: 1721–29.
7. Dominguez SR, Anderson MS, Glodé MP, *et al.* (2006) Blinded case-control study of the relationship between human coronavirus NL63 and Kawasaki syndrome. *J Infect Dis* **194**: 1697–701.
8. Woo PC, Lau SK, Chu CM, *et al.* (2005) Characterization and complete genome sequence of a novel coronavirus, coronavirus HKU1, from patients with pneumonia. *J Virol* **79**: 884–895.
9. Kupfer B, Simon A, Jonassen CM, *et al.* (2007) Two cases of severe obstructive pneumonia associated with a HKU1-like coronavirus. *Eur J Med Res* **12**: 134–38.
10. Snijder EJ, Bredenbeek PJ, Dobbe JC, *et al.* (2003) Unique and conserved features of genome and proteome of SARS-coronavirus, an early split-off from the coronavirus group 2 lineage. *J Mol Biol* **331**: 991–1004.
11. Gorbalenya AE, Enjuanes L, Ziebuhr J, Snijder EJ. (2006) Nidovirales: evolving the largest RNA virus genome. *Virus Res* **117**: 17–37.
12. Lau SK, Woo PC, Li KS, *et al.* (2005) Severe acute respiratory syndrome coronavirus-like virus in Chinese horseshoe bats. *Proc Natl Acad Sci USA* **102**: 14040–45.
13. Tang XC, Zhang JX, Zhang SY, *et al.* (2006) Prevalence and genetic diversity of coronaviruses in bats from China. *J Virol* **80**: 7481–90.
14. Woo PC, Lau SK, Li KS, *et al.* (2006) Molecular diversity of coronaviruses in bats. *Virology* **351**: 180–87.
15. Thiel V, Ivanov KA, Putics A, *et al.* (2003) Mechanisms and enzymes involved in SARS coronavirus genome expression. *J Gen Virol* **84**: 2305–15.
16. Groneberg DA, Hilgenfeld R, Zabel P. (2005) Molecular mechanisms of severe acute respiratory syndrome (SARS). *Respir Res* **6**: 8–23.

17. Stadler K, Masignani V, Eickmann M, *et al.* (2003) SARS — Beginning to understand a new virus. *Nature Rev Microbiol* **1**: 209–18.
18. Alcami A, Koszinowski UH. (2000) Viral mechanisms of immune evasion. *Trends Microbiol* **8**: 410–18.
19. Kopecky-Bromberg SA, Martínez-Sobrido L, Frieman M, Baric RA, Palese P. (2007) Severe acute respiratory syndrome coronavirus open reading frame (ORF) 3b, ORF 6, and nucleocapsid proteins function as interferon antagonists. *J Virol* **81**: 548–57.
20. Sharma K, Surjit M, Satija N, *et al.* (2007) The 3a accessory protein of SARS coronavirus specifically interacts with the 5′UTR of its genomic RNA, using a unique 75 amino acid interaction domain. *Biochemistry* **46**: 6488–99.
21. Sawicki SG, Sawicki DL, Siddell SG. (2007) A contemporary view of coronavirus transcription. *J Virol* **81**: 20–29.
22. Pasternak AO, Spaan WJ, Snijder EJ. (2006) Nidovirus transcription: how to make sense...? *J Gen Virol* **87**: 1403–21.
23. Sawicki SG, Sawicki DL. (2005) Coronavirus transcription: a perspective. *Curr Top Microbiol Immunol* **287**: 31–55.
24. Drosten C, Günther S, Preiser W, *et al.* (2003) Identification of a novel coronavirus in patients with severe acute respiratory syndrome. *N Engl J Med* **348**: 1967–76.
25. Ksiazek TG, Erdman D, Goldsmith CS, *et al.*, SARS Working Group. (2003) A novel coronavirus associated with severe acute respiratory syndrome. *N Engl J Med* **348**: 1953–66.
26. Peiris JS, Lai ST, Poon LL, *et al.*, SARS study group. (2003) Coronavirus as a possible cause of severe acute respiratory syndrome. *Lancet* **361**: 1319–1325.
27. Kuiken T, Fouchier RA, Schutten M, *et al.* (2003) Newly discovered coronavirus as the primary cause of severe acute respiratory syndrome. *Lancet* **362**: 263–70.
28. Marra MA, Jones SJ, Astell CR, *et al.* (2003) The genome sequence of the SARS-associated coronavirus. *Science* **300**: 1399–404.
29. Rota PA, Oberste MS, Monroe SS, *et al.* (2003) Characterization of a novel coronavirus associated with severe acute respiratory syndrome. *Science* **300**: 1394–99.
30. Ruan YJ, Wei CL, Ee AL, *et al.* (2003) Comparative full-length genome sequence analysis of 14 SARS coronavirus isolates and common mutations associated with putative origins of infection. *Lancet* **361**: 1779–1785. Erratum in: *Lancet* **361**: 1832.
31. Anand K, Ziebuhr J, Wadhwani P, *et al.* (2003) Coronavirus main proteinase ($3CL^{pro}$) structure: basis for design of anti-SARS drugs. *Science* **300,** 1763–67.
32. Anand K, Palm GJ, Mesters JR, *et al.* (2002) Structure of coronavirus main proteinase reveals combination of a chymotrypsin fold with an extra alpha-helical domain. *EMBO J* **21**, 3213–24.

33. Hillisch A, Pineda LF, Hilgenfeld R. (2004) Utility of homology models in the drug discovery process. *Drug Discov Today* **9**: 659–69.
34. Xiong B, Gui CS, Xu XY, *et al.* (2003) A 3D model of SARS-CoV 3CL proteinase and its inhibitors design by virtual screening. *Acta Pharmacol Sin* **24**: 497–504.
35. Chen L, Gui C, Luo X, *et al.* (2005) Cinanserin is an inhibitor of the 3C-like proteinase of severe acute respiratory syndrome coronavirus and strongly reduces virus replication *in vitro*. *J Virol* **79**: 7095–103.
36. Scandella E, Eriksson KK, Hertzig T, *et al.* (2006) Identification and evaluation of coronavirus replicase inhibitors using a replicon cell line. *Adv Exp Med Biol* **581**: 609–13.
37. Verschueren KHG, Pumpor K, Anemüller S, *et al.* (2008) A structural view of the inactivation of the SARS-coronavirus main protease by benzotriazole esters. *Chem Biol*, under revision.
38. Hsu WC, Chang HC, Chou CY, *et al.* (2005) Critical assessment of important regions in the subunit association and catalytic action of the severe acute respiratory syndrome coronavirus main protease. *J Biol Chem* **280**: 22741–48.
39. Chen S, Chen L, Tan J, *et al.* (2005) Severe acute respiratory syndrome coronavirus 3C-like proteinase N terminus is indispensable for proteolytic activity but not for enzyme dimerization. Biochemical and thermodynamic investigation in conjunction with molecular dynamics simulations. *J Biol Chem* **280**: 164–73.
40. Shi J, Wei Z, Song J. (2004) Dissection study on the severe acute respiratory syndrome 3C-like protease reveals the critical role of the extra domain in dimerization of the enzyme: defining the extra domain as a new target for design of highly specific protease inhibitors. *J Biol Chem* **279**: 24765–73.
41. Shi J, Song J. (2006) The catalysis of the SARS 3C-like protease is under extensive regulation by its extra domain. *FEBS J* **273**: 1035–45.
42. Yang H, Yang M, Ding Y, *et al.* (2003) The crystal structures of severe acute respiratory syndrome virus main protease and its complex with an inhibitor. *Proc Natl Acad Sci* **100**, 13190–95.
43. Tan J, Verschueren KHG, Anand K, *et al.* (2005) pH-dependent conformational flexibility of the SARS-CoV main proteinase (M^{pro}) dimer: molecular dynamics simulations and multiple X-ray structure analyses. *J Mol Biol* **354**, 25–40.
44. Hsu MF, Kuo CJ, Chang KT, *et al.* (2005) Mechanism of the maturation process of SARS-CoV 3CL protease. *J Biol Chem* **280**: 31257–66.
45. Lee TW, Cherney MM, Liu J, *et al.* (2007) Crystal structures reveal an induced-fit binding of a substrate-like aza-peptide epoxide to SARS coronavirus main peptidase. *J Mol Biol* **366**: 916–32.
46. Hilgenfeld R, Anand K, Mesters JR, *et al.* (2006) Structure and dynamics of SARS coronavirus main proteinase (M^{pro}). *Adv Exp Med Biol* **581**: 585–91.
47. Xu T, Ooi A, Lee HC, *et al.* (2005) Structure of the SARS coronavirus main proteinase as an active C_2 crystallographic dimer. *Acta Cryst* **F61**: 964–66.

48. Xue X, Yang H, Shen W, *et al.* (2007) Production of authentic SARS-CoV M^{pro} with enhanced activity: application as a novel tag-cleavage endopeptidase for protein overproduction. *J Mol Biol* **366**: 965–75.
49. Chen S, Hu T, Zhang J, *et al.* (2008) Mutation of Gly11 on the dimer interface results in the complete crystallographic dimer dissociation of SARS-CoV 3CL^{pro}: crystal structure with molecular dynamics simulations. *J Biol Chem* **283**: 554–64.
50. Snijder EJ, van der Meer Y, Zevenhoven Dobbe J, *et al.* (2006) Ultrastructure and origin of membrane vesicles associated with the severe acute respiratory syndrome coronavirus replication complex. *J Virol* **80**: 5927–40.
51. Anand K, Yang H, Bartlam M, *et al.* (2005) Coronavirus main proteinase: target for antiviral drug therapy. In: A Schmidt, MH Wolf & O Weber (eds), *Coronaviruses with Special Emphasis on First Insights Concerning SARS*, pp. 173–199. Birkhäuser, Basel.
52. Yang H, Xie W, Xue X, *et al.* (2005) Design of wide-spectrum inhibitors targeting coronavirus main proteases. *PLoS Biol* **3**: e324.
53. Lee TW, Cherney MM, Huitema C, *et al.* (2005) Crystal structures of the main peptidase from the SARS coronavirus inhibited by a substrate-like aza-peptide epoxide. *J Mol Biol* **353**: 1137–51.
54. Gosh AK, Xi K, Ratia K, *et al.* (2005) Design and synthesis of peptidomimetic severe acute respiratory syndrome chymotrypsin-like protease inhibitors. *J Med Chem* **48**: 6767–71.
55. Lu IL, Mahindroo N, Liang PH, *et al.* (2006) Structure-based drug design and structural biology study of novel nonpeptide inhibitors of severe acute respiratory syndrome coronavirus main protease. *J Med Chem* **49**: 5154–61.
56. Goetz DH, Choe Y, Hansell E, *et al.* (2007) Substrate specificity profiling and identification of a new class of inhibitor for the major protease of the SARS coronavirus. *Biochemistry* **46**: 8744–52.
57. Yin J, Niu C, Cherney MM, *et al.* (2007) A mechanistic view of enzyme inhibition and peptide hydrolysis in the active site of the SARS-CoV 3C-like peptidase. *J Mol Biol* **371**: 1060–74.
58. Lee CC, Kuo CJ, Hsu MF, *et al.* (2007) Structural basis of mercury- and zinc-conjugated complexes as SARS-CoV 3C-like protease inhibitors. *FEBS Lett* **581**: 5454–58.
59. Schmidt MF, Isidro-Llobet A, Lisurek M, *et al.* (2008) Sensitized detection of inhibitory fragments and iterative development of non-peptidic SARS-CoV protease inhibitors by dynamic ligation screening. *Angew Chem*, in press.
60. Al-Gharabli SI, Shah ST, Weik S, *et al.* (2006) An efficient method for the synthesis of peptide aldehyde libraries employed in the discovery of reversible SARS coronavirus main protease (SARS-CoV M^{pro}) inhibitors. *ChemBioChem* 7: 1048–55.

61. Brockway SM, Lu XT, Peters TR, *et al.* (2004) Intracellular localization and protein interactions of the gene 1 protein p28 during mouse hepatitis virus replication. *J Virol* **78**: 11551–62.
62. Denison MR, Yount B, Brockway SM, *et al.* (2004) Cleavage between replicase proteins p28 and p65 of mouse hepatitis virus is not required for virus replication. *J Virol* **78**: 5957–65.
63. Brockway SM, Denison MR. (2005) Mutagenesis of the murine hepatitis virus Nsp1-coding region identifies residues important for protein processing, viral RNA synthesis, and viral replication. *Virology* **340**: 209–23.
64. Züst R, Cervantes-Barragán L, Kuri T, *et al.* (2007) Coronavirus non-structural protein 1 is a major pathogenicity factor: implications for the rational design of coronavirus vaccines. *PLoS Pathog* **3**: e109.
64a. Wathelet MG, Orr M, Frieman MB, Baric RS. (2007) Severe acute respiratory syndrome coronavirus evades antiviral signaling: role of Nsp1 and rational design of an attenuated strain. *J Virol* **81**: 11620–33.
65. Chen CJ, Sugiyama K, Kubo H, *et al.* (2004) Murine coronavirus nonstructural protein p28 arrests cell cycle in G_0/G_1 phase. *J Virol* **78**: 10410–19.
66. Kamitani W, Narayanan K, Huang C, *et al.* (2006) Severe acute respiratory syndrome coronavirus Nsp1 protein suppresses host gene expression by promoting host mRNA degradation. *Proc Natl Acad Sci USA* **103**: 12885–90.
67. von Brunn A, Teepe C, Simpson JC, *et al.* (2007) Analysis of intraviral protein-protein interactions of the SARS coronavirus ORFeome. *PLoS ONE* **2**: e459.
68. Almeida MS, Johnson MA, Herrmann T, *et al.* (2007) Novel beta-barrel fold in the nuclear magnetic resonance structure of the replicase nonstructural protein 1 from the severe acute respiratory syndrome coronavirus. *J Virol* **81**: 3151–61.
69. Graham RL, Sims AC, Brockway SM, *et al.* (2005) The Nsp2 replicase proteins of murine hepatitis virus and severe acute respiratory syndrome coronavirus are dispensable for viral replication. *J Virol* **79**: 13399–411.
70. Imbert I, Snijder EJ, Dimitrova M, *et al.* (2008) The SARS-coronavirus Nsp3 protein as a replication/transcription scaffolding protein. *Virus Res*, in press.
71. Serrano P, Johnson MA, Almeida MS, *et al.* (2007) Nuclear magnetic resonance structure of the N-terminal domain of nonstructural protein 3 from the severe acute respiratory syndrome coronavirus. *J Virol* **81**: 12049–60.
72. Webster G, Genschel J, Curth U, *et al.* (1997) A common core for binding single-stranded DNA: structural comparison of the single-stranded DNA-binding proteins (SSB) from *E. coli* and human mitochondria. *FEBS Lett* **411**: 313–16.
73. Putics Á, Filipowicz W, Hall J, *et al.* (2005) ADP-ribose-1″-monophosphatases: a conserved coronavirus enzyme that is dispensable for viral replication in tissue culture. *J Virol* **79**: 12721–31.

74. Saikatendu KS, Joseph JS, Subramanian V, *et al.* (2005) Structural basis of severe acute respiratory syndrome coronavirus ADP-ribose-1″-phosphate dephosphorylation by a conserved domain of Nsp3. *Structure* **13**: 1665–75.
75. Egloff MP, Malet H, Putics A, *et al.* (2006) Structural and functional basis for ADP-ribose and poly(ADP-ribose) binding by viral macro domains. *J Virol* **80**: 8493–502.
76. Putics A, Slaby J, Filipowicz W, *et al.* (2006) ADP-ribose-1″-phosphatase activities of the human coronavirus 229E and SARS coronavirus X domains. *Adv Exp Med Biol* **581**: 93–96.
77. Gao G, Guo X, Goff SP (2002) Inhibition of retroviral RNA production by ZAP, a CCCH-type zinc finger protein. *Science* **297**: 1703–06.
78. Tan J, Kusov Y, Mutschall D, *et al.* (2007) The "SARS-unique" domain (SUD) of SARS coronavirus is an oligo(G)-binding protein. *Biochem Biophys Res Commun* **364**: 887–82.
79. Sulea T, Lindner HA, Purisima EO, Ménard R. (2005) Deubiquitination, a new function of the severe acute respiratory syndrome coronavirus papain-like protease? *J Virol* 79: 4550–51.
80. Barretto N, Jukneliene D, Ratia K, *et al.* (2005) The papain-like protease of severe acute respiratory syndrome coronavirus has deubiquitinating activity. *J Virol* **79**: 15189–98.
81. Lindner HA, Fotouhi-Ardakani N, Lytvyn V, *et al.* (2005) The papain-like protease from the severe acute respiratory syndrome coronavirus is a deubiquitinating enzyme. *J Virol* **79**: 15199–208.
82. Barretto N, Jukneliene D, Ratia K, *et al.* (2006) Deubiquitinating activity of the SARS-CoV papain-like protease. *Adv Exp Med Biol* **581**: 37–41.
83. Barrett AJ, Rawlings ND. (2001) Evolutionary lines of cysteine peptidases. *Biol Chem* **382**: 727–33.
84. Lindner HA, Lytvyn V, Qi H, *et al.* (2007) Selectivity in ISG15 and ubiquitin recognition by the SARS coronavirus papain-like protease. *Arch Biochem Biophys* **466**: 8–14.
85. Kim KI, Zhang DE. (2003) ISG15, not just another ubiquitin-like protein. *Biochem Biophys Res Commun* **307**: 431–34.
86. Ritchie KJ, Zhang DE. (2004) ISG15: the immunological kin of ubiquitin. *Semin. Cell Dev Biol* **15**: 237–46.
87. Yuan W, Aramini JM, Montelione GT, Krug RM. (2002) Structural basis for ubiquitin-like ISG 15 protein binding to the NS1 protein of influenza B virus: A protein-protein interaction function that is not shared by the corresponding N-terminal domain of the NS1 protein of influenza A virus. *Virology* **304**: 291–301.
88. Ratia K, Saikatendu KS, Santarsiero BD, *et al.* (2006) Severe acute respiratory syndrome coronavirus papain-like protease: structure of a viral deubiquitinating enzyme. *Proc Natl Acad Sci USA* **103**: 5717–22.

89. Ziebuhr J, Thiel V, Gorbalenya AE. (2001) The autocatalytic release of a putative RNA virus transcription factor from its polyprotein precursor involves two paralogous papain-like proteases that cleave the same peptide bond. *J Biol Chem* **276**: 33220–32.
90. Devaraj SG, Wang N, Chen Z, *et al.* (2007) Regulation of IRF-3-dependent innate immunity by the papain-like protease domain of the severe acute respiratory syndrome coronavirus. *J Biol Chem* **282**: 32208–21.
91. Kiel C, Serrano L. (2006) The ubiquitin domain superfold: structure-based sequence alignments and characterization of binding epitopes. *J Mol Biol* **355**: 821–44.
92. Nassar N, Horn G, Herrmann C, *et al.* (1995) The 2.2 Å crystal structure of the Ras-binding domain of the serine/threonine kinase c-Raf1 in complex with Rap1A and a GTP analogue. *Nature* **375**: 554–60.
93. Hilgenfeld R. (1995) Regulatory GTPases. *Curr Opin Struct Biol* **5**: 810–17.
94. Pacold ME, Suire S, Perisic O, *et al.* (2000) Crystal structure and functional analysis of Ras binding to its effector phosphoinositide 3-kinase γ. *Cell* **103**: 931–43.
95. Huang L, Hofer F, Martin GS, Kim SH. (1998) Structural basis for the interaction of Ras with RalGDS. *Nature Struct Biol* **5**: 422–426.
96. Quilliam LA, Castro AF, Rogers-Graham KS, *et al.* (1999) M-Ras/R-Ras3, a transforming ras protein regulated by Sos1, GRF1, and p120 Ras GTPase-activating protein, interacts with the putative Ras effector AF6. *J Biol Chem* **274**: 23850–57.
97. Chen CJ, Makino S. (2004) Murine coronavirus replication induces cell cycle arrest in G_0/G_1 phase. *J Virol* **78**: 5658–69.
98. Kopecky-Bromberg SA, Martinez-Sobrido L, Palese P. (2006) 7a protein of severe acute respiratory syndrome coronavirus inhibits cellular protein synthesis and activates p38 mitogen-activated protein kinase. *J Virol* **80**: 785–93.
99. Yuan X, Wu J, Shan Y, *et al.* (2006) SARS coronavirus 7a protein blocks cell cycle progression at G_0/G_1 phase via the cyclin D3/pRb pathway. *Virology* **346**: 74–85.
100. Cheng A, Zhang W, Xie Y, *et al.* (2005) Expression, purification, and characterization of SARS coronavirus RNA polymerase. *Virology* **335**: 165–76.
101. Koonin EV. (1991) The phylogeny of RNA-dependent RNA polymerases of positive-strand RNA viruses. *J Gen Virol* **72**: 2197–206.
102. Imbert I, Guillemot JC, Bourhis JM, *et al.* (2006) A second, non-canonical RNA-dependent RNA polymerase in SARS coronavirus. *EMBO J* 25: 4933–42.
103. Zhai Y, Sun F, Li X, *et al.* (2005) Insights into SARS-CoV transcription and replication from the structure of the Nsp7-Nsp8 hexadecamer. *Nature Struct Mol Biol* **12**: 980–86.
104. Ponnusamy R, Mesters JR, Ziebuhr J, *et al.* (2006) Nonstructural proteins 8 and 9 of human coronavirus 229E. *Adv Exp Med Biol* **581**: 49–54.

105. Peti W, Johnson MA, Herrmann T, *et al.* (2005) Structural genomics of the severe acute respiratory syndrome coronavirus: nuclear magnetic resonance structure of the protein Nsp7. *J Virol* **79**: 12905–13.
106. Sutton G, Fry E, Carter L, *et al.* (2004) The Nsp9 replicase protein of SARS-coronavirus, structure and functional insights. *Structure* **12**: 341–53.
107. Bost AG, Carnahan RH, Lu XT, Denison MR. (2000) Four proteins processed from the replicase gene polyprotein of mouse hepatitis virus colocalize in the cell periphery and adjacent to sites of virion assembly. *J Virol* **74**: 3379–3387.
108. Egloff MP, Ferron F, Campanacci V, *et al.* (2004) The severe acute respiratory syndrome-coronavirus replicative protein Nsp9 is a single-stranded RNA-binding subunit unique in the RNA virus world. *Proc Natl Acad Sci USA* **101**: 3792–96.
109. Piotrowski Y, van der Hoek L, Pyrc K, *et al.* (2006) Nonstructural proteins of human coronavirus NL63. *Adv Exp Med Biol* **581**: 97–100.
110. Piotrowski Y, Ponnusamy R, Glaser S, *et al.* (2008) Production of coronavirus non-structural proteins in soluble form for crystallization. In: D Cavenagh (ed.), *Methods in Molecular Biology: SARS- and other coronaviruses.* Humana Press, in press.
111. Theobald DL, Mitton-Fry RM, Wuttke DS. (2003) Nucleic acid recognition by OB-fold proteins. *Annu Rev Biophys Biomol Struct* **32**: 115–33.
112. Bochkarev A, Pfuetzner RA, Edwards AM, Frappier L. (1997) Structure of the single-stranded-DNA-binding domain of replication protein A bound to DNA. *Nature* **385**: 176–81.
113. Brockway SM, Clay CT, Lu XT, Denison MR. (2003) Characterization of the expression, intracellular localization, and replication complex association of the putative mouse hepatitis virus RNA-dependent RNA polymerase. *J Virol* **77**: 10515–27.
114. Bessette PH, Aslund F, Beckwith J, Georgiou G. (1999) Efficient folding of proteins with multiple disulfide bonds in the *Escherichia coli* cytoplasm. *Proc Natl Acad Sci USA* **96**: 13703–08.
115. Parks D, Bolinger R, Mann K. (1997) Redox state regulates binding of p53 to sequence-specific DNA, but not to non-specific or mismatched DNA. *Nucl Acids Res* **25**: 1289–95.
116. You JS, Wang M, Lee SH. (2000) Functional characterization of zinc-finger motif in redox regulation of RPA-ssDNA interaction. *Biochemistry* **39**: 12953–58.
117. Mikhailov VS, Okano K, Rohrmann GF. (2005) The redox state of the baculovirus single-stranded DNA-binding protein LEF-3 regulates its DNA binding, unwinding, and annealing activities. *J Biol Chem* **280**: 29444–53.
118. McBride AA, Klausner RD, Howley PM. (1992) Conserved cysteine residue in the DNA-binding domain of the bovine papillomavirus type 1 E2 protein confers redox regulation of the DNA-binding activity *in vitro. Proc Natl Acad Sci USA* **89**: 7531–35.

119. Sampson DA, Arana ME, Boehmer PE. (2000) Cysteine 111 affects coupling of single-stranded DNA binding to ATP hydrolysis in the herpes simplex virus type-1 origin-binding protein. *J Biol Chem* **275**: 2931–37.
120. Knipe DM, Quinlan MP, Spang AE. (1982) Characterization of two conformational forms of the major DNA-binding protein encoded by herpex simplex virus 1. *J Virol* **44**: 736–41.
121. Dudas KC, Ruyechan WT. (1998) Identification of a region of the herpes simplex virus single-stranded DNA-binding protein involved in cooperative binding. *J Virol* **72**: 257–65.
122. Raghunathan S, Kozlov AG, Lohman TM, Waksman G. (2000) Structure of the DNA binding domain of *E. coli* SSB bound to ssDNA. *Nature Struct Biol* **7**: 648–52.
123. Chen CY, Chang CK, Chang YW, *et al.* (2007) Structure of the SARS coronavirus nucleocapsid protein RNA-binding dimerization domain suggests a mechanism for helical packaging of viral RNA. *J Mol Biol* **368**: 1075–86.
124. Sawicki SG, Sawicki DL, Younker D, *et al.* (2005) Functional and genetic analysis of coronavirus replicase-transcriptase proteins. *PLoS Pathog* **1**: e39.
125. Donaldson EF, Graham RL, Sims AC, *et al.* (2007) Analysis of murine hepatitis virus strain A59 temperature-sensitive mutant TS-LA6 suggests that Nsp10 plays a critical role in polyprotein processing. *J Virol* **81**: 7086–98.
126. Matthes N, Mesters JR, Coutard B, *et al.* (2006) The non-structural protein Nsp10 of Mouse Hepatitis Virus binds zinc ions and nucleic acids. *FEBS Lett* **580**: 4143–49.
127. Joseph JS, Saikatendu KS, Subramanian V, *et al.* (2006) Crystal structure of nonstructural protein 10 from the severe acute respiratory syndrome coronavirus reveals a novel fold with two zinc-binding motifs. *J Virol* **80**: 7894–901.
128. Su D, Lou Z, Sun F, Zhai Y, *et al.* (2006) Dodecamer structure of severe acute respiratory syndrome coronavirus nonstructural protein Nsp10. *J Virol* **80**: 7902–08.
129. Xu X, Liu Y, Weiss S, *et al.* (2003) Molecular model of SARS coronavirus polymerase: implications for biochemical functions and drug design. *Nucl Acids Res* **31**: 7117–30.
130. Azzi A, Lin SX. (2004) Human SARS-coronavirus RNA-dependent RNA polymerase: activity determinants and nucleoside analogue inhibitors. *Proteins* **57**: 12–14.
131. Ivanov KA, Ziebuhr J. (2004) Human coronavirus 229E nonstructural protein 13: characterization of duplex-unwinding, nucleoside triphosphatase, and RNA 5′-triphosphatase activities. *J Virol* **78**: 7833–7838.
132. Ivanov KA, Thiel V, Dobbe JC, *et al.* (2004) Multiple enzymatic activities associated with severe acute respiratory syndrome coronavirus helicase. *J Virol* **78**: 5619–32.

133. Bernini A, Spiga O, Venditti V, *et al.* (2006) Tertiary structure prediction of SARS coronavirus helicase. *Biochem Biophys Res Commun* **343**: 1101–04.
134. Minskaia E, Hertzig T, Gorbalenya AE, *et al.* (2006) Discovery of an RNA virus 3′->5′ exoribonuclease that is critically involved in coronavirus RNA synthesis. *Proc Natl Acad Sci USA* **103**: 5108–13.
135. Eckerle LD, Lu X, Sperry SM, *et al.* (2007) High fidelity of murine hepatitis virus replication is decreased in Nsp14 exoribonuclease mutants. *J Virol* **81**: 12135–44.
136. Freemont PS, Friedman JM, Beese LS, *et al.* (1988) Cocrystal structure of an editing complex of Klenow fragment with DNA. *Proc Natl Acad Sci USA* **85**: 8924–28.
137. Chen P, Jiang M, Hu T, *et al.* (2007) Biochemical characterization of exoribonuclease encoded by SARS coronavirus. *J Biochem Mol Biol* **40**: 649–55.
138. Sperry SM, Kazi L, Graham RL, *et al.* (2005) Single-amino-acid substitutions in open reading frame (ORF) 1b-Nsp14 and ORF 2a proteins of the coronavirus mouse hepatitis virus are attenuating in mice. *J Virol* **79**: 3391–400.
139. Bhardwaj K, Guarino L, Kao CC. (2004) The severe acute respiratory syndrome coronavirus Nsp15 protein is an endoribonuclease that prefers manganese as a cofactor. *J Virol* **78**: 12218–24.
140. Ivanov KA, Hertzig T, Rozanov M, *et al.* (2004) Major genetic marker of nidoviruses encodes a replicative endoribonuclease. *Proc Natl Acad Sci USA* **101**: 12694–99.
141. Bhardwaj K, Sun J, Holzenburg A, *et al.* (2006) RNA recognition and cleavage by the SARS coronavirus endoribonuclease. *J Mol Biol* **361**: 243–56.
142. Ricagno S, Egloff MP, Ulferts R, *et al.* (2006) Crystal structure and mechanistic determinants of SARS coronavirus nonstructural protein 15 define an endoribonuclease family. *Proc Natl Acad Sci USA* **103**: 11892–97.
143. Joseph JS, Saikatendu KS, Subramanian V, *et al.* (2007) Crystal structure of a monomeric form of severe acute respiratory syndrome coronavirus endonuclease Nsp15 suggests a role for hexamerization as an allosteric switch. *J Virol* **81**: 6700–08.
144. Xu X, Zhai Y, Sun F, *et al.* (2006) New antiviral target revealed by the hexameric structure of mouse hepatitis virus nonstructural protein Nsp15. *J Virol* **80**: 7909–17.
145. Kang H, Bhardwaj K, Li Y, *et al.* (2007) Biochemical and genetic analyses of murine hepatitis virus Nsp15 endoribonuclease. *J Virol* **81**: 13587–97.
146. Guarino LA, Bhardwaj K, Dong W, *et al.* (2005) Mutational analysis of the SARS virus Nsp15 endoribonuclease: identification of residues affecting hexamer formation. *J Mol Biol* **353**: 1106–17.
147. Schwarz B, Routledge E, Siddell SG. (1990) Murine coronavirus nonstructural protein ns2 is not essential for virus replication in transformed cells. *J Virol* **64**: 4784–91.

148. von Grotthuss M, Wyrwicz LS, Rychlewski L. (2003) mRNA cap-1 methyltransferase in the SARS genome. *Cell* **113**: 701–02.
149. Ginalski K, Godzik A, Rychlewski L. (2006) Novel SARS unique AdoMet-dependent methyltransferase. *Cell Cycle* **5**: 2414–16.
150. Li W, Moore MJ, Vasilieva N, *et al.* (2003) Angiotensin-converting enzyme 2 is a functional receptor for the SARS coronavirus. *Nature* **426**: 450–54.
151. Li F, Li W, Farzan M, Harrison SC. (2005) Structure of SARS coronavirus spike receptor-binding domain complexed with receptor. *Science* **309**: 1844–46.
152. Probakaran P, Gan J, Feng Y, *et al.* (2006) Structure of severe acute respiratory syndrome coronavirus receptor-binding domain complexed with neutralizing antibody. *J Biol Chem* **281**: 15829–36.
153. Supekar VM, Bruckmann C, Ingallinella P, *et al.* (2004) Structure of a proteolytically resistant core from the severe acute respiratory syndrome coronavirus S2 fusion protein. *Proc Natl Acad Sci USA* **101**: 17958–63.
154. Xu Y, Lou Z, Liu Y, Pang H, *et al.* (2004) Crystal structure of severe acute respiratory syndrome coronavirus spike protein fusion core. *J Biol Chem* **279**: 49414–19.
155. Deng Y, Liu J, Zheng Q, *et al.* (2006) Structures and polymorphic interactions of two hepta-repeat regions of the SARS virus S2 protein. *Structure* **14**: 889–99.
156. Hakansson-McReynolds S, Jiang S, Rong L, Caffrey M. (2006) Solution structure of the severe acute respiratory syndrome-coronavirus heptad repeat 2 domain in the prefusion state. *J Biol Chem* **281**: 11965–71.
157. Beniac DR, Andonov A, Grudeski E, Booth TF. (2007) Architecture of the SARS coronavirus prefusion spike. *Nature Struct Mol Biol* **13**: 751–52.
158. Beniac DR, Devarennes SL, Andonov A, *et al.* (2007) Conformational reorganization of the SARS coronavirus spike following receptor binding: implications for membrane fusion. *PLoS ONE* **2**: e1082.
159. Almazán F, Gálan C, Enjuanes L. (2004) The nucleoprotein is required for efficient coronavirus genome replication. *J Virol* **78**: 12683–88.
160. van der Meer Y, Snijder EJ, Dobbe JC, *et al.* (1999) Localisation of mouse hepatitis virus non-structural proteins and RNA synthesis indicates a role for late endosomes in viral replication. *J Virol* **73**: 7641–57.
161. He R, Leeson A, Andonov A, *et al.* (2003) Activation of AP-1 signal transduction pathway by SARS coronavirus nucleocapsid protein. *Biochem Biophys Res Commun* **311**: 870–76.
162. Surjit M, Liu B, Jameel S, *et al.* (2004) The SARS coronavirus nucleocapsid protein induces actin reorganization and apoptosis in COS-1 cells in the absence of growth factors. *Biochem J* **383**: 13–18.
163. Surjit M, Liu B, Chow VT, Lal SK. (2006) The nucleocapsid protein of severe acute respiratory syndrome-coronavirus inhibits the activity of

cyclin-cyclin-dependent kinase complex and blocks S phase progression in mammalian cells. *J Biol Chem* **281**: 10669–81.

164. Ye Y, Hauns K, Langland JO, *et al.* (2007) Mouse hepatitis coronavirus A59 nucleocapsid protein is a type I interferon antagonist. *J Virol* **81**: 2554–63.
165. Luo C, Luo H, Zheng S, *et al.* (2004) Nucleocapsid protein of SARS coronavirus tightly binds to human cyclophilin A. *Biochem Biophys Res Commun* **321**: 557–65.
166. Luo H, Chen Q, Chen J, *et al.* (2005) The nucleocapsid protein of SARS coronavirus has a high binding affinity to the human cellular heterogeneous nuclear ribonucleoprotein A1. *FEBS Lett* **579**: 2623–28.
167. Luo H, Ye F, Chen K, *et al.* (2005) SR-rich motif plays a pivotal role in recombinant SARS coronavirus nucleocapsid protein multimerization. *Biochemistry* **44**: 15351–58.
168. Luo H, Chen J, Chen K, *et al.* (2006) Carboxyl terminus of severe acute respiratory syndrome coronavirus nucleocapsid protein: self-association analysis and nucleic acid binding characterization. *Biochemistry* **45**: 11827–35.
169. Luo H, Wu D, Shen C, *et al.* (2006) Severe acute respiratory syndrome coronavirus membrane protein interacts with nucleocapsid protein mostly through their carboxyl termini by electrostatic attraction. *Int J Biochem Cell Biol* **38**: 589–99.
170. Huang Q, Yu L, Petros AM, *et al.* (2004) Structure of the N-terminal RNA-binding domain of the SARS CoV nucleocapsid protein. *Biochemistry* **43**: 6059–63.
171. Saikatendu KS, Joseph JS, Subramanian V, *et al.* (2007) Ribonucleocapsid formation of severe acute respiratory syndrome coronavirus through molecular action of the N-terminal domain of N protein. *J Virol* **81**: 3913–3921.
172. Fan H, Ooi A, Tan YW, *et al.* (2005) The nucleocapsid protein of coronavirus infectious bronchitis virus: crystal structure of its N-terminal domain and multimerization properties. *Structure* **13**: 1859–68.
173. Jayaram H, Fan H, Bowman BR, *et al.* (2006) X-ray structures of the N- and C-terminal domains of a coronavirus nucleocapsid protein: implications for nucleocapsid formation. *J Virol* **80**: 6612–20.
174. Nagai K, Oubridge C, Ito N, *et al.* (1995) The RNP domain: a sequence-specific RNA-binding domain involved in processing and transport of RNA. *Trends Biochem Sci* **20**: 235–40.
175. Yu IM, Oldham ML, Zhang J, Chen J. (2006) Crystal structure of the severe acute respiratory syndrome (SARS) coronavirus nucleocapsid protein dimerization domain reveals evolutionary linkage between corona- and arteriviridae. *J Biol Chem* **281**: 17134–39.
176. Chang CK, Sue SC, Yu TH, *et al.* (2005) The dimer interface of the SARS coronavirus nucleocapsid protein adapts a porcine respiratory and reproductive syndrome virus-like structure. *FEBS Lett* **579**: 5663–68.

177. Vennema H, Godeke GJ, Rossen JW, *et al.* (1996) Nucleocapsid-independent assembly of coronavirus-like particles by co-expression of viral envelope protein genes. *EMBO J* **15**: 2020–28.
178. Baudoux P, Carrat C, Besnardeau L, *et al.* (1998) Coronavirus pseudoparticles formed with recombinant M and E proteins induce alpha interferon synthesis by leukocytes. *J Virol* **72**: 8636–43.
179. Lim KP, Liu DX. (2001) The missing link in coronavirus assembly. Retention of the avian coronavirus infectious bronchitis virus envelope protein in the pre-Golgi compartments and physical interaction between the envelope and membrane proteins. *J Biol Chem* **276**: 17515–23.
180. Maeda J, Maeda A, Makino S. (1999) Release of coronavirus E protein in membrane vesicles from virus-infected cells and E protein-expressing cells. *Virology* **263**: 265–72.
181. Ho Y, Lin PH, Liu CY, *et al.* (2004) Assembly of human severe acute respiratory syndrome coronavirus-like particles. *Biochem Biophys Res Commun* **318**: 833–38.
182. Arbely E, Khattari Z, Brotons G, *et al.* (2004) A highly unusual palindromic transmembrane helical hairpin formed by SARS coronavirus E protein. *J Mol Biol* **341**: 769–79.
183. Khattari Z, Brotons G, Akkawi M, *et al.* (2006) SARS coronavirus E protein in phospholipid bilayers: an X-ray study. *Biophys J* **90**: 2038–50.
184. Shen X, Xue JH, Yu CY, *et al.* (2003) Small envelope protein E of SARS: cloning, expression, purification, CD determination, and bioinformatics analysis. *Acta Pharmacol Sin* **24**: 505–11.
185. Ye Y, Hogue BG. (2007) Role of the coronavirus E viroporin protein transmembrane domain in virus assembly. *J Virol* **81**: 3597–607.
186. Liao Y, Lescar J, Tam JP, Liu DX. (2004) Expression of SARS-coronavirus envelope protein in *Escherichia coli* cells alters membrane permeability. *Biochem Biophys Res Commun* **325**: 374–80.
187. Torres J, Wang J, Parthasarathy K, Liu DX. (2005). The transmembrane oligomers of coronavirus protein E. *Biophys J* **88**: 1283–90.
188. Liao Y, Yuan Q, Torres J, *et al.* (2006) Biochemical and functional characterization of the membrane association and membrane permeabilizing activity of the severe acute respiratory syndrome coronavirus envelope protein. *Virology* **349**: 264–75.
189. Torres J, Parthasarathy K, Lin X, *et al.* (2006). Model of a putative pore: the pentameric alpha-helical bundle of SARS coronavirus E protein in lipid bilayers. *Biophys J* **91**: 938–47.
190. Torres J, Maheswari U, Parthasarathy K, *et al.* (2007) Conductance and amantadine binding of a pore formed by a lysine-flanked transmembrane domain of SARS coronavirus envelope protein. *Protein Sci* **16**: 2065–71.
191. Wilson L, McKinlay C, Gage P, Ewart G. (2004) SARS coronavirus E protein forms cation-selective ion channels. *Virology* **330**: 322–31.

192. Krijnse-Locker J, Rose JK, Horzinek MC, Rottier PJ. (1992) Membrane assembly of the triple-spanning coronavirus M protein. Individual transmembrane domains show preferred orientation. *J Biol Chem* **267**: 21911–18.
193. He R, Leeson A, Ballantine M, *et al.* (2004) Characterization of protein-protein interactions between the nucleocapsid protein and membrane protein of the SARS coronavirus. *Virus Res* **105**: 121–25.
194. Fang X, Ye L, Timani KA, *et al.* (2005) Peptide domain involved in the interaction between membrane protein and nucleocapsid protein of SARS-associated coronavirus. *J Biochem Mol Biol* **38**: 381–85.
195. Fang X, Gao J, Zheng H, *et al.* (2007) The membrane protein of SARS-CoV suppresses NF-κB activation. *J Med Virol* **79**: 1431–39.
196. Nelson CA, Pekosz A, Lee CA, *et al.* (2005) Structure and intracellular targeting of the SARS-coronavirus OrF7a accessory protein. *Structure* **13**: 75–85.
197. Hänel K, Stangler T, Stoldt M, Willbold D. (2006) Solution structure of the X4 protein coded by the SARS-related coronavirus reveals an immunoglobulin-like fold and suggests a binding activity to integrin I domains. *J Biomed Sci* **13**: 281–93.
198. Hänel K, Willbold D. (2007) SARS-CoV accessory protein 7a directly interacts with human LFA-1. *Biol Chem* **388**: 1325–32.
199. Tan YJ, Fielding BC, Goh PY, *et al.* (2004) Overexpression of 7a, a protein specifically encoded by the severe acute respiratory syndrome coronavirus, induces apoptosis via a caspase-dependent pathway. *J Virol* **78**: 14043–47.
200. Tan YX, Tan TH, Lee MJ, *et al.* (2007) Induction of apoptosis by the severe acute respiratory syndrome coronavirus 7a protein is dependent on its interaction with the Bcl-XL protein. *J Virol* **81**: 6346–55.
201. Krijnse-Locker J, Ericsson M, Rottier PJ, Griffiths G. (1994) Characterization of the budding compartment of mouse hepatitis virus: evidence that transport from the RER to the Golgi complex requires only one vesicular transport step. *J Cell Biol* **124**: 55–70.
202. Klumperman J, Krijnse-Locker J, Meijer A, *et al.* (1994) Coronavirus M proteins accumulate in the Golgi complex beyond the site of virion budding. *J Virol* **68**: 6523–34.
203. Nguyen VP, Hogue BG. (1997) Protein interactions during coronavirus assembly. *J Virol* **71**: 9278–9284.
204. Tan YJ, Teng E, Shen S, *et al.* (2004) A novel severe acute respiratory syndrome coronavirus protein, U274, is transported to the cell surface and undergoes endocytosis. *J Virol* **78**: 6723–34.
205. Fielding BC, Tan YJ, Shuo S, *et al.* (2004) Characterization of a unique group-specific protein (U122) of the severe acute respiratory syndrome coronavirus. *J Virol* **78**: 7311–18.
206. Ryabova LA, Pooggin MM, Hohn T. (2002) Viral strategies of translation initiation: ribosomal shunt and reinitiation. *Prog Nucleic Acid Res Mol Biol* **72**: 1–39.

207. O'Connor JB, Brian DA. (2000) Downstream ribosomal entry for translation of coronavirus TGEV gene 3b. *Virology* **269**: 172–82.
208. Meier C, Aricescu AR, Assenberg R, *et al.* (2006) The crystal structure of ORF-9b, a lipid binding protein from the SARS coronavirus. *Structure* **14**: 1157–65.
209. Qiu M, Shi Y, Guo Z, *et al.* (2005) Antibody responses to individual proteins of SARS coronavirus and their neutralization activities. *Microbes Infect* 7: 882–89.
210. Fischer F, Peng D, Hingley ST, *et al.* (1997) The internal open reading frame within the nucleocapsid gene of mouse hepatitis virus encodes a structural protein that is not essential for viral replication. *J Virol* **71**: 996–1003.
211. Senanayake SD, Brian DA. (1997) Bovine coronavirus I protein synthesis follows ribosomal scanning on the bicistronic N mRNA. *Virus Res* **48**: 101–105.
212. Robertson MP, Igel H, Baertsch R, *et al.* (2005) The structure of a rigorously conserved RNA element within the SARS virus genome. *PLoS Biol* **3**: e5.
213. Goebel SJ, Hsue B, Dombrowski TF, Masters PS. (2004) Characterization of the RNA components of a putative molecular switch in the 3′m untranslated region of the murine coronavirus genome. *J Virol* **78**: 669–82.
214. Neuman BW, Adair BD, Yoshioka C, *et al.* (2006) Supramolecular architecture of severe acute respiratory syndrome coronavirus revealed by electron cryomicroscopy. *J Virol* **80**: 7918–28.
215. Mizutani T. (2007) Signal transduction in SARS-CoV-infected cells. *Ann N Y Acad Sci* **1102:** 86–95.
216. Canard B, Joseph JS, Kuhn P. (2008) International research networks in viral structural proteomics: again, lessons from SARS. *Antiviral Res.*, in press. Epub ahead of print (Nov 01, 2007).
217. Mesters JR, Tan J, Hilgenfeld R. (2006) Viral enzymes. *Curr Opin Struct Biol* **16**: 776–86.
218. Zeitler CE, Estes MK, Venkataram Prasad BV. (2006) X-ray crystallographic structure of the Norwalk virus protease at 1.5-Å resolution. *J Virol* **80**: 5050–58.

Chapter 17

High-throughput Technologies for Structural Biology: The Protein Structure Initiative Perspective

Andrzej Joachimiak

Macromolecular X-ray crystallography has seen remarkable progress in recent years, contributing significantly to the success of structural genomics and cutting-edge structural biology efforts. These advances were made possible by the development of third-generation synchrotron sources and brought a new dimension to X-ray protein crystallography. It allowed for efficient structure phasing approaches with anomalous signal, the use of very small crystals and studies of very large macromolecular assemblies. Similar progress has been made in the field of NMR that continues to be an important contributor in structural biology. NMR is being applied to large scale projects and new technological advances allow one to address challenging proteins. These advances could not have been exploited fully, were it not for complementary progress in bioinformatics, molecular biology, proteomics, hardware and software for crystallographic data collection, structure determination and refinement, databases, robotics and automation of many processes. Many of these developments were driven by the US-based Protein Structure Initiative and other worldwide structural genomics and proteomics efforts. These developments provide a robust foundation for structural genomics and structural biology programs and assure a productive future. In this

---

Biosciences Division, Midwest Center for Structural Genomics and Structural Biology Center, Argonne National Laboratory, 9700 S Cass Ave. Argonne, IL 60439, andrzejj@anl.gov

chapter, we focus mainly on reviewing X-ray crystallography technologies and their application to large- and small-scale structural genomics and structural biology programs.

Introduction

The Protein Structure Initiative (PSI), the US-based structural genomics program known as the Protein Structure Initiative (PSI), has been inspired by the great success of the Human Genome Project.[1] The long-range goal of the PSI is to make three-dimensional atomic-level structures of most proteins obtainable from a knowledge of their corresponding gene sequences. In the 1990s, it was realized that structural biology was inadequately prepared to provide answers to many questions generated by the Human Genome and other sequencing projects. New, large protein families have been discovered that span all the kingdoms of life or are overrepresented in specific environments. For many of these families, structural and functional information could not be established. The main objective of the PSI is to apply high-throughput technology and determine the structures of large numbers of strategically selected proteins in order to elucidate the protein folding space. The PSI effort is focused on experimental *de novo* structure determination of proteins for which there is currently no structural information available. The PSI makes these data available to the scientific community via the Protein Data Bank (PDB) (www.pdb.org). The PSI also aims to make the process of structure determination more accurate and more efficient, and to disseminate information on methods and technical advances as widely as possible. Similar in scope, structural genomics and proteomics efforts are being pursued world-wide.[2–6]

In the pilot phase, the PSI established several structural genomics centers with the goal of developing tests rigorously and implementing methods and technologies that permit parallel and cost-effective structure determination of many proteins, while allowing for their diverse nature. The PSI approach to this problem is multi-level; targets are characterized and categorized by properties and expected behavior, and processed in parallel by appropriate methods. At its inception, the PSI had to overcome three major challenges: 1) to make high quality protein samples and crystals available for structure determination and reduce the amount of material needed; 2) to identify and establish robust and effective structure phasing approaches; and 3) to increase the speed and efficiency of structure determination.

The PSI effort in methods and technology-development targeted all the steps of the "protein structure determination pipeline" from gene to structure (Fig. 1):

- Family classification and target selection, identification of signal sequences, prediction of transmembrane and disordered regions, domain parsing and optimizing domain constructs;
- Gene cloning and protein expression for microbial, viral and eukaryotic proteins, including new protein expression systems to improve folding and solubility, automation, optimizing vector/strain systems and cell growth conditions;

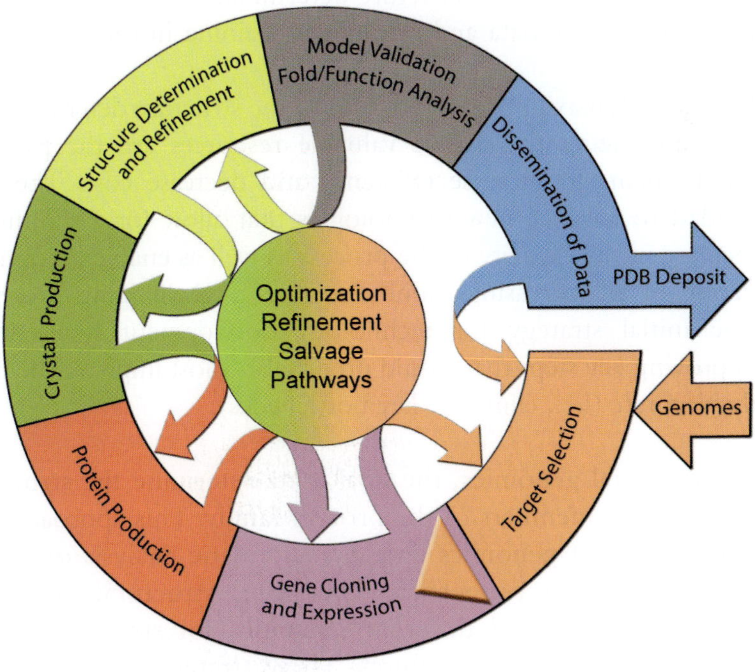

Fig. 1 Schematics of the MCSG structure determination pipeline.

- Protein purification, characterization and production of crystals suitable for synchrotron-based X-ray crystallography, including automation purification for proteins with affinity tags and simplifying purification of proteins with no affinity tags, automation of crystal screening and optimization; improving crystal diffracting properties through salvage approaches using orthologues, surface mutagenesis, chemical modification, and reducing heterogeneity caused by partial ligand occupancy;
- Data collection using cryo-crystallography and synchrotron X-ray sources, automation of crystal handling, optimizing crystal screening, decision making and data collection;
- Structure determination using multi- or single-wavelength anomalous diffraction (MAD/SAD) phasing and automated approaches, automation of model building and structure refinement, verification, fold recognition and function assignment;
- Archiving generated data in dedicated databases and application of LIMs to the integrated structure determination pipeline;
- Dissemination of data and results and training in the use of new technologies;
- For the PSI, maximizing the output of the structure determination and minimizing the use of valuable resources are the primary requirements to increase efficiency and decrease cost. The PSI needed to advance new technologies that allow for a significant increase in the capacity of the process as well as enable accommodation of an increasing number of highly challenging proteins. The initial strategy for technology development focused on improving key steps that would predictably most improve the production rate (i.e., eliminate key bottlenecks).

In structural genomics, the goal is to determine the structure of one or more members of the protein family. This approach can take advantage of genomics data by systematic analysis of family members and their properties. Therefore, structural genomics programs have adopted different strategies for structure determination and dealing with failures. These strategies are described below.

Establishing protein production standards. Strict production standards cull targets with low protein expression levels early in the process because it is important to produce an amount of protein that is sufficient for an extensive crystallization screening and structure determination.

Establishing protein quality standards. A set of procedures is put in place to monitor common problems with protein quality that can lead to a significantly lower crystallization success rate: protein purity (SDS PAGE), protein concentration (UV/VIS spectroscopy, Bradford), cofactor binding (UV/VIS spectroscopy), disorder and misfolding (CD spectroscopy/DXMS/NMR), protein polydispersity (SLS/DLS/SEC), charge homogeneity (IEF), metal binding (XAFS), and protein fingerprinting (tryptic MS). The need for biophysical characterization is motivated primarily by the choice of optimal path and maximization of success rate.

Use of tags to aid protein expression and purification. The use of protease cleavable tags to enhance protein expression, folding and solubility, and aid purification reduces effort and costs associated with protein production and allows for standardization of procedures and automation.

Establishing crystal quality standards. The suitability of crystal quality, size, diffraction limit, mosaicity, unit cell, cryoprotection, selecting optimal cryoprotection conditions and other parameters is assured by testing multiple crystal forms for diffraction properties and crystallization optimization.

Maximize quality at each step. The structure determination process involves so many steps even a small inefficiency at each step can result in significantly reduced overall yield and success rate.

Parallel processing. Structure determination comprises many time-consuming steps (e.g. crystal growth and cloning) that cannot be shortened significantly. In these instances, it is clear that multiple targets must be processed in parallel.

The use of automation and robotics. Efficiency can be improved by applying automation and robotics to the most labor intense steps. It is important to develop dedicated hardware and software, design and engineer multiplexed affinity purification systems, and deploy networked computer-operated systems for execution and documentation of all the steps in the process. As the experimental data accumulates quickly in the databases, new experiments can be planned and executed more efficiently based on past successes and failures.

Automatic process documentation using LIMs and databases. Documentation of the entire experimental process from target selection to structural model is critical and must be automated. The automatic process documentation eliminates the inevitable errors resulting from human documentation of thousands of projects. The databases are linked to many information-rich resources to enhance integration (Fig. 2).

Establishing and applying alternative salvage protocols. A strictly defined set of robust protocols implementing a diverse but limited

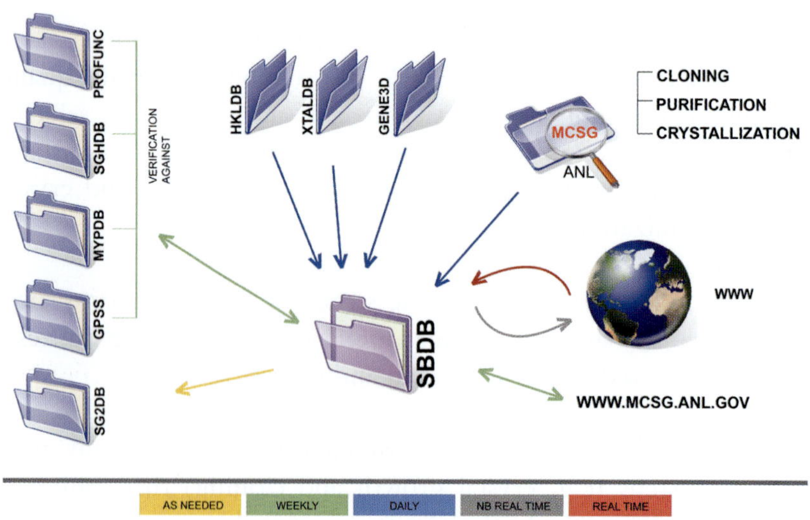

Fig. 2 Schematics of MCSG structural genomics database.

set of experimental approaches is first used without any case-specific modifications. For recalcitrant targets, an alternative set of protocols must be identified and applied. When implemented in a focused way, the alternative methods will rescue some recalcitrant targets.

In structural genomics, the process of solving structures from clone to deposit provides a unique knowledge base. This enables the evaluation of pipeline element effectiveness and the elimination of technologies or software that produce sub-optimal results (at the same time promoting more effective approaches). As the protein progresses through the pipeline and is being evaluated, the probability of success and failure is estimated at each step. This guides the decision making process whether to stop or continue. Similar procedures can be applied to "recalcitrant" targets in that they are moved to an alternative pipeline. This process enables the development of methods that allow continuously redefine valid targets and create new standard operating protocols.

PSI Contributed Many Technological Advances to Structural Biology

The majority of progress in the PSI has been driven by new technology development, process optimization and integration, and parallel processing.[7,8] The PSI, because of its scale and goals and the use of carefully validated genomic information, has been able to rigorously test and comprehensively advance methods and technologies for molecular and structural biology. This program has helped to identify many bottlenecks that were considered "anecdotal" and for the first time provided a large-scale parallel platform to evaluate methods and technologies. The PSI researchers have contributed improvements to existing methods, developed new technologies and have made them available to the biological community. As a result, today we have many robust, efficient and cost effective approaches in molecular biology and protein production, crystal growth, structure determination using X-ray crystallography and NMR, structural model generation, refinement and structure validation. We also better understand their limitations. In a production phase of PSI these methods and technology were rapidly incorporated into the PSI centers pipelines to determine structures of many novel proteins. Moreover, because the PSI protein targets are highly diverse, have low sequence similarity and display very different properties, the methods developed in the PSI have wide applicability in structural biology and many components of the high-throughput pipelines have been already adopted world-wide.

Although in structural genomics, structure determination of a sequence family member is a main goal, identifying effective salvage pathways to increase success rates is important. These approaches are tested on a large set of protein targets under controlled experimental conditions. Large data sets are generated and mined to extract significant trends to identify what works, what does not and what to avoid.

The interaction of structural genomics programs with the biomedical industry has resulted in many technological advances in automation and high-throughput applications in gene cloning, protein expression, purification, crystallization and crystal handling. These advances in instrumentation and software are now available to the biological community and many are freely accessible at the structural biology user facilities.

A comprehensive review of all the technologies developed in the PSI is well beyond the scope of this chapter. Below, we provide several selected examples.

Protein Target Selection Technologies

Tools and Methods for Target Selection

In structural genomics, comprehensive analysis of genomic sequences is essential for target selection. In initiatives such as the PSI, with the primary aim focused on structurally characterizing novel protein families which have no structural representative, selection of the appropriate target protein is critical. However, it is equally important to estimate the number and distribution of these families, and what efforts may be required to achieve a comprehensive structural coverage across all the protein families.[9] For example, it is imperative to know what proportion of structurally uncharacterized families are accessible to high-throughput structural genomics pipelines, specifically those targeting families containing multiple orthologues.[10–13] Orengo's lab analysis suggests that there are important benefits of selecting targets from both structurally uncharacterized domain families while, at the same time, pursuing additional targets from large structurally characterized protein superfamilies.[12] Such a combined approach to target selection is essential for structural genomics to achieve a comprehensive structural coverage of the genomes, leading to greater insights into the structure and mechanisms that underlie protein evolution. The analysis must be dynamically updated with newly sequenced genomic data. A number of new tools and methods have been developed for this purpose. An example of this is provided by the target selection pipeline for the Midwest Center for Structural Genomics (MCSG). The analysis of protein sequences, structures and folds is based on the Gene3D and CATH databases, (www.biochem.ucl.ac.uk/bsm/cath/, www.biochem.ucl.ac.uk/bsm/cath/Gene3D/. CATH provides a hierarchical classification of protein domain structures. The assignments of structures and functions to topology families and homologous superfamilies are made by sequence and structure comparisons. The Gene3D database currently contains >5 million protein sequences from >520 completed genomes classified into ~100,000 protein families. Domain compositions of each multi-domain superfamily are obtained by mapping CATH and Pfam domains onto the genome sequences. The Gene3D web portal provides a combined structural, functional and evolutionary view of the protein world. It is focused on providing structural annotation for protein sequences without structural representatives. The protein sequences have also been clustered into whole-chain families to aid functional prediction.[14] Similar tools have been developed at other PSI structural genomics centers.[15,16]

Analysis of a Large Set of Data Allowed for Optimizing of Protein Targets for Structure Determination

With the availability of large quantities of data from high-throughput structure determination in structural genomics centers, it is possible to recognize certain protein features that correlate with failures; thus, proteins can be identified that are more likely to succeed in the structure determination pipeline. The Joint Center for Structural Genomics (JCSG) has developed methods to identify several protein features that correlate strongly with successful protein production and crystallization and combine them into a single score that assesses "crystallization feasibility." The formula was tested with a jackknife procedure and validated on independent benchmark sets. The "crystallization feasibility" score is being applied to target selection and it contributes towards increasing the success rate, lowering costs, and shortening the time for protein structure determination. Analyses of PDB depositions suggest that very similar features also play a role in non-high-throughput structure determination, suggesting that this crystallization feasibility score would also be of significant interest to structural biology, as well as to molecular and biochemistry laboratories.[17] The database and method is available to the biological community (ffas.burnham.org/XtalPred).

Gene Cloning and Protein Expression

Ligation Independent Cloning (LIC)

To establish high-throughput methods for protein crystallography, all aspects of the production and analysis of protein crystals must be accelerated. Automated, plate-based methods for cloning, expression, and evaluation of target proteins can help researchers investigate the vast numbers of proteins available from sequenced genomes. LIC is well suited to robotic cloning and expression. In the PSI, we have established LIC as an important source of targets. LIC eliminates restriction enzyme screening and DNA ligase components of traditional cloning protocols. LIC provides unique cloning sites, is directional, highly efficient, simple, "generic," rapid, easy, and with a low-background and can be implemented readily in a highly parallel format and with minimal optimization. The LIC approach can generate rapidly multiple constructs from a single template, is compatible with multiple vectors and is well

Vector	Base Vector	Encoded Leader Sequence	Use
pMCSG7	pET21a	N-His-TEV-LICs-	Routine protein production
pMCSG8	pMCSG7	N-His-Sloop-TEV-LICs	Improve solubility
pMCSG9	pMCSG7	N-His-MBP-TEV-LICs	Improve solubility
pMCSG10	pMCSG7	N-His-GST-TEV-LICs	Improve solubility
pMCSG11	pACYCDuet-1	N-His-TEV-LICs	Coexpression
pMCSG12	pACYCDuet-1	N-His-Sloop-TEV-LICs	Coexpression
pMCSG13	pACYCDuet-1	N-His-MBP-TEV-LICs	Coexpression
pMCSG14	pACYCDuet-1	N-His-GST-TEV-LICs	Coexpression
pMCSG17	pMCSG7	N-Stag-TEV-LICs	Coexpression
pMCSG20	pMCSG7	N-Stag-GST-TEV-LICs	Coexpression
pMCSG16	pMCSG7	N-His-AviTag-TEV-LICs	Phage display
pMCSG15	pMCSG7	LICs-TEV-AviTag-His-C	Phage display
pMCSG18	pMCSG7	N-His-TEV-LICs-GFP	Screening
pMCSG19ABC	pMCSG7	N-MBP-TVMV-His-TEV-LICs	Purification
pMCSG21	pDONR/zeo	attL1-TEV-LIC-attL2	Gateway cloning

Fig. 3 The list of LIC vectors developed at MCSG for protein expression.[18–20]

suited to robotic and manual cloning and expression. The PSI has developed a number of cloning vectors for large-scale gene expression that are optimized for structural biology applications. Recently at the MCSG, we developed a new LIC vector, pMCSG19, that produces fusion proteins in which MBP, the His_6-tag and the target protein are separated by highly specific protease cleavage sites in the configuration, MBP-site-His_6-site-protein. *In vivo* cleavage at the first site by the co-expressed protease generates untagged MBP and His_6-tagged target protein. The design and use of new protein expression vectors tested on thousands of gene constructs can also be applied to large proteins and eukaryotic domains[18–20] (Fig. 3). The vectors are available through the PSI.

pCold Vectors

Researchers at the Northeast Structural Genomics Consortium (NESG) took advantage of improved overexpression of proteins in *E. coli* at low temperatures. They developed a series of expression vectors, termed pCold vectors, that drive the high expression of cloned genes upon induction by cold-shock. Proteins can be produced with very high yields, including eukaryotic proteins. The pCold vector system can also be used to selectively enrich target proteins with isotopes to study their properties in cell lysates using NMR spectroscopy. pCold vectors are highly complementary to the widely used pET vectors.[21]

Protein Production by Auto-induction in High Density Shaking Cultures

In the PSI, most proteins are expressed in *E. coli* and are driven by inducible expression systems in which T7 RNA polymerase transcribes coding sequences cloned under control of a T7lac promoter, which efficiently produce a wide variety of proteins in *E. coli.* Systematic analysis of bacterial growth allowed for the development of reliable non-inducing and auto-inducing media in which batch cultures can grow to high densities. Expression strains grown to saturation in non-inducing media retain plasmid and remain fully viable for an extended period of time. Auto-induction allows for the efficient screening of many clones in parallel for expression and solubility, as the cultures have only to be inoculated and grown to saturation, and yields of target protein are typically several-fold higher than that obtained using conventional IPTG induction. Auto-inducing media have been developed for labeling proteins with selenomethionine, 15N or 13C, and for production of target proteins by arabinose induction of T7 RNA polymerase from the pBAD promoter in BL21-AI[22]. The media is commercially available.

Cell-free Protein Expression

Cell-free translation can circumvent a number of limitations of cell-based expression systems for protein production. Among these systems, *E. coli* and wheat germ-based systems are the most widely used.[23,24] The wheat germ system is of special interest for its eukaryotic nature and the advantage of producing eukaryotic multi-domain proteins in a folded state. Several advances in the use of cell-free expression systems have been made in the past few years and successful applications of these systems to produce proteins for functional and structural biology studies have been reported.[23] A wheat germ cell-free platform for protein production that supports efficient NMR structural studies of eukaryotic proteins and offers advantages over cell-based methods has been established at the Center for Eukaryotic Structural Genomics (CESG). The target gene is cloned into a specialized plasmid and screened in a small-scale *in vitro* sequential transcription and translation reaction to ascertain the level of protein production and solubility. A larger scale cell-free translation reaction to incorporate ^{15}N-labeled amino acids into a protein sample is run to test for suitability for NMR structural analysis. For well-behaving proteins, the larger scale cell-free translation reaction with ^{13}C, ^{15}N-labeled amino acids is carried out to prepare a doubly labeled sample for 3D structure determination.[25,26]

High-throughput Protein Purification for Structural Biology

Automation of Protein Purification

A critical issue in structural biology is the availability of high-quality protein samples. 'Structural-biology-grade' proteins must be generated in a quantity and quality suitable for structure determination experiments using X-ray crystallography or NMR. The choice of protein purification and handling procedures plays a critical role in obtaining high-quality protein samples. The purification procedure must yield a homogeneous protein and must be highly reproducible in order to supply milligram quantities of protein and/or its derivative containing marker atom(s). The MCSG has developed protocols for high-throughput automated protein purification using affinity chromatography. These protocols have been implemented on AKTA EXPLORER 3D and AKTAexpress workstations capable of performing multi-dimensional chromatography (Fig. 4). The automated chromatography has been successfully applied to several thousands of proteins of microbial and eukaryotic origin.[27]

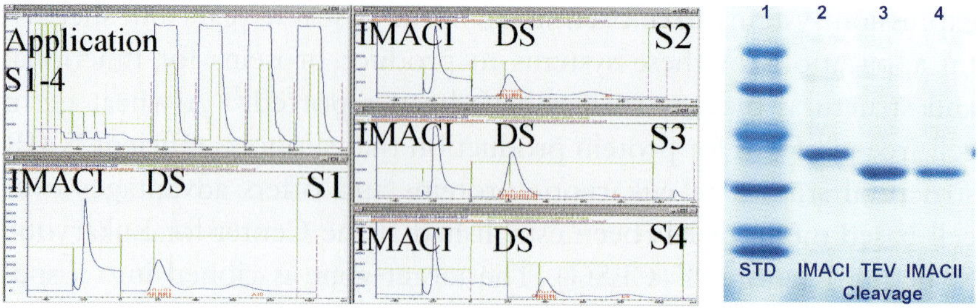

Fig. 4. Automated purification four protein samples (S1–4) on AKTAexpress single. module. Four proteins are being applied to the IMAC column (IMACI) and then eluted and separated on desalting column (DC). The PAGE on the right shows purity of protein after IMACI, cleavage of His_6 tag with TEV protease and a second purification of IMACII as described earlier.[27]

Protein Crystallization

Use Small (nano-liter) Droplets in Protein Crystallization

The main effort in crystallization for structural genomics has been focused on lowering the amount of the sample, discovering the best screening formulations, and automation of high-throughput screening procedures to identify potential crystallization conditions. The use of very small volumes has been explored and extensively tested with over 50,000 proteins in the PSI. Small crystallization volumes tend to reduce equilibration times and increase the success rate.[28–30] However, crystal size is critical for structure determination. Optimization approaches to turn very small or low-quality crystals into useful diffracting ones must be taken into consideration and must be adapted to high throughput.[30] One promising technology involves the use of nanovolume microfluidic crystallization technologies for plug-based and counterdiffusion methods in confined geometries (plastic labcards). These approaches are being developed for *in situ* X-ray screening and data collection. Crystallization screening and optimization in confined microfluidic geometries is expected to enable the crystallization of difficult to produce proteins due to its minimal sample consumption.[31] These promising technologies may revolutionize the crystallization of proteins.

The Use of Synchrotron Facilities for Protein Structure Determination

High-throughput Crystallography Using Synchrotron Radiation

Third generation synchrotron facilities that have been designed to better serve biological research has trasformed macromolecular X-ray crystallography.[32–36] For example, the 19ID undulator beamline of the Structure Biology Center, has been designed and built to take full advantage of the high flux, brilliance and quality of X-ray beams delivered by the Advanced Photon Source.[36] The high flexibility, inherent in the design of the optics, coupled with a kappa-geometry goniometer and beamline control software, allows for optimal strategies to be adopted in protein crystallographic experiments, thus maximizing the chances of their success. The synchrotron beamlines, when combined with crystal cryo-protection and robotic crystal handling, allowed for the optimal use of the "anomalous signal" for phasing structures. Data can be collected from a single crystal and the phases can be extracted semi-automatically.[37] The PSI has significantly contributed to establishing MAD/SAD as a routine method for protein structure determination. Moreover, ultrafast MAD/SAD data collection is now possible on a routine basis and with the widespread use of selenomethionine for phase determination,[35,38] the method has become the most prominent experimental approach in determining structures of novel proteins. Developments in crystallographic software are complementing these advances, paving the way for improving quality and accelerated protein structure determination.[33,34]

New Laboratory X-ray Sources

The Compact Light Source, which is being developed by Lyncean Technologies, Inc. as part of PSI-funded Accelerated Technologies Center for Gene to 3D Structure (ATCG3D), is a novel and unique tunable laboratory X-ray source with peak intensity at X-ray wavelengths that span selenium anomalous absorbance. The ability to efficiently solve new protein crystal structures can be greatly enhanced by the availability of a tunable laboratory X-ray source in the same facility where crystal growth experiments are performed (www.lynceantech.com).

Semi-automated Structure Determination Using X-ray Crystallography and Synchrotron Radiation

A new approach at the MCSG that integrates data collection, data reduction, phasing and model building accelerates significantly the process of structure determination and minimizes both the number of data sets and synchrotron time required for structure solution. The software attempts to solve the structure using different algorithms and approaches, rapidly converting diffraction data into an interpretable electron density map, and for smaller structures, into an initial model. The heuristics for selecting the best computational strategy at different data resolution limits of phasing signal and crystal diffraction are being optimized. The typical end result is an interpretable electron-density map with a partially built structure and, in some cases, an almost complete model. The system is combined with relational databases and linked to external web resources (MCSG, SGPDB, PDB, Swissprot, NCBI and others). The software has been successfully tested on several hundred novel proteins and has resulted in over 300 PDB deposits.[37]

Improving Data Processing for Structure Determination

A novel approach to scaling diffraction intensities has been developed. This method minimizes the variations among multiple measurements of symmetry-related reflections using a stable refinement procedure. The scale factors are described by a flexible exponential function that allows different scaling corrections to be chosen and combined according to the needs of the experiment. The scaling model includes: scale and temperature factor per batch of data; temperature factor as a continuous function of the radiation dose; absorption in the crystal; uneven exposure within a single diffraction image; and corrections for phenomena that depend on the diffraction peak position on the detector. This scaling model can be extended to include additional corrections for various instrumental and data-collection problems.[39]

Improving Structure Deposition

In the PSI, all the protein structures are made available to the public as soon as they are completed and deposited into the PDB. In order to expedite the deposition process, Burley laboratory has developed Deposit3D.[40] This command-line script gathers all the required structure-deposition information and outputs this data into an mmCIF file for subsequent upload through the RCSB PDB ADIT interface. Deposit3D is very useful for structural genomics pipeline projects because it allows workers involved with the various stages of a structure-determination project to pool their different categories of annotation information before starting a deposition session. It also helps individual researchers to standardize files and help in the deposition process.

Structure Determination Using NMR

Structure Determination Using Reduced-dimensionality and G-Matrix FT NMR: GFT-NMR Provides 4D/5D NMR Data in 3D Spectra Reducing Data Collection Times

A standardized protocol enabling rapid NMR data collection for high-quality protein structure determination was developed at NESG that allows one to capitalize on high spectrometer sensitivity: a set of five G-matrix Fourier transform NMR experiments for resonance assignment based on highly resolved 4D and 5D spectral information, is acquired in conjunction with a single simultaneous 3D ^{15}N,^{13}C (aliphatic),^{13}C (aromatic)-resolved [^{1}H,^{1}H]-NOESY spectrum, providing ^{1}H-^{1}H upper distance limit constraints. The protocol was integrated with a methodology for semi-automated data analysis in the NESG pipeline. The protocol effectively removes data collection as a bottleneck for high-throughput structure determination of proteins up to at least approximately 20 kDa, while concurrently providing spectra that are highly amenable to fast and robust analysis.[41]

Structure Determination Using NMR and Microgram Quantities of Protein

Using conventional triple-resonance nuclear magnetic resonance experiments with a 1 mm triple-resonance microcoil NMR probe at NESG, we determined near complete resonance assignments and the 3D structure of the 68-residue *M. mazei* TRAM protein, using only 72 mug (6 microl, 1.4 mM) of protein. This example of a complete solution of an NMR structure, determined using microgram quantities of protein, demonstrates the utility of microcoil-probe NMR technologies for protein samples that can be produced in only limited quantities.[42]

Salvage Approaches in Structural Genomics

Rapid Refinement of Crystallographic Protein Construct Definition Employing Enhanced Hydrogen/Deuterium Exchange MS

Crystallographic efforts often fail to produce suitably diffracting protein crystals. Unstructured regions of proteins play an important role in this problem and considerable advantage can be gained in removing them. The JCSG has developed a number of enhancements to amide hydrogen/high-throughput and high-resolution deuterium exchange MS (DXMS) technology that allows for the rapid identification of unstructured regions in proteins. The utility of this approach for improving crystallization success was tested on proteins with varying crystallization and diffraction characteristics. When compared with targets of known structures, the DXMS method correctly localized even small regions of disorder. DXMS analysis was then correlated with the propensity of such targets to crystallize and was further used to define truncations that improved crystallization. Truncations that were defined solely based on DXMS analysis demonstrated greatly improved crystallization and have been used for structure determination. This approach represents a rapid and generalized method that can be applied to structural genomics or other targets in a high-throughput manner.[43,44]

High-throughput Limited Proteolysis/Mass Spectrometry for Protein Domain Elucidation

Previously published studies have demonstrated that compact globular domains defined by limited proteolysis represent good candidates for production of diffraction quality crystals.[45] Integration of mass spectrometry and proteolysis experiments can provide an accurate definition of domain boundaries at unprecedented rates. The SGX (Structural GenomiX, Inc.) conducted a critical evaluation of this approach with 400 target proteins produced for the New York SGX Research Center for Structural genomics (NYSGXRC). The objectives of this study were to develop parallel/automated protocols for proteolytic digestion and data acquisition for multiple proteins, and to carry out a systematic study to correlate domain definition via proteolysis with outcomes of crystallization and structure determination attempts. Initial results from this work demonstrate that proteins yielding diffraction quality crystals are typically resistant to proteolysis.[46]

In situ *Proteolysis to Aid Crystallization*

In a collaborative effort between the MCSG and SGC, we have systematically study protein *in situ* proteolysis to promote its crystallization. The general applicability of this approach for protein crystallization was evaluated on proteins that were of poor quality or which failed to crystallize. By incubating the proteins in crystallization set ups with a single protease, chymotrypsin, we were able to generate X-ray diffraction quality crystals and determine the structures of several proteins. In all the cases tested, proteolysis removed residues at either the N- or C-termini, or both. The use of *in situ* proteolysis provides a path to significantly increase the success rate of protein structure determination, particularly for recalcitrant proteins.[47]

Reductive Methylation of Proteins to Aid Crystallization

The highest attrition rate in structural biology projects utilizing X-ray crystallography occurs at the step of obtaining X-ray diffraction quality crystals. It is established that protein chemical modification can facilitate crystallization and often crystals of modified proteins diffract to a higher resolution. As part of an effort at the MCSG to increase the success rate of obtaining diffraction quality crystals, reductive methylation of lysine residues in proteins has been tested on several hundred unique protein targets that failed to crystallize or produce X-ray quality crystals. The chemical modification is fast, specific and requires few steps under relatively mild buffer and chemical conditions. Following the method described by I. Rayment,[48] the proteins were methylated and screened using the high-throughput crystallization pipeline. Reductive methylation of lysine residues alters the protein surface properties and crystallization behavior (Fig. 5).

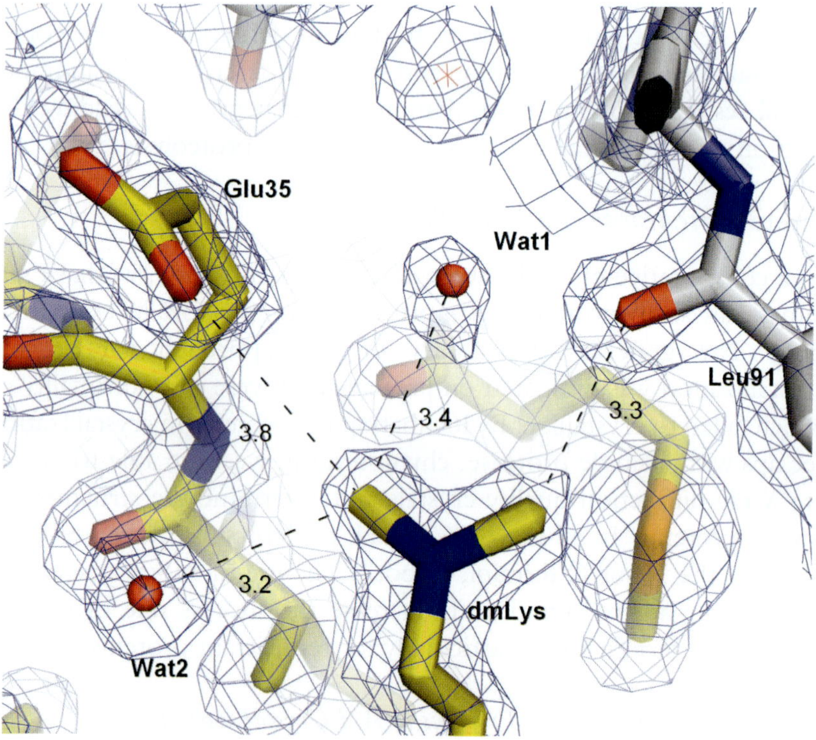

Fig. 5 Electron density map contoured at 1σ level of the double methylated lysine 62 residue making intra- and intermolecular contacts with Glu35, carbonyl of Leu91 and two waters in the structure of HopJ type III effector protein from *Vibrio parahaemolyticus*.[49]

Crystal structures have been obtained for 7% of the screened target proteins. This method is well suited to high-throughput projects as well as regular laboratories.[49]

Surface Entropy Reduction to Promote Protein Crystallization

Derewenda's laboratory has developed a method to engineer surface sequence variants for "stubborn crystallizers" designed to form intermolecular contacts that could support a crystal lattice. This approach relies on the concept of surface entropy reduction (SER), i.e. the replacement of small clusters of two to three solvent-exposed residues, characterized by high conformational entropy, with residues with lower conformational entropy such as alanine. This strategy minimizes the loss of conformational entropy upon crystallization and renders crystallization thermodynamically favorable. This method has been successfully used to crystallize many novel proteins and many stubborn crystallizers. It has proven to be an effective salvage pathway for proteins that are difficult to crystallize.[50] The surface entropy reduction prediction server (SERp), designed to identify mutations that may facilitate crystallization, has been developed at the Integrated Center for Structure and Function Annotation. The server can be accessed at http://www.doe-mbi.ucla.edu/Services/SER.[51]

Incorporation of Selenomethionine into Eukaryotic Proteins in Yeast

S. cerevisiae is an ideal host from which to obtain high levels of post-translationally modified eukaryotic proteins for X-ray crystallography. Malkowski's laboratory at the Center for High-throughput Structural Biology has developed a general method to incorporate selenomethionine into proteins expressed in yeast, based on manipulation of the appropriate metabolic pathways. *sam1*(-) *sam2*(-) mutants, in which the conversion of methionine to S-adenosylmethionine is blocked, and which exhibit reduced selenomethionine toxicity as compared with the wild-type yeast, increased the production of proteins during their growth in selenomethionine, as well as facilitated an efficient replacement of methionine by selenomethionine.[52]

Structure-based Function Prediction

Many proteins selected for structure determination in the PSI do not have biochemical or cellular functions assigned. It has been shown that the 3D structure can often provide important clues about the biochemical function of a protein. In a collaborative effort between the PSI-funded MCSG and the European Bioinformatics Institute, a function prediction server has been developed.[10,11,53,54] The new, fully automated web server (http://www.ebi.ac.uk/thornton-srv/databases/ProFunc) uses a wide range of methods to perform sequence-, structure- and template-based searches on submitted structures for the prediction of the likely functions of proteins. Users submit the coordinates of their structure to the server in PDB format. The results include fold recognition, sequence and structure homologues, surface cleft analysis, active site hits, and ligand and DNA binding sites, as well as reverse templates and genomic location analysis. A summary of the analyses provides an at-a-glance view of what each of the different methods have found. Analysis of data suggests that several of the structure-based methods are very successful and provide examples of local similarity that is difficult to identify using current sequence-based methods. A method based on the Gene Ontology (GO) schema using GO-slims that can allow for the automated assessment of hits, with a success rate approaching that of expert manual assessment, is also included. The server is used in structural genomics where a large proportion of the proteins whose structures are solved, are proteins of unknown function. However, it also finds use in a comparative analysis of members of large protein families. It provides a convenient compendium of sequence and structural information that often hold vital functional clues to be followed up experimentally.

In Summary

The PSI systematically addressed major challenges in structural biology and substantially improved the structure determination of novel proteins. As a result, robust high-throughput structure determination pipelines have been established with the capacity to determine hundreds of protein structures. In the past 7 years, the cost of structure determination has been reduced at least 5 times and the quality of PSI structures has improved significantly. The PSI has also changed the approach to structure determination by identifying bottlenecks, testing salvage approaches and evaluating their efficacy using a large set of novel proteins. Custom and commercial instrumentation integrated into the PSI pipelines are largely available to the scientific community. However, there are a number of challenges that need to be addressed. Some proteins, complexes and cellular assemblies are not compatible with current high-throughput pipelines. Therefore, methods and technology development must continue to address these challenges to advance biomedical research and benefit the scientific community.

Acknowledgments

We thank all members of the Structural Biology Center and the Midwest Center for Structural Genomics at Argonne National Laboratory for their help in conducting experiments. We wish to thank the members of the JCSG, MCSG, NESG, and NYSGXRC, both past and present, for all of their efforts on behalf of the PSI. We also gratefully acknowledge the many contributions of the other ten PSI centers, and the NIGMS that have enhanced our own activities. Finally, we recognize the valuable contributions made by structural genomics researchers throughout the world. We would like to thank Andrea Cipriani for help in preparation of the manuscript for publication and Monica Nocek-Chodkiewicz for help in preparing the figures. This work was supported by the National Institutes of Health Grant GM62414, GM074942 and by the US Department of Energy, Office of Biological and Environmental Research, under contract DE-AC02-06CH11357.

References

1. Norvell JC, Machalek AZ. (2000) "Structural genomics programs at the US National Institute of General Medical Sciences." *Nat Struct Biol* 7(931).

2. Aricescu AR, *et al.* (2006) "Eukaryotic expression: developments for structural proteomics." *Acta Crystallogr D Biol Crystallogr* **62:** 1114–24.
3. Berry IM, DO, Esnouf RM, Harlos K, *et al.*, (2006) "SPINE high-throughput crystallization, crystal imaging and recognition techniques: current state, performance analysis, new technologies and future aspects." *Acta Crystallogr D Biol Crystallogr* **62**(10): 1137–49.
4. Albeck S, AP, Andreini C, Banci L, *et al.* (2006) "SPINE bioinformatics and data-management aspects of high-throughput structural biology." *Acta Crystallogr D Biol Crystallogr* **62**(10): 1184–95.
5. Vedadi M, *et al.* (2007) "Genome-scale protein expression and structural biology of plasmodium falciparum and related apicomplexan organisms." *Mol Biochem Parasitol* **151**: 100–10.
6. Yokoyama S. (2003) "Protein expression systems for structural genomics and proteomics." *Curr Opin Chem Biol* 7: 39–43.
7. Bonanno JB, AS, Bresnick A, Chance MR, *et al.* (2005) "New York-Structural GenomiX Research Consortium (NYSGXRC): a large scale center for the protein structure initiative." *J Struct Funct Genomics* **6**(2–3): 225–32.
8. Lesley SA, KP, Godzik A, Deacon AM, *et al.* (2002) "Structural genomics of the thermotoga maritima proteome implemented in a high-throughput structure determination pipeline." *Proc Natl Acad Sci USA* **99**(18): 11664–69.
9. Marsden RL, Lee D, Maibaum M, *et al.* (2006) "Comprehensive genome analysis of 203 genomes provides structural genomics with new insights into protein family space." *Nucl Acids Res* **34**: 1066–80.
10. Yeats C, MM, Marsden R, Dibley M, *et al.* (2006) "Gene3D: modelling protein structure, function and evolution." *Nucl Acids Res* **34**: D281–84.
11. Watson JD, SS, Ezersky A, Savchenko A, *et al.* (2007) "Towards fully automated structure-based function prediction in structural genomics: a case study." *J Mol Biol.*
12. Marsden RL, Lewis TA, Orengo CA. (2007) "Towards a comprehensive structural coverage of completed genomes: a structural genomics viewpoint." *BMC Bioinformatics* **8**: 86.
13. Todd AE, Marsden RL, Thornton JM, Orengo CA. (2005) "Progress of structural genomics initiatives: an analysis of solved target structures." *J Mol Biol* **348**: 1235–60.
14. Watson JD, *et al.* (2003) "Target selection and determination of function in structural genomics." *IUBMB Life* **55**: 249–55.
15. Slabinski L, *et al.* (2007) "The challenge of protein structure determination — lessons from structural genomics." *Protein Sci* **16**: 2472–82.
16. Liu J, Hegyi H, Acton TB, *et al.* (2004) "Automatic target selection for structural genomics on eukaryotes." *Proteins* **56**: 188–200.
17. Slabinski L, JL, Rychlewski L, Wilson IA, *et al.* (2007) "XtalPred: a web server for prediction of protein crystallizability." *Bioinformatics.*

18. Donnelly MI, ZM, Millard CS, Clancy S, *et al.* (2006) "An expression vector tailored for large-scale, high-throughput purification of recombinant proteins." *Prot Expr Purif* **47**: 446–54.
19. Stols L, GM, Dieckman L, Raffen R, *et al.* (2002) "A new vector for high-throughput, ligation-independent cloning encoding a tobacco etch virus protease cleavage site." *Prot Expr Purif* **25**: 8–15.
20. Dieckman L, GM, Stols L, Donnelly MI, Collart FR. (2002) "High throughput methods for gene cloning and expression." *Prot Expr Purif* **25**: 1–7.
21. Qing G, ML, Khorchid A, Swapna GVT, *et al.* (2004) "Cold-shock induced high-yield protein production in *Escherichia coli*." *Nat Biotech* **22**: 877–82.
22. Studier FW. (2005) "Protein production by auto-induction in high density shaking cultures." *Prot Expr Purif* **41**: 207–34.
23. Endo Y, ST. (2006) "Cell-free expression systems for eukaryotic protein production." *Curr Opin Biotechnol* **17**(4): 373–80.
24. Kigawa T, *et al.* (2004) "Preparation of *Escherichia coli* cell extract for highly productive cell-free protein expression." *J Struct Funct Genomics* **5:** 63–68.
25. Vinarov DA, LB, Peterson FC, Tyler EM, *et al.* (2004) "Cell-free protein production and labeling protocol for NMR-based structural proteomics." *Nat Methods* **1**(2): 149–53.
26. Vinarov DA, MJ. (2005) "High-throughput automated platform for nuclear magnetic resonance-based structural proteomics." *Expert Rev Proteom* **2**(1): 49–55.
27. Kim Y, DI, Zhou M, Wu R, *et al.* (2004) "Automation of protein purification for structural genomics." *J Struct Funct Genom* **5**: 111–18.
28. Stevens RC. (2000) "High-throughput protein crystallization." *Curr Opin Struct Biol* **10**: 558–63.
29. Hui R, Edwards A. (2003) "High-throughput protein crystallization." *J Struct Biol* **142**: 154–61.
30. Chayen NE, Saridakis E. (2002) "Protein crystallization for genomics: towards high-throughput optimization techniques." *Acta Crystallogr D Biol Crystallogr* **58**: 921–27.
31. Gerdts CJ, TV, Yadav MK, Dementieva I, *et al.* (2006) "Time-controlled microfluidic seeding in nL-volume droplets to separate nucleation and growth stages of protein crystallization." *Angew Chem Int Ed Engl* **45**(48): 8156–60.
32. Minor W, TD, Otwinowski Z. (2000) "Strategies for macromolecular synchrotron crystallography." *Struct Fold Des* **8**: R105–10.
33. Walsh MA, Dementieva I, Evans G, *et al.* (1999) "Taking MAD to the extreme: ultrafast protein structure determination." *Acta Crystallogr D Biol Crystallogr* **55**: 1168–73.
34. Walsh MA, Evans G, Sanishvili R, *et al.* (1999) "MAD data collection — current trends." *Acta Crystallogr D Biol Crystallogr* **55**: 1726–32.
35. Dauter Z. (2006) "Current state and prospects of macromolecular crystallography." *Acta Crystallogr D Biol Crystallogr* **62**: 1–11.

36. Rosenbaum G, Alkire R, Evans G, *et al.* (2006) "The Structural Biology Center 19ID undulator beamline: facility specifications and protein crystallographic results." *J Synch Radiat* **13**: 30–45.
37. Minor W, CM, Otwinowski Z, Chruszcz M. (2006) "HKL-3000: the integration of data reduction and structure solution — from diffraction images to an initial model in minutes." *Acta Crystallogr D Biol Crystallogr* **62**: 859–66.
38. Hendrickson WA, Horton JR, LeMaster DM. (1990) "Selenomethionyl proteins produced for analysis by multiwavelength anomalous diffraction (MAD): a vehicle for direct determination of three-dimensional structure." *EMBO J* **9**: 1665–72.
39. Otwinowski Z, BD, Majewski W, Minor W. (2003) "Multiparametric scaling of diffraction intensities." *Acta Crystallogr A* **59**: 228–34.
40. Badger J, HJ, Burley SK, Kissinger CR. (2005) "Deposit3D: a tool for automating structure depositions to the Protein Data Bank." *Acta Crystallogr Sect F Struct Biol Cryst Commun* **61**(9): 818–20.
41. Liu G, SY, Atreya HS, Parish D, *et al.* (2005) "NMR data collection and analysis protocol for high-throughput protein structure determination." *Proc Natl Acad Sci USA* 10487–92.
42. Aramini J, RP, Xiao R, Anklin C, Montelione GT, *et al.* (2007) "Microgram scale protein structure determination by NMR." *Nat Meth* **4**: 491–93.
43. Spraggon G, PD, Klock HE, Wilson IA, *et al.* (2004) "On the use of DXMS to produce more crystallizable proteins: structures of the T. maritima proteins TM0160 and TM1171." *Protein Sci* **12**: 3187–99.
44. Pantazatos D, KJ, Klock HE, Stevens RC, *et al.* (2004) "Rapid refinement of crystallographic protein construct definition employing enhanced hydrogen/deuterium exchange MS." *Proc Natl Acad Sci USA* **101**(3): 751–56.
45. Koth CM, OS, Larson SM, Edwards AM. (2003) "Use of limited proteolysis to identify protein domains suitable for structural analysis." *Meth Enzymol* **368**: 77–84.
46. Gao X, BK, Bonanno JB, Buchanan M, *et al.* (2005) "High-throughput limited proteolysis/mass spectrometry for protein domain elucidation." *J Struct Funct Genom* **6**(2–3): 129–34.
47. Dong A, *et al.* (2007) "*In situ* proteolysis for protein crystallization and structure determination." *Nat Meth.*
48. Rayment I. (1997) Reductive alkylation of lysine residues to alter crystallization properties of proteins. *Meth Enzymol* **276**: 171–79.
49. Kim Y, Quartey P, Volkart L, *et al.* 2007.
50. Cooper DR, BT, Grelewska K, Pinkowska M, *et al.* (2007) "Protein crystallization by surface entropy reduction: optimization of the SER strategy." *Acta Crystallogr D Biol Crystallogr* **63**(5): 636–45.
51. Goldschmidt L, CD, Derewenda ZS, Eisenberg D. (2007) "Toward rational protein crystallization: a Web server for the design of crystallizable protein variants." *Protein Sci* **16**(8): 1569–76.

52. Malkowski MG, *et al.* (2007) "Blocking S-adenosylmethionine synthesis in yeast allows selenomethionine incorporation and multiwavelength anomalous dispersion phasing." *Proc Natl Acad Sci USA* **104**: 6678–83.
53. Laskowski RA, WJ, Thornton JM. (2005) "ProFunc: a server for predicting protein function from 3D structure." *Nucleic Acids Res* **33**: W89–93.
54. Laskowski RA, WJ, Thornton JM. (2005) "Protein function prediction using local 3D templates." *J Mol Biol* **351**: 614–26.

Chapter 18

European Structural Proteomics — A Perspective

Susan Daenke, E. Yvonne Jones and David I. Stuart

Introduction

The concept of structural genomics arose in the mid-to-late 1990s in the USA and Japan as a response to the success of high-throughput (HTP) sequencing methods applied to whole genomes (see http://www.isgo.org). The radical reduction in the unit cost of sequencing, driven by technological advances, raised the intellectual challenge to devise similar advances in the chain of procedures linking gene to protein structure. It was proposed that analogous HTP methods might provide 3-D structures of all the proteins (the "proteome") of an organism to efficiently fill the numerous gaps in observed fold-space. This vision led to the investment of substantial sums of money into a tranche of large-scale structural genomics projects in the USA (e.g. nine projects funded by the NIH/NIGMS Protein Structure Initiative (PSI) from September 2000 to June 2005; see http://www.nigms.nih.gov/psi/) and Japan (e.g. the massive RIKEN project; see http://www.rsgi.riken.go.jp/). These efforts were broadly characterized by the concentration of resources into a small number of large centers, the development of automated technologies to permit

Division of Structural Biology, Wellcome Trust Centre for Human Genetics, University of Oxford, Roosevelt Drive, Oxford OX3 7BN, UK.

an HTP pipeline approach to structure determination, and, especially in the US, a targeting of proteins likely to have novel folds or bring structural light to new protein families.

Europe was slower in implementing HTP approaches to structural biology. The Protein Structure Factory in Berlin, Germany (http://www. proteinstrukturfabrik.de/) led the way, followed by the Oxford Protein Production Facility (OPPF) in Oxford, UK (http://www.oppf.ox.ac.uk/) and the Genopoles in France (notably Gif, Marseille and Strasbourg; http:// rng.cnrg.fr/). However it was not until October 2002 that the first Europe-wide project began. This was a 3-year project funded by the EU FP5 program called SPINE: Structural Proteomics IN Europe (http://www.spineurope.org). Coming later onto the scene, SPINE was deliberately designed to be a second-generation project (indeed, purposefully called a Structural Proteomics project to draw a distinction from the set of structural genomics activities which were by then already well established) and made some radical departures from the earlier initiatives, while at the same time benefiting from the experience and technology development of the earlier projects. SPINE was distinct from first-generation structural genomics efforts established elsewhere in the world, as the intention was to maintain a strong biological drive by close linkage with biomedical/biological collaborators. SPINE was also distinctive in its intention to develop methods that could be made available to the wider structural biology community. SPINE therefore aimed to initiate the development and roll-out of technologies for structural biology around Europe and to define new European standards in areas of HTP methods, LIMS handling and synchrotron technologies.

The challenge set for SPINE was therefore to push forward with cutting-edge technologies while simultaneously generating pan-European integration on biomedically focused structural proteomics. The SPINE consortium comprised 19 leading centers in structural biology distributed throughout Europe (Table 1). The project was coordinated from Oxford and organized into a series of workpackages to each of which various combinations of SPINE laboratories contributed. Eight workpackages covered technology development and implementation, each defining a coherent section of the pipeline leading from gene to structure. These methodological developments were

Table 1

Partner	Institution, Country
1: David Stuart	University of Oxford, UK
2: Par Nordlund	University of Stockholm Sweden
3: Joel Sussman	Weizmann Institute of Science, Israel
4: Matthias Wilmanns	EMBL-Hamburg, Germany
5: Rob Kaptein	Utrecht University, The Netherlands
6: Stephen Cusack	EMBL-Grenoble, France
7: Keith Wilson	University of York, UK
8: Janet Thornton	EMBL-EBI, UK
9: Christian Cambillau	AFMB Marseille, France
10: Dino Moras	CERBM, France
11: Albrecht Messerschmidt	MPI, Martinsried, Germany
12: Lena Gustafsson	Chalmers University, Sweden
13: Titia Sixma	NKI, The Netherlands
14: Udo Heinemann	MDC, Germany
15: Ivano Bertini	CIRMMP, Italy
16: Sine Larsen	ESRF, France
17: Pedro Alzari	Institute Pasteur, France
18: Helena Berglund	Karolinska Institute, Sweden
19: Gunter Schneider	Karolinska Institute, Sweden
20: Alwyn Jones	Uppsala University, Sweden

designed to underpin and be driven by a coherent program of work on biomedical targets, defined in two additional workpackages covering human and pathogen proteins, respectively, and forming the heart of the project. These activities were complemented by a strong training and networking component, which was built into the project with the explicit aim of creating as a European resource a cadre of highly trained structural biologists and technicians. This chapter will review the achievements of SPINE, and set these in the context of further programs in structural proteomics which have been subsequently funded by the EC and are currently in operation.

Much of the information presented here is derived from SPINE reports and publications, most notably the October 2006 special edition of *Acta Crystallographica D* (Vol. 62, pp. 1103–1285) devoted to the achievements of SPINE. The authors acknowledge the contributions of all SPINE partners to the content of this article and to the overall success of the project.

Bioinformatics and LIMS

The principal bottlenecks in structural biology are, at present, the production of soluble protein and its crystallization, so the justification for resourcing bioinformatics and data management must be rooted in the provision of logistic and scientific added value. Nevertheless, the volume of data produced in large-scale structural genomics/proteomics projects has led to the common perception that data management and analysis are of central importance,[1,2] and the distributed nature of the SPINE project suggested that its data management issues would be similarly critical. The first challenge was to define a comprehensive structural biology data model; this was catalyzed by SPINE[3], along with a partial implementation in a form useful for structural biologists. We found real value in recording both failure and success in the structural biology pipeline process such that we could begin an objective analysis of the process bottlenecks.

Figure 1 shows an overview of the bioinformatics and data management requirements for SPINE, and indicates the contributions from the various SPINE partners to this process. One particular area of progress in several labs was in producing simple, yet comprehensive bioinformatics views to enable rational target selection and experimental design.[4] An important feature of such target annotation systems is that the information is updated regularly, to draw on the vast amounts of information being produced by other on-going projects. Since SPINE targeted some difficult and complex target areas, some specific tools were developed, such as VaZyMolO, a viral genome database from the Marseille laboratory, and the native disorder prediction programs RONN (Oxford/Exeter) and FoldIndex© (Weizmann).[5]

Target Selection and Construct Design

Within SPINE, several partners built on their prior experience to assist target selection. Although there was open exchange of ideas, no

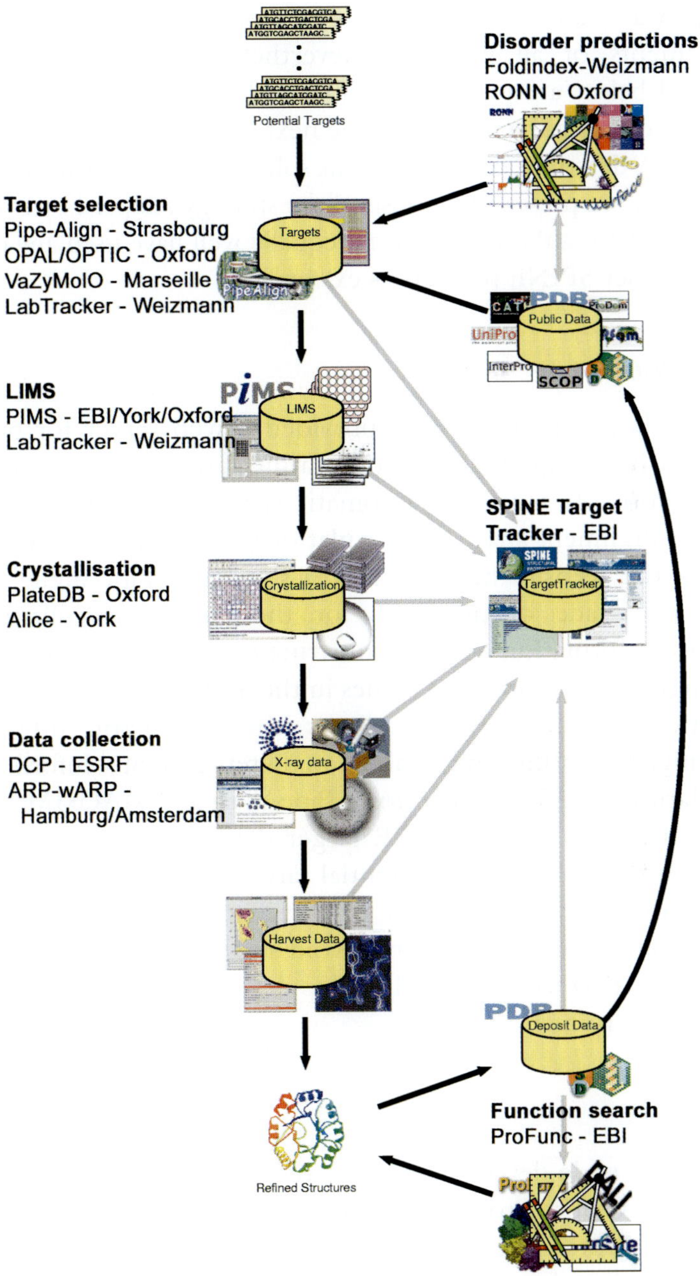

Fig. 1 Overview of bioinformatics and data management for SPINE targets.

attempt was made to construct a single platform incorporating the requirements of all partners. Nevertheless, the powerful tools described below are publicly available and the SPINE-adopted data model[3] provides a framework for the next generation of more unified software. A description of the various solutions is provided by Albeck *et al.*[4] Here, we present a synopsis of the strategy adopted by two representative labs, although we could equally well have used any of half a dozen other SPINE partners as examples.

(i) *Strasbourg*

The Strasbourg partners have provided a web server for protein family analysis (PipeAlign), incorporating target curation and validation protocols as well as automatic structure-based hierarchical multiple alignment analysis.[6] The platform, which is generally available via the website http://igbmc.u-strasbg.fr/PipeAlign/, integrates a cascade of programs for the automatic collation of sequence information and the construction and validation of multiple alignments of protein families in the Strasbourg Gscope bioinformatics platform, which allows the automatic integrated gathering, validation and analysis of heterogeneous information. This platform was used to perform PipeAlign and MACSIM[7] analyses of all targets in the SPINE target database, and an "identity card" was created for each potential target. These "identity cards" are available in a standard XML data exchange format via the project web site. To save time and avoid mistakes, Strasbourg completed this part of the pipeline by developing a combinatorial interface for primer design.

(ii) *Oxford*

The Oxford node developed a single resource for protein and DNA analysis, the Oxford Protein Analysis Linker (OPAL), under which sits the Oxford Protein Target Information Collection (OPTIC), a database for storing the results of these analyses. OPAL incorporates a wide array of both publicly available and bespoke tools developed in-house. The interface to OPAL is through a web form (http://www.oppf.ox.ac.uk/opal/OPAL.php), and the results are returned as a single web page that the user can save to their local machine. The results can also be presented graphically using SEView, a Java applet for browsing molecular sequence data.[8] For authorized users, there is an alternative form which uses locally installed versions of the tools and databases to create and store a full annotation in the OPTIC database; this makes the process easier and faster, and stored annotations are checked regularly and generate user alerts (by email). For genomic-scale annotation, the process of annotation can be trivially scripted. The final phase of target selection, construct design, has been largely automated in the Opine application, which uses the Primer3 tool (http://www-genome.wi.mit.edu/cgi-bin/primer/primer3_www.cgi) customized into a Windows DLL. These tools are routinely used to define sets of constructs of high-value targets.[9]

Wet Lab LIMS

The production of proteins for structural studies includes a number of stages: selection and design of targets, PCR, cloning, recombinant expression (both small-scale and scale-up, using a variety of expression protocols) to produce native and labeled samples, purification and analytical characterization/quality assurance (QA). Furthermore, with high-value targets, it is common to follow a strategy of multiple constructs and expression protocols. This approach generates a cascade of experimental results for which automated tracking, using a laboratory information management system (LIMS), becomes extremely attractive. Commercial LIMS are, at present, poorly adapted to academic research; there is thus a clear need for an easy-to-use LIMS covering the processes of structural proteomics. However, in the absence of such software, the Oxford partners decided in 2001, prior to the start of SPINE, to use as a first implementation an adaptation of a commercially available LIMS, Nautilus (Thermo Electron Corporation; http://www.thermo.com). This has now been developed as far as practically possible, linking to target selection and primer design and covering plate layout, PCR, cloning, small-scale expression trials and crystallization trials. The software is in everyday use, and automated routes exist for the reporting of progress to the OPPF web pages and the SPINE web site. Aspects of the implementation are presented in Albeck *et al.*[4]; suffice it to say that the use of this system was effective in quickly providing a working system and informing the design of the first-generation, purpose-built PIMS software (see below). In contrast, the Weizmann node developed and implemented a purpose-built LIMS to cover its current needs. The system consists of a set of tools for automatic annotation of large-scale data sets covering all aspects from gene to 3D protein structure. This LIMS is intended as a "laboratory notebook" and for tracking and evaluation of different methods, has been based on extensions of the HalX LIMS, in close collaboration with Anne Poupon (University Paris-Sud, Orsay France).[10]

To develop LIMS further, three SPINE partners (Oxford, York and the EBI) have agreed, with CCP4 and the Daresbury laboratory, to jointly develop a free-to-use academic LIMS specially suited to protein production (http://www.pims-lims.org). This development has been funded by the UK BBSRC funding agency and aims to work with additional SPINE partners (including Paris, Amsterdam and Grenoble) to build a system from the bottom up, starting from the Protein Production Data Model mentioned above.[3]

Automation of Cloning and Expression Screening: Increasing Target Throughput

The first experimental links in the chain linking gene sequences to protein structure are cloning and protein expression. The workhorse expression system for structural biology remains *E. coli*, and this was the focus of a SPINE workpackage. The emphasis was on devising and implementing techniques for the parallel processing of constructs. During the course of the project, many SPINE partners implemented molecular cloning and expression screening in 96-plate format. This permitted relatively large numbers of targets to be accommodated, and conversely allowed a relatively large number of different constructs to be investigated for a smaller number of targets at an acceptable cost.

This opened the way to automation, which has been adopted to varying degrees. Most of the groups involved have set up semi-automated liquid handling systems to carry out some or all of their protocols. The motivation to implement automation is largely to enable processes to be scalable and sustainable as error-free operations; however, the protocols can be carried out equally well by hand with appropriate equipment, e.g. multi-channel pipette dispensers. The collaborative nature of the SPINE consortium meant that methods could be developed at a number of European centers and then benchmarked to allow an effective exchange of experience during the three years of the project (see e.g. Berrow *et al.*[11]).

Expression Vectors and Screening Technologies

Amongst SPINE groups, ligation-dependent, ligation-independent and site-specific recombinatorial cloning was used to construct the expression vectors required for protein production (see Table 2 for a synopsis).

The Gateway™ system of recombinatorial cloning and ligation-independent cloning (LIC) was the most widely used (six laboratories). The LIC-PCR procedure is exemplified by the protocol developed by the York Structural Biology Laboratory (SPINE partner York; see Au *et al.*[12]), and is representative of those used by other groups in SPINE. In-Fusion cloning was also used with high success rates (347 PCR products cloned with an efficiency of 89% as assessed by PCR screening of recombinant clones; see Berrow *et al.*[13]). It was notable that LIC methods were taken up increasingly over the course of SPINE and, within these methods, the newer non-Gateway™ methods gained ground later in the project. All SPINE partners adopted the standard His_6-tag for affinity purification, either N-terminal or C-terminal to the protein of interest. The vector design often included the provision for tag removal (N-terminal usually via cleavage with either rhinovirus 3C or tobacco etch virus protease, and C-terminal via carboxypeptidase with a lysine preceding the tag residue to limit cleavage). The large-scale and parallel construction of expression vectors either for multiple targets or multiple versions of

Table 2

Vector Name	Description	Originator
pDESTN-His15	Modifed pDEST15 (InVitrogen) to incorporate N-His6 upstream of GST	Oxford
pET-10AEMBL	pET-22b(+) (Novagen) adapted for Gateway™ incorporates N-His6 and C-His6 tags	Hamburg
pETG-20AEMBL	pET-22b(+) (Novagen) adapted for Gateway™ incorporates N-thioredoxin-His6 and C-His6 tags	Hamburg
pETG-30AEMBL	pET-22b(+) (Novagen) adapted for Gateway™ incorporates N-His6-GST and C-His6 tags	Hamburg
pETG-40AEMBL	pET-22b(+) (Novagen) adapted for Gateway™ incorporates N-MBP and C-His6 tags	Hamburg
pETG-41AEMBL	pET-22b(+) (Novagen) adapted for Gateway™ incorporates N-His6-MBP and C-His6 tags	Hamburg
pETG-50AEMBL	pET-22b(+) (Novagen) adapted for Gateway™ incorporates N-DsbA-His6 and C-His6 tags	Hamburg
pETG-52AEMBL	pET-22b(+) (Novagen) adapted for Gateway™ incorporates N-leaderless-DsbA-His6 and C-His6 tags	Hamburg
pETG-60AEMBL	pET-22b(+) (Novagen) adapted for Gateway™ incorporates N-NusA-His6 and C-His6 tags	Hamburg
pTH10	Gateway adapted from pT7-ZZA N-terminal Z-domain fusion, PreScission™ (3C) cleavage site	Stockholm
pTH18	Gateway adapted from pET21, N-terminal GB1 fusion, PreScission™ cleavage site	Stockholm

(*Continued*)

Table 2 (*Continued*)

Vector Name	Description	Originator
pTH19	Gateway adapted from pET-15b, N-terminal His_6, thrombin cleavage site	Stockholm
pTH1	Gateway adapted from pMAL-c, N-terminal MBP fusion, Factor Xa cleavage site	Stockholm
pTH2	Gateway adapted from pET-43a, N-terminal NusA fusion	Stockholm
pTH3	Gateway adapted from pGb1, N-terminal GB1 fusion	Stockholm
pTH5	Gateway adapted from pT7-ZZA, N-terminal ZZ-domain fusion, Genenase 1 cleavage site	Stockholm
pTH6	Adapted from pDEST15, GST, PreScission™ cleavage site	Stockholm
pTH7	Gateway adapted from pET43a, N-terminal NusA fusion, PreScission™ cleavage site	Stockholm
pTH8	Adapted from pDEST16, N-terminal thioredoxin fusion, PreScission™ cleavage site	Stockholm
pTH24	Adapted from pET-DEST42, C-terminal His_6-tag	Stockholm
pTH27	Gateway adapted from pET-21, N-terminal His_6-tag	Stockholm
pTH28	Gateway adapted from pET-21, N-terminal thioredoxin fusion, PreScission™ cleavage site	Stockholm
pTH29	Gateway adapted from pET-21, N-terminal GST, PreScission™ cleavage site, His_6-tag	Stockholm
pTH30	Gateway adapted from pET-21, N-terminal Z-domain, PreScission™ cleavage site, His_6-tag	Stockholm
pTH31	Gateway adapted from pET11c, C-terminal EGFP fusion	Stockholm

(*Continued*)

Table 2 (*Continued*)

Vector Name	Description	Originator
pTH34	Gateway adapted from pET-21, N-terminal GB1-domain, PreScission™ cleavage site, His_6-tag	Stockholm
pTH35	Gateway adapted from pET-21, N-terminal GST fusion, PreScission™ cleavage site	Stockholm
pTH36	Gateway adapted from pET-21, N-terminal thioredoxin fusion, PreScission™ cleavage site	Stockholm
pTH38	Gateway adapted from pET-43a, N-terminal NusA, PreScission™ cleavage site, His_6-tag	Stockholm
pHGWA	pET-22b adapted for Gateway™ incorporates N-His6 and C-His6 tags	Strasbourg
pHMGGWA	pET-22b adapted for Gateway™ incorporates N-His6-GST and C-His6 tags	Strasbourg
pHMGWA	pET-22b adapted for Gateway™ incorporates N-His6MBP and C-His6 tags	Strasbourg
pHNGWA	pET-22b adapted for Gateway™ incorporates N-His6NusA and C-His6 tags	Strasbourg
pHXGWA	pET-22b adapted for Gateway™ incorporates N-His6 thioredoxin and C-His6 tags	Strasbourg
p0GWA	pET-22b adapted for Gateway™ incorporates C-His6 tag	Strasbourg
p0GGWA	pET-22b adapted for Gateway™ incorporates N-GST and C-His6 tags	Strasbourg
p0MGWA	pET-22b adapted for Gateway™ incorporates N-MBP and C-His6 tags	Strasbourg

(*Continued*)

Table 2 (*Continued*)

Vector Name	Description	Originator
p0NGWA	pET-22b adapted for Gateway™ incorporates N-NusA and C-His6 tags	Strasbourg
p0XGWA	Modified pET-22b for Gateway™ incorporates N-thioredoxin and C-His6 tags	Strasbourg
pG4_casB	Modified pGEX-4T-2 (Pharmacia) for Gateway™ Tac promoter N-GST-thrombin cleavage site	Munich
pG5_casA	Modified pGEX-5X-3 (Pharmacia) for Gateway™ Tac promoter N-GST factor Xa cleavage site	Munich
pI7_casB	Modified pASK-IBA7 for Gateway™ (IBA Institute) tet promoter N-Strep-TagII and factor Xa cleavage site	Munich
pTYB2_casC	Modified pTYB2 (New England BioLabs) for Gateway™ T7/lac promoter C-intein self-cleaving tag	Munich
pTrcHisA_casB	Modified pTrcHisA (InVitrogen) for Gateway™ Trc promoter N-His6 enterokinase cleavage site	Munich
pET-46NKI/LIC	Modified pET-46Ek/LIC vector (Novagen) incorporating 600 bp insert, zero background, N-or C-His 6 tag and enterokinase or HEV3C cleavage site, Amp^R or Kan^R	Amsterdam
pET-22NKI/LIC	Modified pET-22b vector (Novagen) incorporating 600 bp insert, zero background, no-N-term tag, choice of addition of C-term His6 tag, Amp^R or Kan^R	Amsterdam
pET-28NKI/LIC	Modified pET-28a vector (Novagen) incorporating 600 bp insert, zero background,	Amsterdam

(*Continued*)

Table 2 (*Continued*)

Vector Name	Description	Originator
	N-His6 tag and HRV 3C cleavage site, Amp^R or Kan^R	
pET-YSBLIC	pET-28a adapted for LIC incorporates N-His6 tag	York
pET-YSBLIC3C	pET-28a adapted for LIC incorporates N-His6 tag and 3C protease cleavage site	York
pOPINA	pET-28a modified for In-Fusion™ incorporates either N-His6 or C-His tags depending upon site of cloning.	Oxford
pOPINB	pET-28a modified for In-Fusion™ includes N-His6 tag and 3C protease cleavage site or C-His tags depending upon site of cloning	Oxford

fewer targets creates a need for parallel expression screening on a small scale. Over 10 000 expression trials conducted in SPINE laboratories resulted in 70% of targets producing at least one soluble construct. Most interestingly, the bacterial and human targets gave similar results, with viral targets performing somewhat worse. These data are comparable to screening results reported by other large-scale structural proteomics projects largely focused on microbial genomes.[14,15,65]

A large benchmarking study for recombinant protein expression and solubility in *E. coli* was carried out in 2006 as part of the SPINE project.[11] The study compared the solubility profiles of a common set of 96 protein constructs comparing protocols implemented in SPINE laboratories and found that, although 81 of these constructs produced soluble protein, only 25 produced comparable levels of protein in every laboratory. On further analysis, the data demonstrated that different methodologies identified similar groups of best and worst expressing proteins and that small-scale expression can give a good prediction of levels of soluble expression attainable on scale-up. This consistency in prediction is important, since it enables effort downstream of cloning and expression screening to be invested in the most tractable targets. The study also highlighted the variability between protocols to detect proteins that fell in the mid-range of expression, indicating that for a substantial subset of targets, the parameters chosen for screening have a major effect on the expression outcome. This benchmarking study formed the basis for refinement of a set of guidelines for the most effective strategy for the production, for a particular protein, of sufficient material for structural studies. These guidelines include (i) construct optimization, (ii) the use of homologous proteins from different species and (iii) the exploration of expression space in both prokaryotic and eukaryotic systems.[16]

A collaborative pilot project by the York and Oxford partners on proteins from a biomedically important prokaryote, *Bacillus anthracis* (*B. anthracis*), provided a vehicle for the development of robust technologies for protein expression, crystallization and structure determination, to stress-test the pipeline under true HTP conditions and to investigate more automated methodologies.[12] Overall, 44 *B. anthracis* and *B. subtilis* structures were solved by X-ray crystallography in the pilot study, plus two by NMR. Three biologically important protein structures, BA4899, BA1655, and BA3998, involved in tRNA modification, sporulation control and carbohydrate metabolism, respectively, exemplify these achievements (Table 3).

Within the *B. anthracis* study, target analysis by biophysical clustering based on pI and hydropathy provided useful information for future target selection strategies. An analysis of the correlates of success with physico-chemical properties has previously been made for the *T. maritima* project,[17] and some of the findings applied also to the SPINE *B. anthracis* cohort. In particular, when analyzed in terms of the calculated pI and hydropathy, the open reading frames of the *B. anthracis* group were in a similar manner to those of *T. maritima* (Fig. 2). In line with the findings of Canaves *et al.*, the targets that led to successful structure determinations almost always came from cluster A. Such plots are probably useful to direct target selection and construct design. It is intriguing to note that two structures were solved from cluster B by NMR analysis.

Table 3

Bacillus anthracis	YPLATE1 G1; phosphoribosylglycinamide cyclo-ligase OXG2 David; Thil tRNA thiolase BA5489 YPLATE1 B6; UDP glucose 4-epimerase BA5505 Transcription regulator CodY N-terminal domain YPLATE1 D3; alkenesulfonate-mono-oxygenase EDPLATE1 A4; glycosyl transferase YPLATE1 A4; phosphoribosylaminoimidazole-succinocarboximide Alanine dehydrogenase Ald-1 BA0592 Cytidine deaminase Cdd2 BA4525 YPLATE2 E5; inosine-uridine preferring nucleoside hydrolase OPTIC6701; glyceraldehyde 3-phosphate dehydrogenase 2 OPTIC6189; glyceraldehyde 3-phosphate dehydrogenase 1 OPPF1288; light pyruvate kinase OPPF1317; heavy pyruvate kinase OPPF1314; 5-formyl tetrahydrofolate cycloligase OPTIC5808; translation elongation factor P OPTIC3381; uridylate kinase SpoOE phosphatase OPTIC1859; methionine aminopeptidase OPTIC5390; 3-oxo-acyl (carrier protein) reductase OPTIC5399; ribulose phosphate epimerase OPTIC6009; histidyl tRNA synthetase 2 OPTIC6697; enolase YNMR1; YisI conserved domain protein BA1655 Natalia 5; dihydrodipicolinate synthetase YR4; purine nucleoside phosphorylase deoD YPLATE1 B1; phosphoribosylaminoimidazole carboxylase purE YPLATE1 A2; guanosine monophosphate reductase guaC YPLATE1 F3; ferrochelatase hemH1 Neil IV; endonuclease IV

(*Continued*)

Table 3 (*Continued*)

	YPLATE1 F5; superoxide dismutase Mn sodA2
	YPLATE1 C6; superoxide dismutase Mn sodA1
	YMB13; ribulose-phosphate-3-epimerase
Bacillus subtilis	BCBRG102; CsrA
	YVLAD1; appA
	YTHAY1; forespore regulator of sigmaK checkpoint Bsu2771
	Neil 6; dUTPase YncF
	Neil 7; dUTPase YosS
	IGBMC-1123-000; PhoP
	CIRMMP02; Sco1
	CIRMMP05; S46VCopAa
	CIRMMP06; S46VCopAab
	CIRMMP08; oncomodulin
	CIRMMP10; SOD like protein

Eukaryotic Expression Technologies

SPINE was committed to work on high-value human and viral protein targets, despite the observation that many of these targets are intractable in prokaryotic expression systems. A variety of eukaryotic expression systems were developed during the SPINE tenure, most successfully in insect and mammalian cells (in general, yeast fared less well). The development and implementation of HTP eukaryotic expression methodologies was the subject of one SPINE workpackage, resulting in medium- or high-throughput protein expression in baculovirus and the use of transient or stable expression in mammalian cells. Methods designed specifically for HTP expression were adopted in the baculovirus system; for example, an alternate method of recombinant isolation (BaculoDirect, Invitrogen) to integrate baculovirus expression systems into Gateway or Gateway-based recombination facilitated the quick generation of recombinant viruses. Similarly, the development and streamlining of protocols for the transient expression of proteins in mammalian cells offered a methodology which was compatible with HTP approaches.

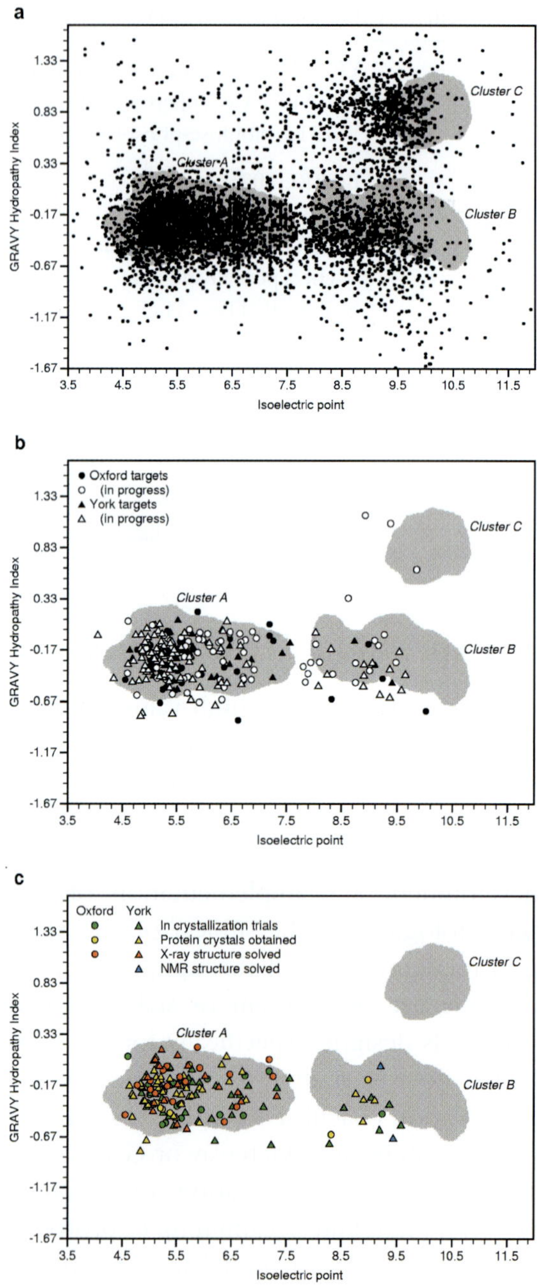

Fig. 2 Biophysical target clustering based on PI and hydropathy can be used to inform target selection strategies.

Transient expression in mammalian cells proved a particularly successful strategy for the production of secreted glycoproteins.[18] The applicability of this approach for structural biology targets was enhanced by the development of methodologies to manipulate the nature and level of glycosylation to yield protein samples that were more amenable to crystallization.[19] One highlight resulting from these developments was the structure of the entire extracellular adhesive complex of the receptor protein tyrosine phosphatase mu. The extracellular region of this cell surface receptor, comprising six domains bearing a total of 12 predicted N-linked glycosylation sites, was crystallized as a trans-dimer yielding a 3 Å-resolution structure (Fig. 3; see Aricescu *et al.*[20]).

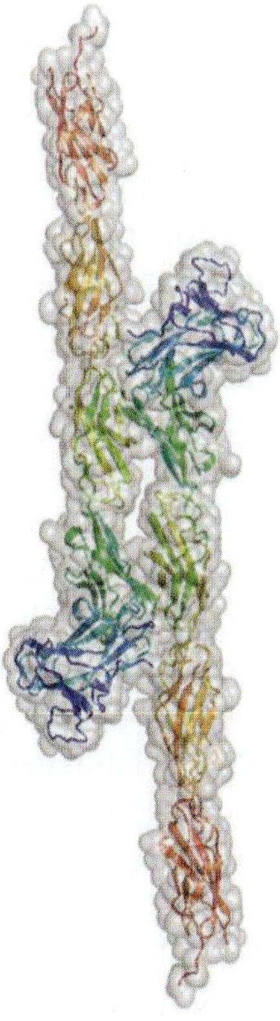

Fig. 3 Crystal structure of RPTP mu.

Protein Characterization

The current protein structure determination pipeline contains two major bottlenecks, protein production and crystallization (or formation of a suitable sample for NMR). Protein production presented a particular challenge for the SPINE partners because many of the biomedically important proteins targeted for structure determination were eukaryotic, whereas the workhorse *E. coli* expression system was prokaryotic. The overall statistics for the SPINE project testify to the severity of this problem: some 30% of constructs for eukaryotic targets yielded soluble protein,[66] in contrast to some 89% for the expression of carefully chosen bacterial proteins.[21] The toll on success taken at the second bottleneck is also acute: less than 30% of the total set of SPINE targets which expressed as soluble protein yielded suitable crystals or NMR samples,[66] although the success rates for subsets of targets selected for potential tractability (e.g. anthrax proteins[12]) rose to 60%. Given this attrition rate, careful optimization of the quality of samples taken forward to crystallization or preliminary NMR analysis is vital. Empirical observations suggest that there is a correlation between certain biophysical properties of a protein preparation and the probability of obtaining crystals; for example, it was suggested that monodispersity measured by dynamic light scattering correlates well with the ability to crystallize.[67–69]

The breadth of experience of the SPINE consortium provided an opportunity to assess the efficiency of a variety of quality assessment (QA) tools. To investigate which QA methods were being used routinely, a survey was carried among the SPINE partners. In almost all purification protocols, size exclusion chromatography (SEC) was used as a polishing step and the SEC elution profiles provided a first, and very useful, indication of protein purity, homogeneity and oligomeric state. In all SPINE laboratories, the purity was also checked by SDS-PAGE, which yields the approximate subunit mass and hence probable identity of the purified protein (details of the survey are reported in Geerlof *et al.*[70]).

As discussed above, anecdotal evidence has suggested correlation between success in crystallization and various biophysical properties of a protein preparation. The SPINE survey provided some (but not overwhelming) support for some of these correlations.[70] The routine use of dynamic light scattering (DLS) to determine the monodispersity of the protein was widespread in SPINE. Mass spectrometry (MS) was also widely used (~50% of SPINE laboratories) to check the quality of the preparation, to confirm the identity of the purified protein by accurate determination of its molecular mass and to obtain information about post-translational modifications such as phosphorylation and glycosylation.

Methods to determine the folding state of the sample, such as circular dichroism (CD) and NMR, were less routinely employed. The main reason for the limited application of these techniques was the lack of access to the necessary equipment; however, this presumably reflects a prioritization of resources driven by the relatively low information content of a technique such as CD. Other QA methods that were mentioned in the survey but were not routinely used were analytical SEC, native PAGE, and static light scattering (SLS). Of these, SLS in particular is a potentially very powerful technique.[70] A notable development during the course of SPINE was the uptake by several laboratories of the fluorescence-based thermal shift assay for protein stability (commonly known as thermofluor); this technology was implemented in the context of structural proteomics by the Stockholm partners, and is emerging as a QA tool and probe function of great value.[22,70]

HTP Crystallization, Imaging and Crystal Handling

The implementation of automated, miniaturized HTP protein crystallization was a keynote achievement of the SPINE project that has had a major impact on structure determination.[23] This methodology was made possible by developments in the implementation of small-volume (nanodrop) crystallization methods, algorithms for automated crystal image recognition, systematic protocols for optimizing crystal growth and quality, and novel procedures for membrane–protein crystallization. The success of this program led to an increase in the number of European laboratories using automated nanoliter crystallization methods by roughly an order of magnitude during the course of the project.

Accurate dispensing of small drops is challenging, not least because of problems with sample cross-contamination and drop evaporation, although a number of different solutions to these problems have been discovered. Accurate dispensing is complicated further by the differences in viscosity, volatility and surface tension of solutions to be dispensed; however, the Cartesian Microsys and TTP Mosquito are two instruments that are now widely used to dispense nanoliter droplets, requiring only about 15 μl of protein material for a full 96-well plate screen of crystallization conditions. Sitting-drop vapor diffusion experiments are still most commonly used for crystallization, in part because this is the simplest method to automate. Fluidigm R have developed a microfluidic free-interface diffusion technique whereby screening 96 crystallization conditions requires only 1.5 μL of protein solution, and this has been successfully applied within SPINE, mainly to targets of a particular value.

Several systems for crystallization tray storage, integrated with an image acquisition system that allows automated and scheduled imaging of the crystallization experiments have been used by SPINE partners. Indeed, purpose-built software has been developed by the partners (see e.g. Fig. 4, Mayo *et al.*[24] and Berry *et al.*[23]). The systematic imaging of crystallization experiments offered by these systems has been found to be of real scientific value, allowing a proper evidence-based evaluation of results. The imaging schedule defined in each laboratory is determined by several factors: the amount of storage available, the importance of the tray's contents (laboratories may wish to keep high-value samples for longer periods and image trays more regularly in these late stages to monitor for any nucleation or slow growth of crystals) and the speed of the imaging device. The result is that the time period during which trays are imaged at regular intervals varies between one month and a year, depending on the laboratory. Interestingly, the experience of the Oxford partners is that although most crystals grow within a few days in nanodrops, there is a small number of tough problems where a small number of crystals

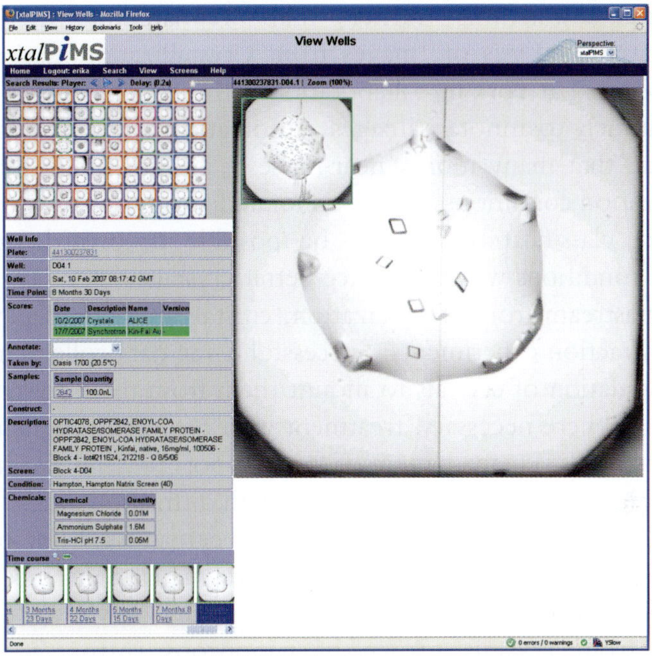

Fig. 4 An image of a crystal trial shown in the Xtal PIMS crystallization management system.

grow much more slowly, often over a period of months. Experience gained with these image viewers is now going into the design of the next generation of viewers, as a follow-on from the SPINE work, primarily funded by an EU FP6 project, BIOXHIT (and forming part of the PIMS project described above).

One important consideration for viewing systems is integration with image analysis and human annotation tools. This integration creates the prospect of useful data mining from the huge databases of images that are being generated (the largest single database amongst the SPINE partners now comprises over 53 million images). The aim of analysis software is simply the detection of crystals; but if this can be automated, it will provide a massive database with reliable classification outcomes that can feed back into the design of screening and optimization protocols. Within the SPINE project, the aim was to produce software that could be adapted to multiple systems through retraining algorithms rather than reprogramming. ALICE (Automated anaLysis of Images from Crystallization Experiments) was developed at York in collaboration with Oxford, where it is now used routinely to annotate images. Automatic classification of images will mean that many more "near-misses", i.e. potentially promising crystallization conditions, will be annotated; and so even with a rather imperfect classification, it may be possible to provide guidance towards conditions which produce better crystals.

Downstream from crystallization, crystals must be deployed for X-ray diffraction experiments. Successful X-ray data collection entails (i) manipulation of crystals to mount them from their crystallization drops; (ii) where necessary, treatment with cryo-protectants to allow cryo-cooling; (iii) management of the transport to X-ray facilities; and (iv) linking of protein production data with diffraction data. The handling of crystals grown in nanodrops has been an issue for some laboratories, but the problems can be overcome; for instance, in the Oxford laboratory, the vast majority of structures solved are currently derived from crystals grown in nanodrops (http://www.oppf.ox.ac.uk/OPPF/public/technologies/crystal.jsp). This pipelining between crystallization and (automated) data collection has been addressed by collaborative work between SPINE, the e-HTPX project, synchrotrons and others to produce systems such as DNA and ISPyB, which exploit recent developments in automated beamline sample changers (see below, Beteva *et al.*[25] and Cipriani *et al.*[71]). An immediate benefit of this has been the ability to view diffraction images from the home lab within a few seconds of collection.

NMR Spectroscopy

At present, the process of structure determination is less automated and slower for NMR than X-ray crystallography, and SPINE therefore supported the development of HTP NMR pipelines (reviewed by Ab *et al.*,[26] some highlights from which are presented here). Attention was given to bottlenecks from sample preparation, data acquisition, data processing and data analysis through to structure determination, leading to improvements in sensitivity, automation, speed, robustness and validation. Specific highlights were protonless ^{13}C direct detection methods and inferential structure determination (ISD). In addition, NMR was applied to deliver over 60 NMR structures of proteins including five that failed to crystallize — compelling evidence that NMR spectroscopy has a role to play in structural proteomics pipelines.

The integration of NMR into a HTP pipeline requires HTP NMR screening of samples for solubility, monodispersity and foldedness. SPINE co-developed, with Bruker BioSpin, a novel flow cell for use in cryoprobes that provided high sensitivity and rapid sample handling. Faster data acquisition was also achieved by improving the polarization recovery for deuterated proteins, namely using recovered H^N anti-TROSY polarization in queued TROSY,[27] resulting in a 50% time saving. For non-deuterated proteins, where HSQC-type transfer is the method of choice, the implementation of an extended flip-back (EFB) scheme[28] recovered >50% of the previously wasted non-amide proton polarization, affording a sensitivity enhancement of 40% or more from accelerated recovery of H^N polarization.

^{13}C direct detection benefits from the fact that the contribution to relaxation from a paramagnetic center on ^{13}C is 16-fold smaller than from ^{1}H, so avoiding ^{1}H transfers can improve signal detection in paramagnetic systems. ^{13}C NMR can also be useful for protein regions showing high conformational/chemical exchange that often co-localize with functional domains. In SPINE, these ^{13}C methods have been used for both paramagnetic systems[29–31] and diamagnetic systems.[32,33,72]

Developments improved the power and speed of resonance assignments. The application of wavelet-denoised NOESY spectra, using software that coupled the automated peak identification and structure determination[34], was integrated into existing ARIA software. The advantages of this strategy are most significant for noisy NOESY spectra. The method requires little additional computational time. The software tools provided de-noising, peak picking and modules to export different file formats, and now incorporates programs for combined NOESY assignment and structure calculation.

ISD is a novel objective approach to determine the probability distribution of an unknown structure (and its precision) based on prior information and experimental data. It has been implemented in a Markov chain Monte Carlo algorithm and demonstrated to improve the quality of structures. The Bayesian inference method has also been applied to solve the problem of assigning relative weights to experimental evidence and physical restraints at little or no computational cost.

Other developments were also driven by SPINE. These included, firstly, the use of paramagnetic RDCs for metalloproteins containing paramagnetic ions. Paramagnetic RDCs provide information on domain conformational freedom in multi-domain proteins with a paramagnetic metal in one domain; this was exploited to study the relationship between the N- and C-terminal domains of calmodulin.[35] Secondly, improvements in AUREMOL improved the simulation of multidimensional NOESY spectra. Thirdly, a method using the unassigned portion of NOE spectra was implemented for structure calculation in ARIA and CANDID. SPINE tests showed that this allowed correct structures to be found in CANDID, with up to 30% of assignments missing (cf. 10% previously).

X-ray Diffraction Data Collection: Automation for HTP Samples

Although major strides have been made in synchrotron data collection in recent years, bottlenecks arise with the increasing number of crystals requiring testing, especially in sample mounting and centering in the X-ray beam. This is relatively slow if done manually, but automation can significantly improve efficiency. SPINE therefore put considerable effort into this area, resulting in the SPINE standard sample holder and vial, developed for use in a robotic sample changer.[36] The sample holder was based on the Hampton Reasearch CrystalCap Magnetic system, and bridged the gap between existing manual sample handling and automated loading whilst still being broadly compatible. The SPINE standard incorporates a ECC200 Data Matrix code on the vial cap base for sample tracking and a human-readable code on the rim. The sample code can be scanned by hand-held readers directly into the ISPyB LIMS. SPINE standard sample holders and vials are now commercially available from major suppliers of crystallography materials.

The European Synchrotron Radiation Facility (ESRF) provided a testbed for beamline automation within SPINE. Prototype sample changers (SC1 and SC2) were installed on two beamlines (ESRF ID29 and ID14-3), followed by the production model (SC3) which was installed on eight beamlines (ID14-1, ID14-2, ID14-3, ID14-4, ID29, ID23-1, ID23-2 and BM14 at ESRF). To encourage uptake, a basic starter kit was made generally available in February 2006, comprising five cassettes, forceps for handling cassettes and vials, a vial releaser and a cassette canister compatible with dry shipping dewars. The technology allows 50 samples to be screened automatically in 2.5 hours. Considerable effort was devoted to the software, which integrates directly with applications such as DNA, ISPyB and the EXES system, to form the core of a data collection pipeline.

The ESRF data collection pipeline combines the automation in sample handling and data collection with the automatic alignment of optical elements and improved beamline diagnostics.[25] The overall pipeline developments were the result of a collaboration between SPINE and BIOXHIT, eHTPX, the BM14 beamline and DNA. The pipeline now provides a workable system for remote control of the beamline and remote data collection. We expect that this next stage in the application of automation will mark a major change, providing a better user experience, reducing costs for access and allowing for a more efficient use of beamlines.

Crystallographic Structure Determination

To improve throughput, macromolecular crystallography procedures must be streamlined; and work in a number of laboratories in Europe, including several SPINE partners, directly addressed this during the period of SPINE. Scripting now links the various stages in the analysis, and better algorithms continue to be formulated in key areas. However, SPINE had only limited resources to contribute to the development of HTP crystallographic computing; whilst some of these were devoted to method development in specific software platforms such as ARP/wARP, particular efforts were made to bring together the major users and providers of crystallographic software in order to gain access to the latest tools and to provide feedback and input to inform the ongoing developments. Thus, SPINE held two early workshops, which helped focus attention on the needs and possibilities for automation, and a third workshop towards the end of the project, where the current methodology was tested against a set of targets selected from Oxford and York to assess the current state of the art applied to real problems. We will provide a synopsis of the lessons of this workshop (for a full analysis, see Bahar *et al.*[37]).

Twenty-three data sets were selected from a group of bacterial targets under study in Oxford and York. Data were provided as merged structure factors and sequences, and these were the basis of the major efforts of the workshop. In addition, for a different set of targets, raw images were made available for assessment of an automated processing protocol; in the workshop, this work was essentially restricted to a single-problem data set (which was thereby solved).[37]

It was realized very early in the workshop that the experimental data as provided often did not carry all of the necessary quality assessment information in a form accessible by user or computer. Some of the information such as wavelength should have been recorded in the reflection file header. We reiterate here the suggestions of Bahar *et al.*[37] for a set of core data to be recorded in an accepted exchange format. The major requirements for decision making in automatic procedures fall into four categories:

1. Sample parameters, e.g. the sequence, molecular weight and expected numbers of "heavy" atoms.
2. Details of the X-ray experiment: (i) direct parameters such as wavelength, beam line and temperature; (ii) derived parameters including unit cell, point group, the likely number of molecules in the asymmetric unit and the presence of any non-crystallographic translational operators; and (iii) quality indicators including nominal resolution, estimated B factor and anisotropy plus completeness, multiplicity, $I/\sigma(I)$ and merging R factor, all as functions of resolution.

3. Intensity statistics to be tested against expectation values, including cumulative intensity distributions and moments. These are sensitive indicators of problems in the experiment, such as twinning or local errors in the processing (e.g. saturation of substantial numbers of low-resolution terms).
4. Identification of special features of the crystal, such as pseudo-symmetry or potential alternative indexing.

The XIA-DPA pipeline for automatic data processing[37] was tested extensively and found to perform adequately for many data sets, but further work is required to ensure that the process detects and flags aberrant cases. A final assessment of data quality at the end of an automated procedure may be valuable. Automated procedures have two features to offer, reproducibility and standardization, which should allow for more objective summary statistics. Thus, they may perform the tedious transformations between different packages, enabling all statistics to be calculated on the same basis.

In summary, it was found that the success rate of the overall pipeline was reasonable. It is clear that the HTP agenda has improved the usability and effectiveness of crystallographic software for the general user. As data processing programs improve in reliability and more sophisticated algorithms become available, there is no doubt that the overall success rate of the pipeline will improve further.

Structure Annotation

Structural annotation is essential if the increase in data from the HTP technologies is to impact fully on biomedical research. Within the SPINE consortium, the EBI focused particularly on this issue. ProFunc (Laskowski *et al.*[38]; http://www.ebi.ac.uk/thornton-srv/databases/ProFunc/) was developed to help identify the likely biochemical function of a protein from its three-dimensional structure. This software makes use of a range of methods, some novel, to identify functional motifs or close relationships to functionally characterized proteins. A summary of the analyses provides an at-a-glance view, with more detailed results available on separate pages. ProFunc should be of particular use in large-scale projects where structures are determined for hypothetical proteins of unknown function and in the comparative analysis of large protein families. The system was made available to SPINE members, and a large amount of testing led to the inclusion of new algorithms for finding structural motifs and similarities, better default parameters for many of the individual tasks and a clearer graphical presentation of the results. Importantly, the system has been extended to ligand annotation, since ligand prediction will play a crucial role in elucidating function from structure.

A second approach to protein function prediction developed by the EBI exploited HTP ligand screening.[39,40] This method aims to exclude those ligands whose shape does not match the predicted binding pocket. A new protein structure is analyzed with the program SURFNET[41], which identifies surface clefts and indentations, with the aid of residue conservation scores. Harmonic expansion is performed to describe the shape of these pockets, and orientation based on the moments of the coordinates allows resulting coefficients to be used for extremely rapid shape comparison. All ligands in the PDB were clustered based on their shape coefficients, and tests showed that, in many cases, shape alone is sufficient to identify the correct ligand type.

Summary of Structural Results

The aim of SPINE was to develop and exploit technological advances in structural biology to tackle difficult problems related to human health and disease. The targets chosen for study in SPINE were human proteins plus proteins from bacterial and viral pathogens. The majority of these represented *bona fide* high-value targets; however, as part of the strategy to evaluate and implement HTP technologies, some "low-hanging fruit" targets were also selected for study.

The human protein targets fell into four broad groups: those involved in cancer, immune defense mechanisms, neuronal development and neurodegenerative disease. Of the 2000 targets selected in SPINE, around 800 were within these categories. More than 200 novel protein or protein complex structures were determined from the human target pool.

Within the cancer targets, four important kinesin structures were solved. These included (i) the structure determination of the human kinetochore-associated protein CENP-E,[42,43] essential for some aspects of kinetochore microtubule attachments; (ii) the pseudo-atomic structure of the MT-CENP-E complex, solved by a combination of X-ray crystallography and 3D image reconstruction[44]; (iii) the structure of human mitotic Eg5 in complex with a new potent antimitotic inhibitor, as well as the production of a new crystal form for native Eg5[73,74] (Garcia-Saez and Kozielski, patent pending); and (iv) the structure of a human kinesin-associated protein in two different nucleotide states.[45] In addition, an atypical protein kinase C (PKCι) in complex with the bis(indolyl)maleimide inhibitor BIM1[46] was determined. This structure revealed, for the first time, the architecture of the turn motif phosphorylation site, which is characteristic of PKCs and PKB/AKT but is completely different to that in PKA. The PKCι/BIM1 complex constitutes the basis for rational drug design for selective PKCι inhibitors.

More than 20 crystal structures (novel structures or primary examples of protein–ligand complexes) were determined from a target group of classical nuclear receptors (NRs) such as RAR, vitamin D receptor, oestrogen receptor and orphan receptors (e.g. RXR, ERR and ROR). A resulting structure-based sequence analysis of the ligand-binding domain of NRs led to the identification of two classes of the NR superfamily, distinguished by their oligomeric behavior.[47]

SPINE targets involved in the human immune response centered largely on the MHC–T-cell receptor (TCR) complexes, which mediate T-cell recognition of host and pathogen antigens. The crystal structures of a series of MHC Class 1 peptide–TCR complexes were determined, including complexes of TCRs recognizing two different tumor antigens (Chen *et al.*[48]) and TCRs recognizing three different AIDS virus antigens. Such structures are potentially informative for therapeutic strategies; for example, an MHC tumor peptide–TCR structure characterized the location of TCR binding on the antigenic peptide and suggested methods for improving the peptide shape complementarity to both MHC and TCR to improve affinity, providing a basis for the rational optimization of peptides for use in clinical trials to boost anti-tumor immune responses.

Several proteins important in neurological diseases were targeted for structural analysis. One key target group was the copper-binding ATPases 7A and 7B, also called the Menkes and Wilson proteins. The solution structures of four single domains (2, 3, 5 and 6) of the ATPase7A membrane protein and two domains (5 and 6) of ATPase 7B were determined, including the soluble copper chaperone. These structures provided insights into the effects of the disease-linked mutation A629P, which reduces the affinity for copper(I) and makes the protein more susceptible to proteolytic cleavage and/or results in a reduced capability for copper(I) translocation.[49]

The crystal structure of GlcCerase, an acid-β-glucosidase which contains a disease-linked mutation in Gaucher disease[50], was determined. The subsequent structure of GlcCerase conjugated with the irreversible inhibitor conduritol-B-epoxide revealed an active site conformation, which is consistent with the observed reduction in catalytic activity of the mutated protein.[51] The native and complex structures may therefore act as the basis for engineering an improved GlcCerase protein for enzyme replacement therapy and for the rational design of drugs aimed at restoring the activity of defective GlcCerase.

The human pathogen targets included proteins of both bacterial and viral origin. The bacteria targeted represented a range of threats to human health, including food poisoning, respiratory diseases and potential bioterrorism agents, all of which manifest antibiotic resistant strains. The targets, in addition to those already noted, were *Mycobacterium tuberculosis* (*MTB*), *Neisseriae* (*N. meningitidis* and *N. gonorrhoeae*; Oxford), *E. coli* (e.g. 0157:H7; Stockholm, Weizmann, Florence), *Klebsiella pneumoniae* (Lisbon) and *Streptomycetes* (Stockholm, Weizmann). The expected tractability of these targets varied widely. Some bacterial targets were selected to provide a cohort of proteins to validate the pipelines being assembled, especially the *B. anthracis* proteins mentioned above (and described in more detail in Au *et al.*[12]), where 42 structures were determined, and some targets from *Campylobacter jejuni*. On average, two or three constructs were produced for each target and each construct was subjected to an average of two expression trials, with 29% of constructs producing soluble protein. However, there was significant variability between target organisms, with 45% of all *B. anthracis* constructs expressed in soluble form, but only 33% for MTB and 15% for the other bacteria.

The major effort on TB involved six partners (Paris, Hamburg, Uppsala, Stockholm, York and the Weizmann Institute), primarily targeting proteins from *MTB* (H37Rv). We select here a single snapshot from this major body of work to illustrate the SPINE efforts to tackle the problem of insoluble protein expression. *MTB* proteins are notoriously difficult to express in *E. coli*.[21,52] To address this, the Hamburg partners developed an MTB expression system based on the faster-growing close relative *Mycobacterium smegmatis*.[53] Proteins expressed in this system were also correctly post-translationally modified (e.g. glycosylation and methylation). The starting protocol[53], based on an inducible acetamidase promoter, was modified to make it compatible with the EMBL pETM vector system, allowing the *MTB* LipB enzyme to be expressed as a native enzyme without a covalently bound ligand (*E. coli* expression produced a mixture of native and ligand-bound enzymes). The structure was subsequently solved by X-ray crystallography at a spacing of 1.08 Å.[54]

A range of viruses were targeted by SPINE groups, from large dsDNA viruses such as Epstein Barr virus, causing infectious mononucleosis and carcinomas (Grenoble partners), and *Vaccinia* virus (VACV), the smallpox vaccine (Oxford partners), to somewhat simpler ssRNA viruses such as the newly emergent human pathogen, the severe acute respiratory syndrome coronavirus (SARS-CoV) (Marseille and Oxford). In addition, some viruses of less direct relevance to human health have been investigated, for instance bacteriophage P2 (Marseille). We found that, as for other targets, the bottleneck in producing viral structures is the production of soluble protein. This problem is particularly severe for viral proteins, which are very insoluble compared to those from bacterial or mammalian sources, with only 163 out of ~600 selected targets yielding useful soluble expression, resulting in a haul of only ~30 crystal structures. As for the bacterial targets, not all viruses are equal in their difficulty; thus, we had a significantly higher success rate in the production of viral proteins in *E. coli* for poxviruses and flaviviruses compared to herpes viruses and SARS-CoV.

For the bacterial targets, we will give snapshots of technology helping difficult projects, in this case the implementation of a simple protocol for side chain methylation[55] and the much more ambitious development of a library-based method to derive soluble fragments of intractable proteins.[56] The reductive methylation of lysine residues is not new[75], and the reduction of surface entropy has been reported to facilitate crystallization. To test the effectiveness of our protocol, we carried out a study on ten, otherwise intractable, eukaryotic and viral targets. Upon methylation, several of these produced diffraction quality crystals.[55] One of these, MHV NSP6 (Meier *et al.*, unpublished), a replication protein from a MHV that is a related virus to SARS-CoV, did not crystallize in its unmodified form. After reductive methylation (where the mass of the protein increased by 310 Da, equivalent to 11 residues being modified), the protein crystallized readily in commercially available screening solutions. After optimization, large crystals were grown and the structure was solved at 2.3 Å. The ESPRIT methodology was developed by the Grenoble partners as a library-based method to find soluble, expressable fragments of difficult targets[56] and has been applied to several viral proteins, with particular success with a fragment of a subunit of influenza virus polymerase, providing a first glimpse of a very difficult but important target.[57]

Spine2, Specialist European Programs and the European Outlook

As part of the first wave of European-based HTP structural proteomics initiatives, which also included the Réseau National des Génopoles in France and the Berlin Protein Structure Factory, SPINE yielded products that are widely used beyond this community and developed solutions that are broadly applicable to many proteins. SPINE made HTP technologies accessible to many protein production laboratories, and enabled or at least facilitated their adoption by groups that would otherwise have struggled to acquire HTP capabilities. The portfolio of European Commission (EC)-funded integrated programs nucleated by SPINE has expanded over the subsequent few years to include BIOXHIT (synchrotron technologies), VIZIER (structural virology of RNA viruses), SPINE2-COMPLEXES (see below), E-MEP (membrane protein structures) AND 3D-Repertoire (yeast complexes) and a number of European networks (e.g. 3D-EM).

The momentum generated by SPINE provided the opportunity for the further development and streamlining of HTP technologies and their application to more challenging structure/function problems; the next-generation project, SPINE2-COMPLEXES, was funded by the EC to do just that. It is timely to attack more difficult biological systems by combining the knowledge of genomes with HTP methods for structural proteomics. With a remit to further improve HTP procedures for cloning, protein expression and purification, biophysical characterization, crystallization, data collection and structure solution, SPINE2-COMPLEXES — which began in July 2006 — added electron microscopy technologies. The aim is to utilize the SPINE approach to provide key structural insights into proteins and protein complexes involved in a common theme of "signaling pathways from receptor to gene". The target focus includes proteins presenting particular difficulties for structure determination, e.g. glycosylated proteins, proteins expressed in eukaryotic cells, co-expressed proteins, intrinsically unstable proteins and large protein complexes. Significant achievements have already resulted from the project, such as the publication of new technologies for controlling glycosylation in secreted proteins, a key step towards crystallization.[19] The success of SPINE2-COMPLEXES will be measured in terms of the scientific impact of its output rather than simply counting PDB depositions. The structures of target complexes will be central to understanding the corresponding fundamental cellular processes as well as the molecular basis of cancer, auto-inflammatory and neurological diseases, which form the specific biological areas of focus for SPINE2-COMPLEXES.

Some of the ambitious aims of SPINE2-COMPLEXES are currently far beyond the capacities of individual laboratories working in isolation, but the project consortium will provide the infrastructure and critical mass to tackle these challenges collectively. Although there are very limited resources available in SPINE2-COMPLEXES for pure technology development and infrastructure, it will be a test platform for the existing infrastructure for structural biology in Europe. The results generated will help inform a major infrastructure development program, funded initially by the EC as part of the ESFRI (European Strategy Forum on Research Infrastructures) Roadmap (http://www.cordis.europa.eu/esfri), which will aim to build a truly competitive integrated infrastructure for structural biology in Europe.

Conclusions

In summary, SPINE was genuinely productive, with 308 novel structures determined. However, it is hard to quantify the cost-effectiveness of the process, since many labs received funds from additional sources; indeed, one of the strengths of SPINE was that it allowed many of the partners to leverage additional resources. Furthermore, the attribution of structures in the PDB to SPINE was also incomplete and inconsistent; the best source of precise information on the structures solved is therefore www.spineurope.org/page.php?page=scoreboard. Other metrics may be more telling, for instance the number of SPINE papers published, which is now >215 (with more than 100 further papers being SPINE-related). However, these metrics do not fully address the aims of SPINE as stated above, i.e. to establish standards and technologies/methods for roll-out across Europe. Success here can be judged by the dramatic uptake in key enabling technologies for various stages in the pipeline; some highlights include parallel ligation-independent cloning methods, small-scale expression screening, improved protein QA (e.g. rapid dissemination of the thermofluor method), and robotic methods for nanoliter drop crystallization (e.g. the spread of Cartesian dispensers from 2 SPINE partners at the start of SPINE to ~30 labs around Europe by its end, with a similar number of Mosquito robots also being sold). On a global scale, a highly polarized discussion about the value of structural genomics still rages[58–64]; however, we believe that the experience of SPINE demonstrates the value of excellent groups coming together for collaborations where shared technical and methodological developments can mutually benefit scientific programs in biomedical structural biology.

References

1. Burley SK, Almo SC, Bonanno JB, *et al.* (1999) "Structural genomics: beyond the human genome project." *Nat Genet* **23**: 151–57.
2. Stevens RC, Yokoyama S, Wilson IA. (2001) "Global efforts in structural genomics." *Science* 294: 89–92.
3. Pajon A, Ionides J, Diprose J, *et al.* (2005) "Design of a data model for developing laboratory information management and analysis systems for protein production." *Proteins* **58**: 278–284.
4. Albeck S, Alzari P, Andreini C, *et al.* (2006) "SPINE bioinformatics and data-management aspects of high-throughput structural biology." *Acta Crystallogr D Biol Crystallogr* **62**: 1184–95.
5. Esnouf RM, Hamer R, Sussman JL, *et al.* (2006) "Honing the *in silico* toolkit for detecting protein disorder." *Acta Crystallogr D Biol Crystallogr* **62**: 1260–66.
6. Plewniak F, Bianchetti L, Brelivet Y, *et al.* (2003) "PipeAlign: a new toolkit for protein family analysis." *Nucl Acids Res* **31**: 3829–32.
7. Thompson JD, Muller A, Waterhouse A, *et al.* (2006) "MACSIMS: multiple alignment of complete sequences information management system." *BMC Bioinformatics* 7: 318.
8. Junier T, Bucher P. (1998) "SEView: a Java applet for browsing molecular sequence data." *In Silico Biol* **1**: 13–20.
9. Siebold C, Berrow N, Walter TS, *et al.* (2005) "High-resolution structure of the catalytic region of MICAL (molecule interacting with CasL), a multidomain flavoenzyme-signaling molecule." *Proc Natl Acad Sci USA* **102**: 16836–41.
10. Prilusky J, Oueillet E, Ulryck N, *et al.* (2005) "HalX: an open-source LIMS (Laboratory Information Management System) for small- to large-scale laboratories." *Acta Crystallogr D Biol Crystallogr* **61**: 671–78.
11. Berrow NS, Büssow K, Coutard B, *et al.* (2006) "Recombinant protein expression and solubility screening in *Escherichia coli*: a comparative study." *Acta Crystallogr D Biol Crystallogr* **62**: 1218–26.
12. Au K, Berrow NS, Blagova E, *et al.* (2006) "Application of high-throughput technologies to a structural proteomics-type analysis of *Bacillus anthracis*." *Acta Crystallogr D Biol Crystallogr* **62**: 1267–75.

13. Berrow NS, Alderton D, Sainsbury S, *et al.* (2007) "A versatile ligation-independent cloning method suitable for high-throughput expression screening applications." *Nucleic Acids Res* **35**: e45.
14. Christendat D, Yee A, Dharamsi A, *et al.* (2000) "Structural proteomics: prospects for high throughput sample preparation." *Prog Biophys Mol Biol* **73**: 339–45.
15. Lesley SA, Kuhn P, Godzik A, *et al.* (2002) "Structural genomics of the *Thermotoga maritima* proteome implemented in a high-throughput structure determination pipeline." *Proc Natl Acad Sci USA* **99**: 11664–69.
16. Aricescu AR, Assenberg R, Bill RM, *et al.* (2006) "Eukaryotic expression: developments for structural proteomics." *Acta Crystallogr D Biol Crystallogr* **62**: 1114–24.
17. Canaves JM, Page R, Wilson IA, Stevens RC. (2004) "Protein biophysical properties that correlate with crystallization success in *Thermotoga maritima*: maximum clustering strategy for structural genomics." *J Mol Biol* **344**: 977–91.
18. Aricescu AR, Lu W, Jones EY. (2006) "A time- and cost-efficient system for high-level protein production in mammalian cells." *Acta Crystallogr D Biol Crystallogr* **62**: 1243–50.
19. Chang VT, Crispin M, Aricescu AR, *et al.* (2007) "Glycoprotein structural genomics: solving the glycosylation problem." *Structure* **15**: 267–73.
20. Aricescu AR, Siebold C, Choudhuri K, *et al.* (2007) "Structure of a tyrosine phosphatase adhesive interaction reveals a spacer-clamp mechanism." *Science* **317**: 1217–20.
21. Alzari PM, Berglund H, Berrow NS, *et al.* (2006) "Implementation of semi-automated cloning and prokaryotic expression screening: the impact of SPINE." *Acta Crystallogr D Biol Crystallogr* **62**: 1103–13.
22. Ericsson UB, Hallberg BM, Detitta GT, *et al.* (2006) "Thermofluor-based high-throughput stability optimization of proteins for structural studies." *Anal Biochem* **357**: 289–98.
23. Berry IM, Dym O, Esnouf RM, *et al.* (2006) "SPINE high-throughput crystallization, crystal imaging and recognition techniques: current state, performance analysis, new technologies and future aspects." *Acta Crystallogr D Biol Crystallogr* **62**: 1137–49.
24. Mayo CJ, Diprose JM, Walter TS, *et al.* (2005) "Benefits of automated crystallization plate tracking, imaging, and analysis." *Structure* **13**: 175–82.
25. Beteva A, Cipriani F, Cusack S, *et al.* (2006) "High-throughput sample handling and data collection at synchrotrons: embedding the ESRF into the high-throughput gene-to-structure pipeline." *Acta Crystallogr D Biol Crystallogr* **62**:1162–69.
26. Ab E, Atkinson AR, Banci L, *et al.* (2006) "NMR in the SPINE Structural Proteomics project." *Acta Crystallogr D Biol Crystallogr* **62**: 1150–61.

27. Diercks T, Orekhov V. (2005) "qTROSY — a novel scheme for recovery of the anti-TROSY magnetisation." *J Biomol NMR* 32: 113–27.
28. Diercks T, Daniels M, Kaptein R. (2006) "Extended flip-back schemes for sensitivity enhancement in multidimensional HSQC-type out-and-back experiments." *J Biomol NMR* **33**: 243–59.
29. Arnesano F, Banci L, Bertini I, *et al.* (2003) "A strategy for the NMR characterization of type II copper(II) proteins: the case of the copper trafficking protein CopC from *Pseudomonas Syringae.*" *J Am Chem Soc* **125**: 7200–08.
30. Bermel W, Bertini I, Felli IC, *et al.* (2003) "13C direct detection experiments on the paramagnetic oxidized monomeric copper, zinc superoxide dismutase." *J Am Chem Soc* **125**: 16423–29.
31. Babini E, Bertini I, Capozzi F, *et al.* (2004) "Direct carbon detection in paramagnetic metalloproteins to further exploit pseudocontact shift restraints." *J Am Chem Soc* **126**: 10496–97.
32. Bertini I, Felli IC, Kümmerle R, *et al.* (2004) "13C-13C NOESY: an attractive alternative for studying large macromolecules." *J Am Chem Soc* **126**: 464–65.
33. Arnesano F, Balatri E, Banci L, *et al.* (2005) "Folding studies of Cox17 reveal an important interplay of cysteine oxidation and copper binding." *Structure* 13: 713–22.
34. Dancea F, Günther U, (2005) "Automated protein NMR structure determination using wavelet de-noised NOESY spectra." *J Biomol NMR* **33**: 139–52.
35. Bertini I, Del Bianco C, Gelis I, *et al.* (2004) "Experimentally exploring the conformational space sampled by domain reorientation in calmodulin." *Proc Natl Acad Sci USA* 101: 6841–46.
36. Cipriani F, Felisaz F, Launer L, *et al.* (2006) "Automation of sample mounting for macromolecular crystallography." *Acta Crystallogr D Biol Crystallogr* **62**: 1251–59.
37. Bahar M, Ballard C, Cohen SX, *et al.* (2006) "SPINE workshop on automated X-ray analysis: a progress report." *Acta Crystallogr D Biol Crystallogr* **62**: 1170–83.
38 Laskowski RA, Watson JD, Thornton JM. (2005) "ProFunc: a server for predicting protein function from 3D structure." *Nucleic Acids Res* **33**(Web Server issue): W89–93.
39. Morris RJ. (2004) "Statistical pattern recognition for macromolecular crystallographers." *Acta Crystallogr D Biol Crystallogr* **60**: 2133–43.
40. Morris RJ, Najmanovich RJ, Kahraman A, Thornton JM. (2005) "Real spherical harmonic expansion coefficients as 3D shape descriptors for protein binding pocket and ligand comparisons." *Bioinformatics* **21**: 2347–55.
41. Laskowski RA. (1995) "SURFNET: a program for visualizing molecular surfaces, cavities, and intermolecular interactions." *J Mol Graph* **13**: 323–30, 307–08.

42. Garcia-Saez I, Blot D, Kahn R, Kozielski F. (2004) "Crystallization and preliminary crystallographic analysis of the motor domain of human kinetochore-associated protein CENP-E using an automated crystallization procedure." *Acta Crystallogr D Biol Crystallogr* 60: 1158–60.
43. Garcia-Saez I, Yen T, Wade RH, Kozielski F. (2004) "Crystal structure of the motor domain of the human kinetochore protein CENP-E." *J Mol Biol* 340: 1107–16.
44. Neumann E, Garcia-Saez I, DeBonis S, *et al.* (2006) "Human kinetochore-associated kinesin CENP-E visualized at 17 Å resolution bound to microtubules." *J Mol Biol* **362**: 203–11.
45. Garcia-Saez I, DeBonis S, Lopez R, *et al.* (2007) "Structure of human Eg5 in complex with a new monastrol-based inhibitor bound in the R configuration." *J Biol Chem* **282**: 9740–47.
46. Messerschmidt A, Macieira S, Velarde M, *et al.* (2005) "Crystal structure of the catalytic domain of human atypical protein kinase C-iota reveals interaction mode of phosphorylation site in turn motif." *J Mol Biol* **352**: 918–31.
47. Brelivet Y, Kammerer S, Rochel N, *et al.* (2004) "Signature of the oligomeric behaviour of nuclear receptors at the sequence and structural level." *EMBO Rep* **5**: 423–49.
48. Chen JL, Stewart-Jones G, Bossi, G, *et al.* (2005) "Structural and kinetic basis for heightened immunogenicity of T cell vaccines." *J Exp Med* 201: 1243–55.
49. Banci L, Bertini I, Cantini F, *et al.* (2005) "An atomic-level investigation of the disease-causing A629P mutant of the Menkes protein, ATP7A." *J Mol Biol* **352**: 409–17.
50. Dvir H, Harel M, McCarthy AA, *et al.* (2003) "X-ray structure of human acid-beta-glucosidase, the defective enzyme in Gaucher disease." *EMBO Rep* **4**: 704–09.
51. Premkumar L, Sawkar AR, Boldin-Adamsky S, *et al.* (2005) "X-ray structure of human acid-beta-glucosidase covalently bound to conduritol-B-epoxide. Implications for Gaucher disease." *J Biol Chem* 280: 23815–19.
52. Bellinzoni M, Riccardi G. (2003) "Techniques and applications: the heterologous expression of *Mycobacterium tuberculosis* genes is an uphill road." *Trends Microbiol* **11**: 351–58.
53. Daugelat S, Kowall J, Mattow J, *et al.* (2003) "The RD1 proteins of *Mycobacterium tuberculosis*: expression in *Mycobacterium smegmatis* and biochemical characterization." *Microbes Infect* **5**: 1082–95.
54. Ma Q, Zhao X, Nasser Eddine A, *et al.* (2006) "The *Mycobacterium tuberculosis* LipB enzyme functions as a cysteine/lysine dyad acyltransferase." *Proc Natl Acad Sci USA* **103**: 8662–67.

55. Walter TS, Meier C, Assenberg R, *et al.* (2006) "Lysine methylation as a routine rescue strategy for protein crystallization." *Structure* **14**: 1617–22.
56. Hart DJ Tarendeau F, (2006) "Combinatorial library approaches for improving soluble protein expression." *Acta Crystallogr D Biol Crystallogr* **62**: 19–26.
57. Tarendeau F, Boudet J, Guilligay D, *et al.* (2007) "Structure and nuclear import function of the C-terminal domain of influenza virus polymerase PB2 subunit." *Nat Struct Mol Biol* **14**: 229–33.
58. Petsko GA. (2007) "An idea whose time has gone." *Genome Biol* **8**: 107.
59. Banci L, Baumeister W, Heinemann U, *et al.* (2007) "An idea whose time has come." *Genome Biol* **8**: 408.
60. Blundell T (2007) "New dimensions of structural proteomics: exploring chemical and biological space." *Structure* **15**: 1342–43.
61. Harrison SC. (2007) "Comments on the NIGMS PSI." *Structure* **15**: 1344–46.
62. Janin J. (2007) "Structural genomics: winning the second half of the game." *Structure* **15**: 1347–49.
63. Moore PB. (2007) "Let's call the whole thing off: some thoughts on the Protein Structure Initiative." *Structure* **15**: 1350–52.
64. Gerlt JA. (2007) "A protein structure (or function?) initiative." *Structure* **15**: 1353–56.
65. Chance MC, Fisor A, Sali A, *et al.* (2004) "High throughput computational and experimental techiniques in structural genomics." *Genome Res* **14**: 2145–54.
66. Banci L, Bortini I, Cusack S, *et al.* (2006) "First proteomics effective methods in exploiting high-throughput technologies for the determination of human.
67. D'Aray A. (1994) "Crystallizing protein — a rational approach." *Acta Crystallogr D Biol Crystallogr* **50**: 1469–71.
68. Ferré-D' Amaré AR, Burley SK .(1997) "Dynamic light scattering in evaluating crystallizability of macromolecules." *Meth Enzymol* **276**: 157–66.
69. Ferré-D'Amaré AR, Burley SK. (1997) "Way and means of dynamic light scattering to assess crystalizability of macromolecules and micromolecular assemblies." *Structure* **2**: 357–59.
70. Geerlof A, Brown J, Coutard B, *et al.* (2006) " The impact of protein characterization on structural proteomics." *Acta Crystallogr D Biol Crystallogr* **62**: 1125–36.
71. Cipriani F, Felisaz F, Launer L, *et al.* (2006) "Automation of sample mounting for macromolecular crystallography." *Acta Crystallogr D Biol Crystallogr* **62**: 1251–59.
72. Bermel W, Bertini I, Felli IC, *et al.* (2005) "A selective experiment for the sequential protein backbone assignment from 3D heteronuclear spectra." *J Magn Reson* **172**: 324–328.

73. Garcia-Saez I, DeBowis S, Lopez K, *et al.* (2007) "Structure of human Eg5 in complex with a monastrol-based inhibitor bound in the *R* conformation." *J Biol Chem* **282**: 9740–47.

74. Lopez R, Rousseau B, Kozielski F, *et al.* FR Patent 05-02518-(2005), World patent No 2006097617 (2006).

75 Rayment I, Rypniewski WR, Schmidtbase K, *et al.* (2003) "Three-dimensional structure of myosin subfragment-1: a molecular motor." *Science* **261**: 50–58.

Chapter 19

Structural Genomics and Structural Proteomics: A Global Perspective

Lucia Banci*, Wolfgang Baumeister†, Udo Heinemann‡, Gunter Schneider§, Israel Silman¶ and Joel L. Sussman||

The concept of Structural Genomics (SG) arose towards the mid-1990s as a consequence of the availability of whole-genome information and the success of high-throughput (HTP) methods in DNA sequencing. It was envisaged that similar HTP methods could be applied to determining the 3-D structures of "all" the proteins (the "proteome") of an organism. As a part of a general research strategy for functional genomics, systematic, genome-driven and high-throughput crystal and NMR structure determination projects were planned. The rationale was that these data would significantly advance our understanding, at the

* Centro Risonanze Magnetiche, University of Florence, Via Luigi Sacconi 6, Sesto Fiorentino, Florence 50019, Italy, Email: banci@cerm.unifi.it.

† Max Planck Institute of Biochemistry, Am Klopferspitz 18a, Martinsried D-82152, Germany.

‡ Max-Delbruck-Center for Molecular Medicine, Robert-Roessle-Str 10, Berlin D-13125, Germany.

§ Karolinska Institutet, Scheelevägen 2, Stockholm S-171 77, Sweden.

¶ Department of Neurobiology, Weizmann Institute of Science, Rehovot 76100, Israel

|| Department of Structural Biology, Weizmann Institute of Science, Rehovot 76100, Israel.

molecular, and eventually, at higher levels, of the functional processes underlying function and dysfunction of the cell and the organism. An interim objective was to provide an efficient way of filling existing gaps in "fold-space," i.e. to try to determine at least one structure for every existing sequence family, so as to provide suitable templates for modeling the structures of all the proteins present in a given genome. Till now, other gene products such as regulatory RNAs and ribozymes have remained outside the focus of SG projects.

Until quite recently, many structural biologists and protein chemists would have questioned the value of the use of homology modeling, in accurately predicting novel protein structures or for their use in drug design. But there are now an increasing number of examples where predicted structures have proved invaluable in both contexts,[1–3] and indeed in engineering proteins capable of performing novel functions.[4,5]

The SG "vision" led to the investment of very large sums of money in large scale projects, both in the USA (~$300 million invested by the NIH/NIGMS Protein Structure Initiative (PSI) in nine large centers from 2000 to 2005,[6] (http://www.nigms.nih.gov/psi), and in Japan (~US$70 million per annum invested in the Protein 3000 national project from 2002 to 2007,[7] with the bulk going to the RIKEN Research Institute, http://www.rsgi.riken.go.jp). Both these national programs were characterized by the concentration of large resources in a small number of big centers, by the concomitant development of novel, automated technologies for implementing a HTP pipeline approach to structure determination; a focus on novel folds as the major criterion for success; and for the US initiative, a policy requiring immediate public deposition of structural data.

In June 2005, the USA NIH/NIGMS activity moved into Phase 2, which involved the large-scale funding of four production units, and funding on a smaller scale of several other centers focussed on the development of complementary new technologies. Phase 2 will run through 2005–2010, again, with a total investment of ~$300M.[6]

Japan was one of the first countries to embrace SG — Japanese-led projects oriented towards such an approach were conceptualized as early as 1995. Officially, the SG program in Japan began with the Protein Folds Project, which was initiated at the RIKEN Institute in 1997, and in the following year was transferred to the newly established RIKEN Genomic Sciences Center GSC (http://www.riken.go.jp). Another project, known as the Structurome Project began in October, 1999 at the RIKEN Harima Institute of the SPring-8 synchroton; this project focussed on proteins of the extremophile bacterium, *Thermus thermophilus*.[8] Although the structurome project used mainly X-ray crystallography for structure determination, the Protein Folds project at the GSC was intimately linked from its inception to GSC's large new nuclear magnetic resonance facility. Research activities within the Protein Folds project focussed on structure determination of mouse and plant proteins, being synergistically aligned with work on DNA libraries developed by scientists at the GSC.

Europe proceeded more slowly than both the USA and Japan in implementing large-scale SG programs. The Protein Structure Factory in Berlin, Germany (http://www.proteinstrukturfabrik.de) led the way, followed by the OPPF at Oxford (http://www.oppf.ox.ac.uk), England, and the Genopoles in France (notably at Gif-sur-Yvette, Marseille and Strasbourg, http://rng.cnrg.fr). However, it was not until October 2002 that the first Europe-wide project began. This was a three-year Integrated Project, funded by the EU FP5 program, which bore the acronym SPINE, standing for Structural Proteomics in Europe (http://www.spineurope.org). SPINE was a second-generation project with respect to the evolution of the concept of SG projects, and was deliberately called a Structural Proteomics (SP) project. This name was chosen to make a distinction from the earlier SG initiatives, from which it radically departed, while obviously benefiting from their experience and from the technologies already developed by other projects. Additional SG-related integrated projects were subsequently established and funded by the EC, which had either specific methodological (e.g. BIOHXIT) or thematic (e.g. 3D Repertoire, VIZIER, Interaction PROTEOME) aims, as well as related smaller scale projects. It is worth noting that, even taken together, these projects were, in terms of financial investment, on a much smaller scale than the corresponding Japanese and US initiatives.

The worldwide activities briefly surveyed above led the SG/SP community to establish an organization called the *International Structural Genomics Organization* (ISGO), so as to exchange and coordinate views and information. ISGO is now a well-established body, which, among other activities, publishes, as its official journal, the *Journal of Structural and Functional Genomics* (JSFG), and organizes a biannual international SG conference.

The Differing Approaches of SG/SP and Classical Structural Biology (SB)

The strategy and implementation of the first SG projects launched in the USA were at the center of a major and thorough debate,[6] similar to that which preceded the funding and launching of the Human Genome Project a decade earlier. It was realized that implementation of SG programs required even more demanding technological developments than those required for the Human Genome Project. It was necessary to develop HTP procedures for a series of stages, from gene cloning through expression, protein purification, crystallization, data collection to structure analysis and refinement. It is a tribute to the efforts of the various SG projects taken together that automatized procedures have been developed for all these steps, although, hardly surprisingly, there is still much scope for improvement.

Although all the SG projects share the common objective of contributing to the "fold space," which will permit structural modeling of any protein with a known sequence, the individual SG consortia differ in the criteria for selecting their protein targets. Thus, for example, even within the framework of the US PSI initiative, some consortia chose family-based criteria for target selection, whereas others focussed on the genome of a given organism.

Much discussion was devoted to the productivity that might be expected from the various SG projects in terms of the number of structures determined. The first round of the PSI set a goal of determining about 10 000 structures. But it soon became clear that this initial goal was unrealistic; by the end of the first five years, ~1300 structures had been solved. The first round of the PSI was, however, successful in developing automatized technologies at a level that permitted the second round to enter into a "production" phase. It was also anticipated that, after an initial peak in generation of novel structures, a decline would occur after the easiest structures, the so-called 'low-hanging fruit,' had been determined. Moreover, to quote John Norvell, Director of the PSI at NIGMS/NIH, "... the fact remains that some proteins are not amenable to high-throughput approaches." Nevertheless, SG projects have already made, and are continuing to make, significant contributions to the determination of new folds and new domains, thus providing the various databases, such as CATH and SCOP,[9,10] with a substantially larger number of unique new domains than had been provided by the standard structural biology (SB) approach. SG centers have indeed contributed to about half of the new SCOP families, superfamilies and folds in the two and a half years since January 1, 2004. Moreover, the structures solved by SG projects are ~4-fold less sequence-redundant than typical PDB structures.[11]

Extensive discussions were also directed towards the comparison between the approaches and impact of "SG/SP" versus "Structural Biology (SB)" endeavors. A significant proportion of the structures generated by SG/SP centers have lower citation levels than those generated by SB studies,[12] suggesting that the biological/functional characterization of a protein performed in the context of a classical SB study has a broader impact on the biochemical/biological community. Ultimately, however, the cumulative impact of SG/SP, by providing comprehensive structural data applicable to the majority of proteins, will most certainly excede the sum of the impacts of the individual structures solved. SG/SP projects aim to achieve as broad a coverage of the proteome as possible. As a consequence, target selection has, in general, been directed towards unique proteins, defined as proteins whose sequence has <30% identity with structures already present in the PDB. In contrast, a SB approach is usually devoted to the detailed study of a limited number of proteins, often already well characterized in terms of mechanism, specificity and biological role. This may result in the deliberate choice of a number of closely related proteins, or of complexes of a given protein with a number of ligands, in order to address in depth certain aspects of its mode of action and biological function.

It is now becoming apparent that the number of folds is quite limited, and that quite different sequences can assume a similar fold.[13] An awareness is also emerging that the classical SB approach and the SG/SP approach are, in fact, complementary, as the structure of a given protein is essential for understanding its function; but such an isolated snapshot does not suffice to provide complete functional knowledge.

Another major issue that has attracted the attention of the scientific community, and has promoted an ongoing debate, concerns both the size of the proteins studied and the quality of the structures determined, within the various SG/SP projects, as compared with the individual SB projects. Some scientists and officers of funding agencies had indeed expressed concern that, due to the HTP approaches adopted, the structures determined in SG/SP projects would be of lower quality than of those determined in the individual SB projects. It is widely accepted that the quality of the structures determined in the framework of SP/SG projects is quite comparable to, or even better than that of structures determined in SB projects.[14]

If one compares the efforts of the PSI centers with those of traditional SB laboratories in terms of cost/structure, it has been calculated that novel structures solved by PSI centers are significantly less costly than those solved by traditional SB laboratories, whether the structures involved are individual novel structures per se, or new PFAM families or new SCOP superfamilies or folds. Futhermore, there is significant evidence that the cost for solution of structures at the PSI Centers is decreasing quite substantially. However, for large high-impact structures, like the ribosome, which contains a significant number of non-identical polypeptide chains, this is not; in fact, the cost per individual polypeptide chains is significantly lower.[14]

The Goals and Policies of International SG/SP Projects

A compilation of the worldwide SG/SP initiatives, which updates progresses on the basis of the Target Registration Database (TB) of the PDB,[15,16] is presented in Table 1, which also lists the main focus of each project and its website. The first round of the PSI adopted an almost "pure" SG approach, which favored a high production rate for protein structures, and oriented target selection towards the principal goal of completing "fold space." This focus has been revised in PSI-II, where the focus has also been put on function in target selection, through the funding of specialized centers devoted to specific classes of proteins.

The PSI also invested major efforts, especially in PSI-1, in methodological developments essential for implementation of HTP approaches, and major technological advances were made as a consequence. These advances resulted in a dramatic reduction in the cost per structure. It has been estimated that the average cost per structure at the PSI centers, during the period from 1 February 2004 to 31 January 2005, was US$138 000.

PSI-2 is more oriented towards structure "production," exploiting the technical advances obtained during PSI-1. The total number of structures solved during PSI-1 (September 2000 to June 2005) was just over 1100 (http://www.nigms.nih.gov/Initiatives/PSI/Background/PilotFacts.htm). During PSI-2, which is still ongoing, ~1200 structures have been solved so far, still far short of the 10 000 structures envisaged at the onset of PSI-1.

In Japan, the SG initiative at the RIKEN Institute focussed on the "fold" approach, i.e. aiming at the determination of the structures of a large number of distinct protein domains. To select the target proteins, mouse and plant genomes were clustered into families on the basis of amino acid sequences, and families for which no experimental structure was yet available were selected. Then, families of particular biological interest were prioritized. For protein production, the cell-free protein production method pioneered at RIKEN[17] has been used

Table 1 Structural Genomics and Proteomics Project List: — Worldwide Initiatives — 2002–2007

Acronym.	Coordinator	Short Description	URL
BSGC PSI-1	Sung-Hou Kim, Lawrence Berkeley Natl. Lab.	The main focus of this initiative is an integrated SG effort on minimal organisms, *Mycoplasma genitalium* and *Mycoplasma pneumoniae*, to study proteins essential for life. The goals include classification of fold families, obtaining representative proteins from each family, inferring molecular functions of proteins of unknown function, and optimizing key steps for structure determination. Structures are determined by X-ray crystallography.	http://www.strgen.org
CESG PSI-2	John Markley Univ. Wisconsin Madison	The center is elucidating 3D structures of proteins encoded by the genome pf *Arabidopsis thaliana*, an important model plant. The initial focus of the center is to develop HTP methods for protein production, characterization and structure determination, using X-ray crystallography and NMR spectroscopy.	http://www.uwstructuralgenomics.org

(*Continued*)

Table 1 (*Continued*)

Acronym.	Coordinator	Short Description	URL
CHTSB PSI-2	Michael Malkowski	The broad goal of this center is to overcome the most significant obstacles to structure determination by focusing on technology development in areas related to sample preparation for X-ray diffraction studies.	http://www.chtsb.org
CSMP PSI-2	Suzan Betheil	Atomic structure determination of both bacterial and human membrane proteins. Human membrane proteins encode the targets for ~40% of all therapeutic drugs currently used, but understanding of their mechanisms of action at the atomic level is still lacking. Many of the human protein structures sought have therapeutic importance, and their solution will provide atomic-level templates for drug design/discovery.	http://csmp.ucsf.edu/index.htm
JCSG PSI-2	Ian Wilson	This center is developing HTP methodologies for target selection, protein production, crystallization, and structure determination by	http://www.jcsg.org

(*Continued*)

Table 1 (*Continued*)

Acronym.	Coordinator	Short Description	URL
		X-ray crystallography. Initial focus is on novel structures from *Thermotoga maritime*, *C. elegans* and on human proteins thought to be involved in cell signaling. It will also cover the structures of similar proteins from other organisms to ensure the inclusion of the greatest number of different protein folds. The 5-year goal is to generate 3D structures of approximately two thousand proteins.	
ISFI PSI-2	Thomas C. Terwilliger	This is an NIH PSI Specialized Center focussed on developing and applying a set of synergistic technologies designed to overcome recognized bottlenecks in structure determination at the key steps of production of soluble protein and protein crystallization.	http://techcenter.mbi.ucla.edu
MCSG PSI-2	Andrzej Joachimiak	The group will select protein targets from Eukarya, Archaea, and Bacteria, with an emphasis on previously unknown folds and on proteins from	http://www.mcsg.anl.gov/index.html

(*Continued*)

Table 1 (*Continued*)

Acronym.	Coordinator	Short Description	URL
		disease-causing organisms. Another focus of this group is to establish methodologies for highly cost-effective protein production, crystallization, structure determination by X-ray crystallography, and refinement, with to the objective of reducing the average cost per structure from $100 000 to $20 000.	
NESG PSI-2	Gaetano Montelione	This consortium is targeting proteins from eukaryotic model organisms, which are subjects of extensive functional genomics research, including *S. cerevisiae*, *C. elegans*, and *D. melanogaster*, as well as homologues from the human genome. Its aim is to develop integrated key technologies such as protein expression and structure determination by both X-ray crystallography and NMR spectroscopy. By developing HTP and cost-effective platforms, it plans to solve >180 protein structures per year at a cost, excluding capital equipment, of $10 000–$20 000 per structure.	http://bioinfo5.mbb.yale.edu/nesgc

(*Continued*)

Table 1 (*Continued*)

Acronym.	Coordinator	Short Description	URL
NYSGXRC PSI-2	Stephen Burley	The consortium expects to solve several hundred protein structures from organisms ranging from bacteria to humans, with an emphasis on developing leads for drug discovery. The consortium is also focusing on development of key HTP technologies such as computational methods for protein family classification and target selection, protein production, purification, and structure determination by X-ray crystallography. Its long-term goal is to determine >10 000 3D structures.	http://www.nysgrc.org
SECSG PSI-1	Bi-Cheng Wang	This consortium will analyze part of the human genome and the entire genomes of two representative organisms, the eukaryotic microorganism, *Caenorhabditis elegans*, and its more primitive prokaryotic ancestor, *Pyrococcus furiosus*. There is an emphasis on technology developments, especially automation of various X-ray crystallography and NMR spectroscopy data collection techniques.	http://www.secsg.org

(*Continued*)

Table 1 (*Continued*)

Acronym.	Coordinator	Short Description	URL
SGPP PSI-1	Wim Hol	The focus of this consortium is the development of methods and technologies for determining structures of proteins from pathogenic protozoans, many of which cause deadly diseases such as sleeping sickness (*Trypanosoma brucei*), Chagas' disease (*Trypanosoma cruzi*), leishmaniasis (*Leishmania*) and malaria (*Plasmodium falciparum* and *Plasmodium vivax*). Using X-ray crystallography, the consortium plans to discover novel folds and templates for drug design.	http://www.sgpp.org
TBSGC PSI-1	Thomas Terwilliger	The consortium plans to determine and analyze the structures of over 400 proteins from *Mycobacterium tuberculosis*, including ~40 novel folds and 200 representatives of new protein families, and to analyze these structures in the context of functional information. This will be strongly directed to the design of new and improved drugs and vaccines for tuberculosis. HTP methodology developments have also been carried out as a pilot project using a hyperthermophile. The consortium uses X-ray crystallography for structure determination.	http://www.doe-mbi.ucla.edu/TB

(*Continued*)

Table 1 *(Continued)*

Acronym.	Coordinator	Short Description	URL
BSGI	Mirek Cygler	The aim of this project is to allow researches to investigate the function and structure of genes and proteins that can be used in developing new drugs. The facility will emphasize protein mapping, identification and characterization. The project will bring together investigators who use biochemical assays, cell biology methodologies, genomics, protein engineering, DNA chip technology, protein sequence analysis, and X-ray crystallography, among other tools.	http://euler.bri.nrc.ca/brimsg/bsgi.html
S2F	John Moult and Osnat Herzberg	The project determines structures of hypothetical proteins, i.e. those whose structures cannot be related to any previously characterized proteins and whose functions are thus, as yet, unknown. The initial targets have been selected from *Hemophilus influenzae*. Structure determination utilizes both X-ray crystallography and NMR spectroscopy.	http://s2f.umbi.umd.edu/families.php

(Continued)

Table 1 (*Continued*)

Acronym.	Coordinator	Short Description	URL
RSGI	Shigeyuki Yokoyama	It focusses on the "fold" approach, i.e. aiming to determine the structures of a large number of distinct protein domains. It has established a high-throughput pipeline for protein sample preparation for structural genomics and proteomics by using cell-free protein synthesis. This center has had a very high success rate, i.e. as of 15 Jan 2008, determining 1343 crystal structures, and 1373 NMR structures.	http://www.rsgi.riken.jp/rsgi_e
KSPRO	Se Won Suh	Its major focus is on proteins from organisms such as *Mycobacterium tuberculosis* and *Helicobacter pylori* that may result in novel targets for drug discovery. X-ray crystallography and NMR are being used for structure determination.	http://kspro.org

extensively, being particularly advantageous for producing isotope-labeled proteins for NMR structure solution. Mission-oriented infrastructures were established which exploited an impressive park of NMR spectrometers in Yokohama, as well as the Spring-8 synchrotron at Harima. As of Sep 2007 *ca.* 1914 structures had been released in the PDB, of which 1040 had been solved by NMR.

In China, the Structural Genomics Consortium of the Chinese Academy of Sciences was established in the spring of 2001. Five universities and institutions have joined together to form this consortium, *viz.* the University of Science and Technology of China; the Institute of Biophysics, CAS; the Shanghai Institute for Biological Sciences, and the Shanghai Second Medical University. Five X-ray crystallography groups, three NMR groups, one bioinformatics group and four molecular biology/biochemistry groups are involved in these SG activities. The consortium is focusing on proteins expressed in human hematopoietic stem/progenitor cells, and on proteins related to blood diseases.[18,19]

In Taiwan, the new synchrotron-based Protein Crystallography Facility at the NSRRC was inaugurated in November 2005 (http://www.nsrrc.org.tw). With the NSRRC's protein crystallography beamlines having become operational, Taiwan is a new player in the fields of proteomics and structural genomics.

The Korean Structural Proteomics Research Organization was established in February 2002 to promote and coordinate proteomics research activities in Korea (http://xtalg.gist.ac.kr). Its major focus is on proteins of bacteria such as *Mycobacterium tuberculosis* and *Helicobacter pylori*, which may lead to discovery of new drugs for treatment of tuberculosis and ulcers, respectively. Both X-ray crystallography and nuclear magnetic resonance (NMR) spectroscopy are being used for structure determination.

The Israel Structural Proteomics Center (ISPC) (http://www.weizmann.ac.il/ISPC) was established by scientists from the Weizmann Institute of Science, Rehovot, ISRAEL to increase the efficiency of all stages of 3D protein structure determination.[20] Targets submitted to the ISPC are primarily related to human health and disease. The center has a unique combination of scientific expertise and state-of-the-art instrumentation for high-throughput production and crystallization of proteins. Each target is cloned into multiple vectors, using ligation independent cloning. Expression is extensively screened in several bacterial strains with different fusion proteins. Proteins which are not soluble are expressed either in bacterial cell free extracts or in yeast (*Pichia pastoris*). Parallel purification of up to six proteins can be performed using an AKTA3D. Purified proteins are screened for crystallization using a Douglas Instruments ORYX6 robot, which employs the batch method under oil, and a TTP-Labtech MOSQUITO robot for sitting and hanging drops crystallization. This has yielded a high-percentage of high quality diffracting crystals. All the different stages are manipulated by a laboratory information management system (LIMS) in which several bioinformatics tools have been incorporated to facilitate the analysis of our targets. The ISPC now receives targets from scientists both in academia and industry. The ISPC believes that making structural information accessible to the entire scientific community will stimulate novel studies and developments related to health and disease.

The Taiwanese, Korean and Israeli projects show that even relatively small countries are capable of developing domestic SG/SP projects, evidence for the worldwide relevance and impact of the SG/SP endeavor.

In Australia, three SG projects are at the planning stage. They will focus on microbial virulence factors, macrophage proteins, and cold-adapted organisms (http://www.isgo.org/list/index.php#Australia).

In addition to the SP/SG projects ongoing in the USA, Europe and Japan, transnational consortia are also being established. The most prominent, to date, is the Structural Genomics Consortium (SGC), headed by Aled Edwards, which was established in 2004, and maintains research centers in Toronto, Oxford and Stockholm. It focuses on human proteins of medical relevance, and is the first consortium to have solved the structures of a large number of human proteins, which are far harder to produce than prokaryotic proteins.[21]

In order to coordinate the efforts of the multiple SG/SP projects currently functioning, the SG/SP community, in particular the publicly funded projects, have agreed on a series of actions directed towards making all the targets public, ensuring the prompt release of all structures analyzed, and facilitating the open exchange of new technologies as they come "on-line." A direct outcome of this policy has been the establishment of web sites and repository databases, which are providing the scientific community as a whole with open access to a wealth of data. Particularly relevant are the databases of selected targets, which allow researchers to avoid duplication in target selection (http://targetdb.pdb.org). This approach has indeed proved successful, as a recent survey[22] reported that only 14% of the structures determined by the various consortia have close homology (>30% sequence identity) with structures analyzed by other consortia. Databases containing information on methodological issues, such as cloning, expression, and purification, are also available. For example, the Protein Expression Cloning and Purification Database, PepcDB (http://pepcdb.pdb.org), was established to collect detailed status information and experimental details of each step in the protein production pipeline.[16]

Achievements of SG/SP Projects

The few years during which the various SG/SP projects have produced data and results can be used to measure their effectiveness and their impact. A simple way to measure their effectiveness is to count the number of experimentally determined structures that they have generated in terms of their absolute number, the fraction of the total structures deposited in the PDB, and, perhaps more importantly, the fraction of unique structures (defined as such on the basis of the sequence identity being <30% of that of any other structure deposited in the PDB).

In a recent paper, Chandonia and Brenner[12] reviewed the results and impact of SG efforts worldwide, and presented extensive statistics, with particular emphasis on structural novelty. Their analysis showed that the numbers of new structures or, more importantly, of the first structure reported for a PFAM family, came far more often from an SG/SP project than from a classical SB project. SG centers worldwide now account for about half of all new structurally characterized families. For PSI centers, for example, the percentage of domains representing a new SCOP fold or superfamily was 16%, significantly higher than for the non-SG average, which was 4%. For non-SG/SP structures, >70% of those solved in the past 10 years were related to proteins which had already been structurally characterized in a different state, i.e. with mutations, with bound ligands, or in a different complex.[12]

The analysis of the achievements of SG projects and of the advancements in structural knowledge only in terms of the number of structures, of novel structures and reduced cost per structure is quite reductive. An additional major outcome has been the development of pioneering HTP technologies in the fields of protein production, purification and crystallization, as well as structure determination, using both X-rays and NMR. These achievements have fall-out well beyond the SG projects themselves, also contributing significantly to SB and to life science studies in general.

The structural knowledge provided by the SG/SP projects can suggest functional properties or a biological role for proteins of unknown function. Indeed, in a few cases, newly analyzed structures have been used to infer the functional properties and the mechanism of action of a given protein.

Finally, SG/SP projects, as a spin-off of their HTP approach, which involves screening for expression of several constructs of a number of orthologous genes for each of tens of thousands of targets worldwide, have produced huge archives of cloned and expressed proteins. The vast majority have not resulted in crystals suitable for X-ray data collection, or in samples suitable for NMR spectroscopy. Nevertheless, these archives contain a wealth of precious information for other biochemical and biological studies.

The European Structural Genomics Project SPINE

Europe, which tackled the SP/SG scientific challenge later than both the USA and Japan, has developed an approach combining features of both SP/SG and SB, and exploiting the positive aspects of the two disciplines. In particular, this has been the approach of SPINE, which was the first Structural Proteomics project to be funded at the European level.

SPINE developed an approach that combines technical and methodological development with the generation of protein structures of high medical relevance, selected from pathogens (as was done in the TB Structural Genomics Consortium (http://www.doe-mbi.ucla.edu/TB)) or from human proteins involved in diseases.

A principal contribution of SPINE has been to serve as a catalyst for the development of a pan-European network of laboratories with HTP SG/SP capabilities. SPINE has contributed to the spread of novel technologies (e.g. affordable nano-crystallization and expression screening robotics), rather than establishing large central facilities. It has taken advantage of the diversity of European laboratories so as to generate novel ideas or to benchmark alternative strategies, the best of which have then been more widely adopted.

SPINE has pushed the development of European standards in several areas of HTP technology, notably the development of LIMS systems and automatization of the handling of frozen crystals at synchrotrons (http://www.spineurope.org/page.php?page=protocol_vials), which is already progressing towards courier mail transfer of crystals from the users to synchrotrons, and thus to monitoring of data collection by scientists from their home laboratories. Furthermore, it has been driven by the notion of selecting "high-value targets for human health" rather than by "filling fold space" by solving many of the structures of an entire small proteome, or even by selecting "low-hanging fruit" in the context of development of techniques and methodologies. By so doing, it has provided a pragmatic working definition of the term "structural proteomics." Surprisingly, despite the fact that many of the targets selected by SPINE were difficult ones, the success rate that it has achieved in the structure determination of human proteins compares favorably with the success rates of other major SG programs focussed on bacterial proteins. Thus, the Joint Center for Structural Genomics (JCSG; http://www.jcsg.org), which started in 2000, is one of the most effective large US projects, and has focussed mainly on the proteome of the bacterial thermophile,

Thermotoga maritima (with annual funding substantially greater than that allocated to SPINE). The scoreboard for this project, after seven years of operation, was (on 7/9/07): targets selected: 19 749; cloned: 16 213; expressed: 14 819; crystallized: 1082; solved: 465 (X-ray), 15 (NMR); deposited in PDB: 468. The corresponding output of SPINE after ~5 years operation is highly encouraging and on a par with the US projects: targets selected: 2395; cloned: 1534; expressed: 1177; soluble: 687; solved: 252 (X-ray), 56 (NMR); deposited in PDB: 122 (Jul, 2006). These figures also conceal considerable parallel work on many targets, with the total number of expression trials being ~14 000. The SPINE statistics, showing a total of 308 structures solved, reflect novel structures only; the number including ligand- and metal ion-bound isoforms is >370, with more than 200 being human proteins. To put this in perspective, the total number of new human structures (with <95% identity to prior structures) deposited into the PDB during the first 11 months of 2005 was 337. It should be stressed, however, that the funding for structures that have been "counted" as SPINE targets, has not always been exclusively funded by SPINE only, as was the case for the PSI.

By its policy of maintaining an open decentralized network, together with a focus on high-value targets, SPINE has overcome the potentially divisive dichotomy between the "traditional" way of doing SB ("one post-doc/one project" with in-depth complementary functional investigations) and "factory-style" SG (multiple parallel projects, abandoning of failures, target proteins of often unknown function). The SPINE mode of work, whereby HTP techniques are exploited for high-value targets, is likely to become the norm for SB. SPINE has put in place strong links with a number of companies that have stimulated technology transfer to SMEs, and encouraged beta-testing of new products in SPINE laboratories. Furthermore, the output of SPINE in terms of published papers is outstanding with, to date, 219 publications citing SPINE support.

The current and earlier SG/SP projects have revolutionized the way in which structural biology is now being done worldwide, through the introduction of novel automated, systematic and methodological strategies at each step of the structure determination pipeline (although their cost-effectiveness, particularly as a method for discovering new folds, is open to discussion). Of perhaps greater importance than the numbers of structures delivered, SPINE has had a remarkable impact in Europe, acting as the springboard for the second generation of FP6 Integrated Projects, such as VIZIER and BIOXHIT, which apply and further hone the appropriate technologies for specific target areas, as well as SPINE2-Complexes, initiated in the summer of 2006 (http://www.spine2.eu/index.php), which represents a step forward with respect to the classical, as one might say "old style" structural biology approach, as these new projects exploit a HTP "factory style" approach to address functional processes at the cellular level in their complexity and in their entirety. SPINE2-Complexes has moved on from the goals of SPINE, which were to advance technologies and solve structures of single proteins, to developing approaches for solving structures of protein complexes, with the eventual challenging objective of integrating such complexes into higher-order cellular structures. The measure of the success of the project will not be the number of structures solved but rather their biological impact.

The Scientific Advisory Board of SPINE, in their final review of its achievements during the three-year term for which it was funded, wrote to the European Commission as follows: "*The SPINE impact on the European Community has been very significant and there is no other funding mechanism to accomplish what they have done. SPINE has been a tremendous success as the catalyst for structural biology throughout Europe. This model programme should be duplicated for other EU projects.*"

Highlights of SPINE'S Achievements

The following provides a snapshot of some of the major achievements of the SPINE project that have laid significant foundations on which future SG/SP research can build.

1. Efficient small-scale automated HTP pipelines for protein cloning, expression and purification in prokaryotes, now utilized by many European laboratories both within and outside SPINE.

2. New mammalian expression technologies and refinement of procedures for optimization of expression in eukaryotic systems.
3. Incorporation of quality assurance (QA) into the HTP protein production pipeline, including technologies such as mass spectrometry, ThermoFluor analysis and small-angle X-ray scattering.
4. Methods for achieving soluble expression of protein domains and subdomains, suitable for structural analysis, from previously intractable proteins.
5. Dissemination of nanoliter crystallization technologies.
6. Progress in crystal imaging and image recognition testing.
7. Development of ^{13}C protonless NMR spectroscopy methodology that provides a significant breakthrough in structure solution, particularly for larger proteins.
8. Establishment and testing of an expert system for crystal diffraction data collection from user laboratory to synchrotron; this involves utilization of automated procedures from sample loading through crystal alignment, to data collection and reporting.
9. Development of a SPINE sample holder standard has been adopted across Europe and, more recently, also in China (http://www.spineurope.org/page.php?page=protocol_vials).
10. An integrated protein information server for SG/SP, providing a comprehensive resource for protein selection, annotation and data collection: including PipeAlign, OPAL, OPTIC, FoldIndex, SeqAlert, SeqFacts, RONN, BestPrimers, OPINE, eHTPX hub, ISPyB, DNA automated data collection, ProFunc server, SURFNET, and many others.
11. Solution of the structures of 30 *Bacillus anthracis* structures out of 361 target proteins selected.
12. Analysis of more than 50 high-impact structures, including pathogen and human proteins (see http://www.spineurope.org).
13. Contributed to benchmarking definition in SG via a series of multi-lab comparisons of the various stages of expression and protein production.
14. SPINE played a major role in providing credibility for the consideration of structural biology as a research area whose requirements

for infrastructure were eventually incorporated into the ESFRI Roadmap. This resulted in the funding of the preparatory phase of the new infrastructure INSTRUCT at the beginning of 2008.

The Legacy of SPINE

In large part due to large-scale EU support, SPINE has given visibility and identity to European scientists engaged in SP/SG, and has achieved an international stature comparable to that attained by equivalent large-scale projects in the USA and Japan. This provided an effective mechanism by which worldwide opportunities for scientific exchange in the field could be funnelled through SPINE to individual European laboratories. In a similar way, SPINE, and now SPINE2-COMPLEXES, due to the extensive network developed, can serve as a natural contact point for companies wishing to beta-test new technologies relevant to SG/SP, as positive results can be rapidly disseminated.

SPINE has been exemplary in combining the expertise of the consortium members with that of related consortia, both inside and outside the EC, to pioneer benchmark procedures (e.g. for constructs, expression vectors, folding protocols, crystallization screens and their visual analysis, data collection and rapid structure determination), all of which may result in the adoption of pan-European standards. The establishment of such standards will be greatly facilitated through maintenance of careful quantitative records of both successes and failures, at all stages of the HTP pipeline, by means of the LIMS being built around the PIMS initiative, which arose largely out of preparatory work within SPINE. PIMS is destined to become a *de facto* standard in the area of SP/SG, for which such a standard is sorely lacking.

In parallel to work on structural analysis of the component proteins of the proteome, major efforts are now underway to map the interactions of human proteins (the so-called human "interactome"). This requires that the definition of human complexes be placed on a more systematic and complete basis, and European laboratories are playing an important role in this effort, building on the strong platform of achievement established by SPINE, and with the FP6 Integrated Project SPINE2-COMPLEXES positioned to play a leading role in this endeavor.

Other European Structural Genomics Projects 2002–2006

Following the success of SPINE, the EC funded a series of other programs in the area of SG/SP, focussing on specific SG/SP technologies, targets, standardization of methods, and plans for the future. The EC has several different instruments to fund projects, all of which require partners from at least three member countries (or associated countries, such as Switzerland and Israel). These programs, together with a brief summary of their activities, are listed in Table 2.

Perspectives

The functional perspective is becoming increasingly relevant both to target selection and prioritization. Analysis of the entries in the PDB has shown that approximately 70% of the human genes with a Gene Ontology annotation (molecular function, biological process or cell component) are not yet structurally characterized by even one identifiable domain. The structural coverage of the human genome is even lower with respect to sequence space; there is approximately 10% coverage by structures with >40% sequence identity.[23]

Indeed PDB content, not surprisingly, is significantly enriched in terms of functional coverage in "low-hanging fruit" and validated drug targets. Accordingly, SG projects are beginning to turn their attention from coverage of fold space to that of functional space. This includes individual proteins that are often hard to study, such as membrane proteins; however, attention is increasingly being turned to higher order structures, starting with functional complexes and the long term objective of obtaining structures of organelles and cellular structures.

Table 2 EC Funded SG/SP Projects 2002–2007

Acronym.	Coordinator	Short Description	URL
BIOXHIT	Victor Lamzin	Coordinates scientists at all European synchrotrons, together with leading software developers, in an unprecedented joint effort to develop, assemble and provide a highly effective technology platform for SG.	http://www.bioxhit.org
Vizier	Bruno Canard	Aims to have a groundbreaking impact on the identification of potential new drug targets in RNA viruses through comprehensive structural characterization of the replicative machinery of a carefully selected and diverse set of viruses.	http://www.vizier-europe.org
IMPS	Jean-Luc Popot	Aims to develop broad-range tools for SP of membrane proteins.	http://cordis.europa.eu/fetch? CALLER=FP6_PROJ&ACTION=D&DOC=117& CAT=PROJ&QUERY=1179147785074&RCN=78727
Opticryst	Roslyn Bill	Development, implementation and exploitation of new technologies to overcome bottlenecks in optimization of protein crystallization.	http://www.opticryst.info

(Continued)

Table 2 (*Continued*)

Acronym.	Coordinator	Short Description	URL
thera-cAMP	Enno Klußmann	Identification of lead compounds that specifically modulate protein-protein interactions in cAMP signaling networks.	http://www.thera-camp.eu
SPINE	David Stuart	Development of new methodologies and technologies for HTP structural biology.	http://www.spineurope.org
SPINE2-Complexes	David Stuart	Structure determination of protein complexes associated with signaling pathways involved in human health and disease, and concomitant development of cutting edge technologies for the production and structure determination of such complexes.	www.spine2.eu
E-MeP	Roslyn Bill	Development and implementation of new technologies to overcome bottlenecks that preclude the HTP determination of high-resolution structures of membrane proteins and membrane protein complexes.	http://www.e-mep.org
3D Repertoire	Luis Serrano	This project brings together the top European structural biology institutions in a collaboration aimed at solving the structures of a large number of functional protein complexes in yeast.	www.3drepertoire.org

(*Continued*)

Table 2 (*Continued*)

AcroAcronym.	Coordinator	Short Description	URL
NMR-Life	Ivano Bertini	Development of cutting-edge NMR technologies for studying functional protein complexes *in vitro* and *in situ*.	http://www.postgenomicnmr.net
EXTEND-NMR	Ernest D. Laue	Development of novel computational tools that extend the scope of NMR spectroscopy and make possible functional and structural studies on large proteins and biomolecular complexes.	http://www.biocompetence.eu/index.php/kb_5/io_3577/io.html
UPMAN	Harald Schwalbe	Use of NMR to understand protein misfolding and aggregation.	http://schwalbe.org.chemie.uni-frankfurt.de/upman
FSB-V-RNA	Sybren Wijmenga	The structural, functional and virological analysis of RNA and RNA-protein complexes from viruses.	http://www.fsgvrna.nmr.ru.nl
NDDP	Rolf Boelens	Use of cutting-edge NMR techniques to develop a fast, integrated approach for support of structure-based drug design.	http://projects.bijvoet-center.nl/nddp
3D-EM	Andreas Engel	3D-EM aims to establish a standardized platform of advanced technology and methodology that.	http://www.3dem-noe.org

(*Continued*)

Table 2 (*Continued*)

Acronym.	Coordinator	Short Description	URL
		will allow Europe to maintain the lead in structural research. It will allow the coordination of research, training activities, research-industry collaboration, and the transfer of knowledge, via publications and focused scientific meetings, in the field of electron microscopy.	
HT-3DEM	Andreas Engel	To enhance European leadership in 3D EM, this project proposes the development of an automated platform permitting HTP screening and analysis of native protein complexes and protein crystals using EM.	http://www.ht3dem.org
MSGP	Christian Cambillau	The project is conducted by a joint CNRS and industrial consortium aimed at the discovery of new anti-bacterial and antiviral targets. The targets include proteins from *Escherichia coli* and *Mycobacterium tuberculosis* as well as viral proteins.	http://www.afmb.univ-mrs.fr/rubrique93.html
BIGS	Chantal Abergel	Focuses on the discovery of new antibacterial gene targets among evolutionary conserved genes of uncharacterized function.	http://igs-server.cnrs-mrs.fr

(*Continued*)

Table 2 (*Continued*)

Acronym.	Coordinator	Short Description	URL
OPPF	David Stuart, Ernest Laue	Explores the biomedical relevance of human pathogens, in particular of herpes viruses.	http://www.oppf.ox.ac.uk/OPPF/
XMTB	Matthias Wilmanns	Is focussed on the identification of lead compounds against *Mycobacterium tuberculosis* (TB), using a structure-based approach.	http://xmtb.org
YSG	Herman van Tilbeurgh	A lab-scale platform for the systematic production and structure determination of proteins is being tested on 250 yeast non-membrane proteins of unknown structure. Strategies and final statistics are evaluated.	http://genomics.eu.org/spip
PSF	Udo Heinemann	Target proteins are human proteins relevant to health and disease.	http://www.proteinstrukturfabrik.de
ISPC	Joel Sussman	Aims to increase the efficiency of protein structure determination. Targets submitted to the ISPC are primarily related to human health and disease.	http://www.weizmann.ac.il/ISPC
SGC	Aled Edwards	The SGC operates out of the Universities of Oxford and Toronto and Karolinska Institutet, Stockholm.	http://www.sgc.ox.ac.uk

(*Continued*)

Table 2 (*Continued*)

Acronym.	Coordinator	Short Description	URL
		The primary focus of the Oxford laboratory is the study of human proteins involved in phosphorylation and integral membrane proteins, as well as of enzymes associated with metabolic pathways. The Toronto group seeks to determine the 3D structures of human proteins of therapeutic relevance to diseases such as cancer, diabetes, and metabolic disorders.	
FESP	Joel L. Sussman	Has the role of developing a strategy for SG/SP in the broader context of anticipated developments in biological research.	http://www.ec-fesp.org

Acknowledgements

This work was supported, in part, by European Commission Grant for the Forum for European Structural Proteomics (FESP), contract number: LSSG-CT-2005-018750.

References

1. Okumoto S, Looger LL, Micheva KD, *et al.* (2005) Detection of glutamate release from neurons by genetically encoded surface-displayed FRET nanosensors. *Proc Natl Acad Sci USA* **102**(24): 8740–45.
2. Becker OM, Dhanoa DS, Marantz Y, *et al.* (2006) An integrated in silico 3D model-driven discovery of a novel, potent, and selective amidosulfonamide 5-HT1A agonist (PRX-00023) for the treatment of anxiety and depression. *J Med Chem* **49**(11): 3116–35.
3. Dooley AJ, Shindo N, Taggart B, *et al.* (2006) From genome to drug lead: identification of a small-molecule inhibitor of the SARS virus. *Bioorg Med Chem Lett* **16**(4): 830–33.
4. Looger LL, Dwyer MA, Smith JJ, Hellinga HW. (2003) Computational design of receptor and sensor proteins with novel functions. *Nature* **423**(6936): 185–90.
5. Deuschle K, Okumoto S, Fehr M, *et al.* (2005) Construction and optimization of a family of genetically encoded metabolite sensors by semirational protein engineering. *Protein Sci* **14**(9): 2304–14.
6. Norvell J, Berg JM. (2008) Policies in structural genomics/structural proteomics — PSI. In: Sussman JL, Silman I (eds). *Structural Proteomics and its Impact on the Life Sciences* (in press): World Scientific Publishing.
7. Cyranoski D. (2006) "Big science" protein project under fire. *Nature* **443**(7110): 382–.
8. Yokoyama S, Hirota H, Kigawa T, *et al.* (2000) Structural genomics projects in Japan. *Nat Struct Biol* 7 (Suppl): 943–45.
9. Lo Conte L, Ailey B, Hubbard TJ, *et al.* (2000) SCOP: a structural classification of proteins database. *Nucl Acids Res* **28**(1): 257–79.
10. Greene LH, Lewis TE, Addou S, *et al.* (2007) The CATH domain structure database: new protocols and classification levels give a more comprehensive resource for exploring evolution. *Nucl Acids Res* **35**(Database issue): D291–97.
11. Levitt M. (2007) Growth of novel protein structural data. *Proc Natl Acad Sci USA* **104**(9): 3183–88.
12. Chandonia J-M, Brenner SE. (2006) The impact of structural genomics: expectations and outcomes. *Science* **311**(5759): 347–51.

13. Fogg MJ, Alzari P, Bahar M, *et al.* (2006) Application of the use of high-throughput technologies to the determination of protein structures of bacterial and viral pathogens. *Acta Cryst* **62**(Pt 10): 1196–207.
14. Brown EN, Ramaswamy S. (2007) Quality of protein crystal structures. *Acta Crystallogr D Biol Crystallogr* **63**(Pt 9): 941–50.
15. Westbrook J, Feng Z, Chen L, *et al.* (2003) The Protein Data Bank and structural genomics. *Nucl Acids Res* **31**(1): 489–91.
16. Kouranov A, Xie L, de la Cruz J, *et al.* (2006) The RCSB PDB information portal for structural genomics. *Nucl Acids Res* **34**(Database issue): D302–D5.
17. Kigawa T, Yabuki T, Yoshida Y, *et al.* (1999) Cell-free production and stable-isotope labeling of milligram quantities of proteins. *FEBS Lett* **442**(1): 15–19.
18. Shi Y, Wu J. (2007) Structural basis of protein-protein interaction studied by NMR. *J Struct Funct Genomics*: (in press).
19. Gong WM, Liu HY, Niu LW, *et al.* (2003) Structural genomics efforts at the Chinese Academy of Sciences and Peking University. *J Struct Funct Genomics* **4**(2–3): 137–39.
20. Albeck S, Burstein Y, Dym O, *et al.* (2005) 3D structure determination of proteins related to human health in their functional context at the Israel Structural Proteomics Center (ISPC). *Acta Cryst* D61: 1364–72.
21. Gileadi O, Knapp S, Lee WH, *et al.* (2007) The scientific impact of the Structural Genomics Consortium: a protein family and ligand-centered approach to medically-relevant human proteins. *J Struct Funct Genomics* **8**: 107–19.
22. Todd AE, Marsden RL, Thornton JM, Orengo CA. (2005) Progress of structural genomics initiatives: an analysis of solved target structures. *J Mol Biol* **348**(5): 1235–60.
23. Xie L, Bourne PE. (2005) Functional coverage of the human genome by existing structures, structural genomics targets, and homology models. *PLoS Comput Biol* **1**(3): e31.

Chapter 20

Policies in Structural Genomics/ Structural Proteomics

A. The Protein Structure Initiative: Policies and Update

John Norvell and Jeremy Berg
National Institute of General Medical Sciences,
National Institutes Health,
Bethesda, MD USA.

Building on the successes of structural biology and the genome sequencing projects, scientists around the world proposed in the late 1990s the establishment of large-scale high throughput structural genomics projects to extend the structural coverage of the sequenced genes. The National Institute of General Medical Sciences (NIGMS), a component of the National Institutes of Health (NIH), organized three workshops and held discussions with the Institute's advisory council in the planning of one of these projects, the Protein Structure Initiative (PSI). The PSI's goal is ambitious: to make three-dimensional atomic level structures of most proteins easily available from a knowledge of their corresponding DNA sequences by focusing on high-throughput structural determination of unique protein structures. The PSI began as a 5-year project, consisting of a pilot centers program and a methodology and technology development grants program, and nine pilot research centers were established in 2000 and 2001. These pilot centers developed methodology and technology that increased the success rates and lowered the costs of structural determination, leading to the construction and automation of the protein production and structural determination pipeline. From the beginning of this project, the NIGMS staff and advisors were concerned about policy issues in this field, including data release, patents, and international coordination. The NIH policy for all structural biology research grants requires that coordinates and related information be deposited and released at the time of publication of structural results. But the requirements for the PSI centers are more demanding as the PSI project is designed to be a public resource and a large amount of funds are invested in a few centers. Therefore, the policy must ensure that results and products must be made available to the scientific community quickly and structural coordinates and related information were required to be deposited and released promptly after completion of the structures. This ensures that all scientists will have the opportunity to pursue follow-up studies on these structures. In addition, each center was required to develop a public web page to provide information on their research, including new technologies and methodologies, target selection, high throughput approaches, data management, and related material. The pilot centers were also required to list their targets on a publicly available web page with weekly updates on the progress toward structural determination. This information is essential to avoid duplication of efforts within the PSI centers and with the structural biology community. Another requirement is the dissemination of information on protein production, purification, and crystallization. The NIGMS supports two databases for the centralization of these data (TargetDB and PepcDB). As expression clones and other materials generated from PSI research are useful to scientists interested in pursuing detailed functional studies that are beyond the scope of the PSI, the PSI centers are required to make these materials readily available.

With the experiences of the Human Genome Project in mind, many scientists in structural genomics began discussions on a global approach to these policies, especially data release and international coordination. The NIGMS and the Wellcome Trust, UK co-sponsored the first international structural genomics policy meeting in Hinxton, UK in April 2000. These issues were discussed further at a conference in Florence, Italy in June 2000 held by the Organization for Economic and Cooperative Development and at a workshop held before the International Conference on Structural Genomics in Yokohama, Japan

in November, 2000. Following these meetings, a final policy meeting was held in Airlie, VA, US, sponsored by the NIGMS, Wellcome Trust, and RIKEN, Japan. As reported elsewhere in this chapter, this meeting focused on international cooperation and policies such as data release, publication, coordinate deposition, and intellectual property. Agreements were reached at Airlie on many of these issues and the international organization, ISGO, was created. Much of the discussion focused on the policy for deposition and release of coordinates. The agreed policy on this topic permitted flexibility, as explained elsewhere in this chapter. The NIGMS policy for PSI was more stringent than the ISGO policy and required deposition and release of coordinates within six weeks of completion of the structures, with no exceptions. The PSI centers do retain intellectual property rights, but only those consistent with the PSI data release policy.

The pilot centers program, PSI-1, ended in mid-2005. Numerous new methods, automated and parallel procedures, robotic instruments, and salvage procedures were developed during this period and incorporated into the structural genomics pipelines. About 1300 structures were solved during these five years, with about 65% of these unique, that is no structure for a protein with a sequence more than 30% identical was present in the Protein Data Bank (PDB) at the time of deposition. By the fifth year of PSI-1, the cost per structure was reduced dramatically — to $138K. For each of the last two years, the PSI structures have represented about 40% of all the unique structures deposited into the PDB worldwide.

Following an examination of the progress of PSI-1 and the lessons learned from this pilot project, the NIGMS designed and then initiated a second phase, PSI-2, which began July 2005. As a major goal, PSI-2 aims to increase the number of structural families with structural representatives, including families with high biological impact, and to continue methodology and technology development, especially for challenging classes of proteins such as membrane proteins. In addition, PSI-2 should facilitate the use of structures by the broad scientific community. To achieve these goals, the PSI-2 centers program was designed with five separate components: 1) four large-scale research centers focused on the production of a

large number of unique protein structures that, combined with computational models, will permit broad structural coverage; 2) six specialized centers focused on technical problems associated with pipeline bottlenecks and challenging proteins (such as membrane proteins, small complexes, and proteins from eukaryotes); 3) two homology modeling centers to improve the accuracy of comparative modeling; 4) a materials repository center to store and distribute expression clones; and 5) a knowledgebase to serve as a centralized information, analysis, and dissemination center. One of the specialized centers is co-funded by the NIH National Center for Research Resources (NCRR).

As a public resource, the PSI-2 has special regulations and policies similar to those of PSI-1 and the centers are required to release all results promptly. This includes the deposition and release of structures and related information into the PDB within four weeks of completion. The overall PSI-2 goal of providing broad structural coverage and the determination of unique protein structures from large protein families was built into the PSI-2 project, but the details of target selection were left to be worked out by the PSI-2 researchers. The target selection strategy has focused on coarse sampling of large families (including Pfam) with no structural representatives in PDB for broad structural coverage and moderate sampling of very large families with limited structural representatives in PDB. Other PSI-2 targets focus on structural coverage of single organisms, metagenomes, and microbiomes.

During the first and second year of PSI-2 (July 2005–June 2007), the four large-scale centers determined 426 and 617 protein structures, developed additional new methods, and jointly devised a target selection process to maximize structural coverage and the biomedical relevance of the structures. Over 70% of these PSI-2 structures are unique, and the cost per structure has been reduced to $65,000.

The target selection strategy is intended to increase substantially the probability that investigators who have their attention drawn to relatively uncharacterized proteins will find proteins, identifiable by sequence comparison methods, for which structures are available in

the PDB. For example, genome-wide association studies are revealing genes in which variation contributes to differences in disease susceptibility or other traits. A significant number of the protein-coding genes identified in such studies are likely to encode proteins that have not been structurally characterized. Similarly, a variety of proteomic studies in a wide range of organisms are identifying proteins for which no structural information is available. In these cases, the availability of even rudimentary homology models will contribute to hypothesis generation and target prioritization. The development of more powerful homology modeling and other computational tools that are underway will increase the utility of the structures derived from structural genomics efforts.

B. Structural Genomics in European Framework Programs

Josefina Enfedaque, Saša Jenko Kokalj
and Jacques Remacle
European Commission, Research Directorate General,
BE-1049 Brussels, Belgium.

Over the last 23 years, the European Union (EU), via the use of subsequent EU Framework Programmes (FPs) (Fig. 1), has supported trans-European collaborative research projects. European collaborative research is highly successful and functions extremely well due to the autonomy and flexibility given to researchers through the proposal, contract and project management policies of the European Commission (EC). This extensive multi-laboratory collaboration is often essential for assembling the scale of resources needed to advance in a wide range of research areas.

A significant part of these different Framework Programme budgets was dedicated to support collaborative research in life sciences and medical research. The overall budget of the new FP7 program is €50.5 billion, with €6 billion dedicated to health-related research.

The sequencing of the human genome and many other genomes heralded a new age in human biology, offering unprecedented opportunities to improve human health and to stimulate industrial and economic activity.

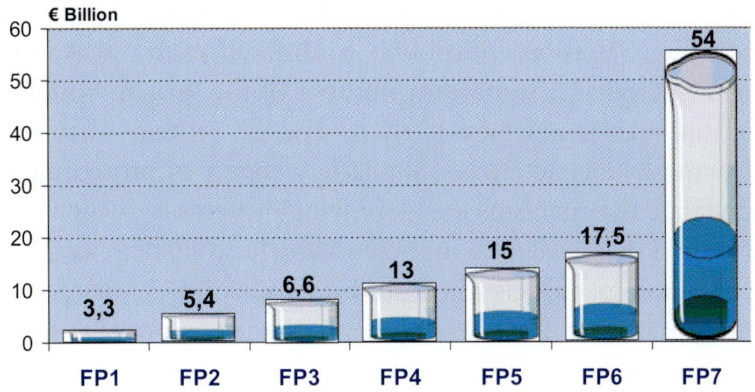

Fig. 1 Budgets and anticipated budgets of the EU Framework Programmes (FP1–FP7, 1984–2013).

The EC identified quite early on the importance of genomics research. Indeed, between 1990 and 2002 (FP3–FP5), over €120 million was invested into genomic research. Several major discoveries were accomplished with these investments, including the sequencing of the first eukaryotic genome (yeast); the sequencing of the first plant genome (*Arabidopsis thaliana*); and the assembling of the physical and genetic maps of the human genome, important and necessary tools for the sequencing of the human genome.

Even though "the book of life" is now deciphered, the main challenges remain. A global understanding of the complete function of approximately 22 000 human genes and their interactions between themselves and with the environment constitutes a major challenge to understand normal and pathological situations. To tackle this challenge, the EC made genomics and post-genomics research a research priority in FP6 (2002–2006). In FP6, approximately €589 million was invested into fundamental genomics, with the overall aim to foster the basic understanding of genomic information by developing the knowledge base, tools and resources needed to decipher the function of genes and gene products relevant to human health, and to explore their interactions with each other and with their environment. The FP6 Fundamental Genomics section involved the following sub-areas: gene expression and proteomics, structural genomics,

comparative genomics and population genetics, bioinformatics and multidisciplinary functional genomics approaches to basic biological processes.

The concept of structural genomics arose in the mid-to-late 1990s in the USA and Japan as a response to the success of high-throughput (HTP) sequencing methods applied to whole genomes (see www.isgo.org). It was imagined that similar HTP methods could be applied to obtain three-dimensional structures of all the proteins (the "proteome") of an organism, which would in particular be an efficient way of filling in the gaps in observed fold-space. This vision led to the investment of substantial funds into large-scale structural genomics projects in the USA (e.g. nine projects funded by the NIH/NIGMS Protein Structure Initiative (PSI) from September 2000 to June 2005; www.nigms.nih.gov/psi/) and Japan (e.g. the massive RIKEN project; www.rsgi.riken.go.jp/). In Europe, the first large initiative in implementing HTP approaches to structural biology was launched in 2002 (FP5) with SPINE: Structural Proteomics IN Europe (www.spineurope.org). The challenge set for SPINE was to push forward with cutting-edge technologies aimed at biomedically relevant targets, while at the same time generating a pan-European integrated effort directed towards biomedically focused structural proteomics. The project produced 308 novel protein structures and a further 61 derivative structures. SPINE also developed European standards in protein crystal handling for X-ray crystallography. These developments are now being used to tackle protein complexes involved in a number of signaling pathways in human health and disease (Stuart *et al.*, 2006).

The effort in structural genomics was further increased in the EU FP6 Programme (2002–2006) with the objectives to enable researchers to determine, more effectively and at a higher rate than was currently feasible, the 3D structure of proteins and other macromolecules, which is important for elucidating protein function and is essential for drug design. Whereas, in general, projects elsewhere have tended towards a high-throughput approach that covers an organism, an organelle or a category of proteins, most of the European projects are oriented towards technology development or high-value targets, in most cases associated with diseases (specifically, drug targets, viral pathogens, membrane receptors, neuronal development and degeneration, immunology, and cancer).

The FP6 projects can be classified into two categories: projects that are generating high-value 3D structures of proteins and complexes of fundamental and biomedical importance, using high-throughput X-ray crystallography, NMR and/or electron microscopy (VIZIER, FSG-V-RNA, SPINE-2-COMPLEXES, E-MeP, 3D-REPERTOIRE, CAMP); and projects that are developing new and/or improving existing methods for the production, characterization and structure determination of proteins and complexes as well as new and improved technological and bioinformatics tools (3D-EM, BIOXHIT, Opticryst, IMPS, Extend-NMR, NDDP, UPMAN, NMR-Life, HT-3DEM, GENEFUN) (Fig. 2).

- **VIZIER:** The aim of this large integrated project (23 partners, EU contribution €13 million) is to gather the background knowledge

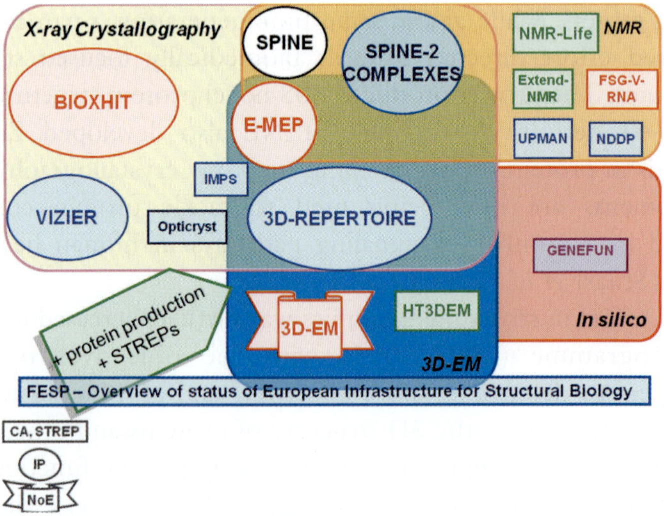

Fig. 2 EC-funded projects in SG/SP 2002–2007. The various EC instruments to fund the research shown above include CA (Coordinated Activity), STREP (Specific Targeted Research Projects), IP (Integrated Project), and NOE (Network of Excellence).

on viral replication that is needed to develop new drugs which could help prevent new viral outbreaks. This project, unprecedented in its size, has set out the sequence of the genomes of hundreds of viruses, defined the proteins essential to their replication, and through a major 3D structural effort is identifying the crucial common sites of these proteins that could be the target for new anti-viral drugs with large spectrum of action. VIZIER is the first large-scale research initiative in the world to address the challenging concept of anticipating the unexpected emergence of RNA viruses (www.vizier-europe.org/).

- **FSG-V-RNA** is a more targeted project (10 partners, EU contribution €2.4 million) that complements VIZIER and aims at developing/improving tools and approaches for the rapid and efficient structural analysis of RNA and RNA–protein complexes which are vital for the function of HBV, HCV and HIV viruses (www.fsgvrna.nmr.ru.nl/).
- **SPINE-2-COMPLEXES** (18 partners, EU contribution €12 million) is the second generation of the SPINE project, and aims to investigate signaling pathways from receptor to gene by combining the knowledge of genomes with HTP methods for structural proteomics. The project targets the development and application of HTP methods for an efficient determination of atomic-resolution structures of protein–protein and protein–ligand complexes (www.spine2.eu/).
- **E-MeP** is a large integrated project (18 partners, EU contribution €10.3 million) that aims at solving bottlenecks which preclude the determination, at high throughput, of high-resolution structures of membrane proteins and membrane protein complexes. Optimized tools and methods for expression, refolding, solubilization, purification and crystallization have been developed and efforts are being made towards the standardization of refolding and solubilization techniques. An integrated database cataloging E-MeP's results, protocols and other pertinent data has been established as well (www.e-mep.org/).
- **3D-REPERTOIRE** (26 partners, EU contribution €12 million) aims at determining the structures of all amenable complexes from

the budding yeast *S. cerevisiae* at medium or high resolution by electron microscopy, X-ray crystallography, and *in silico* methods; there structures will later serve to integrate toponomic and dynamic analyses of protein complexes in a cell. The project's deliverables include improved protocols and vectors for the expression and purification of large complexes, development of software to automatically build protein complexes using structural information regarding the complex components or related proteins, and innovative software to automatically fit modeled complexes into low-resolution structures (www. 3drepertoire.org/).

- **CAMP** (10 partners, EU contribution €2.7 million) is a specific targeted research project aimed at developing protease substrates and inhibitors with fluorescent molecules to monitor proteases in tissue culture and live animals in both healthy and pathological situations, and thereby control proteases in cardiovascular diseases and cancers. The ultimate goal is the design and development of drugs to control proteases involved in inflammations, cardiovascular diseases, cancer and neurodegeneration (camp.bioinfo.cipf.es/node/1).
- **3D-EM** (18 partners, EU contribution €10 million) is bringing together European excellence in electron microscopy approaches for studying protein complexes and cellular supramolecular architecture. This network of excellence addresses the field of electron microscopy (single particle analysis, electron crystallography, cryo-electron tomography) to boost research activities and transfer of knowledge. A standardized comparison and exchange of experimental data obtained with differing EM methods facilitates these efforts, and contributes to the understanding of cellular function in molecular detail (www.3dem-noe.org/).
- **BIOXHIT** (21 partners, EU contribution €10 million) is bringing together scientists working at all European synchrotron light sources and key software developers from the field of X-ray crystallography, thereby creating an unprecedented collaborative environment for high-throughput structure determination. It has facilitated the introduction of automation, standardization and

information management aspects into the X-ray crystallography of biological macromolecules for rapid and accurate 3D structure determination of biologically and medically relevant proteins. BIOXHIT enabled the European structural biology community to remain competitive with the US Protein Structure Initiative and the Japanese Protein-3000 program (www.bioxhit.org).

- **Opticryst** (11 partners, EU contribution €2.3 million) is addressing the critical post-protein production bottleneck area in the field of structural genomics. Moving away from current approaches and applying methods based on understanding the fundamental principles of crystallization, the OptiCryst project focuses on designing techniques to actively control the crystallization environment. It is aiming at increasing the success rate of the production of diffraction-quality crystals from the current rate of 21% to at least 40% (www.opticryst.org/).
- **IMPS** (9 partners from 3 different EU member states, budget €1.9 million) is an important part of a range of research projects funded by the EU to understand membrane proteins within the body that are important for medical knowledge and drug design. IMPS is exploring innovative approaches to membrane proteins such as overexpression, stabilization and micro-crystallography in collaboration with the European Synchrotron Facility.
- **Extend-NMR** (8 partners, EU contribution €2 million) is aimed at developing novel computational tools that extend the scope of NMR spectroscopy, that make possible functional and structural studies of larger proteins and biomolecular complexes which are not amenable to crystallization, and that facilitate studies of the conformations of excited states and studies of molecular dynamics (www.ccpn.ac.uk/ccpn/projects/extendnmr/extend-nmr-project-information).
- **NDDP** (6 partners, EU contribution €1 million) is using cutting-edge NMR techniques for dynamic characterization of drug–receptor interactions at atomic resolution to develop a fast, integrated approach to support structure-based drug design using phosphatases,

a major class of drug targets for a broad range of medical indications (6 partners, EU contribution: €1 Million) (http://projects.bijvoet-center.nl/nddp/).

- **UPMAN** (9 partners, EU contribution €1.9 million) is studying structural states of proteins, ranging from highly flexible unfolded monomers to soluble oligomers and precursors of fibrillar aggregates, that are relevant to understanding protein misfolding and aggregation. Novel NMR and computational tools are being developed to investigate the disorganized ensembles characteristic of some of the most interesting and biologically relevant species (schwalbe.org.chemie.uni-frankfurt.de/upman/).
- **NMR-Life** (10 partners, EU contribution €1.07 million) promotes the networking and coordination of NMR research in structural genomics by the exchange of personnel and good practices, the implementation of a virtual laboratory and the organization of meetings (www.postgenomicnmr.net).
- **HT-3DEM** (5 partners, EU contribution €1.8 million) has an objective to develop an innovative technology platform for high-throughput screening and analysis of native protein complexes and protein crystals by EM that will reduce processing time and cost and will greatly increase the probability of obtaining high-quality 2D crystals of membrane proteins (www.ht3dem.org/).
- **GENEFUN** (10 partners, EU contribution €1 million) is developing improved bioinformatics tools for reliably assigning functions to genes. The developed function prediction methods will make a significant contribution towards improving the *in silico* annotation of gene function (www.genefun.org/).

In total, the investment in structural proteomics increased considerably in FP6 to reach over €100 million (Fig. 3). This investment has allowed European research to achieve an international stature comparable to large-scale projects in the USA and Japan.

While it is still premature to predict the success of the FP6 structural genomics projects since many of them are still running today, one could conclude that the EC is funding top-class, ambitious and state-of-the-art projects networking the excellence in Europe. Many

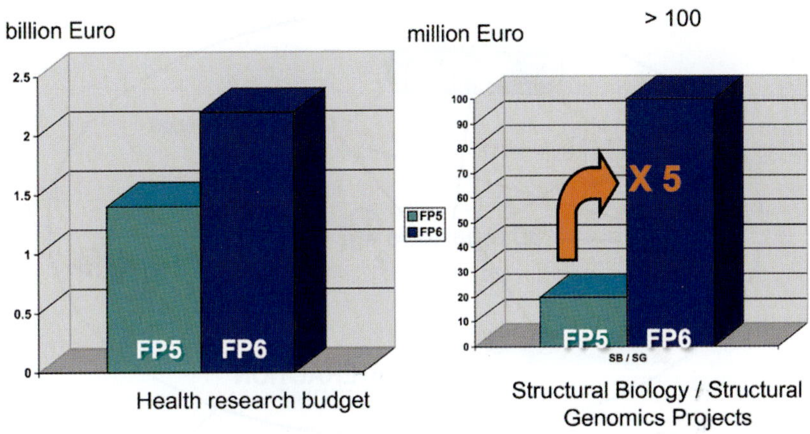

Fig. 3 Funding for health research budget and specifically for structural biology collaborative projects in FP6 (2002–2006).

of these projects have already generated major discoveries that resulted in high-level publications. Most importantly, these projects have played an important role in integrating the research community in Europe, thereby increasing their visibility at the national, European and international levels and improving their capacity to tackle in a collaborative manner ambitious challenges in research (Fig. 4). By doing so, these projects have also substantially contributed to reduce the fragmentation of research in Europe and to implement the concept of the European research area in these important research fields (Fig. 4).

A thorough assessment of the existing structural genomics and structural proteomics projects and infrastructures at the national, European and worldwide levels is one of the major tasks of the Forum for European Structural Proteomics (FESP), resulting in strategic planning and goal setting for future European policy in the area of structural genomics and proteomics. (Banci *et al.*, 2007)

The future of this field will rely on combining integrated structural biology with cell biology so that an atomic-level dissection of the cell can be reconstituted into a functional system (3D cellular structural biology). For FP7, structural genomics projects will not only be devoted to large-scale data gathering initiatives, but will participate in more integrated, systems biology approaches. In addition, FP7 will

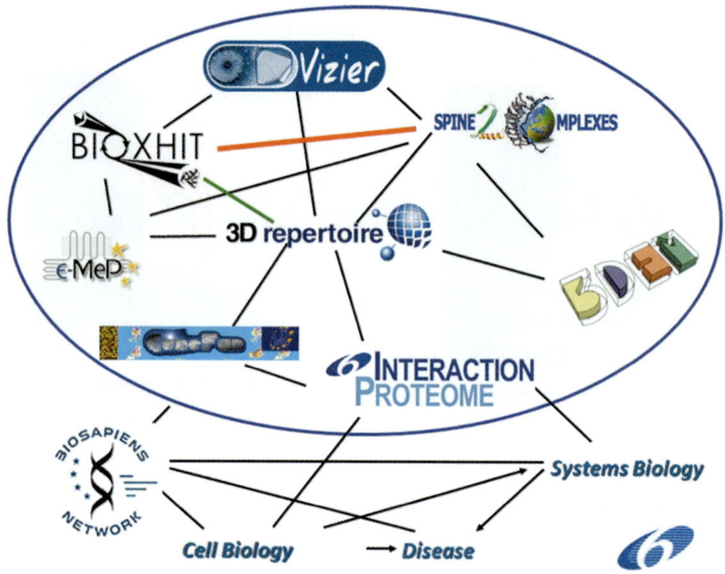

Fig. 4 FP6-funded projects reduced the fragmentation of research in Europe by integrating the European research community in the field of structural genomics.

continue to support projects aiming at developing new and/or improving existing tools and technologies to facilitate protein and protein complex structure determination.

Understanding membrane proteins remains one of the frontier areas of cellular biology. Membrane proteins comprise about a third of all proteins in human cells. However, in the current protein structure database, membrane protein structures only represent a small fraction (less than 0.3%) of the structure coordinates deposited in the Protein Data Bank. To increase our knowledge on this important family of proteins, the EC gave priority to the structure-function analysis of membrane transporters and channels for the identification of potential drug target sites in the first FP7 call for proposal. From this call, two complementary large integrated projects were selected for funding: (Europa Press Release)

- **EDICT:** European Drug Initiative on Channels and Transporters project aims at characterizing the structure/function of several membrane superfamilies in human and pathogenic microorganisms, covering a wide variety of human disorders or diseases and addressing global health issues. The main strength of EDICT (where two of the partners are Nobel Prize winners in the field of membrane proteins) is its powerful methodological structural genomics pipeline for different families of membrane proteins.

- **NeuroCypres:** This project is tackling channels (Cys-loop receptors) of the central and peripheral nervous system involved in severe neurodegenerative diseases and disorders. Its main strength is its multidisciplinary approach to understand the biology behind and the link between their dysfunction and the disease.

Both the EDICT and NeuroCypres projects are driven by a consortium of top European researchers in the field of membrane proteins in Europe. Competitive data will be generated, and importantly these projects will also spend considerable effort in training the new generation of European scientists to sustain European competitiveness in this important research area. They will provide excellent research to improve the health of European citizens, and increase and strengthen the competitiveness and innovative capacity of European health-related industries and businesses. The EC will create interactions and catalyze possible collaborations between these two flagship projects to assure Europe a leader position in the field of membrane proteins.

Finally, to meet the future challenges of 3D cellular structural biology, it will be important to invest in the infrastructure. To address the needs in terms of infrastructure for structural biology in Europe, the EC started with feasibility studies in collaboration with member states to establish an integrated pan-European structural biology infrastructure based on new centers or major upgrades of existing ones (http://cordis.europa.eu/esfri/roadmap.htm). Europe is extraordinarily well-placed to meet this challenge. The centers will develop complementary expertise and instrumentation depending on their focus areas, but all will maintain a set of core technologies. Each center will have a specific biological focus (e.g. viruses, membrane proteins, ion channels, large transient complexes, enzymes, filamentous proteins). The centers will thus develop complementarities, taking into account the originality; the importance of the biological questions being addressed; and the relevance to European priorities such as human health, the environment, therapeutic innovation and biotechnologies. The centers will be open to the European academic and industrial world and will provide, on a project basis, access to production and experimental facilities. The Integrated Structural Biology Infrastructure will provide a central framework for 21st-century biology and pharmaceuticals. (ESFRI Roadmap Report.)

References

Banci L, Baumeister W, Enfedaque J, Heineman U, Schneider G, Silman I, Sussman JL. (2007) "Structural proteomics: from the molecule to the system." *Nat. Struct. Mol. Biol. Vol 14*, a meeting report.

Europa Press Release (26/10/2007) EU funded scientists decode proteins with potential for new medicines.

European Strategy Forum on Research Infrastructures: European Roadmap for Research Infrastructures, Report 2006, p. 51.

Stuart DI, Jones EY, Wilson KS, Daenke S. (2006) "SPINE: Structural Proteomics IN Europe — the best of both worlds." *Acta Cryst. D. Vol. 62*, preface.

C. Policy Aspects in Structural Genomics/Proteomics

Barbara Skene
Formerly Head of Department, Molecules, Genes and Cells
The Welcome Trust
London
UK

The Wellcome Trust expects the researchers that it funds to maximize the availability of research data with as few restrictions as possible and has a long history of promoting the early release of data in the public domain.

The Bermuda Principles

The Trust's role in promoting early data release began with its involvement in the funding of the International Human Genome Sequencing Project. In 1996, the Wellcome Trust sponsored and organized the first International Strategy Meeting on Human Genome Sequencing which was held in Bermuda. This meeting gave rise to the famous "Bermuda Principles" (http://www.wellcome.ac.uk/doc_wtd002751.html) in which it was agreed that:

- Primary genomic sequence should be in the public domain
  It was agreed that all human genomic sequence information, generated by centers funded for large-scale human sequencing, should be freely available and in the public domain in order to encourage research and development, and to maximize its benefit to society.
- Primary genomic sequence should be rapidly released
  Sequence assemblies should be released as soon as possible; in some centers, assemblies of greater than 1 KB would be released automatically on a daily basis. Finished annotated sequence should be submitted immediately to the public databases. It was further agreed that these principles should apply for all human genomic sequence generated by large-scale sequencing centers, funded for the public good, in order to prevent such centers from establishing a privileged position in the exploitation and control of human sequence information.
- Coordination
  In order to promote the coordination of activities, it was agreed that large-scale sequencing centers should inform HUGO (Human Genome Organization) of their intention to sequence particular regions of the genome. HUGO presented this Sequence Index on their world wide web page and directed users to the web pages of individual centers for more detailed information regarding the current status of sequencing in specific regions. This mechanism enabled centers to declare their intentions in a general framework while also allowing

more detailed interrogation at the local level. This arrangement lasted only one year; at the Second International Strategy Meeting on Human Genome Sequencing in 1997, it was decided that the NCBI would take over the Sequence Index created by HUGO in order to coordinate the collection and presentation of this data more efficiently.

Community Resource Projects

In 2002, it became clear that new strategies and other advances in large-scale DNA sequencing necessitated a re-examination and updating of the data release policies originally developed to implement the Bermuda Principles for pre-publication sequence data.

To address this question, the Wellcome Trust sponsored a meeting in January 2003 on the subject of sharing data from large-scale biological research projects (http://www.wellcome.ac.uk/doc_wtd 003208.html). The meeting attendees recommended that the 1996 Bermuda Principles should be reaffirmed and that the agreement be extended to apply to all sequence data, including both the raw traces and whole genome shotgun assemblies.

The attendees further recommended that the principle of rapid pre-publication release should apply to other types of data from other large scale production centers specifically established as "community resource projects." Community Resource Projects were defined as research projects specifically devised and implemented to create a set of data, reagents or other material whose primary utility will be as a resource for the broad scientific community. Examples of such projects include the International Human Genome Sequencing Consortium, the Mouse Genome Sequencing Consortium, the Mammalian Gene Collection, the SNP Consortium, the Structural Genomics Consortium and the International HapMap project.

The Structural Genomics Consortium

The Structural Genomics Consortium (SGC) was established in 2003 as a charitable organization to determine the three-dimensional structures of proteins of medical relevance and place them in the public domain without restriction. Operations commenced in July 2004 with funding from Canadian and British sponsors from both the public and private sectors — the Wellcome Trust, GlaxoSmithKline, Genome Canada, the Ontario Research and Development Challenge Fund, the Ontario Innovation Trust, Canadian Institutes of Health Research and the Canada Foundation for Innnovation. In 2005, a consortium of Swedish sponsors (VINNOVA, Swedish Foundation for Strategic Research, Knut and Alice Wallenberg Foundation and the Karolinska Institute) joined the initiative and the SGC has now established laboratories with ~180 scientists worldwide at the Universities of Oxford and Toronto as well as at the Karolinska Institute in Stockholm.

The goal of this undertaking is to develop the infrastructure and processes necessary for rapid, parallel structural determination of proteins of relevance to human health, with the aim of having the capability to determine 200 protein structures per year. In Phase I (2004–2007), the SGC has exceeded its goal to deposit 386 structures of proteins from a defined set of ~2500 genes that have relevance to human health and disease, such as those associated with diabetes, cancer and malaria. At the end of Phase I, ~450 novel human and pathogen proteins had been deposited in the Protein Data Bank (PDB- www.rcsb.org) (with an overall contribution of around a quarter (~23%) of novel human protein structures during 2006.)

The SGC research is focused on protein families (such as protein kinases, GTPases and oxidoreductases) with a goal to provide a complete structural description of all its members to allow detailed description of distinct structural features (e.g. active sites) as well as enabling comparative intra-family structural analysis. The structural information generated by the SGC is expected to have a tremendous impact on human health by furthering our understanding of the relevant proteins and by supplying structural information on new targets for therapeutic intervention. It will also provide a structural framework for the rational design of new or improved drugs that can modulate the function of proteins suitable as intervention points in disease treatment.

The SGC has annual and monthly milestones agreed by its Board of Directors and these act as a major incentive to deposit structures in PDB as quickly as possible. All the structures are required to meet data and structure quality criteria as defined by the SGC's Scientific Committee of independent scientists. In addition, all deposited structures are quality controlled independently by the PDB (www.rcsb.org). In certain pertinent cases, the SGC scientists have an opportunity to request an 8-week hold on data release if they are involved in a research collaboration with another group in order to obtain additional data or write up the results for a research publication prior to the coordinates being released. However, in most (>95%) of the cases, structural data is released in advance of publication and most often publication of results has taken place after deposition in the public domain, sometimes more than a year after the initial release. It appears, therefore, that the early release of structural data does not inhibit the ability to publish high quality analyses of the data at a later date.

Summary

The Trust has played a major role in the development and funding of resources for genomics and structural biology, both in the creation of the data (e.g. the Human Genome Sequencing Project and the Structural Genomics Consortium) and in facilitating the dissemination of the data through the support of projects such as ENSEMBL http://www.ensembl.org/index.html and the EBI Macromolecular Structure Database http://www.ebi.ac.uk/msd/). The Trust has also played a major role in the coordination of policy development through the sponsorship and organization of a number of international meetings; for example, the Bermuda meetings on Human Genome Sequencing, the Fort Lauderdale meeting on large-scale biological research projects and the Hinxton and Airlie House meetings on Structural Genomics. These meetings had the strong support of the scientific community and the funding agencies around the world to maximize the availability of research data with as few restrictions as possible. The Trust is committed to extending this policy to other types of data that could be shared for added benefit and in 2006, it published its Policy On Data Management And Sharing http://www.wellcome.ac.uk/doc_WTX035043.html.

D. Policies and Updates of the RIKEN Structural Genomics/Proteomics Initiative

Shigeyuki Yokoyama
Protein Research Group, Genomic Sciences Center,
RIKEN Yokohama Institute, 1-7-22 Suehiro-cho,
Tsurumi, Yokohama 230-0045, Japan.

The RIKEN Structural Genomics/Proteomics Initiative (RSGI) (http://www.rsgi.riken.go.jp/rsgi_e/) was organized by the Protein Research Group, Genomic Sciences Center, RIKEN Yokohama Institute (http://protein.gsc.riken.jp) and SPring-8 Center, RIKEN Harima Institute in 2001, with the aim of unifying on-going structural genomics/proteomics efforts in the RIKEN. The RIKEN started a structural genomics/proteomics project (the Protein Folds Project) on a smaller scale in 1997, and expanded it as the Protein Research Group of the Genomic Sciences Center in 1998. The Structurome Project, a structural genomics project focused on *Thermus thermophilus* HB8, was started by the RIKEN Harima Institute in 1999. The RIKEN, together with its funding agency, sent a delegation to the First International Structural Genomics Meeting in 2000 in Hinxton, UK, which was co-sponsored by the National Institute of General Medical Sciences (NIGMS), USA and the Wellcome Trust, UK. In order to promote international discussion on structural genomics policies, the RIKEN and its funding agency, the Ministry of Education, Culture, Sports, Science, and Technology of Japan (MEXT), co-sponsored, with the NIGMS and the Wellcome Trust, the Second International Structural Genomics Meeting held on April 4–6, 2001, in Airlie, Virginia, USA. Intensive discussions on policies resulted in an agreement and the International Structural Genomics Organization (ISGO) was formed. The document of the Airlie Agreement (Agreed Principles and Procedures: Coordination of International Programs in

Structural Genomics) is available at the ISGO web site (http://www.isgo.org/home/).

In 2002, the MEXT started a five-year national project, the National Project on Protein Structural and Functional Analyses (NPPSFA or "Protein 3000." The RSGI was selected as one of the Protein 3000 centers and was responsible for overseeing the program. In accordance with the Airlie Agreement, coordinates determined by the RSGI were to be deposited in the Protein Data Bank (PDB) and released within six months, with a progress report including the target list to be provided every week in XML format (the Task Force on Target Tracking, http://www.isgo.org/organization/index.php) to the TargetDB database (http://targetdb.pdb.org/), the target proteins being selected through collaboration with private companies.

Proteins involved in important biological phenomena were selected as targets. Proteins from small-genome microorganisms, such as the eubacterial extreme thermophile, *Thermus thermophilus* HB8 (http://www.thermus.org/e_index.htm), were selected, as they constitute the "minimum protein sets" for cells. In particular, proteins involved in replication, recombination, transcription, and translation of the fundamental genetic system were intensively studied. For human, mouse, and plant proteins, proteins involved in signal transduction and/or nucleic acid binding were selected. Disease-related proteins deemed important for an understanding of the mechanisms of the diseases and may be drug target candidates, were studied in collaboration with pharmaceutical companies. To transfer the research results to the pharmaceutical companies, the RSGI carried out a collaborative program, the "Partnership Program" (http://www.rsgi.riken.go.jp/rsgi_e/partnerG/) from January 2004 to March 2007. Under the basic collaborative research agreement, protein research progress reports were to be immediately provided to the partner companies, which then selected target proteins for the next specific collaboration. After an individual collaboration research agreement between RIKEN and a partner company had been signed, RIKEN would provide protein samples for research as well as information to the partner company.

A large number of cDNA clones were expressed to obtain samples for structure and function studies. For protein sample preparation, the cell-free protein synthesis method was used in addition to *in vivo* expression systems. The RSGI organized international hands-on workshops on cell-free protein synthesis techniques, at which more than 80 scientists from 14 different countries participated. For protein complexes and integral membrane proteins, the cell-free protein synthesis method is combined with cell-based expression using yeast, insect, and human cells. The cell-free systems have also been developed from human and insect cells. We used both X-ray crystallography and NMR spectroscopy for the 3D structural analyses of proteins, developing high-throughput technologies. In the five years from fiscal 2002 to 2006, we determined 2675 structures (1342 NMR structures and 1333 X-ray structures). The NMR structures are mostly of Pfam domains (http://www.sanger.ac.uk/Software/Pfam/) from human/mouse proteins, whereas many of the X-ray structures are of bacterial and archaeal proteins. The RSGI has provided large numbers of templates that can be used to model other members of the protein families. On average, each NMR structure from the RSGI has contributed to about 300 new homology models, and each X-ray structure about 200 new models at a level of 30% sequence identity (S. Yokoyama *et al.* (2007), *Nature* **445**(7123): 21.). The Protein 3000 Project ended successfully in March, 2007.

The RIKEN organized the International Conference on Structural Genomics (ICSG) 2000 in Yokohama in 2000, and the second ICSG was organized in Berlin in 2002 as the scientific meeting of the ISGO. The RIKEN also joined the organization of the ISGO ICSG 2006 in Beijing and Yokohama. The *Journal of Structural and Functional Genomics* was initiated by Prof. Yoji Arata of the RIKEN as an electronic journal, and is now published as the official journal of the ISGO (http://www.isgo.org/jsfg/).

E. The International Structural Genomics Organization: Policies for Structural Genomics

Thomas C. Terwilliger[1], Shigeyuki Yokoyama[2], Udo Heinemann[3], Ian Wilson[4], Dino Moras[5],

David Stuart[6], Seiki Kuramitsu[7], Edward N. Baker[8], Stephen Burley[9] and Joel Sussman[10]

1 Los Alamos National Laboratory, Los Alamos, NM 87545, USA. (terwilliger@LANL.gov)
2 RIKEN Genomic Sciences Center, 1-7-22 Suehiro-cho, Tsurumi, Yokohama 230-0045, Japan. (yokoyama@biochem.s.u-tokyo. ac.jp)
3 Max-Delbrück-Centrum für Molekulare Medizin Berlin-Buch Robert-Rössle-Str. 10, 13125 Berlin, Germany. (heinemann@mdc-berlin.de)
4 The Scripps Research Institute, 10550 N. Torrey Pines Rd, La Jolla, CA 92037, USA. (wilson@scripps.edu)
5 Laboratoire de Biologie et de Génomique Structurales I.G.B.M.C., 1, rue Laurent Fries 67404 Illkirch cedex, France. (moras@titus.u-strasbg.fr)
6 Wellcome Trust Centre for Human Genetics, University of Oxford, Roosevelt Drive Oxford OX3 7BN, UK. (dave@strubi.ox.ac.uk)
7 Department of Biological Sciences, Graduate School of Science, Osaka University, 1–1 Machikaneyama-cho, Toyonaka, Osaka 560-0043, Japan. (kuramitu@bio.sci.osaka-u.ac.jp)
8 School of Biological Sciences, University of Auckland, Private Bag 92019, Auckland, New Zealand.
9 SGX Pharmaceuticals Inc, 10505 Roselle Street, San Diego, CA 92121, USA.
10 Department of Structural Biology, Weizmann Institute of Science, Rehovot 76100, Israel.

International collaboration and cooperation have been important elements of structural genomics since the inception of the field. The successes of the human genome project and the lessons about the importance of data release and international cooperation learned from sequencing projects provided a powerful example of how an international project could be organized to benefit humanity. In 2000 and 2001 researchers embarking on projects in structural genomics met at the First and Second International Structural Genomics meetings in

Hinxton, UK and Airlie, VA, USA to discuss the goals of structural genomics and how organizations involved in structural genomics could best attain them. These workshops were organized by the Wellcome Trust (UK), the National Institutes of Health (USA) and RIKEN/MEXT (Japan), and the international nature of the sponsorship of these workshops was important in underscoring the importance of their outcomes.

A set of Agreed Principles were developed as a result of the Hinxton and Airlie meetings. These Agreed Principles emphasize data sharing and collaboration in structural genomics. Additionally, the structural genomics researchers at the Hinxton and Airlie meetings formed the International Structural Genomics Organization (see http://www.isgo.org) to help the structural genomics community attain its goals based on the principles of community involvement, openness, and international agreement.

The Agreed Principles from the Airlie meeting (see http://www.nigms.nih.gov/news/meetings/airlie.html) describe the goals of structural genomics and principles of international cooperation in structural genomics. The overall goal of structural genomics is, "the discovery, analysis and dissemination of three-dimensional structures of protein, RNA and other biological macromolecules representing the entire range of structural diversity found in nature." The participants in the Hinxton and Airlie meetings discussed the importance of this overall goal and drew up the Agreed Principles in order to, "encourage harmonious cooperation among a broad range of public and private sector institutions" in this international effort.

The Agreed Principles of structural genomics emphasize that the success of an international effort to determine macromolecular structures would require not only efforts to determine and analyze structures, but also efforts to develop all aspects of methodology for structure determination. The Principles further underscore the need for international support and coordination of structural genomics as well as the need for systematic archiving and dissemination of the resulting data.

It was widely felt that projects with public funding have the highest responsibilities for openness and sharing of data, and a key set of policies in the Agreed Principles are those that were agreed upon to apply to all structural genomics efforts with public funding. The policies for structural genomics projects with public funding include: 1) free exchange of data and materials, 2) deposition of coordinates and other mandatory data in the PDB immediately on completion of structure determination, with public release in a short time and always less than 6 months, and 3) open exchange of targeting information.

The ideas of free exchange of data and materials followed upon policies developed in the genome sequencing projects and upon the successes of open-source computing projects. It was generally felt that far more could be achieved by structural genomics as a whole if each effort would share their data at an early stage and if materials created by one effort could be shared with others. The prompt deposition in the Protein Data Bank of three-dimensional coordinates of the structures that are determined was a central element of the discussions at the Hinxton and Airlie meetings and is a critical part of the Agreed Principles. It was widely recognized that immediate deposition and prompt release of coordinates and other key information was required to achieve the maximum possible impact of structural genomics. The open exchange of targeting information was a new idea that was developed to allow structural genomics groups to have independent targeting, yet to avoid unintentional overlap of effort.

At the Airlie and Hinxton meetings it was generally felt that projects with private funding did not have the same responsibility for sharing as those with public funding. Instead it was felt that efforts should be made to encourage the eventual deposition of structures determined with private funding in public databases for the benefit of everyone.

The Agreed Principles note that the rapidity of structure determination might make publishing full journal articles on each structure impractical, and set in motion the current practice of rapid publication of short articles describing key features of new structures from structural genomics groups. The Principles also note the importance of technology exchange among structural genomics laboratories, a process that has been of high importance in developing the field as well as in sharing its success with the rest of the biological community.

At the inception of the structural genomics field, there was some concern about possible lower standards for structures determined by structural genomics. The Agreed Principles attempted to address this with a statement that, "quality is not to be sacrificed in the interests of quantity" and with guidelines for numerical criteria to be deposited along with structural data. Recent analyses of several measures of the quality of structures from structural genomics efforts have shown them to be of a quality that is comparable to those from non structural genomics efforts (Bhattacharya *et al.*, 2007).

The importance of archiving and dissemination of the data obtained from international structural genomics efforts was recognized in the Agreed Principles. An important outgrowth of this recognition has been a long-term effort to collect not simply the final products of each structure determination, but rather information about the detailed protocols and materials used in the process, so that future efforts could benefit from a full understanding of what was done and could analyze these data to identify optimal procedures.

The Agreed Principles address the important issues of patenting of structures and of information involving structures. The consensus guiding policy statement agreed upon was that, "Raw fundamental data on the shape of natural protein molecules, including 3D positional coordinates, should be made freely available to researchers everywhere. However, intellectual property protection for inventions based on these can play an important role in stimulating the development of important new health care projects." This was accompanied by a statement encouraging strengthening the utility requirement for patentability.

To help frame and define the Agreed Principles, ISGO commissioned a series of Task Forces to examine specific aspects of structural genomics. These Task Forces have included: Informatics; Numerical Criteria for Evaluating and Assuring Structure Quality; Tracking and Registration of Targets; Deposition, Archiving, and Curation of the Primary Information; Mechanisms for Publication and Recording of Methods; and Intellectual Property. The Task Force reports are all available at the web site http://www.nigms.nih.gov/news/meetings/airlie.html#tasks).

The Agreed Principles in structural genomics from the Airlie meeting have had a major effect on the field of structural genomics. They have given individual researchers and research groups in the field guidelines for sharing of data and targeting information. Perhaps even more importantly, the Agreed Principles have provided a framework within which organizations and national funding agencies could develop specific policies for data deposition and sharing. Having the international agreements embodied in the Agreed Principles in place has meant that some of the overall decisions of what data should be shared, what should be patented, and how soon data should be deposited were worked out very early and did not have to be reinvented by each group carrying out structural genomics.

Reference

1. Bhattacharya A, Tejero R, Montelione GT. (2007) Evaluating protein structures determined by structural genomics consortia. *Proteins* **66**: 778–95.

Index